Modelling Mathematical Methods and Scientific Computation

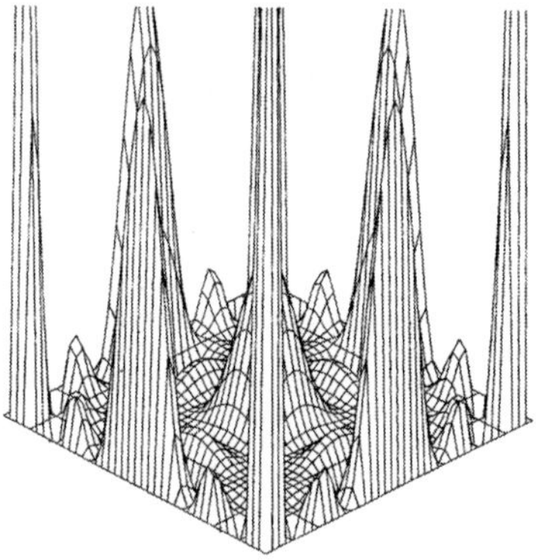

CRC Mathematical Modelling Series

Titles included in the series:

Movchan: ***Mathematical Modelling of Solids with Nonregular Boundaries***

Mittnik: ***System Theoretic Methods in Empirical Economic Modelling***

Villa/Brandimarte: ***Advanced Models for Manufacturing Systems Management***

Kusiak/Bielli: ***Designing Innovation in Industrial Logistics***

Kliemann/Namachchivaya: ***Nonlinear Dynamics and Stochastic Mechanics***

Bellomo: ***Modelling Mathematical Methods and Scientific Computation***

Modelling Mathematical Methods and Scientific Computation

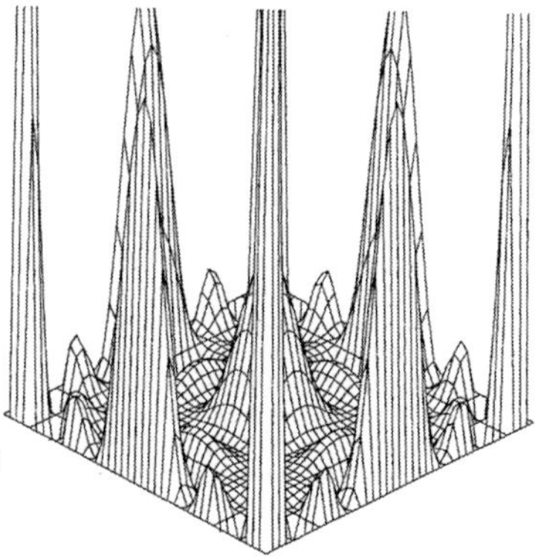

Nicola Bellomo
Luigi Preziosi
Politecnico di Torino, Torino, Italy

CRC Press
Boca Raton   Ann Arbor   London   Tokyo

FIRST INDIAN REPRINT, 2014

Library of Congress Cataloging-in-Publication Data

Bellomo, N.
Modelling mathematical methods and scientific computation / by Nicola Bellomo and Luigi Preziosi.
p. cm. — (Mathematical modelling series)
Includes bibliographical references and index.
ISBN 978-0-8493-8331-1
1. Mathematical models. I. Preziosi, Luigi. II. Title. III. Series.
QA401.B44 1994
501'.1—dc20 94-38722
CIP

Direct all inquiries to CRC Press, Inc., 2000 Corporate Blvd., N.W., Boca Raton, Florida 33431.

International Standard Book Number 978-0-8493-8331-1
Library of Congress Card Number 94-38722
Printed and bound in India by Bhavish Graphics.

CONTENTS

PREFACE

Mathematical Modelling as a Science

Physical systems can be observed, studied, and measured using, if necessary, experimental devices, which may be especially designed for such a purpose. After such a phenomenological analysis, the description of the behavior of the observed system and its inner interpretation is generally delivered to mathematical models.

The relevance of mathematical modelling does not need to be emphasized. As a matter of fact, the conceptual framework of all modern sciences is the classical mathematical physics developed in the last century as the natural development of Newtonian and Lagrangian mechanics. Technological sciences need the classical models of fluid and solid mechanics, transport phenomena, and electromagnetism. Biological, life, and medical sciences use models that are the natural evolution of the classical ones, developed in the framework of mechanical sciences.

Mathematical modelling is now generally accepted in all applied sciences: economy, medical research, and so on. However, despite this recognized relevance, the science of mathematical modelling is not yet a fully well established discipline.

This volume makes an effort to cover such a field with reference to mathematical modelling by the equation of classical mathematical analysis. It will be the first of a Series devoted to mathematical modelling in applied sciences.

This first part of the Preface provides a preliminary description of the main features of mathematical modelling as a science. It is important, for this aim, to list the sequential steps generally followed to ***create*** a mathematical model.

- The physical system to be mathematically modelled is observed and phenomenologically studied.
- The state variable, namely a suitable collection of physical variables, is selected. This variable is charged, in the mathematical model of the description and evolution of the physical system.
- An evolution equation is derived in order to define the time and space evolution of the state variable. Such an equation will be called the ***mathematical model.***

- The mathematical model generates a mathematical problem when suitable initial- and/or boundary conditions are implemented.
- The simulation of the physical system in the description of the model is obtained solving the mathematical problem. The solution will be based upon suitable mathematical methods.
- Suitable comparisons between the simulation provided by the model and the behavior of the real system can validate or reject the model. In the second case, the modelling procedure needs to be revised or even completely reorganized.

This brief sketch clearly shows that the ***science of mathematical modelling*** also refers to the study of mathematical methods with a deep analysis of the interplay between modelling and mathematical methods.

It is worth pointing out that mathematical modelling is a creative science, which requires observation, initiative, and invention. Modelling motivates applied mathematics, which, on the other hand, is required to support modelling and contributes to addressing the invention along mathematically reasonable paths.

Another important point to be stressed is that mathematical modelling can often contribute to a deeper understanding of physical reality. Indeed, the construction of a mathematical model contributes to discovering the organized structures of physical systems. Moreover, the simulation can point out behaviors that have not been or even cannot be observed.

Contents and Style of Presentation

Having all this in mind, the ***contents*** of this volume can now be described.

The first chapter provides the general framework needed by mathematical modelling: definitions, classifications, general modelling procedures, and validation methods.

This chapter organizes the content of the rest of the book which follows the classification proposed in Chapter 1. The second chapter deals with the analysis of discrete models: modelling methods and related mathematical methods. The relevant content of this chapter is the analysis of models defined in terms of ordinary differential equations.

The third chapter deals with the analysis of continuous models, and, in particular, with models defined in terms of partial differential equations.

The fourth chapter deals with inverse type problems and stochastic modelling. Inverse type problems are stated when some features either of the mathematical model or of the mathematical problem are not known. On the other hand, some information is given about the solution of the problem, being directly obtained by measurements on the real physical system.

The second part of the chapter tackles the problem of dealing with the stochastic behavior of some physical systems and of treating the uncertainty of the mathematical model with respect to the real system.

These uncertainties may be caused, among other reasons, by the fact that some of the variables that characterize the real system are "hidden" in the mathematical model, which takes into account only some aspects of the real system.

Referring now to the ***style*** of presentation of the contents of the book, we must mention that such a style is essential and concise. Use of remarks is frequently made in order to point out crucial aspects. Some useful mathematical methods are reported in the various appendixes at the end of the book.

This book is addressed to readers who possess a mathematical education at the level of bachelor's degree and are interested in further mathematical education towards mathematical modelling and related mathematical analysis. When the contents leave the ground of mathematical modelling and move towards mathematical methods, most of the attention is addressed to the applications. Theorems, which are typical of pure mathematics, are dealt with only occasionally although some essential bibliographical indication is given.

An Advanced Course in Mathematical Modelling and Applied Mathematics

This volume is proposed as an advanced textbook for higher courses of mathematics referred to as master courses in technological and applied sciences. The proposal is essentially founded on the idea that ***mathematical methods suitable to solve mathematical problems cannot be separated from the various aspects of mathematical modelling.***

As a matter of fact, solution of problems in applied sciences essentially means analysis of mathematical problems. Moreover, the formulation of a mathematical model requires the knowledge not only of suitable mathematical methods finalized to the construction of the model itself, but also of the methods for the solution of the related mathematical problems.

In fact, in the construction of a mathematical model, one cannot forget the methods to solve the related problems: a sophisticated model, however fascinating, may generate hard, or unsolvable, mathematical problems. Therefore such a model may be practically useless. On the other hand, an oversimplified model may be unrealistic and the simplification of the related mathematical problem may be, this time, useless.

As already mentioned, the modelling of some physical system, must necessarily be followed by the analysis of the prediction of the model, i.e., by the solution of some mathematical problem, in order to validate the model by comparing its prediction with the behavior of the real system. If the validation gives unsatisfactory results, then the model must be revisited so that modelling and methods will be, once more, linked together.

After having given some motivations on the need of linking modelling and mathematical methods, the main lines to be followed in the organization of a course in applied mathematics and mathematical modelling can be given. We assume that the course is addressed to an audience who possess the fundamental knowledge of differential calculus, simple differential equations with constant coefficients, and basic elements of probability theory. With this in mind, the following guidelines are listed:

- Mathematical equations need to be constantly related to modelling of physical systems. Different classes of equations will lead to different classes of models and vice versa.
- The reader needs to reach a broad knowledge on the various classes of mathematical equations and related problems. In fact mathematical modelling cannot be addressed to a class of equations stated "a priori."
- Nonlinear problems are of relevant importance. Linearity has to be regarded as a special situation. Methods of nonlinear analysis are necessary to deal with the analysis of models.
- Mathematical methods must be effective. If a mathematical problem cannot be effectively solved, then the mathematical model that has generated the problem is useless.
- Scientific computation is necessary to the solution of problems related to the analysis of models. Programs must be simple and readable so that their use is not personalized.

This book provides the contents of a university course according to the various requirements, listed above.

Scientific Programs and Problems

It has been already mentioned that scientific computing plays an important role in applied mathematics. In fact, the analysis of models requires the solution of problems, often nonlinear, such as initial-value problems for ordinary differential equations or initial- and/or boundary-value problems for partial differential equations. The solution can be obtained, except in some very special cases, only by computational methods.

As a matter of fact, analytic methods must be exploited, as far as it is possible, in order to obtain qualitative information. However, almost always, quantitative information can be obtained only by numerical techniques and scientific programs. This book provides, at the end of each chapter, besides this first introductory one, some scientific programs related to the solution of sample problems.

The style of the scientific programs fulfills the requirement of being simple and readable also to those not familiar with programming. The programs are addressed to users of applied mathematics. The expert can certainly improve and optimize them. However this is not the important aspect compared with the requirement of simplicity we have explained above.

Some problems are proposed with reference to the contents of the various chapters. Sometimes, the reader is addressed towards the solution of these problems. The proposed problems are not always simple exercises. In fact modelling cannot be reduced to exercises. Some of the problems refer to a rather complete analysis of a system in terms of mathematical modelling and/or to the analysis of related mathematical problems. However an effort is made throughout the book, and consequently also in the proposed problems, to decompose a complex and apparently difficult problem into a suitable sequence of simple problems.

In principle this should be one of the aims of applied mathematics and it is certainly an effort, hopefully an output, of this book.

Nicola Bellomo and Luigi Preziosi

Chapter 1

MATHEMATICAL MODELLING

1.1 Introduction

A physical system can be observed and studied phenomenologically in order to reach a knowledge, as deep as possible, of the inner structure and of the behavior of the system itself.

For such a purpose, experiments often need to be organized, and suitable experimental devices need to be designed. These may be, in some cases, quite sophisticated inventions of the human brain. The history of science reports about the scientific organization of experiments aimed at understanding physical reality.

When, through the collection and organization of the experimental data, the systematic observation is developed into the analysis, interpretation, and prediction of the behavior of the observed system, then a ***mathematical model*** is produced.

In past times, mathematical modelling was essentially related to mathematical physics. Such a science has provided, mainly through the last century, all fundamental models of the modern mechanical and technological sciences.

Mathematical modelling is nowadays a discipline that plays a crucial role in all applied sciences, from natural to technological ones. This volume covers the relevant aspects of this discipline:

- Model classification;
- Derivation techniques;
- Mathematical methods dealing with problems related to the model analysis;
- Model validation and optimization.

We concentrate, in this book, on mathematical models, which can be represented by evolution equations in time and/or space. We also deal with methods of treating some of the mathematical problems, which are generated by the model analysis.

The reader should be aware of the fact that other types of modelling are possible: Computer simulations, geometrical constructions, and so on. However, modelling by mathematical equations still can be regarded as the fundamental way to simulate physical reality.

A detailed description of the content of the whole book was given in the preface. Therefore, we shall simply mention here that the chapters which follow are organized to deal with modelling and model analysis with reference to specific classes of equations (or models) or problems (or model analysis). In fact, both modelling techniques and related mathematical methods are strongly referred to the class of equations.

This chapter is a general introduction which provides the necessary definitions and classifications. In addition, some ideas — still general ones — are given about modelling techniques and model validation. Here, these two topics are only introduced. In fact, as already mentioned, a detailed treatise can be given only when the class of problems under study is chosen.

In particular, after this introduction, the second section provides the definition of a mathematical model and of all the ingredients needed to obtain a model. The third section deals with the classification of mathematical models and of the related mathematical methods. The fourth section provides a description of the main guidelines for modelling. The fifth section develops some general ideas towards model validation. Finally, the last section reports some free speculations on mathematical modelling as a science.

In order to focus the attention of the reader on the crucial points, we use a concise style and use "Remarks" extensively.

1.2 Definition of Mathematical Model

The Preface and Section 1.1 have already introduced some general ideas on the concept of mathematical model. We now present detailed definitions, in particular, a definition of ***mathematical model***.

In order to do this, we need some ingredients. Among them, a ***state variable***, able to describe the model in terms of the ***independent variables*** through a suitable ***evolution equation***, which may depend on some ***parameters***.

More precisely the following definitions can be given:

DEFINITION 1.2.1 ***Independent variables***
Time

$$t \in [0, T] \subseteq \mathbb{R}_+ \tag{1.2.1}$$

and space

$$\mathbf{x} = \{x, y, z\} \in \mathcal{D} \subseteq \mathbb{R}^3 . \tag{1.2.2}$$

are the ***independent variables****. A physical system can be observed in a time interval* $[0, T] \subseteq \mathbb{R}_+$ *and in a volume* $\mathcal{D} \subseteq \mathbb{R}^3$.

REMARK 1.2.1 In order to define the physical volume occupied by the observed system, a fixed system of orthogonal axes $Oxyz$ with unit vectors **i**, **j**, **k** can be used. ■

REMARK 1.2.2 The ***past*** is represented by the negative values of the time, $t \in \mathbb{R}_-$; the ***future*** by positive values $t \in \mathbb{R}_+$. A system can be observed for positive times, however, the mathematical model can refer to negative times. In this case $t \in \mathbb{R}$. ■

DEFINITION 1.2.2 ***State variable***
The ***state variable*** *is a vector valued variable, generally a function of the independent variables,*

$$\mathbf{u} = \mathbf{u}(t, \mathbf{x}) \; : \; [0, T] \times \mathcal{D} \longmapsto \mathbb{R}^n , \tag{1.2.3}$$

where

$$\mathbf{u} = \{u_1, \ldots, u_i, \ldots, u_n\} , \tag{1.2.4}$$

such that **u** *is the set of the variables in charge of describing (in the mathematical model) the physical state of the observed system.*

REMARK 1.2.3 The ***state variable*** is also called ***dependent variable***. It can take values either for all values or for discrete values of the ***independent variables***. ■

DEFINITION 1.2.3 ***Parameters***
The ***parameters*** *are physical quantities that characterize the physical system to be modelled. These quantities can be defined either in a dimensional or in a dimensionless form and can be obtained either by direct measurements on the system itself or by comparisons between the predictions of the model and the behavior of the real system. In the latter, the definition of* ***identification parameters*** *is also used.*

We have now all the tools needed to provide a definition of mathematical model.

DEFINITION 1.2.4 Mathematical model
A ***mathematical model*** *is an equation, or a set of equations, whose solution provides the time-space evolution of the* ***state variable****, that is the physical behavior, in the framework of the mathematical model, of the related physical system. The equation that defines the mathematical model can also be called the* ***state equation****.*

The structure of the state equation is not, and cannot be fixed *a priori*. It depends upon the adopted modelling technique.

The flow chart in Table 1.2.1 shows the main guidelines to be followed in the invention of a mathematical model. The content of this section, in particular, refers to block B. Modelling techniques refer to block C. Mathematical methods refer to block D.

As shown in the flow chart, an analysis of the mathematical model can bring to state the ***validity*** of the model itself. Then, if the model is valid, the simulation can be organized. Otherwise, if its prediction is far from real behavior, then a revision of the model is necessary. The starting point is again the selection of the state variable.

Although nothing has yet been said about modelling techniques, we can already point out at least some general aspects of mathematical modelling.

1. The choice of the state variable is somewhat pragmatic and can be approved only after comparisons between the prediction of the model and behavior of the real system.

2. Modelling involves a compromise between the need of a model suitable to perfectly simulate physical reality and the need of an effective model suitable to "approximate" physical reality in a convenient way. The first model may be too expensive or even impossible to run. The second one may not be able to provide the required accuracy and generality.

3. One may also have several simple models, each one of them being suitable for a certain range of the independent variables, parameters, and/or state variable or under certain circumstances and conditions. The full picture then can be obtained by linking the pieces of the puzzle. If, through this procedure, it is possible to cover every desired aspect, then this collection of models might be preferred to having a complex, expensive model comprehensive of all details.

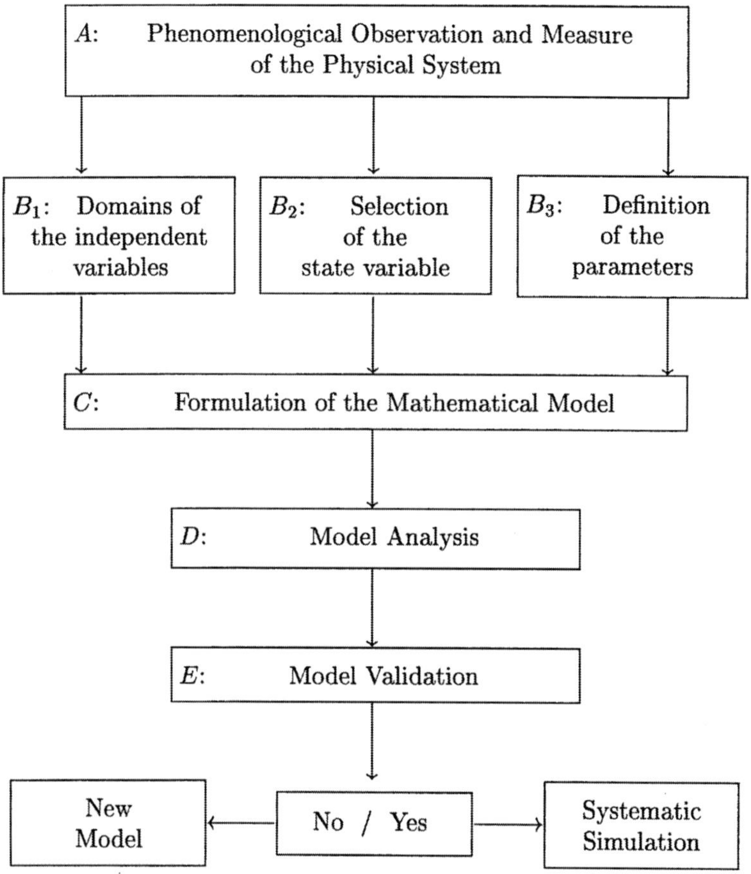

Table 1.2.1 — Flow chart: Modelling and model analysis.

These aspects of mathematical modelling will be developed in the last section of this chapter in a more general framework.

A simple example is now given in order to explain the notations used in this section. The example refers to a simple mathematical model, written without providing full details on its construction. Modelling methods will be dealt with in later sections.

Example 1.1 ***Diffusion of pollutant***
Assume we want to describe the diffusion of the concentration of a pollutant in a river. According to detailed rules, which will be dealt with in Chapter 3, the following simple model can be proposed

$$\frac{\partial u}{\partial t} + c\frac{\partial u}{\partial x} = d\frac{\partial^2 u}{\partial x^2}, \tag{1.2.5}$$

where, according to the general rules previously given, the following hypotheses have been made:

- The state (dependent) variable is the concentration, u, of the pollutant;
- The river is schematized as a one-dimensional region. The longitudinal position along the river is defined by the independent space variable x;
- Besides x, the state variable depends on the time t (the other independent variable);
- The physical system will be obviously influenced by a parameter c, which defines the convectional velocity, and by a parameter d, which defines the diffusion of u along the space variable.

The reader is referred to Chapter 3 in order to have some information about the derivation of this type of model. We will accept, for the moment, the mathematical model as it is, restricting our attention to some comments on its structure. In particular, we note the following:

- The model provides a simplified description of physical reality. For instance, the other two space coordinates are not taken into account and the terms c and d are assumed to be constant, whereas these quantities can be functions, at least, of the independent variables.
- The model can be written, in compact form, as follows

$$\mathcal{F}u = 0, \tag{1.2.6}$$

where

$$\mathcal{F} = \frac{\partial}{\partial t} + c\frac{\partial}{\partial x} - d\frac{\partial^2}{\partial x^2}. \tag{1.2.7}$$

- Even if c and d are assumed to be functions of the independent variables, for instance of the space variable x, the model can still be written as in Eq.(1.2.6), where the operator $\mathcal{F}$ is now

$$\mathcal{F} = \frac{\partial}{\partial t} + c(x)\frac{\partial}{\partial x} - d(x)\frac{\partial^2}{\partial x^2}. \tag{1.2.8}$$

□

1.3 Classification of Mathematical Models

As we have already seen, a mathematical model is an evolution equation in which the state variable is the dependent variable, and time and/or space are the independent variables. Of course, solving such an equation, which refers to the analysis of the model, essentially consists of the solution of a mathematical problem. For this reason, a classification of mathematical models is important. In fact, such a classification will suggest both modelling techniques and mathematical methods to handle the problems related to the ***modelling*** $\rightarrow$ ***solution*** $\rightarrow$ ***validation*** procedure.

The classifications that will be proposed in this section are based on one of the following titles:

- Analysis of the structure of the state variable;
- Analysis of the type of the state equation;
- Analysis of the structure of the evolution equation;
- Analysis of the characteristics of the parameters.

Each title will lead to different classifications: There may be, however, common features. For simplicity of notation, the classification is organized for scalar systems, namely for systems in which the state variable has only one component. The generalization to systems of equations is straightforward.

1.3.1 Classification by state variable

This type of classification is based upon the analysis of the structure of the state variable and, in particular, of its arguments (the independent variables). In detail, the following definitions can be given:

DEFINITION 1.3.1 Dynamic and static models
A mathematical model is ***dynamic*** *if the state variable u depends on the time variable t. Otherwise the mathematical model is* ***static***.

DEFINITION 1.3.2 Discrete and continuous models
A mathematical model is ***discrete*** *if the state variable does not depend on the space variables. Otherwise the mathematical model is* ***continuous***.

The classification can then be represented by the following scheme:

	Static	Dynamic
Discrete	$u = u_0$	$u = u(t)$
Continuous	$u = u(\mathbf{x})$	$u = u(t, \mathbf{x})$

Table 1.3.1 — Classification of mathematical models.

For instance, the state variable in Example 1.1 is dynamic and continuous. The same example can be considered in the stationary case if the time derivative is suppressed. In the latter, the state variable is static and continuous. An example of a discrete state variable can be found later on in Example 1.3.

1.3.2 Classification by type of state equation

This type of classification refers to the analysis of the type of state equation and is strongly related, as we shall see, to that proposed in Section 1.3.1. In detail, the following definitions can be given:

DEFINITION 1.3.3 Algebraic model
A mathematical model is ***algebraic*** *if the state equation is an algebraic equation over the state variable and the set of parameters* $\mathbf{r}$*, which characterize the system*

$$f(u; t, \mathbf{x}, \mathbf{r}) = 0\,, \tag{1.3.1}$$

where t *and* $\mathbf{x}$ *behave as parameters in the same way as* $\mathbf{r}$.

DEFINITION 1.3.4 Models and ODE
A mathematical model is expressed in terms of ordinary differential equations if the state variable satisfies an ordinary differential equation of the type

$$\frac{du}{dt} = f(t, u; \mathbf{r})\,. \tag{1.3.2}$$

DEFINITION 1.3.5 Models and PDE
A mathematical model is expressed in terms of partial differential equations if the state equation involves partial derivatives of the state variable

with respect to the independent variables. A formal example in one space variable is

$$\frac{\partial u}{\partial t} = f\left(t, x, u, \frac{\partial u}{\partial x}, \frac{\partial^2 x}{\partial x^2}; \mathbf{r}\right). \tag{1.3.3}$$

REMARK 1.3.1 If the state variable is a multidimensional vector, then Eqs.(1.3.1–1.3.3) are replaced by systems of equations: One for each component u_i of the state variable $\mathbf{u}$. ■

REMARK 1.3.2 The classifications proposed in this section are strictly related to those of Section 1.3.1. In fact, discrete models generally correspond to Eq.(1.3.2), continuous models to Eq.(1.3.3). When time does not explicitly appear in the model, then the equation corresponds to a static model. Of course, the classification can be further specialized within each class of equations. Various examples will be given in the next chapters. ■

REMARK 1.3.3 Equations (1.3.1–1.3.3) do not cover, of course, all possible types of equations. For instance, one can have integral and integrodifferential or, in general, operator equations. Examples will be given in later chapters. ■

REMARK 1.3.4 If the state variable is multidimensional, then the mathematical model can be expressed by equations of a different type, for instance ordinary differential equations coupled with partial differential equations. ■

1.3.3 Classification by structure of the state equation

This type of classification applies to all types of models, either discrete or continuous, and refers to the linearity or nonlinearity of the model.

DEFINITION 1.3.6 Linear and nonlinear models
A mathematical model is ***linear*** *if the state equation can be written in the form*

$$\mathcal{L}\mathbf{u} = \mathbf{0}, \tag{1.3.4}$$

where $\mathcal{L}$ is a linear operator that satisfies the following condition

$$\mathcal{L}(\mathbf{u}_1 + \mathbf{u}_2) = \mathcal{L}\mathbf{u}_1 + \mathcal{L}\mathbf{u}_2. \tag{1.3.5}$$

A mathematical model is ***nonlinear*** *if the state equation can be written as*

$$\mathcal{N}\mathbf{u} = \mathbf{0}, \tag{1.3.6}$$

where $\mathcal{N}$ is a nonlinear operator that does not satisfy condition (1.3.5). A mathematical model is **semilinear** *if the state equation can be written as*

$$\mathcal{L}\mathbf{u} + \varepsilon\mathcal{N}_\varepsilon\mathbf{u} = \mathbf{0}\,, \tag{1.3.7}$$

where ε is a parameter, and the operator $\mathcal{N}_\varepsilon$ may depend on ε. In some cases ε can be small.

In what follows, we will indicate if necessary by $\mathcal{L}_i$ and $\mathcal{N}_i$, respectively, the i-th component of the operators $\mathcal{L}$ and $\mathcal{N}$.

As we know, if the model is linear, then it is reasonable to hope that analytic solutions can be obtained. In the case of a weakly perturbed system, the solution of the linear problem can be exploited in order to look for asymptotic expansions in powers of the parameter ε. Otherwise the analysis of the model needs methods of nonlinear analysis.

Linearity, however, has to be regarded as a "special" situation. Physical systems are generally nonlinear and show linear behaviors only in particular situations. For this reason, this textbook is mainly devoted to nonlinear methods.

Example 1.2 Nonlinear diffusion of pollutant

The mathematical model introduced in Example 1.1 can be generalized to the case of a diffusion parameter depending on the concentration u. In this case the model is

$$\frac{\partial u}{\partial t} + c\frac{\partial u}{\partial x} = \frac{\partial}{\partial x}\left[d(u)\frac{\partial u}{\partial x}\right]\,.$$

Such a model can be written in the form

$$\mathcal{L}u + \mathcal{N}u = 0\,,$$

where

$$\mathcal{L} = \frac{\partial}{\partial t} + c\frac{\partial}{\partial x}$$

is the linear part and

$$\mathcal{N}u = \frac{\partial}{\partial x}\left[d(u)\frac{\partial u}{\partial x}\right]$$

is a nonlinear term. □

1.3.4 Classification by parameters and stochasticity

Classification by parameters and stochasticity also applies to all models, either discrete or continuous, and refers to the deterministic or stochastic structure of the model. In other words, a model with deterministic parameters is a deterministic one; a model whose parameters are random is stochastic.

As we shall see in Chapter 4, stochasticity in mathematical modelling can be generated in several ways: Initial and boundary conditions, perturbation noise, and so on.

Chapter 4 will provide a detailed analysis of stochasticity aspects in mathematical modelling from the fundamental elements of probability theory to the classification of models in mathematical methods.

For the moment, we shall simply recall that stochasticity in mathematical models is often a natural feature. In fact, we have already seen that the definition of the state variable ***selects*** some variables and ***hides*** others. Therefore, the behavior predicted by the model may show some random uncertainties with respect to the real system.

In addition, features of the model, parameters, or conditions needed to deal with the mathematical problems generated by the model, i.e., initial and boundary conditions, may be known only with some uncertainty or may be affected by noise.

As a matter of fact, the various quantities mentioned above are generally obtained by measurements. Some uncertainty can often affect the result of the measurements because of the reliability of the instruments or of the influence of the outer environment upon the system that is being studied.

The analysis developed in the next two chapters will be devoted to deterministic models. The analysis of stochastic models is confined to Chapter 4. As we shall see, several difficulties have to be handled: From the identification of the stochasticity, to the development of suitable mathematical methods, and, finally, to the representation of the output, i.e., of the random behavior of the state variable in time and space.

Example 1.3 Population dynamics

Consider the classical n-species Lotka-Volterra model, which defines the time-evolution of a system of n species, each characterized by the number u_i of individuals belonging to the i-th species. The model describes a physical situation such that the rate of growth of the i-th species is determined by the reproduction capability of the individuals of the species, i.e., it is proportional to u_i. On the other hand such a rate of growth is decreased by competition among individuals of the same species, proportional to u_i^2, and by competition among individuals of different species, proportional to $u_i u_j$.

The model, in its general form, can then be written as follows

$$u_i = u_i(t) \; : \quad \frac{du_i}{dt} = a_i u_i - u_i \sum b_{ij} u_j \, , \tag{1.3.8}$$

where $i = 1, \ldots, n$ and the terms a_i and b_{ij} are assumed, for the moment, to be constant. In particular,

a_i is the intrinsic rate of increase,

b_{ij} is the competition coefficient regarding interaction between the i-th and the j-th species.

For a one-population dynamic

$$\frac{du}{dt} = au - bu^2 \, . \tag{1.3.8'}$$

For a two-population dynamic

$$\begin{cases} \dfrac{du_1}{dt} = a_1 u_1 - b_{11} u_1^2 - b_{12} u_1 u_2 \\ \dfrac{du_2}{dt} = a_2 u_2 - b_{22} u_2^2 - b_{21} u_2 u_1 \, . \end{cases} \tag{1.3.8''}$$

More detailed information of this type of modelling will be given in what follows. For the moment, with reference to model (1.3.8), and having in mind what has been said up to now, we note the following:

- The ***state variable*** is $\mathbf{u} = \{u_i\}_{i=1}^n$, that is, the set of the number of individuals belonging to each species.
- It can be useful to normalize each u_i with respect to

$$u_0 = \sum_{i=1}^{n} u_i(t = 0) \, , \tag{1.3.9}$$

  that is, to the total number of individuals at the initial time $t = 0$.
- The set of ***parameters*** that characterize the model is given by the set of all a_i and all b_{ij}, $\mathbf{r} = \{a_i, b_{ij}\}$.

- The mathematical model is a ***discrete dynamic model*** and is described by a set of ***nonlinear ordinary differential equations***, which in vector form can also be written as

$$\frac{d\mathbf{u}}{dt} = \mathbf{f}(\mathbf{u}; \mathbf{r}), \tag{1.3.10}$$

where $\mathbf{f} = \{f_i\}_{i=1}^n$ and

$$f_i = a_i u_i - u_i \sum_{j=1}^{n} b_{ij} u_j . \tag{1.3.11}$$

Moreover, the linear and nonlinear part can be separated as follows

$$\mathcal{L}_i = \frac{du_i}{dt} - a_i u_i \tag{1.3.12}$$

and

$$\mathcal{N}_i = u_i \sum_{j=1}^{n} b_{ij} u_j . \tag{1.3.13}$$

- If one or more parameters in $\mathbf{r}$ are random variables, then the system is stochastic. Otherwise, if all parameters are deterministic, the model is deterministic as well.
- If $du_i/dt = 0$ for all $i = 1, \ldots, n$, then the model is ***static discrete*** and is defined by a ***nonlinear algebraic system*** of the type

$$\mathbf{f}(\mathbf{u}; \mathbf{r}) = \mathbf{0} . \tag{1.3.14}$$

- If dependence on space is included, i.e., $u = u(t, \mathbf{x})$, a model governing the evolution of $\mathbf{u}$ can be written as

$$\frac{\partial u_i}{\partial t} + \frac{\partial u_i}{\partial x}\frac{dx}{dt} + \frac{\partial u_i}{\partial y}\frac{dy}{dt} + \frac{\partial u_i}{\partial z}\frac{dz}{dt} = f_i(\mathbf{u}; \mathbf{r}), \tag{1.3.15}$$

which corresponds to a ***continuous dynamic model.***

Note that Eq.(1.3.15) can also be written as

$$\left(\frac{\partial}{\partial t} + \mathbf{v} \cdot \nabla\right) u_i = f_i(\mathbf{u}; \mathbf{r}), \tag{1.3.16}$$

where

$$\mathbf{v} = \frac{dx}{dt}\mathbf{i} + \frac{dy}{dt}\mathbf{j} + \frac{dz}{dt}\mathbf{k}$$

and

$$\nabla = \frac{\partial}{\partial x}\mathbf{i} + \frac{\partial}{\partial y}\mathbf{j} + \frac{\partial}{\partial z}\mathbf{k}$$

is the gradient operator. □

Example 1.4 ***Small vibration of a string***
Consider a mathematical model that defines the small vibration of a perfectly elastic uniform string suspended at both extremes and vibrating in a plane. In particular, the string is suspended at $x = a$ and $x = b$.

The mathematical model is

$$u = u(t,x) : \quad \frac{\partial^2 u}{\partial t^2} = c^2 \frac{\partial^2 u}{\partial x^2} - h\frac{\partial u}{\partial t}, \tag{1.3.17}$$

where u is the displacement of the string, $t \in [0, T]$, $x \in [a, b]$, c^2 is the ratio between the elastic constant and the mass per unit length of the string, and h is the fluid-dynamic drag coefficient.

The model expresses the local equilibrium of each element of the string under applied and inertial forces. The equilibrium is seen in the framework of the Newtonian mechanical model. The applied forces are the fluid-dynamic drag and the elastic forces applied to its edges according to the phenomenological model of linear elasticity. Details on this model will be given in Chapter 3, where continuous models are dealt with.

Let us first rewrite Eq.(1.3.17) in terms of a system of two equations of first-order with respect to time derivatives

$$\begin{cases} \dfrac{\partial u_1}{\partial t} = u_2 \\ \dfrac{\partial u_2}{\partial t} = c^2 \dfrac{\partial^2 u_1}{\partial x^2} - h u_2 \,. \end{cases} \tag{1.3.18}$$

Referring to the model defined in Eq.(1.3.18), we note the following:

- The state variable is $\mathbf{u} = \{u_1, u_2\}$, namely the position and the velocity of the displacement of the string. The set of parameters that characterizes the model is $\mathbf{r} = \{c, h\}$.
- The model is a ***dynamic continuous*** one and is described by a set of two coupled, linear, partial differential equations.

- The static behavior of the model, which is obtained by setting

$$\frac{\partial u_1}{\partial t} = \frac{\partial u_2}{\partial t} = 0\,, \tag{1.3.19}$$

is defined by

$$u_2 = 0\,, \qquad c^2\,\frac{d^2u_1}{dx^2} = -hu_2\,. \tag{1.3.20}$$

The solution of such a static continuous system is often referred to as ***steady state*** or ***stationary solution***.

- As usual, the deterministic model (1.3.18) becomes a stochastic one if the parameters c or h are simulated by random variables.

We shall finally comment that the linearity of the model is simply due to the hypothesis of small displacements of the string and of linear action of the fluid-dynamic drag. □

The various examples we have proposed should give an idea of the fact that all systems are dynamic, nonlinear, and continuous. Linear and/or discrete models are almost always a simplification of physical reality. On the other hand, we should presume that real systems have a deterministic behavior. Stochasticity is then a feature of the model due to the uncertainty in measuring all the parameters that characterize the model itself.

1.4 Modelling Methods

The derivation of specific models, referred to a given physical system, is the core of the mathematical modelling process. The process can start after the phenomenological analysis of the physical system and after the selection of the parameters, the state variable, and its domain of definition in time and space. In other words, referring to the flow chart of Table 1.2.1, we are dealing with block C, after having completed the analysis mentioned in blocks A and B.

As we have already mentioned, a unified theory of the mathematical modelling of physical systems cannot be found in the existing literature. Nevertheless, some general rules organized in a reasonably logic pattern can be given. As a matter of fact this book also covers this topic.

A general methodology is proposed in this section. Examples of applications are given throughout the book.

Assume that the state variable has already been selected. Now, the mathematical model consists of a number of evolution equations, linearly independent, equal to the number of components of the state variable.

Before approaching such a problem we need to define a general framework consisting of:

- Selecting the interactions between the inner system and the outer environment, which should be modelled according to their effects on the system;
- Modelling the cause-effect relations of the interactions mentioned in the item above;
- Modelling the phenomenological behaviors of the physical quantities that characterize the inner system;
- Defining the general mechanical model, or physical laws, which must be satisfied both by the inner and the outer system.

Referring to the first point, it is necessary to determine if the inner system (which is the one to be modelled) is fully isolated by the outer environment or not. In the latter, the action of the outer environment upon the state variable has to be modelled.

Referring to the second point, we mean all phenomenological behaviors of the physical quantities that characterize the system, i.e., viscosity, elasticity, diffusion and transfer properties, and so on.

Finally we must state the general mechanical framework to be followed: Newtonian mechanics, chemical laws, local interactions, and so on.

Once such a general framework has been stated, the evolution equations can be derived. These equations are essentially of two types:

- Conservation equations, namely the equations that ***state the conservation*** of physical quantities along the evolution of the system: i.e., mass, momentum, energy, and so on;
- Equilibrium equations, which state the connection between ***cause*** and ***effect***.

Some simple (rough) models will be proposed as examples in order to understand the modelling procedure.

Example 1.5 Population dynamics

Consider the mathematical model of population dynamics introduced in Section 1.3 and defined in Eq.(1.3.8). The mathematical model is derived with reference to the framework, which has been described in this section, according to the following sequential steps:

1. The system is fully isolated, i.e., it is assumed that there is not interaction with the outer environment.

2. A first phenomenological law (or model) is stated so that the population growth of the i-th species linearly depends upon the number of the individuals belonging to the said species.
3. A second phenomenological law (or model) is stated such that the interaction between individuals of different species produces a decrease proportional to the product of individuals of the interacting species.
4. The mathematical model is simply obtained equating the ***cause***

$$a_i u_i - u_i \sum_{j=1}^{n} b_{ij} u_j \tag{1.4.1}$$

to the ***effect***

$$\frac{du_i}{dt} \,. \tag{1.4.2}$$

□

Example 1.6 Heat transfer along a bar

Consider the heat transfer along a uniform bar and let the state variable be defined by the temperature $u(t,x)$, which depends on time denoted by t and on the space coordinate denoted by x.

Having in mind that a more rigorous and detailed derivation of such a mathematical model will be left to Chapter 3, one may approach, as a first attempt, the heat transfer model following these sequential steps:

1. The system is not isolated from the outer environment. One may assume that the outer environment subtracts from the element dx a heat flux proportional to the difference between u and the environment temperature u_a

$$q_{\text{out}} = c\,\big(u - u_a(t)\big)\,dx\,, \tag{1.4.3}$$

where q_{out} is the outgoing heat flux and u_a is assumed to be a given function of time.

2. A second phenomenological law is stated assuming that the heat flow along the x-axis is proportional to the temperature gradient by a heat diffusion coefficient, which may depend on the temperature

$$q = -A\kappa(u)\frac{\partial u}{\partial x}\,, \tag{1.4.4}$$

where q is the heat flux, A is the given section area of the bar, and $\kappa(u)$ is the heat conductivity coefficient.

3. A third phenomenological model is stated such that κ weakly decreases with temperature

$$\kappa = \kappa_0 \left[1 - \frac{\varepsilon}{u_0}(u - u_0) \right] , \tag{1.4.5}$$

where ε is a small dimensionless parameter and $\kappa_0 = \kappa(u_0)$.

4. A fourth phenomenological model is stated assuming that the temperature increase is proportional to the volume and to the total heat flux through a coefficient c_v (the heat capacity), which is assumed to be constant.

5. Finally, the mathematical model is obtained by a conservation equation, which states that the total heat flux absorbed by the element with volume $A\,dx$ is transferred into an increase of temperature as stated in item 4.

Transferring the sequential procedure into the actual derivation of the model is a matter of technical calculations. In fact the heat flux entering the element of length dx is

$$q_T = \left(q + \frac{\partial q}{\partial x}\, dx \right) - q - q^- = A \frac{\partial}{\partial x} \left[\kappa(u) \frac{\partial u}{\partial x} \right] dx - c(u - u_a)\, dx \,. \tag{1.4.6}$$

Then equating q_T to the increase of heat capacity

$$q_T = A\, dx c_v \frac{\partial u}{\partial t} , \tag{1.4.7}$$

yields the model

$$\frac{\partial u}{\partial t} = \frac{\kappa_0}{c_v} \frac{\partial^2 u}{\partial x^2} - \frac{\varepsilon \kappa_0}{u_0} \left(\frac{\partial u}{\partial x} \right)^2 - \frac{c}{A c_v}(u - u_a) \,. \tag{1.4.8}$$

□

The classification of such a model as indicated in Section 1.3 is left to the reader as an exercise. Further examples will be developed throughout this textbook.

1.5 Validation of Mathematical Models

Once a mathematical model has been "invented," then the evolution of the state variable can be simulated by solving the state equation joined to suitable initial and/or boundary conditions. What is obtained can be compared with the behavior of the real system and if the comparison is satisfactory, the model can be regarded as a ***valid*** one. Otherwise it needs to be revised and, if necessary, rejected.

It is convenient to deal separately with discrete and continuous systems. It is also useful to refer to the first two Appendixes, which deal, respectively, with functional spaces and polynomial approximation.

In what follows, the behavior of the real system will be denoted by $\mathbf{v}$ and that predicted by the model by $\mathbf{u}$. Generally measurements on the real system are obtained for discrete values of time and space. However, it will be assumed that $\mathbf{v}$ is continuous with respect to the independent variables. In fact, the continuous behavior can be achieved by suitable interpolation techniques such as those summarized in Appendix 2.

1.5.1 Validation of discrete dynamic models

Let

$$\mathbf{u} = \mathbf{u}(t) \; : \; [0, T] \longmapsto \mathbb{R}^n$$

be the state variable in the mathematical model and

$$\mathbf{v} = \mathbf{v}(t) \; : \; [0, T] \longmapsto \mathbb{R}^n$$

the behavior of the variable $\mathbf{u}$ in the real system.

In order to determine if the model is ***valid***, we need a tool that enables us to decide when simulation and solution are satisfactorily close to each other. That is, we need a definition of the ***size*** S of $\mathbf{v}$ and of the ***distance*** d between $\mathbf{u}$ and $\mathbf{v}$. The ***size*** S is the norm of $\mathbf{v}$

$$S = \|\mathbf{v}\| \, ; \tag{1.5.1}$$

the ***distance*** d is the norm of the difference

$$d = \|\mathbf{u} - \mathbf{v}\| = \|\mathbf{v} - \mathbf{u}\| \, . \tag{1.5.2}$$

We will then say that the mathematical model is valid if

$$d \leq \varepsilon S \, , \tag{1.5.3}$$

where ε is the maximum admissible gap related to the size S. Such a gap must be fixed *a priori* according to the level of accuracy required by the model.

As known and with reference to Appendix 1, different norms can be selected and different results can be achieved according to such a selection. The selection of the norm should be based upon the kind of ***approximation*** needed by the model. The notations used in what follows refer to $\mathbf{u}$. The same notations can be generalized to $\mathbf{v}$.

If an ***average*** approximation is needed, the following norm, among others, can be used

$$\|\mathbf{u}\| = \left\{ \sum_{i=1}^{n} \|u_i\|_1^2 \right\}^{\frac{1}{2}}, \tag{1.5.4}$$

where

$$\|u_i\|_1 = \frac{1}{T} \int_0^T |u_i|(s)\, ds\,, \tag{1.5.5}$$

where, of course, to perform the integration, the solution has to satisfy integrability properties. In this case the average is made over time and then among all components of $\mathbf{u}$.

Distance in average can be also obtained in mean square

$$\|u_i\|_2 = \frac{1}{T} \left\{ \int_0^T u_i^2(s)\, ds \right\}^{\frac{1}{2}}, \tag{1.5.6}$$

which requires integrability of the square power of the solution.

An alternative is the following

$$\|\mathbf{u}\| = \max_{i=1,\dots,n} \|u_i\|_1\,. \tag{1.5.7}$$

In this case, the average is still calculated over time, but the norm is evaluated over the component of $\mathbf{u}$ where the situation is less favorable.

On the other hand, if a ***uniform*** approximation is needed, then the following norm can be used

$$\|\mathbf{u}\| = \sup_{t\in[0,T]} \max_{i=1,\dots,n} |u_i|\,. \tag{1.5.8}$$

It is plain that the norm (1.5.8) is more severe than that defined in (1.5.4–1.5.6). In fact, a large norm value of the difference between $\mathbf{u}$ and $\mathbf{v}$ can be determined simply by a localized gap. Additional details are given in Appendix 1 as well as in the literature cited therein.

The selection of one norm or the other is a delicate mathematical problem. In fact, one has to select a norm consistent with all requirements of the physics of the problem and, at the same time, the norm must be that of the function spaces for which the solution has been sought. This project cannot always be performed, and sometimes an acceptable compromise must be pragmatically selected.

1.5.2 Validation of continuous dynamic models

Consider, for the moment, the simplest case of a scalar state variable in one space variable only

$$u = u(t, x) \, : \, [0, T] \times [a, b] \longmapsto \mathbb{R} \,. \tag{1.5.9}$$

The validation procedure is the same as that presented in Section 1.5.1 in Eqs.(1.5.1–1.5.3). However, it is necessary to properly select the norm of u taking into account the fact that there are two independent variables. The procedure consists of taking the norm with respect to x and then with respect to t. The norm with respect to x is indicated with a double bar; the norm with respect to both time and space variables is indicated with a triple bar.

As before, we can have mean or uniform evaluation, respectively, by

$$|||u||| = \frac{1}{T} \int_0^T \|u\|(s)\, ds \tag{1.5.10}$$

and

$$|||u||| = \sup_{t \in [0,T]} \|u\|(t) \,. \tag{1.5.11}$$

The norm (1.5.10) can also be taken in mean square

$$|||u||| = \frac{1}{T} \left\{ \int_0^T \|u\|^2(s)\, ds \right\}^{\frac{1}{2}} . \tag{1.5.12}$$

On the other hand, the norm with respect to x can be specialized, under suitable integrability properties, as follows

$$\|u\|_1 = \frac{1}{|b-a|} \int_a^b |u|(x)\, dx \,, \tag{1.5.13}$$

$$\|u\|_2 = \frac{1}{|b-a|} \left\{ \int_a^b u^2(x)\, dx \right\}^{\frac{1}{2}} , \tag{1.5.14}$$

or uniformly

$$\|u\|_\infty = \sup_{x\in[a,b]} |u|(x)\,. \qquad (1.5.15)$$

Once more, we have to say that the indications given above are only the most commonly used and that several other choices are possible.

If the system is in more than one space variable, then integration over all space variables must be taken in the norms (1.5.13) and (1.5.14) and, similarly, one has to take the *supremum* over all variables in the norm (1.5.15). If the system is multidimensional, i.e., the state variable is a vector, then analogously to the discrete system one may define

$$|||\mathbf{u}||| = \max_{i=1,\dots,n} |||u_i|||\,, \qquad (1.5.16)$$

or

$$|||\mathbf{u}||| = \left\{\sum_{i=1}^{n} |||u_i|||^2\right\}^{\frac{1}{2}}\,. \qquad (1.5.17)$$

1.5.3 Validation of static models

In the case of static systems, the dependence on time does not exist. The validation procedure is still the same, however the selection of the norm is simplified. In particular, for ***static discrete models*** the norm is

$$\|\mathbf{u}\| = \max_{i=1,\dots,n} |u_i|\,, \qquad (1.5.18)$$

or

$$\|\mathbf{u}\| = \left\{\sum_{i=1}^{n} u_i^2\right\}^{\frac{1}{2}}\,. \qquad (1.5.19)$$

In the case of continuous discrete systems it is sufficient to take the double bars norm as norm of $\mathbf{u}$, obtained by generalizing the expressions defined in Eqs.(1.5.13–1.5.15) and operating as in Eqs.(1.5.18–1.5.19).

1.6 Mathematical Modelling as a Science

Mathematical modelling constantly supports the development of technological and natural sciences by providing the essential contribution of the mathematical methods. As already mentioned, it is historically recognized that the contribution of mathematical modelling was given primarily by the development of mathematical physics.

We have also seen that mathematical modelling starts playing a relevant role as soon as the investigation of a physical system leaves pragmatic methods and looks for a well formalized theory, suitable to define the properties and the behavior of the physical system to be studied.

It has been shown in the preceding sections that mathematical modelling can be regarded as a well structured science, which follows organized logic lines aimed at the modelling of physical systems in all fields of applied sciences.

The main lines to be followed in the modelling process are summarized and visualized in the flow chart of Table 1.2.1. Each block of the flow chart needs to be adapted and specialized to the various classes of mathematical models reported in Section 1.3. A detailed description of the flow chart of Table 1.2.1 and, in particular, of the content of each block, can be regarded as a conceivable presentation of the various features of mathematical modelling as a science.

In particular, we point out that the modelling process has to be considered as a loop (see Table 1.6.1) that might be interrupted when there is a satisfactory agreement between simulation and observation of the phenomena. This loop, however, may never be stopped if the formulation of the mathematical model or the choice of simulation/validation technique is improperly made.

The contents of this book are already described in the Preface, therefore such a description will not be repeated here. However, some additional comments will be given in order to explain its organization.

- Modelling not only leads to a simulation of physical reality, but also can contribute to a deep understanding of physical systems. Therefore, a mathematical model, even if it ***approximates*** physical reality, should not hide important features. In particular, it should not hide nonlinear behaviors or nonlinear features of the phenomena being modelled.

- Analysis of mathematical models essentially means solution of mathematical problems obtained by joining to the state equation suitable initial and/or boundary conditions. This type of analysis needs the development of mathematical methods that can be organized for classes of models. As a matter of fact, the mathematical methods applied to the analysis of discrete systems are somewhat different from those applied to the analysis of continuous models. Mathematical methods, according to what we have said in the preceding item, should be those of nonlinear analysis. Linearity should be regarded as a particular situation.

- Mathematical problems are not generally as mathematicians wish. In other words, real situations are not such that existence, uniqueness, and regularity of the solution can be proved. Often, mathemat-

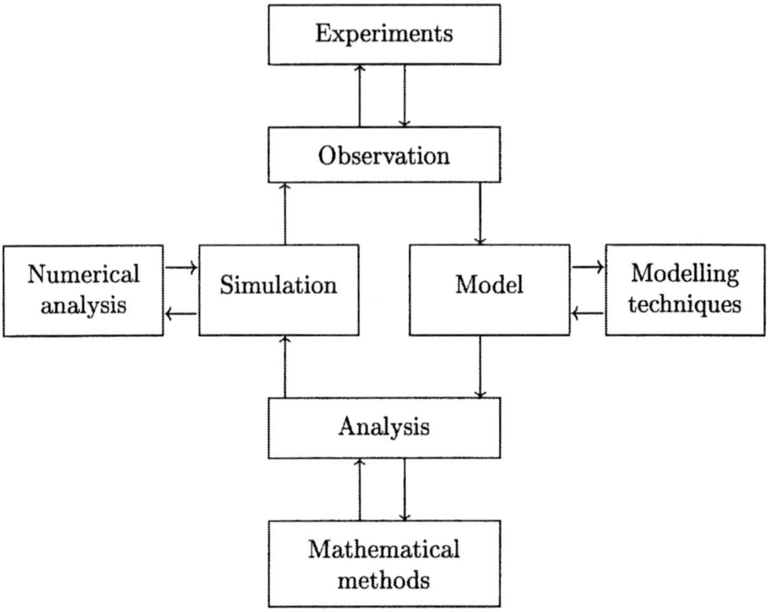

Table 1.6.1 — Modelling loop.

ical problems are imposed by physical reality. To clarify this concept we shall recall that the analysis of inverse-type problems is very important in mathematical problems. In fact, it may often happen that although some information on the solution is given, some features of the model (the parameters) or of the mathematical problem (initial or boundary conditions) cannot be measured. Inverse problems are almost always ill posed. On the other hand, it is then plain that the solution of inverse-type problems is of relevant importance in the construction of mathematical models.

- Physical systems sometimes show stochastic behaviors. In some situations, even if the mathematical model is of a deterministic type, the related mathematical problem may be of a stochastic type. In fact, initial or boundary conditions cannot be measured precisely and this type of information may be affected by some stochastic noise. Stochasticity in mathematical models may be an unavoidable feature, and, consequently, suitable mathematical methods need to be developed in order to deal with stochastic problems.

This book is organized to cover all features mentioned in the above items. The contents refer to the relevant aspects of applied mathematics and attempt to provide effective methods.

There are several interesting books, generally concentrated to particular classes of models, available in the mathematical literature. In this textbook we provide an overall, as complete as possible, treatise of the topic.

Referring to the pertinent literature, we shall first cite the interesting article by Hernandez [HER], which contributes to several fundamental aspects of the science of mathematical modelling. The collection of papers edited by Klamkin [KLA] provides examples of several interesting models. A useful guide is published by Edwards and Hamson [EaM].

The book by Lin and Segel [LaS] is an important contribution to the literature and certainly has inspired some of the features of this book. Interesting books, among others, are those by Beltrami [BEL], Dim and Ivey [DaI], and Haberman [HAB]. Relevant examples of modelling applied to industrial problems are given in the volumes edited by Friedmann [FRI]. Stochastic aspects in mathematical modelling are discussed in the book by Bellomo, Brzezniak and de Socio [BBD].

Additional citations will be given throughout the book and will be directly related to specific problems.

We conlude this chapter by emphasizing, once more, the fact that the construction–invention of a successful model is an important result. However, this does not conclude the investigation over physical reality, as we have already mentioned. In fact, after the simulation procedure by the model, one has to look at the physical system with enriched knowledge, look for a deeper understanding of it, and, if necessary, look for a new model.

In principle, this iterative approach may never be stopped. In fact, it can be regarded as one of the mathematical aspects of human thinking.

Problems for Chapter 1

Problem 1.1

Consider a mathematical model suitable to describe the competition between predators P and prey p homogeneously distributed in a certain territory. In particular, suppose that the model is the following

$$\begin{cases} \dfrac{dP}{dt} = c_1 P + c_2 p \\ \dfrac{dp}{dt} = c_3 p + c_4 P \,. \end{cases}$$

1. Define the dependent and the independent variables of such a model.
2. Define the parameters.
3. Is the model a discrete or a continuous one?
4. Is the model a static or a dynamic one?
5. Are the constants c_i, $i = 1, \ldots, 4$, which characterize the model, positive quantities? Why?
6. Assume that the following term

   $$c_5 P^2 \,, \qquad c_5 < 0 \,,$$

   is added to the first equation. What is the physical meaning of such a term?
7. Assume that the term $g(t) \geq 0$ is added to the second equation. What is the physical meaning of such a term?

Note to Problem 1.1 *The mathematical model defines the dynamics of the number of predators P and prey p. The presence of P contrasts the growth of p. The reader should consider this note with reference to questions 5 and 6.*

Problem 1.2

Consider the mathematical model for the dynamics of predators P and prey p nonhomogeneously distributed in a two-dimensional territory. Let the model be defined by the following evolution equation

$$\begin{cases} \dfrac{\partial P}{\partial t} + v_1 \dfrac{\partial P}{\partial x} + v_2 \dfrac{\partial P}{\partial y} = c_1 P + c_2 p \\[2ex] \dfrac{\partial p}{\partial t} + v_3 \dfrac{\partial p}{\partial x} + v_4 \dfrac{\partial p}{\partial y} = c_3 p + c_4 P + s(t,x,y)\,. \end{cases}$$

1. Do the questions posed in Problem 1.1 also have some meaning in this problem? Provide answers to those that have meaning.
2. What is the qualitative difference between the term g defined in question 7 of Problem 1.1 and the term s defined in the model dealt with in this problem?

Note to Problem 1.2 *The terms s and g are source terms due to the inlet of elements p.*

Problem 1.3
Consider the one-dimensional convection–diffusion model

$$\frac{\partial u}{\partial t} + c(u)\frac{\partial u}{\partial x} = \frac{\partial}{\partial x}\left(h(u)\frac{\partial u}{\partial x}\right) + g(t,x,u)\,.$$

1. Explain why this model is a dynamic continuous one.
2. Explain why the model is nonlinear.
3. What are the conditions required to characterize such a model as a static one?
4. What are the conditions required to characterize such a model as a discrete one?
5. What are the conditions required to characterize such a model as a linear one?

Note to Problem 1.3 *The reader should simply go carefully through the definitions and the examples of Chapter 1.*

Problem 1.4
Suppose that the validation of the models mentioned in Problems 1.1–1.3 should be organized.

1. Provide, for each model, the proper definition of size and distance.

2. Indicate which is a suitable norm for each case and explain the reason of such a choice.
3. Explain why more than one reply can be accepted in answer to question 2.

Note to Problem 1.4 *The reader is referred to Appendix 1 and should reply to the questions of such a problem bearing in mind that for the discrete models a uniform approximation should be preferred to approximation in the mean. Approximation in the mean can be accepted for the continuous models.*

Problem 1.5

Identify, referring to the models of Problems 1.1–1.3, the various steps of the modelling procedure reported in Section 1.4 of Chapter 1.

Problem 1.6

Consider the following competition model

$$\frac{du}{dt} = \alpha u - \beta v$$

$$\frac{dv}{dt} = \gamma v^2 ,$$

where α and β are positive constants and γ may take both positive and negative values.

Find the analytic solutions for initial conditions $u_0 = v_0 = 1$ and report this solution on a graphic for given values of α, β, and γ.

1. What is the qualitative behavior of the solutions? Do these solutions exist for all times?
2. How is the qualitative behavior modified if γ is negative? What is the main difference with respect to the previous situation?
3. Can one deal with the questions posed in Problem 1.4 in this specific problem?

Note to Problem 1.6 *The second equation can be integrated independently from the first, simply by separation of variables.*

Chapter 2

DISCRETE MODELS

2.1 Plan of Chapter 2

The analysis developed in Chapter 1 has shown that the mathematical models of ***discrete dynamic systems*** are organized in equations that define the ***time-evolution*** of a discrete state variable

$$\mathbf{u} = \mathbf{u}(t) : \quad [0, T] \longmapsto \mathbb{R}^n , \tag{2.1.1}$$

where $\mathbf{u}$ is the set of dependent variables

$$\mathbf{u} = \{u_1, \ldots, u_i, \ldots, u_n\} = \{u_i\}_{i=1}^n \tag{2.1.2}$$

defined over the domain $[0, T]$ of the independent variable t.

As already seen in Chapter 1, the state variable $\mathbf{u}$ is in charge of describing, in the framework of the mathematical model, the physical state of the system under consideration. Generally, mathematical models related to discrete dynamic systems are structured in terms of ordinary differential equations, which can be written in terms of a system of n equations in n unknowns

$$\begin{cases} \dfrac{du_1}{dt} = f_1(t, u_1, \ldots, u_n; \mathbf{r}) \\ \vdots \\ \dfrac{du_n}{dt} = f_n(t, u_1, \ldots, u_n; \mathbf{r}) , \end{cases} \tag{2.1.3}$$

where $\mathbf{r}$ is a given set of parameters

$$\mathbf{r} = \{r_1, \ldots, r_i, \ldots, r_p\} . \tag{2.1.4}$$

The same equation can be written in a compact vector form

$$\frac{d\mathbf{u}}{dt} = \mathbf{f}(t, \mathbf{u}; \mathbf{r}), \tag{2.1.5}$$

where $\mathbf{u}$ was defined in (2.1.2) and

$$\mathbf{f} = \{f_1, \ldots, f_i, \ldots, f_n\} = \{f_i\}_{i=1}^n. \tag{2.1.6}$$

The formulation (2.1.3) is called ***normal form***. Higher-order equations

$$\frac{d^n u}{dt^n} = f\left(t, u, \frac{du}{dt}, \ldots, \frac{d^{n-1}u}{dt^{n-1}}; \mathbf{r}\right) \tag{2.1.7}$$

can be easily reported in terms of a normal system of n equations in n unknowns by the change of variables

$$\begin{aligned} u_1 &= u, \\ u_2 &= \frac{du}{dt}, \\ &\vdots \\ u_n &= \frac{d^{n-1}u}{dt^{n-1}}, \end{aligned}$$

which yields

$$\begin{cases} \dfrac{du_1}{dt} = u_2 \\ \dfrac{du_2}{dt} = u_3 \\ \quad\vdots \\ \dfrac{du_n}{dt} = f(t, u_1, \ldots, u_n; \mathbf{r}). \end{cases} \tag{2.1.8}$$

Discrete mathematical models are not formulated only in terms of ordinary differential equations. In fact, we can also have integral equations, which may involve integration over a parameter or over time.

In the first case the state variable depends upon a parameter and there is a quantity that is obtained by a suitable average over this parameter. For instance, in a scalar system

$$u = u(t; \mathbf{r}, \alpha), \tag{2.1.9}$$

the structure of the model can be formally indicated as follows

$$\frac{du}{dt} = \int_a^b \phi\big(t, u(t; \mathbf{r}, \alpha); \mathbf{r}, \alpha\big)\, d\alpha\,. \tag{2.1.10}$$

In a similar way the model can be characterized by memory terms, which can depend upon time, in particular on the past evolution of the dependent variable, and have the following structure

$$\frac{du}{dt} = f\big(t, u(t; \mathbf{r}); \mathbf{r}\big) + \int_0^{+\infty} \phi\big(t - s, u\big((t - s); \mathbf{r}\big); \mathbf{r}\big)\, ds\,. \tag{2.1.11}$$

Specific examples will be given, in Chapters 3 and 4 in order to provide "realistic" applications to the "abstract" statements and framework given above.

The contents of this chapter are directly related to the line that must be followed in mathematical modelling of physical systems.

In particular, the sequential steps of the modelling procedure and of the related analysis (already stated in Chapter 1) are, for sake of completeness, also reported here.

- Selection of the state variable, which will describe the physical state of the system under consideration;
- Derivation of the mathematical model, i.e., the evolution equation for the state variable;
- Formulation of the mathematical problem, i.e., implementation of the necessary conditions (initial or limit conditions) for its solution;
- Qualitative analysis of the mathematical problem: existence and qualitative behavior of the solutions;
- Solution of the mathematical problem;
- Analysis of the results produced by the model and validation of the model.

The flow chart of Table 2.1.1 indicates the main steps of modelling procedure and model analysis as proposed in this chapter. In particular, this chapter covers several aspects of mathematical modelling and related methods for the analysis of discrete dynamic systems.

In detail, the second section deals with modelling techniques and provides some examples of mathematical models. The third section deals with the formulation of mathematical problems and with the related definition of solution. The fourth section deals with the qualitative analysis

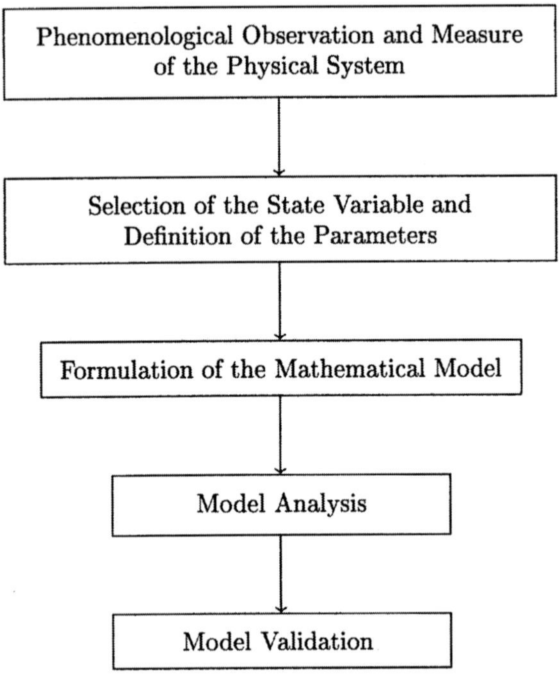

Table 2.1.1 — Flow chart.

of the mathematical problem. The fifth section deals with mathematical methods for the analysis of linear systems. The sixth section is about stability and linearization. The seventh section deals with bifurcation and chaos. The eighth section provides some numerical methods for the analysis of initial-value problems. Finally, the ninth section provides a collection of scientific programs useful for the simulation and quantitative analysis of discrete models.

In this chapter, we cover every aspect of modelling, model analysis, and scientific computation with attention both to analytic and numerical methods. However, greater attention is paid to methods that apply to nonlinear systems. As a matter of fact, if a mathematical model is to provide a realistic picture of physical reality, then nonlinear aspects cannot be neglected. In other words, a model must include all relevant nonlinear aspects, and the related analysis will need all relevant methods of nonlinear analysis.

The literature on mathematical methods for ordinary differential equations is certainly vast, both from the theoretical and the applicative point of view. The following bibliographical references represent some of the available books that are close to the intellectual line of this volume.

The topics dealt with in Section 2.4 are reported, among others, in Birkhoff and Rota [BRO], while an applicative approach to the study of linear system is given by Benton [BEN] and by Jeffrey [JEF].

As far as the theory of stability is concerned, the book by Beltrami [BEL] proposes several models, paying particular attention to their stability properties. It is a good starting point for studying the subject with an applicative approach. The book by Seydel [SEY] is slightly more technical and develops a detailed study of the topic, always applying the results to physical models. Finally, the book by Iooss and Joseph [IaJ] and the book by Keller [KEL] present the subject from the mathematical and the numerical point of view, respectively.

The theory on limit cycles is throughly dealt with by Yan-Qian [YAN], while Stoker [STO] looks at the phenomenon from a more physical point of view. Hopf bifurcation is deeply examined by Hassard, Kazarinoff, and Wan [HKW] and by Marsden and Mc Cracken [MaM], both of which give the proofs of the main theorems, their practical application, and an extensive study of several models. The book by Marsden and Mac Cracken [MaM] is more on the mathematical side but the first chapter explains in a very understandable way the bifurcation mechanisms.

An applicative textbook, which presents the basic concepts and methods involved in the numerical solution of ordinary differential equations, is the one by Sewell [SEW]. For a deeper insight, the book by Butcher [BUT] specifically focus on Runge–Kutta methods. The numerical solution of stiff ordinary differential equations is dealt with by Aiken [AIK] and by Gear [GEA] while boundary-value problems are carefully examined in the classical book by Fox [FOX] and in the more recent one by Ascher, Mattheij, and Russell [AMR].

2.2 About Mathematical Modelling

As we have already seen in Chapter 1, the way of producing a mathematical model for the simulation of a real system is not unique. However, some general rules can still be given. The basic idea consists in relating causes and effects or in modelling the conservation equations of the quantities that refer to the physical system and that are preserved during its evolution.

In more detail, we present the following steps for modelling a discrete dynamic systems.

Step 1 Observe the physical system and select the discrete state variable $\mathbf{u} = \mathbf{u}(t)$ suitable to describe the physical state of the system.

Step 2 Model the causes that may generate the effects and that determine the temporal evolution of the physical state of the system.

Step 3 Model the interplay between causes and effects.

Step 4 Model the conservation equation for the quantities that depend on $\mathbf{u}$, say $\varphi_i = \varphi_i(\mathbf{u})$, which are preserved during the evolution of the system.

REMARK 2.2.1 In general, one needs a number of equations (either of equilibrium or of conservation) equal to the dimension of $\mathbf{u}$. ∎

REMARK 2.2.2 The modelling of the interplay between causes and effects or of the conservation equations often requires a ***background model*** to support the behavior of the physical system. Say, in mechanics we have Newton's laws to model the interplay between forces (cause) and momentum dynamics (effect). Still in mechanics we can have conservation of mass, momentum, and energy. ∎

The ***background model*** mentioned in Remark 2.2.2 is often regarded as a physical theory. On the other hand, we have to look at it as a model with validity only in certain physical situations. For instance, Newtonian mechanics may be replaced by relativistic mechanics or quantum mechanics.

The examples of mathematical models and modelling that will be reported in what follows correspond to different ways of organizing the interplay between causes and effects or conservation of physical quantities. In detail, we treat the following situations, indicating some references for a deeper insight on the background model and on the modelling techniques and hypothesis.

- Mass-point dynamics: Direct application of Newtonian dynamics;
- Rigid-body dynamics: Application of Lagrangian dynamics;
- Population dynamics: Direct cause-effect-source-sink interplay (see Maynard Smith [MSM] and May [MAY]);
- Nonlinear electric circuit: Application of Kirkoff laws (see Hasler and Nelrynck [HaN]);
- Biological systems: Direct simulation of the system by using known properties of mathematical equations and of simpler physical systems (see Cronin [CRO] and Jack, Noble, and Tsien [JNT]).

Although the examples listed above do not cover all conceivable types of modelling, the reader should still receive from them a good deal of information before being introduced to the mathematical techniques for treating mathematical problems.

2.2.1 Mass-point dynamics

Consider a particle with mass m moving on a straight line, subject to a force F. Our aim is to foresee the motion of this particle.

To derive a mathematical model, follow these steps:

Step 1 Select the state variable.

Step 2 Model the cause-effect mechanics.

Step 3 Model the force acting on the particle.

The mathematical model can be defined choosing, with reference to the first step, the state variable as position and velocity of the particle

$$\mathbf{u} = \left\{ u_1 = x\,, \quad u_2 = v = \frac{dx}{dt} \right\}, \tag{2.2.1}$$

where x is the distance of the particle from a reference point on the straight line.

This implies an ***approximation*** of physical reality, as the dimension of the particle is not taken into account: The particle is a point mass and rotational dynamic is not taken into account.

Referring to the second step, we can observe that the applied force causes a change in velocity. The application of Newton's model yields

$$m\frac{dv}{dt} = F\,, \tag{2.2.2}$$

where F must be given or modelled on the basis of experimental information.

In general, F is a function of time, position, and velocity of the particle

$$F = F(t, x, v)\,. \tag{2.2.3}$$

The use of position (2.2.1) yields the following mathematical model

$$\begin{cases} \dfrac{du_1}{dt} = u_2 \\[2mm] \dfrac{du_2}{dt} = \dfrac{1}{m} F(t, u_1, u_2)\,. \end{cases} \tag{2.2.4}$$

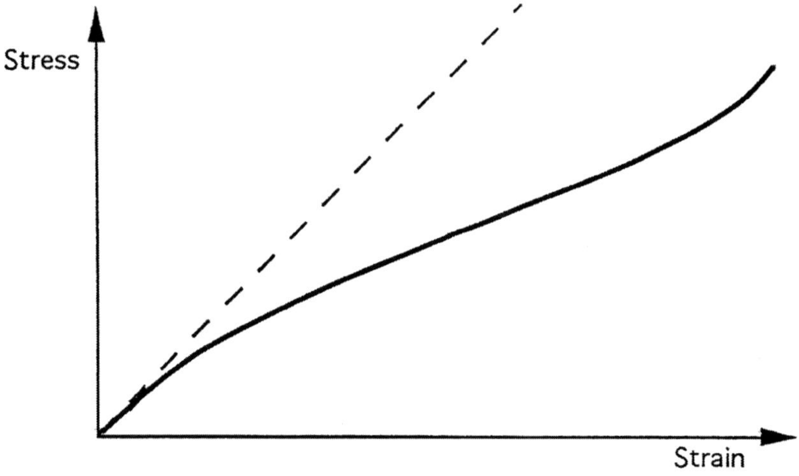

Figure 2.2.1 — Typical stress-strain relation for cross-linked natural rubber (an elastic material) at room temperature.

If, for instance, the particle is connected to the origin with a massless spring with vanishing rest length (which of course are all simplifying assumptions), its action can be modelled by

$$F = -k(x)x\,. \tag{2.2.5}$$

If the rigidity $k(x)$ is assumed constant, one has Hooke's law, but this is only a linear approximation. In Fig. 2.2.1 a typical experimental stress-strain diagram is plotted.

If, moreover, the particle moves in a fluid environment or gas environment, there is a drag force acting on the particle which may depend on the position of the particle and on its velocity and can be modelled as

$$D = -h(x, |v|)v\,, \tag{2.2.6}$$

where h is a positive function.

Equations (2.2.5) and (2.2.6) are called ***constitutive relations*** and essentially have to be found experimentally. According to the previous assumption, one has the following nonlinear model

$$\begin{cases} \dfrac{du_1}{dt} = u_2 \\ \dfrac{du_2}{dt} = -\dfrac{h(u_1, |u_2|)}{m} u_2 - \dfrac{k(u_1)}{m} u_1 \,, \end{cases} \tag{2.2.7}$$

which becomes linear if h and k are assumed to be constant.

2.2.2 Rigid-body dynamics

Consider a rigid cylindrical rod hinged at point O (the center of one base) with a horizontal pin, and assume that this hinge is free to rotate about a vertical axis ζ, as shown in Fig. 2.2.2, so that the centre of gravity can move on an imaginary spherical surface. The state variables are the angle θ formed by the axis x of the rod and the vertical axis ζ through O, the angle φ formed by the vertical plane containing the axis x of the cylinder with a reference plane, and their time derivatives. A spring, which, as usual, is assumed to be massless, with vanishing rest length and satisfying Hooke's law, connects the center of the other basis with the vertical axis ζ through O, so that the spring always keeps its horizontal position, as shown in Fig. 2.2.2.

Assuming that there is no friction, the kinetic and potential energies of the system are, respectively,

$$T = \frac{1}{2}\left(I_x \cos^2\theta\dot{\varphi}^2 + I_y \sin^2\theta\dot{\varphi}^2 + I_z\dot{\theta}^2\right) \tag{2.2.8}$$

and

$$U = -mg\frac{\ell}{2}\cos\theta + \frac{k}{2}\ell^2\sin^2\theta \,, \tag{2.2.9}$$

where $\dot{\theta} = d\theta/dt$, $\dot{\varphi} = d\varphi/dt$, I_x, I_y and I_z are the principal momenta of inertia about the axes x, y, and z represented in Fig. 2.2.2, ℓ is the length of the cylinder and k is the spring rigidity.

The Lagrange equations

$$\begin{cases} \dfrac{d}{dt}\dfrac{\partial \mathcal{L}}{\partial \dot{\theta}} - \dfrac{\partial \mathcal{L}}{\partial \theta} = 0 \\ \dfrac{d}{dt}\dfrac{\partial \mathcal{L}}{\partial \dot{\varphi}} - \dfrac{\partial \mathcal{L}}{\partial \varphi} = 0 \end{cases} \tag{2.2.10}$$

with $\mathcal{L} = T - U$ yield

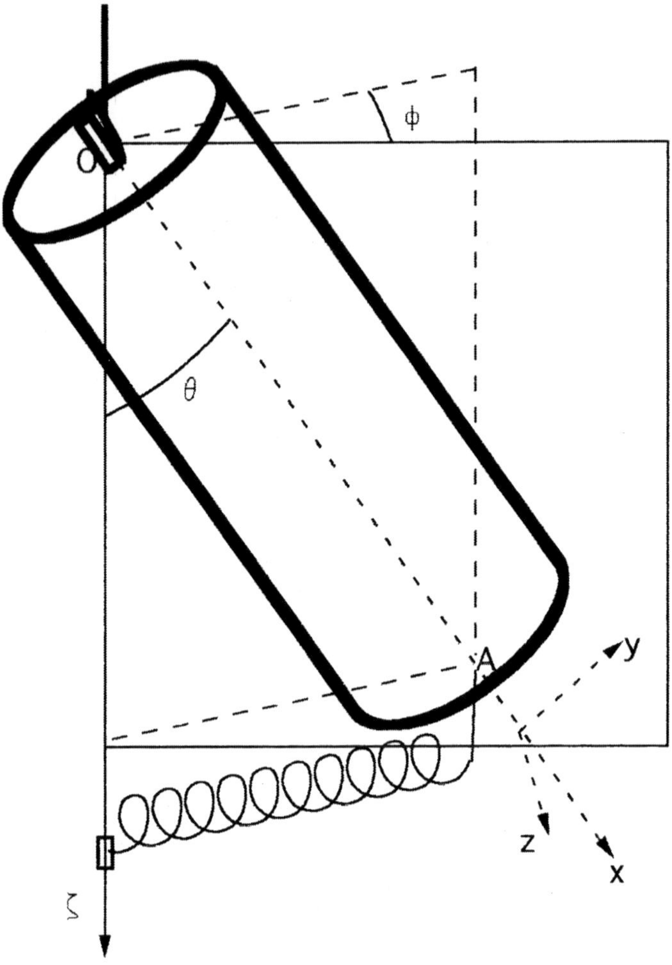

Figure 2.2.2 — Rotating pendulum.

$$I_z \frac{d^2\theta}{dt^2} + (I_x - I_y)\sin\theta\cos\theta\left(\frac{d\varphi}{dt}\right)^2 + mg\frac{\ell}{2}\sin\theta + k\ell^2\sin\theta\cos\theta = 0 \tag{2.2.11}$$

and

$$\frac{d}{dt}\left[\left(I_x\cos^2\theta + I_y\sin^2\theta\right)\frac{d\varphi}{dt}\right] = 0\,. \tag{2.2.12}$$

Equation (2.2.12) represents the conservation of angular momentum about the vertical rotation axis.

REMARK 2.2.3 Observe that if one neglects I_x, as one usually does if one assumes that the rod has a negligible cross section, Eq.(2.2.12) becomes singular ($\sin\theta$ may vanish) causing problems in its integration. This difficulty derives from the nonphysical assumption of a null cross section. ■

We end this example by considering the case in which we steadily rotate the rod about the vertical axis ζ with a constant angular velocity $d\varphi/dt = \Omega$ and analyze at the motion of the pendulum from the uniformly rotating frame. In this case the system has only one degree of freedom, θ, and its evolution equation can be written as

$$\frac{d^2\theta}{dt^2} + \beta \sin\theta\big(1 + \alpha\cos\theta\big) = 0\,, \tag{2.2.13}$$

where

$$\alpha = \frac{2}{mg\ell}\big[k\ell^2 + (I_x - I_y)\Omega^2\big] \tag{2.2.14}$$

and

$$\beta = \frac{mg\ell}{2I_z}\,.$$

Introducing the state variable

$$\mathbf{u} = \left\{u_1 = \theta\,,\ u_2 = \frac{d\theta}{dt}\right\}\,,$$

Eq.(2.2.13) can be rewritten as

$$\begin{cases} \dfrac{du_1}{dt} = u_2 \\[2ex] \dfrac{du_2}{dt} = -\beta\sin u_1\big(1 + \alpha\cos u_1\big)\,. \end{cases} \tag{2.2.15}$$

Observe that β is always positive, while α can have both signs, according to the magnitude of I_x, I_y, Ω, k, and ℓ.

2.2.3 Population dynamics

Consider a population of self-reproducing living organisms, such as a species of bacteria, of animals, or even mankind. We want to model its evolution. In order to do that the following simplifications will be made:

- Age and sex differences will not be taken into account, so that the evolution of the population can be described counting the number N of individuals.
- The population is uniformly distributed in a closed habitat in which migration is not allowed. Hence, the state variable N will only depend on time.
- The number of individuals is so large that $N(t)$ can be considered a continuous function.
- Reproduction is carried out by all individuals, continuously in time.

The growth of the population is due to the balance of births and deaths

$$\frac{dN}{dt} = B(t,N)N\,, \tag{2.2.16}$$

where $B(t,N)$ is the per capita rate at which each member of the population reproduces and dies. As a first approximation one may assume that B is a constant. This leads to an exponential growth of the population, which, by experience, is valid only at the very initial stages of growth of a population in an ideal environment (plenty of nutrients and no crowding problem). In more realistic situations the habitat can support a maximum population level M, known as ***carrying capacity***. The growth rate B is then positive only if $N < M$. As N gets bigger than M, the death rate, due, for instance, to lack of food, becomes larger than the birth rate causing a decrease in population. The simplest function with these properties is

$$B(t,N) = B_0\left(1 - \frac{N}{M}\right)\,. \tag{2.2.17}$$

Equation (2.2.16) can be rewritten as

$$\frac{dN}{dt} = B_0 N\left(1 - \frac{N}{M}\right)\,, \tag{2.2.18}$$

where both B_0 and M usually depend on time. In fact, they may, for instance, fluctuate with the seasons. In this case, they can be modelled as periodic functions of a one-year period.

Model (2.2.18) can be also justified as follows. Assume that the population is limited by the amount of food available, and that its growth rate $B(t, N)$ is proportional to the amount of extra food available. From the daily available food $F(t)$ a quantity proportional to the population is consumed, so that the extra food supply is given by $F(t) - \alpha N$, which again gives Eq.(2.2.17).

Model (2.2.18) represents the basis used for several different problems. For instance, one may introduce migration, still keeping the population uniformly distributed. This may be a voluntary periodic migration or related to fishing or hunting. If then $H(t)$ is the percentage of the population that is hunted, fished, or simply leaves the habitat, Eq.(2.2.18) is modified to

$$\frac{dN}{dt} = B_0(t)N\left(1 - \frac{N}{M(t)}\right) - H(t)N\,. \qquad (2.2.19)$$

There are cases, however, in which Eq.(2.2.17) is still not accurate enough. For instance, fishes are better protected by predators when swimming in school. This implies that the death rate decreases for moderate populations and therefore that B is an increasing function of N for N less than a certain value $\widehat{N}$. The carrying capacity may also depend, as we shall see in May's model (2.2.21–2.2.22), on other quantities such as the quantity of available food.

Consider now a habitat with two interacting species $\mathbf{N} = \{N_1, N_2\}$. In general, the presence of a second species has either a positive or a negative effect on the first one and *vice versa*.

- If the species enhance each other, then the interaction is called ***mutualism*** or ***symbiosis***.
- If the species negatively affect each other, then they are said to be in ***competition***, for instance, for the same food supply.
- A system in which one species, called ***predator***, feeds on the other, called ***prey*** (so that the predator population is enhanced by the presence of the prey population and, of course, the prey population, because of the presence of predators, decreases) is called a ***predator-prey system***.

We will here derive some evolution equations for the predator-prey system, giving at the end some hints on their modifications to study the other cases.

The prey population can be modelled by Eq.(2.2.19), where the hunting function H can be assumed proportional to the number N_2 of predators

$$\frac{dN_1}{dt} = B_1 N_1\left(1 - \frac{N_1}{M_1}\right) - H_1 N_1 N_2\,, \quad B_1,\, M_1,\, H_1 > 0\,. \qquad (2.2.20)$$

A perhaps more satisfactory model is the one in which even though the predator eats as much as it can when food is scarce, i.e., N_1 is small, it satiates in periods of abundant food supply, after having eaten a quantity equal to H_M. In this case, Eq.(2.2.20) can be replaced, for instance, by

$$\frac{dN_1}{dt} = B_1 N_1 \left(1 - \frac{N_1}{M_1}\right) - \frac{H_1 H_M N_1 N_2}{H_M + H_1 N_1}, \quad B_1,\, M_1,\, H_1,\, H_M > 0. \tag{2.2.21}$$

Coming to the predator, one may assume that its growth is given by Eq.(2.2.18) with a carrying capacity proportional to the prey population

$$\frac{dN_2}{dt} = B_2 N_2 \left(1 - \frac{N_2}{M_2 N_1}\right), \quad B_2,\, M_2 > 0. \tag{2.2.22}$$

We will refer to Eqs.(2.2.21–2.2.22) as ***May's model*** [MAY].

Alternatively, one may assume that in absence of prey the predator population drops, while the presence of prey enhances it with a per capita rate proportional to the prey population. These hypotheses yield the following ordinary differential equation

$$\frac{dN_2}{dt} = -B_2 N_2 + H_2 N_1 N_2, \quad B_2,\, H_2 > 0, \tag{2.2.23}$$

which, together with Eq.(2.2.20), forms the so-called ***Lotka–Volterra model***.

We shall end this section by observing that the modelling equations for the mutualism or the competition cases can be obtained by similar reasoning. For instance, a pair of equations like (2.2.20), one for N_1 and one for N_2, may provide an example of a competition model if H_1 and H_2 are both positive and of a symbiosis model if H_1 and H_2 are both negative.

In general, we have seen several models each corresponding to a certain phenomenological analysis. This gives an idea of how the modelling procedure evolves from the selection of the state variables, goes through a suitable phenomenological analysis and its modelling, and finally is concluded by the derivation of the actual model.

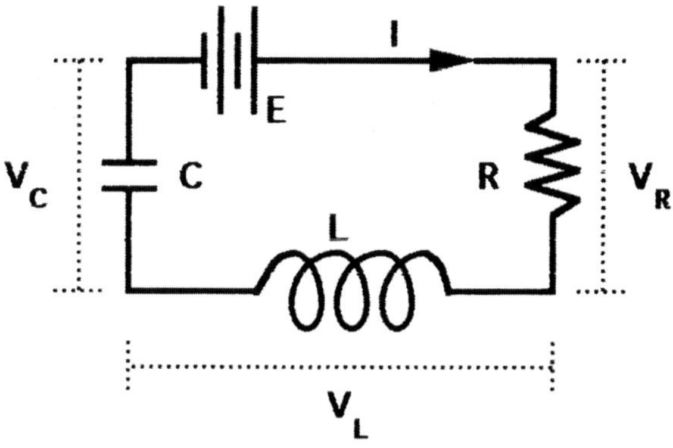

Figure 2.2.3 — RLC circuit.

2.2.4 Nonlinear electrical circuits

Consider a standard RLC circuit constituted by a resistor R, an inductance L, and a capacitor C in series, as shown in Fig. 2.2.3.

Under the assumption that the circuit elements are linear, we have that the tension drop in each of them is

$$V_R = RI$$

$$V_L = L\frac{dI}{dt}$$

$$V_C = \frac{Q}{C},$$

where Q is the charge and $I = dQ/dt$ is the current in the circuit. Due to Kirkhoff's law, their sum has to be equal to the electromotive force E

$$L\frac{dI}{dt} + RI + \frac{Q}{C} = E,$$

or

$$L\frac{d^2Q}{dt^2} + R\frac{dQ}{dt} + \frac{Q}{C} = E. \tag{2.2.24}$$

We note that Eq.(2.2.24) is obtained under the assumption that the circuit elements are linear. In many circuits that use semiconductor components, however, this is not true and nonlinear effects have to be taken into account. In this way, for instance, Ohm's law

$$V_R = RI_R \qquad \text{or} \qquad I_R = GV_R,$$

where G is the conductance, has to be replaced with

$$V_R = R(I_R)I_R \qquad \text{or} \qquad I_R = G(V_R)V_R\,.$$

A simple example is given by the tunnel diode in which the current flowing across the component is not even a monotone function of the voltage, as shown in Fig. 2.2.4a, and can be approximated by

$$I_R = a_1 V_R - a_2 V_R^2 + a_3 V_R^3 - a_4 V_R^4 + a_5 V_R^5\,, \qquad \text{with} \quad a_i > 0\,.$$

The simplest circuit we can consider is shown in Fig. 2.2.4b, where R is the internal resistance of the battery, and L and C are introduced to take into account second-order effects, such as fringing electric and magnetic fields, parasitic inductances, capacitances, and resistance due to the unavoidable wire connections.

By applying Kirkhoff's first law in A one has

$$I_L = I_C + I_D = C\frac{dV_D}{dt} + f(V_D) \tag{2.2.25}$$

and then

$$V_C = E - RI_L - L\frac{dI_L}{dt}\,. \tag{2.2.26}$$

Hence, choosing $\mathbf{u} = \{u_1 = I_L\,, u_2 = V_C = V_D\}$ as the state variable, we rewrite Eqs.(2.2.25–2.2.26) as

$$\begin{aligned} \frac{du_1}{dt} &= \frac{1}{L}(E - Ru_1 - u_2) \\ \frac{du_2}{dt} &= \frac{1}{C}\big[u_1 - f(u_2)\big]\,. \end{aligned}$$

In the same way, there are nonlinear capacitors, for instance the varicap diodes, for which

$$V_C = \frac{Q}{C(Q)} \qquad \text{or} \qquad Q = C(V_C)V_C\,,$$

and nonlinear inductors, for instance the Josephson junction, for which

$$\Phi = \Phi(I_L) \qquad \text{or} \qquad I_L = I_L(\Phi)\,,$$

where $d\Phi/dt = V_L$.

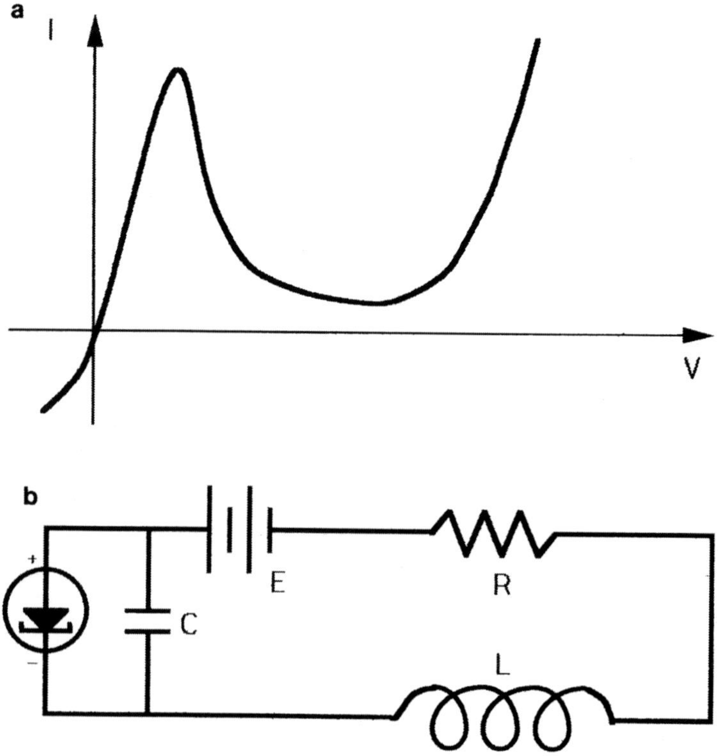

Figure 2.2.4 — (a) Sample current-voltage relation for tunnel diodes. (b) Sample circuit with a tunnel diode.

Considering again the RLC circuit shown in Fig. 2.2.3 but with nonlinear components in series, and operating as before, we have

$$E(t) = V_L + V_R + V_C = \frac{d\Phi}{dI}(I)\frac{dI}{dt} + R(I)I + \frac{Q}{C(Q)},$$

which yields the nonlinear ordinary differential equation

$$\frac{d^2Q}{dt^2} + f\left(\frac{dQ}{dt}\right)\frac{dQ}{dt} + g\left(Q, \frac{dQ}{dt}\right)Q = h\left(t, \frac{dQ}{dt}\right). \qquad (2.2.27)$$

Observe that if the inductance function $\Phi(I_L)$ is linear in I_L, then the dependence of g and h on dQ/dt drops, yielding

$$\frac{d^2Q}{dt^2} + f\left(\frac{dQ}{dt}\right)\frac{dQ}{dt} + \widetilde{g}(Q)Q = \widetilde{h}(t)\,.$$

If, furthermore, $\widetilde{g}$ is constant, Eq.(2.2.27) can be derived with respect to time and rewritten as

$$\frac{d^2I}{dt^2} + F(I)\frac{dI}{dt} + GI = H(t)\,. \qquad (2.2.28)$$

The following cases can be particularly interesting:

$$f\left(\frac{dQ}{dt}\right) = a\,, \quad \widetilde{g}(Q) = \alpha + \beta Q^2\,, \quad \widetilde{h}(t) = \delta\cos\Omega t \qquad (2.2.29)$$

and

$$F(I) = -(\alpha + \beta I^2)\,, \quad G = 1\,, \quad H(t) = \delta\cos\Omega t\,, \qquad (2.2.30)$$

where a, α, β, δ, and Ω are constants. The equations obtained using these relations are

$$\frac{d^2Q}{dt^2} + a\frac{dQ}{dt} + \alpha Q + \beta Q^3 = \delta\cos\Omega t \qquad (2.2.31)$$

and

$$\frac{d^2I}{dt^2} - (\alpha + \beta I^2)\frac{dI}{dt} + I = \delta\cos\Omega t\,. \qquad (2.2.32)$$

We will refer to Eqs.(2.2.31) and (2.2.32) as the ***modified forced Duffing's equation*** and the ***Van der Pol equation***, respectively.

Particularly interesting are the cases in which a negative current corresponds to a positive voltage (Fig. 2.2.5b), which can be obtained with a twin tunnel diode circuit (Fig. 2.2.5a), and in which three different currents correspond to the same value of the voltage (Fig. 2.2.5a).

As a generalization of Eqs.(2.2.31) and (2.2.32), we will also consider the so-called ***Liénard's equation***

$$\frac{d^2u}{dt^2} + a(u)\frac{du}{dt} + b(u) = \delta\cos\Omega t\,. \qquad (2.2.33)$$

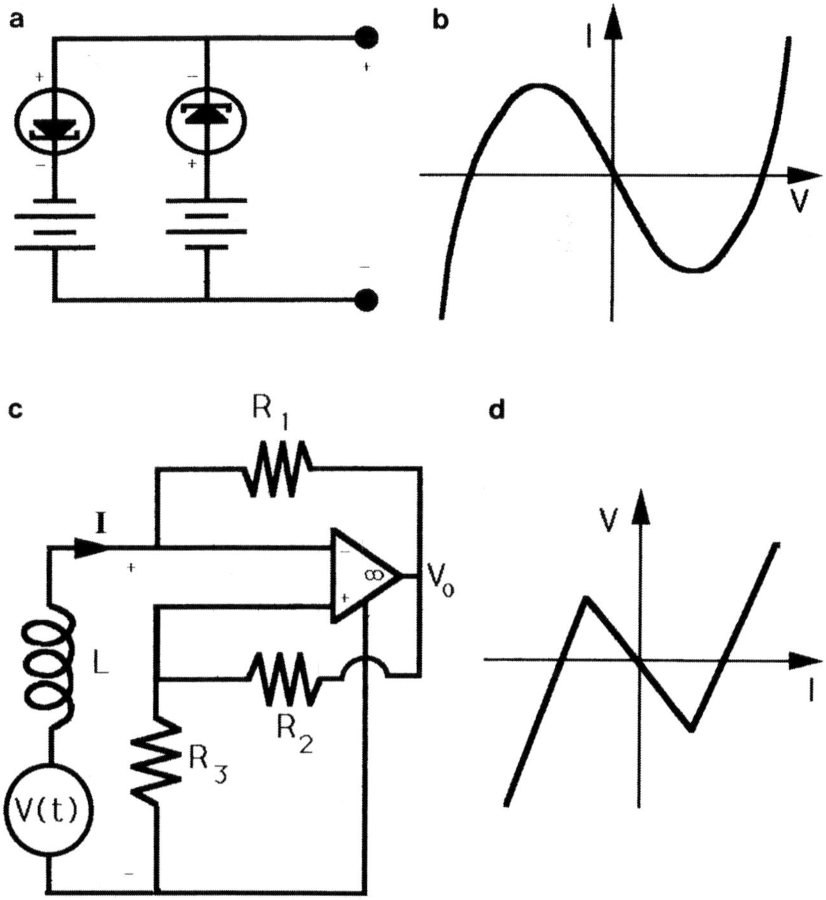

Figure 2.2.5 — Sample current-voltage relation (b) and (d), respectively, for twin tunnel diodes (a) and for a bistable (flip-flop) circuit (c).

2.2.5 Nerve impulse propagation models

The nerve cells are usually made of a central body from which a wire-like ***axon*** departs. This axon is in charge of transmitting information to the neighborhood cells. Suitable electrical events, which will not be considered here, generate an electric wave called ***action potential*** that travels along the axon and constitutes the signal to be transmitted to the other nerve cells.

We will here briefly consider the mechanisms that permit the propagation of this electric wave along the axon. This is essentially related to the properties of the axonal membrane, a double layer of fatty molecules, which coats the axon. Both inside and outside the axon there is a strong accumulation of ions, and, according to physiological mechanisms, these concentrations change because of the migration of ions through the membrane. At rest there is a higher concentration of sodium ions outside and of potassium ions inside the axon yielding a tension drop across the membrane. This situation can be regulated by several mechanisms. In fact, scattered in the lipid layer there are giant protein molecules, which act as ion selective channels. Among these the most important one is the ***sodium channel***, which allows the passage of sodium ions, Na^+, only. In the same way there are ***potassium channels*** (K^+) and secondary channels, which govern the passage of other ions, mostly chlorine ions Cl^-. There are also ion pumps, such as the ***sodium-potassium exchange pump***, which, by consuming metabolic energy, redistributes ions, for instance, pushing the sodium ions outside the axon and the potassium ions inside.

In order to describe the movements of ions across the membrane Hodgkin and Huxley [HaH] modelled the membrane as an electric circuit formed by a capacitor (the lipid layer) and three batteries with internal resistance representing the sodium, potassium, and secondary chlorine channels in parallel, as shown in Fig. 2.2.6.

The mathematical model can now be obtained. In fact, the first step is taken by proceeding as in Section 2.2.4. In detail, we have

$$C\frac{dV}{dt} = g_{Na}(V_{Na} - V) + g_K(V_K - V) + g_L(V_L - V) + I\,, \qquad (2.2.34)$$

where I is the injected current, V_K, $V_L < 0$ and $V_{Na} > 0$ are the tension drops across the batteries (the channels) and g_K, g_{Na}, and g_L are the voltage-dependent conductance of the channels per unit area. All voltages are measured with respect to the rest value, i.e., $V = 0$ at rest.

The next step is to find the relation linking the conductances to the voltage across the membrane. From experimental observations and curve-fitting procedures it was successfully argued that the channels are made up of four subunits that act independently. The channel is open when all subunits are active, as shown in Fig. 2.2.7.

The four potassium channel subunits are closed at rest and open as the voltage drop increases. Therefore, if n_K is the fraction of open subunits in the potassium channel, the conductance can be modelled as proportional to n_K^4

$$g_K = \widehat{g}_K n_K^4\,, \qquad (2.2.35)$$

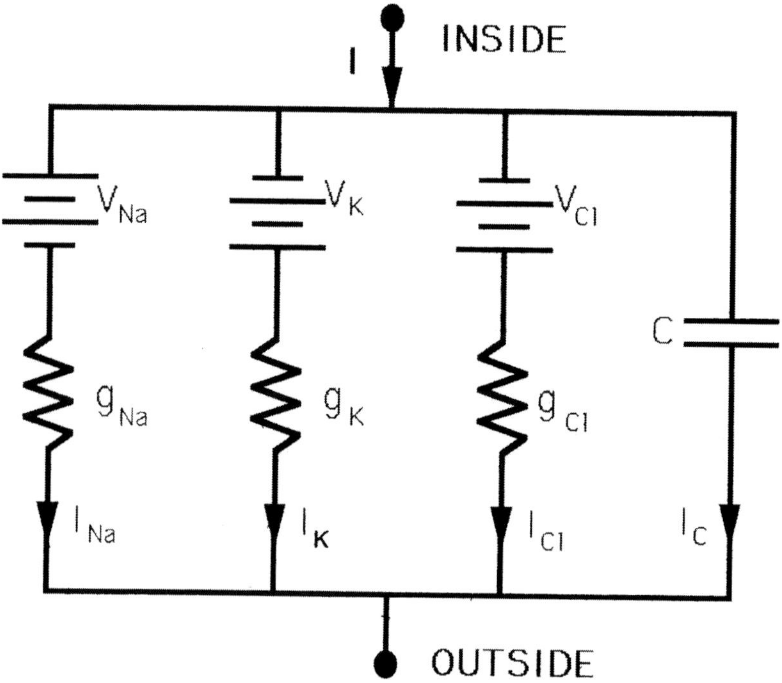

Figure 2.2.6 — Electric circuit modelling the action of the axon membrane.

with

$$\frac{dn_K}{dt} = -F'_K(V)n_K + F''_K(V)(1 - n_K), \tag{2.2.36}$$

where $F'_K(V)$ and $F''_K(V)$ are known functions that will be specified later on.

As far as the sodium channel is concerned, three of its subunits are closed and the fourth is open at rest. As V increases the three subunits open, while the fourth one closes at a smaller time scale. This reasoning gives

$$g_{Na} = \widehat{g}_{Na} n_{Na}^3 p_{Na}\,, \tag{2.2.37}$$

with

$$\frac{dn_{Na}}{dt} = -F'_{Na}(V)n_{Na} + F''_{Na}(V)(1 - n_{Na}) \tag{2.2.38}$$

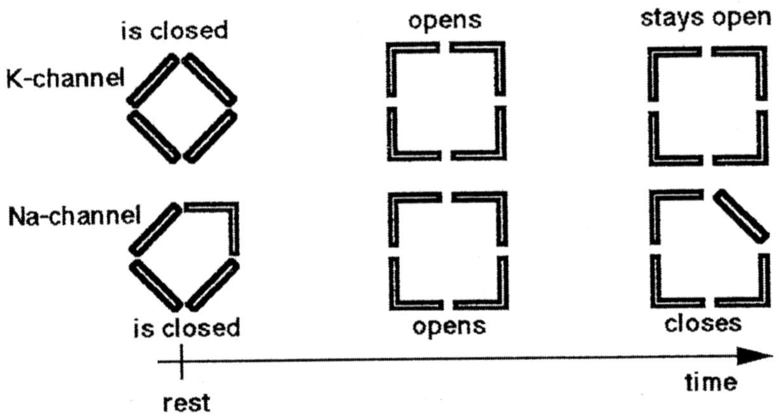

Figure 2.2.7 — Behavior of channel subunits.

$$\frac{dp_{Na}}{dt} = -G'_{Na}(V)p_{Na} + G''_{Na}(V)(1 - p_{Na})\,. \tag{2.2.39}$$

Finally, the conductance g_L can be reasonably assumed constant. Collecting Eqs.(2.2.34–2.2.39) and using several experimental data, Hodgkin and Huxley proposed the following model

$$\begin{cases} C\dfrac{dV}{dt} = 120n_{Na}^3 p_{Na}(115 - V) - 36n_K^4(12 + V) + \\ \qquad\qquad + 0.3(10.6 - V) + I \\ \dfrac{dn_K}{dt} = -\dfrac{e^{-V/80}}{8}n_K + \dfrac{10 - V}{100\left[e^{(10-V)/10} - 1\right]}(1 - n_K) \\ \dfrac{dn_{Na}}{dt} = -4e^{-5V/90}n_{Na} + \dfrac{25 - V}{10\left[e^{(25-V)/10} - 1\right]}(1 - n_{Na}) \\ \dfrac{dp_{Na}}{dt} = -\dfrac{p_{Na}}{1 + e^{3-V/10}} + \dfrac{7}{100}e^{-V/20}(1 - p_{Na})\,. \end{cases} \tag{2.2.40}$$

REMARK 2.2.4 This complex system of ordinary differential equations still models a very simple situation. In fact, it deals with a space independent phenomenon, while, of course, nerve impulse propagation is typically a space dependent problem. Furthermore, in order to study the action of agglomerates of nerve cells, further simplifications are needed.

A classical one is obtained by assuming that n_K and p_{Na} are constants and equal to their rest value. In this way the second and fourth equations of (2.2.40) are eliminated. ■

Keeping in mind the previous remark and in order to simplify the model, FitzHugh [FIT] used a completely different approach. In fact, he looked for a smaller number of simpler differential equations able to model the characteristics and the phenomena made evident by experiments and by the quantitative results obtained using the full Hodgkin and Huxley equations (2.2.40).

The basic ideas are the following:

- There are phenomena occurring at different time scales. One state variable, say u_2, may then represent the evolution of slowly varying quantities, such as n_K and p_{Na} qualitatively, and one state variable, say u_1, may represent the rapidly varying quantities, such as n_{Na} and V.
- The simplest evolution equation, that is, with a linear right-hand side, is assumed to govern the slowly varying variable u_2.
- The evolution equation for the rapidly varying variable u_1 is made of a nonlinear term chosen according to some stability consideration, which will be studied in Section 2.7.4, a coupling linear term, and a constant α, which represents the effect of an applied current.

Following these considerations FitzHugh proposed the following model

$$\begin{cases} \dfrac{du_1}{dt} = \dfrac{1}{\varepsilon}\left(u_1 - \dfrac{u_1^3}{3} - u_2 + \alpha\right) \\ \dfrac{du_2}{dt} = \varepsilon(u_1 - \beta u_2 + \gamma)\,, \end{cases} \tag{2.2.41}$$

where

$$\beta \in (0,1)\,, \quad \gamma \in \left(1 - \frac{2}{3}\beta, 1\right)\,, \quad \text{and} \quad \varepsilon \tag{2.2.42}$$

are parameters; ε is of a smaller order compared with unity.

2.3 Mathematical Formulation of Problems

As we have seen in the various examples proposed in Section 2.2, the modelling procedure leads to an evolution equation for the state variable. The study of the model essentially consists in the qualitative and quantitative analysis of the mathematical problem generated by associating with the evolution equation the conditions necessary for its solution.

Bearing this in mind, the following concepts need to be clarified:

- Consistency of the mathematical model;
- Good formulation of the mathematical problem;
- Well-posedness of the mathematical problem.

We need to be precise and remark that we will deal with systems of ordinary differential equations that may be written as a system of first-order equations in normal form as in Eq.(2.1.5). Referring now to the consistency and to some qualitative properties of the mathematical model, the following definitions can be introduced.

DEFINITION 2.3.1 Consistency of mathematical models
The mathematical model (2.1.5) is ***consistent*** *if the number of unknowns equals the number of linearly independent equations.*

DEFINITION 2.3.2 Autonomous model
The mathematical model (2.1.5) is ***autonomous*** *if the time t does not explicity appear as an argument of $\mathbf{f}$. On the other hand, if t appears in $\mathbf{f}$, the model is called* ***nonautonomous.***

DEFINITION 2.3.3 Linear model
A system of ordinary differential equations is ***linear*** *if it can be written as*

$$\frac{d\mathbf{u}}{dt} = \mathbf{A}(t)\mathbf{u} + \mathbf{b}(t)\,, \tag{2.3.1}$$

where $\mathbf{A}(t)$ is an $n \times n$ matrix and $\mathbf{b}(t)$ is a vector with n components.

REMARK 2.3.1 Let $A_{ij}(t)$ be the coefficients of $\mathbf{A}(t)$ and $b_i(t)$ be the components of the vector $\mathbf{b}(t)$, then Eq.(2.3.1) can be written explicity as

$$\begin{cases} \dfrac{du_1}{dt} & = A_{11}(t)u_1 + \cdots + A_{1n}(t)u_n + b_1(t) \\ & \vdots \\ \dfrac{du_n}{dt} & = A_{n1}(t)u_1 + \cdots + A_{nn}(t)u_n + b_n(t) \,. \end{cases} \tag{2.3.2}$$

If the system is autonomous, then **A** and **b** are, respectively, a constant matrix and a constant vector, therefore A_{ij}, $i, j = 1, \ldots, n$ and b_i, $i = 1, \ldots, n$ in (2.3.2) are constant. ■

REMARK 2.3.2 It is important to distinguish between linear and nonlinear systems. In fact, in the former case we can attempt analytic solutions (see Section 2.5), while, in the latter case, a qualitative analysis can only give partial answers to the behavior of the system and we are likely to need numerical methods to obtain quantitative results. ■

Consider Eq.(2.1.3) in the time interval $[0, T] \subset \mathbb{R}$. We call ***initial condition*** the value

$$u_i(t = 0) = u_{i0} \tag{2.3.3}$$

of the variable u_i at $t = 0$. In the same fashion we call ***limit condition*** the value

$$u_j(t = T) = u_{jT} \tag{2.3.4}$$

of the variable u_j at the time $t = T$.

Bearing this in mind, the following definitions can be given.

DEFINITION 2.3.4 Initial-value problem
*The **initial-value problem** is well formulated if Eq.(2.1.3) is associated with n **initial conditions** u_{i0} for $i = 1, \ldots, n$. The initial-value problem can be written, in integral vector form, as*

$$\mathbf{u}(t) = \mathbf{u}_0 + \int_0^t \mathbf{f}\big(s, \mathbf{u}(s)\big)\, ds \,. \tag{2.3.5}$$

DEFINITION 2.3.5 Boundary-value problem
The ***boundary-value problem*** *is well formulated if Eq.(2.1.3) is associated with $p < n$* ***initial conditions,*** *each associated with a different component and to $(n-p)$* ***limit conditions,*** *that is*

$$u_{k_i}(0) = u_{i0}\,, \quad i = 1, \ldots, p \quad \text{with} \quad k_i \neq k_j \quad \text{if} \quad i \neq j\,,$$
$$u_{k_i}(T) = u_{iT}\,, \quad i = p+1, \ldots, n \quad \text{with} \quad k_i \neq k_j \quad \text{if} \quad i \neq j\,.$$

The set of initial and limit conditions can be also called ***boundary conditions.***

Finally, we adopt the following terminology:

DEFINITION 2.3.6 Well-formulated problem
A problem is ***well formulated*** *if the evolution equation is associated with the correct number of initial or boundary conditions for its solution.*

DEFINITION 2.3.7 Well-posed problem
A problem is ***well posed*** *if it is characterized by existence, uniqueness, and continuous dependence of the solution on the initial data.*

As we shall see in the next section, a problem that is well formulated does not necessarily admit a solution in a prescribed time interval.

2.4 On Existence, Uniqueness, and Continuity

The main purpose of a model formulated for a physical system is to predict for a certain time interval its behavior starting from the knowledge of the state at $t = 0$. We have seen, in the preceding section, that an evolution equation linked to proper initial conditions generates a mathematical problem called the initial-value problem. The predictions of the model are obtained by solving such a problem. There are, then, some basic requirements a problem should satisfy:

1. The solution must exist, at least for the period of time desired.
2. The solution must be unique.
3. The solution must depend continuously on the initial data, so that if a little error is made in describing the present state, one has an estimate on the effect of this error in the future.

As stated in the previous section, if these requirements are satisfied, then the initial-value problem is said to be ***well posed.***

Furthermore, in the modelling procedure, physical reality is often simplified or evaluation errors are unconsciously committed. For this reason, it is often useful to have an estimate on the difference between the solutions of two similar models.

Bearing in mind these expectations, in this section we give some sample theorems for systems of ordinary differential equations.

In order to do that, we have to introduce a norm in $\mathbb{R}^n$, which might be, for instance, the Euclidean one

$$\|\mathbf{u}\| = \left(\sum_{i=1}^{n} u_i^2\right)^{1/2}, \tag{2.4.1}$$

and the following definition.

DEFINITION 2.4.1 Lipschitz condition
A vector function $\mathbf{f}(t,\mathbf{u})$ satisfies a ***Lipschitz condition*** *in a region $\mathcal{D}$ of the $(t,\mathbf{u})$-space if there exists a constant L (called* ***Lipschitz constant****), such that, for any $(t,\mathbf{u})$ and $(t,\mathbf{v})$ in $\mathcal{D}$*

$$\left\|\mathbf{f}(t,\mathbf{u}) - \mathbf{f}(t,\mathbf{v})\right\| \leq L\,\|\mathbf{u}-\mathbf{v}\|. \tag{2.4.2}$$

REMARK 2.4.1 Note that both terms on the left-hand side of (2.4.2) involve the same instant of time. ■

REMARK 2.4.2 The Lipschitz condition is a property between continuity and differentiability. In fact, it can be proved that if $\mathbf{f}(t,\mathbf{u})$ is defined in a bounded, closed, and convex(*) domain $\mathcal{D}$ and if the partial derivatives of $\mathbf{f}$ with respect to u_i exist with

$$\max_{i,j=1,\dots,n}\ \sup_{(t,\mathbf{u})\in\mathcal{D}} \left|\frac{\partial f_i}{\partial u_j}\right| = M\,,$$

then $\mathbf{f}$ satisfies a Lipschitz condition in $\mathcal{D}$ with Lipschitz constant equal to M. It also can be proved that if $\mathbf{f}(t,\mathbf{u})$ satisfies a Lipschitz condition in $\mathcal{D}$, then for each fixed t, $\mathbf{f}$ is a continuous function of $\mathbf{u}$ in $\mathcal{D}$. As is well known, the opposite is not true. ■

(*) A domain is convex if any segment joining two points of the domain lies entirely within the domain.

Considering that this textbook is not devoted to qualitative analysis, but to modelling and mathematical methods, some fundamental theorems on initial-value problems will be simply reported and commented without details on their proofs. We point out that boundary-value problems are more difficult to deal with in a simple way, and their treatment deserves particular attention, which falls beyond the scope of this section. However, in the book by Bailey, Shampine, and Waltman [BSW] the problem is proposed in a very understandable language. Almost every theorem has its own application and/or counter example, so the reader is actually driven with care along the path to the core of the problem.

We have to recall that, before applying any model, it is useful to lay its foundations on good mathematical ground, seeking the best theorems of existence, uniqueness, and continuity the mathematical problem allows. In this framework we recall here the following fundamental theorems.

THEOREM 2.4.1 ***(Existence)***
If the initial-value problem

$$\begin{cases} \dfrac{d\mathbf{u}}{dt} = \mathbf{f}(t, \mathbf{u}) \\ \mathbf{u}(t_0) = \mathbf{u}_0 \end{cases} \tag{2.4.3}$$

has $\mathbf{f}(t, \mathbf{u})$ *continuous in the rectangle*

$$\mathcal{R} = \big\{(t, \mathbf{u}) : \|\mathbf{u} - \mathbf{u}_0\| \leq K\,,\ |t - t_0| \leq T\big\}\,, \tag{2.4.4}$$

then there exists at least one solution of the initial-value problem defined and continuous for $|t - t_0| \leq \widehat{T}$, *where*

$$\widehat{T} = \min\left\{T, \frac{K}{M}\right\} \qquad \text{and} \quad M \geq \big\|\mathbf{f}(t, \mathbf{u})\big\| \quad \forall\,(t, \mathbf{u}) \in \mathcal{R}\,. \tag{2.4.5}$$

■

THEOREM 2.4.2 ***(Uniqueness)***
If besides the continuity condition given in the existence Theorem 2.4.1, the function $\mathbf{f}$ *of the initial-value problem (2.4.3) satisfies a Lipschitz condition in the rectangle* $\mathcal{R}$ *defined in Eq.(2.4.4), then there exists a unique solution of the initial-value problem defined and continuous for* $|t - t_0| \leq \widehat{T}$ *with* $\widehat{T}$ *defined in (2.4.5) for any initial condition. Furthermore,* $\|\mathbf{u}(t) - \mathbf{u}_0\| \leq M\widehat{T}$. ■

THEOREM 2.4.3 ***(Continuous Dependence on Initial Data)***
Let $\widehat{\mathbf{u}}$ and $\widetilde{\mathbf{u}}$ be the two solutions of the ordinary differential equation in (2.4.3) with initial data $\mathbf{u}(t_0) = \widehat{\mathbf{u}}_0$ and $\mathbf{u}(t_0) = \widetilde{\mathbf{u}}_0$, respectively. If $\mathbf{f}$ is continuous and satisfies the Lipschitz condition, then

$$\left\|\widehat{\mathbf{u}}(t) - \widetilde{\mathbf{u}}(t)\right\| \leq \left\|\widehat{\mathbf{u}}_0 - \widetilde{\mathbf{u}}_0\right\| e^{L|t-t_0|},$$

where L is the Lipschitz constant. ∎

THEOREM 2.4.4 ***(Continuous Dependence on f)***
Consider for $t \in [t_0, t_0 + T]$ the initial-value problems

$$\begin{cases} \dfrac{d\mathbf{u}}{dt} = \widehat{\mathbf{f}}(t, \mathbf{u}) \\ \mathbf{u}(t_0) = \widehat{\mathbf{u}}_0 \,, \end{cases} \qquad \text{and} \qquad \begin{cases} \dfrac{d\mathbf{u}}{dt} = \widetilde{\mathbf{f}}(t, \mathbf{u}) \\ \mathbf{u}(t_0) = \widetilde{\mathbf{u}}_0 \,, \end{cases} \tag{2.4.6}$$

with $\widehat{\mathbf{f}}$ and $\widetilde{\mathbf{f}}$ defined and continuous in a common domain $\mathcal{D}$.

If one of them satisfies a Lipschitz condition with constant L and if

$$\left\|\widehat{\mathbf{f}}(t, \mathbf{u}) - \widetilde{\mathbf{f}}(t, \mathbf{u})\right\| \leq \varepsilon\,, \qquad \forall\,(t, \mathbf{u}) \in \mathcal{D}\,,$$

then

$$\left\|\widehat{\mathbf{u}}(t) - \widetilde{\mathbf{u}}(t)\right\| \leq \left\|\widehat{\mathbf{u}}_0 - \widetilde{\mathbf{u}}_0\right\| e^{L|t-t_0|} + \frac{\varepsilon}{L}\left[e^{L|t-t_0|} - 1\right],$$

where $\widehat{\mathbf{u}}(t)$ and $\widetilde{\mathbf{u}}(t)$ are the solutions of the two initial-value problems defined in (2.4.6). ∎

REMARK 2.4.3 It should be noted that the Lipschitz condition is not needed to assure the existence of a solution of the initial-value problem; instead, it is essential in the uniqueness proof. Actually, the previous theorems can be slightly improved, especially by specializing the proofs to the particular case. For instance, existence results can be obtained for $\mathbf{f}(t, \mathbf{u})$ with a limited number of finite discontinuities [INC]. ∎

Although the previous theorems answer the questions posed at the beginning of this section and can be applied to most of the initial-value problems occurring in nature, they leave an important question unanswered: How large is the domain of the solution? Theorems 2.4.1 and

2.4.2 assure existence and uniqueness of the solution for $|t - t_0| \leq \widehat{T}$. But the evaluation of $\widehat{T}$ in (2.4.5) can lead to useless estimates on the existence interval. Of course, the domain might be larger, possibly extending to all $t \geq t_0$. One then needs criteria to determine the largest possible domain of existence. This question gives rise to the class of so-called ***extension theorems***, which can be found in classical textbooks on differential equations.

Of course, all the previous theorems state sufficient conditions to assure the thesis. A solution may exist even if $\mathbf{f}$ blows up to infinity, or may be unique even if $\mathbf{f}$ does not satisfy a Lipschitz condition. To understand this point the following example will be helpful.

Example 2.1 Particle dynamics

Consider a particle with mass m modelled as a mass point, which moves along a straight line with velocity u subject to a force proportional to u^α, $\alpha \in \mathbb{R}$, starting at $t = 0$ with velocity u_0.

The mathematical model is then

$$\begin{cases} \dfrac{du}{dt} = \beta u^\alpha \\ u(0) = u_0 \, . \end{cases} \tag{2.4.7}$$

Observe that the function u^α has the following properties:

- If $\alpha < 0$, it blows up to infinity when $u = 0$;
- If $\alpha = 0$, it satisfies the Lipschitz condition with $L = 0$;
- If $0 < \alpha < 1$, it is continuous but not Lipschitz (it is Lipschitz away from 0);
- If $\alpha = 1$, it satisfies the Lipschitz condition with $L = 1$;
- If $\alpha > 1$, it is continuous but not Lipschitz (it is locally Lipschitz).

Hence according to Theorems 2.4.1 and 2.4.2, existence is assured only for $\alpha \geq 0$ and uniqueness only for $\alpha = 0, 1$.

A second aspect to remember is that, in the above theorems, existence and uniqueness are proved for any initial condition u_0.

If $u_0 \neq 0$, the solution to (2.4.7) can be easily obtained by separation of variables

$$u(t) = \begin{cases} u_0 \left[1 + \dfrac{\beta(1-\alpha)}{u_0^{1-\alpha}} t \right]^{\frac{1}{1-\alpha}} & \text{if } \alpha \neq 1 \\ u_0 e^{\beta t} & \text{if } \alpha = 1 \, . \end{cases} \tag{2.4.8}$$

Things are more complex if $u_0 = 0$, as will become evident.

In order to better understand the differences occurring as α varies, we will focus our attention on some specific examples. We develop these examples keeping in mind the specific system (particle dynamics) modelled by the ordinary differential equation in Eqs.(2.4.7). The reader can verify the physical consistency of the results.

The solutions are plotted in Fig. 2.4.1.

1. $\alpha = -2$
   One can check easily that in this case the solution to (2.4.7) is
   $$u(t) = \sqrt[3]{u_0^3 + 3\beta t}\,,$$
   even for $u_0 = 0$ and therefore exists and is unique for all $t \geq 0$ and for any initial condition.

2. $\alpha = -1$
   In this case, if $u_0 \neq 0$, the solution (2.4.8) reduces to
   $$u(t) = u_0\sqrt{1 + \frac{2\beta t}{u_0^2}}\,. \tag{2.4.9}$$
   At this point several cases must be distinguished.
   • If $\beta > 0$, Eq.(2.4.9) defines a unique solution, which exists for any $t \geq 0$, but if $\beta < 0$, this unique solution exists only for
   $$0 \leq t \leq \frac{u_o^2}{2|\beta|}\,.$$
   • What happens if $u_0 = 0$? In this case, if $\beta > 0$ the initial-value problem looses uniqueness, since both $u = \sqrt{2\beta t}$ and $u = -\sqrt{2\beta t}$ satisfy the initial-value problem. If $\beta < 0$, the initial-value problem has no solution at all.
   • The difference between the cases $\alpha = -2$ and $\alpha = -1$ can be roughly explained as follows: In both (1) and (2), the function **f** satisfies the Lipschitz condition and is continuous away from $u = 0$, then when the solution approaches zero it may loose uniqueness or even existence. In (1) this does not occur. Instead, in (2) it occurs as soon as u vanishes. Actually, u never vanishes if $\beta > 0$. In fact, referring to (2.4.7), if, for instance, $u_0 > 0$, then β/u_0 is positive. Therefore, the solution increases initially, stays positive, and keeps

increasing. On the contrary, if $\beta < 0$ and, for instance, $u_0 > 0$, then $u(t)$ is always a decreasing function and crosses the t-axis, loosing existence.
• Hence, in summary, if $\beta > 0$ the solution exists, but uniqueness is lost if $u_0 = 0$. If, instead, $\beta < 0$, the solution does not even exist if $u_0 = 0$.

3. $\alpha = 2/3$
In this case, if $\beta u_0 > 0$, the solution

$$u(t) = u_0 \left(1 + \frac{\beta t}{3\sqrt[3]{u_0}}\right)^3 \tag{2.4.10}$$

is unique and defined in $[0, +\infty)$. If, instead, $u_0 = 0$, then trivially $u = 0$ and $u = (\beta t/3)^3$ are solutions, but there is actually an infinite set of solutions

$$u(t) = \begin{cases} 0\,, & \text{if } t \in [0, \hat{t}] \\ \left[\dfrac{\beta(t - \hat{t})}{3}\right]^3, & \text{otherwise} \end{cases}$$

for any choice of $\hat{t} > 0$.
Finally, if $\beta u_0 < 0$, then Eq.(2.4.10) is the unique solution in the interval $[0, 3\sqrt[3]{|u_0/\beta^3|}]$, i.e., till $u(t) \neq 0$, after that something similar to the case $u_0 = 0$ may occur, and uniqueness is lost.

4. $\alpha = 2$
In this case, the solution to the initial-value problem (2.4.7)

$$u(t) = \frac{u_0}{1 - \beta u_0 t}$$

is unique and exists for all $t \geq 0$ if $\beta u_0 \leq 0$ and blows up for $t = 1/\beta u_0$ if $\beta u_0 > 0$.

In these examples we encountered a fairly wide class of possibilities. The expected and actual results are summarized in Table 2.4.1. We point out, once again, that the results stated in the theorems are independent of u_0. The number of solutions and the maximal time of existence is indicated for $|\beta| = 1$.

The nonexistence results correspond to situations that are physically not admissible. □

	Theoretical prediction	$u_0 \neq 0$	$u_0 = 0$
$\alpha = -2$		1, $[+\infty]$	1, $[+\infty]$
$\alpha = -1$		$(\beta > 0)\ 1, [+\infty]$ / $(\beta < 0)\ 1, \left[\frac{u_0^2}{2}\right]$	$(\beta > 0)\ 2, [+\infty]$ / $(\beta < 0)\ 0$
$\alpha = 0$	uniqueness	1, $[+\infty]$	1, $[+\infty]$
$\alpha = \frac{2}{3}$	existence	$(\beta u_0 > 0)\ 1, [+\infty]$ / $(\beta u_0 < 0)\ \infty, [+\infty]$	∞, $[+\infty]$
$\alpha = 1$	uniqueness	1, $[+\infty]$	1, $[+\infty]$
$\alpha = 2$	existence	$(\beta u_0 > 0)\ 1, \left[\frac{1}{u_0}\right]$ / $(\beta u_0 < 0)\ 1, [+\infty]$	$1, [+\infty]$

Table 2.4.1 — Theoretical existence and uniqueness results, number of solutions, and, in square brackets, maximal existence time for $|\beta| = 1$.

2.5 Linear Systems

We have seen in Section 2.3 that some models are linear. However, in spite of their simplicity, even in this case it is not possible to give a general procedure to construct the analytic solution.

In spite of this, the study of linear systems

$$\frac{d\mathbf{u}}{dt} = \mathbf{A}(t)\mathbf{u} + \mathbf{b}(t)\,, \quad \mathbf{u} \in \mathbb{R}^n\,, \tag{2.5.1}$$

represents a fundamental starting point in the study of ordinary differential equations.

Even if one has to deal with a nonlinear ordinary differential equation, a useful way to understand some of its general features is to study its linearized version. Furthermore, the solution of some nonlinear problems can be approximated by a sequence of solutions of linear systems. In fact, several problems can be solved by suitable linearization methods [CAR], [JEF].

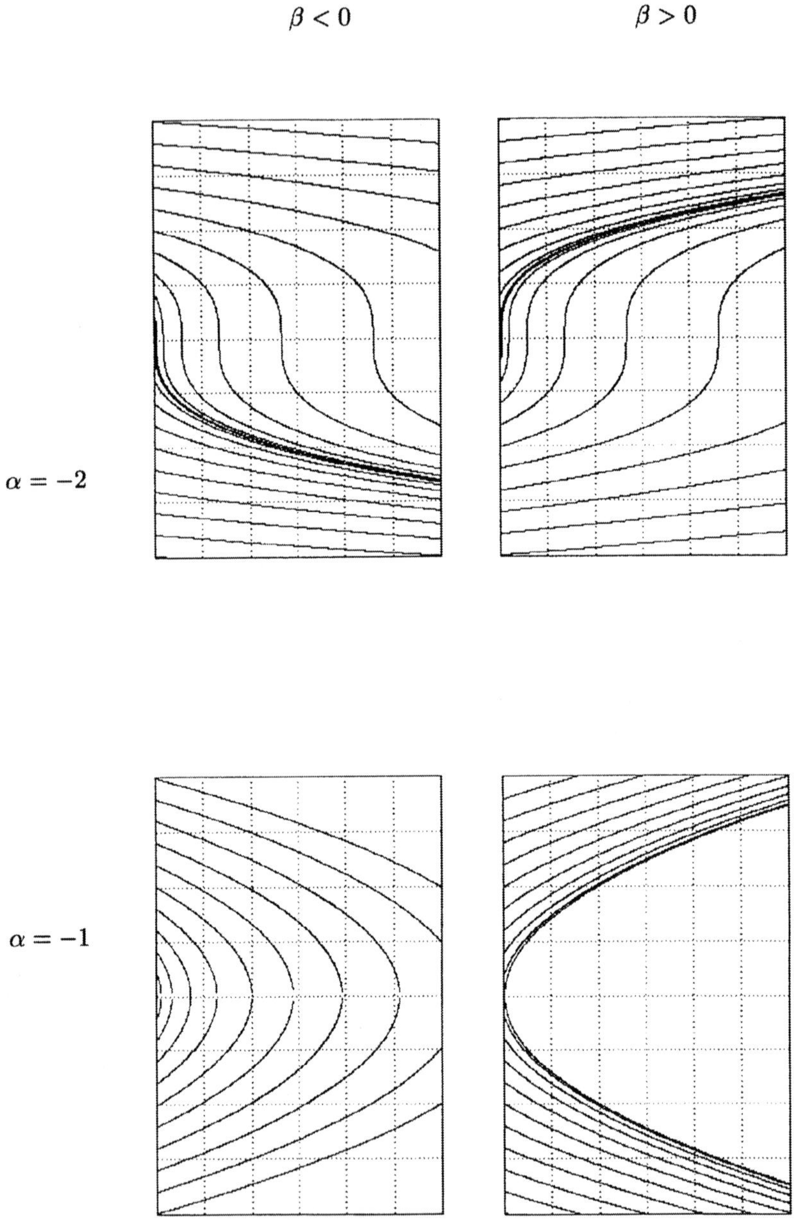

$\beta < 0$
$\beta > 0$
$\alpha = -2$
$\alpha = -1$

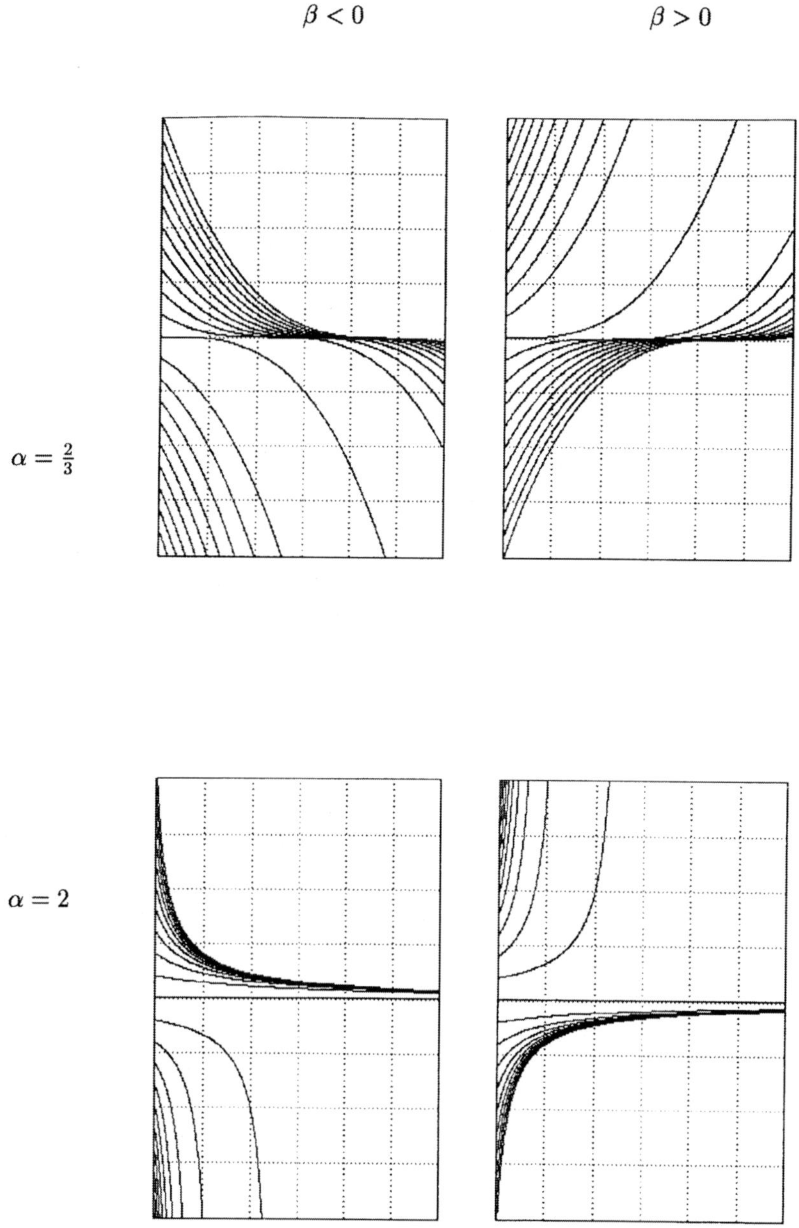

Figure 2.4.1 — Solutions of the particle dynamic problem (2.4.7) for several initial conditions and values of α and β.

The qualitative analysis of linear systems is widely dealt with in the literature and can be approached in several ways. First, we recall the following:

Proposition 2.5.1
Any linear system with continuous coefficients on a closed interval I satisfies the Lipschitz condition with

$$L = \sum_{i,j=1}^{n} \sup_{t \in I} |A_{ij}(t)| \,. \tag{2.5.2}$$

■

Hence, by the theorems stated in the previous section, the following results hold:

THEOREM 2.5.1 ***(Solutions of Linear Systems)***
The initial-value problem

$$\begin{cases} \dfrac{d\mathbf{u}}{dt} = \mathbf{A}(t)\mathbf{u} + \mathbf{b}(t) \\ \mathbf{u}(t_0) = \mathbf{u}_0 \,, \end{cases} \tag{2.5.3}$$

with $A_{ij}(t)$ and $b_i(t)$ defined and continuous for $|t-t_0| \leq T$, has a unique (and continuous) solution on $|t - t_0| \leq T$. ■

A peculiar characteristic, which makes easier and more interesting the study of linear systems, is the ***principle of superposition***. This property assures that if $\mathbf{u}_1, \ldots, \mathbf{u}_m$ are m solutions of the homogeneous linear system

$$\frac{d\mathbf{u}}{dt} = \mathbf{A}(t)\mathbf{u}\,, \qquad \mathbf{u} \in \mathbb{R}^n \,, \tag{2.5.4}$$

then any linear combination

$$\sum_{i=1}^{m} C_i \mathbf{u}_i \,, \quad C_1, \ldots, C_m \in \mathbb{R} \tag{2.5.5}$$

is still a solution of (2.5.4).

At this point, it is fundamental to introduce the following definition.

DEFINITION 2.5.1 Linearly independent solutions
A function $\mathbf{u}$ *is said to be* ***linearly dependent*** *on* $\mathbf{u}_1, \ldots, \mathbf{u}_m$ *if it can be written as*

$$\mathbf{u} = \sum_{i=1}^{m} C_i \mathbf{u}_i \,, \tag{2.5.6}$$

with $C_1, \ldots, C_m \in \mathbb{R}$. *On the other hand,* $\mathbf{u}_1, \ldots, \mathbf{u}_m$ *are* ***linearly independent,*** *if for all* $(C_1, \ldots, C_m) \in \mathbb{R}^m - \{\mathbf{0}\}$ *there exists a* $\hat{t}$ *such that*

$$\sum_{i=1}^{m} C_i \mathbf{u}_i(\hat{t}) \neq \mathbf{0} \,. \tag{2.5.7}$$

It can be proved that for (2.5.4) there exist n linearly independent solutions and that any $(n+1)$-th solution linearly depends on them. This means that to solve (2.5.4), one has to look for n linearly independent solutions. The general solution is then a linear combination of them. This set of independent solutions for the homogeneous problem is called the ***fundamental set.***

Going back to the nonhomogeneous linear initial-value problem stated in (2.5.3), its solution consists in the following steps:

Step 1 Find a fundamental set $\{\mathbf{u}_1, \ldots, \mathbf{u}_n\}$ for the homogeneous system

$$\frac{d\mathbf{u}}{dt} = \mathbf{A}(t)\mathbf{u}$$

related to (2.5.1), which will be called the ***reduced equation.***

Step 2 Find a particular solution $\tilde{\mathbf{u}}$ of the nonhomogeneous system (2.5.1), so that the general solution can be written as

$$\mathbf{u}(t) = \sum_{i=1}^{n} C_i \mathbf{u}_i(t) + \tilde{\mathbf{u}}(t) \,.$$

Step 3 Impose the n initial (or boundary) conditions to find the values of the coefficients $C_1, \ldots, C_n$.

The first step is the stumbling block. The second step can be classically completed by using the so-called ***Lagrange method of variation of parameters,*** while the third step is usually a matter of algebraic calculation.

The situation is much simpler in the case of systems of ordinary differential equations with constant coefficients

$$\frac{d\mathbf{u}}{dt} = \mathbf{A}\mathbf{u}\,, \qquad \mathbf{u} \in \mathbb{R}^n\,. \tag{2.5.8}$$

In this case, it is possible to give a general procedure to complete Step 1. In fact, one can look for solutions in the form

$$\mathbf{u}(t) = \mathbf{v}e^{\lambda t}\,. \tag{2.5.9}$$

Substituting (2.5.9) back into (2.5.8) indicates that λ and $\mathbf{v}$ have to satisfy the matrix equation

$$(\mathbf{A} - \lambda\mathbf{I})\mathbf{v} = \mathbf{0}\,, \tag{2.5.10}$$

which must be an eigenvalue and the corresponding eigenvector of the matrix $\mathbf{A}$.

If $\mathbf{A}$ has n linearly independent eigenvectors $\mathbf{v}_1, \dots, \mathbf{v}_n$ corresponding to the eigenvalues $\lambda_1, \dots, \lambda_n$ (which need not all be distinct), then the solution of (2.5.8) is

$$\mathbf{u} = C_1\mathbf{v}_1e^{\lambda_1 t} + \dots + C_n\mathbf{v}_ne^{\lambda_n t}\,, \tag{2.5.11}$$

where $C_1, \dots, C_n$ are arbitrary constants.

The situation becomes more complex if there are eigenvalues with multiplicity $r > 1$ with less that r corresponding eigenvectors.

In the simplest case, $\mathbf{A}$ has $n-2$ distinct eigenvalues $\lambda_1, \dots, \lambda_{n-2}$ with eigenvectors $\mathbf{v}_1, \dots, \mathbf{v}_{n-2}$ and a double eigenvalue $\lambda_{n-1} = \lambda_n = \lambda$ corresponding to a single eigenvector $\mathbf{v}$. In this case, the general solution of (2.5.8) is

$$\mathbf{u} = C_1\mathbf{v}_1e^{\lambda_1 t} + \dots + C_{n-2}\mathbf{v}_{n-2}e^{\lambda_{n-2} t} + \big[C_{n-1}\mathbf{v} + C_n(\mathbf{w} + \mathbf{v}t)\big]e^{\lambda t}\,, \tag{2.5.12}$$

where $C_1, \dots, C_n$ are arbitrary constants and $\mathbf{w}$ is a solution of

$$(\mathbf{A} - \lambda\mathbf{I})\mathbf{w} = \mathbf{v}\,. \tag{2.5.13}$$

If, instead, for instance, $\mathbf{A}$ has $n-3$ distinct eigenvalues $\lambda_1, \dots, \lambda_{n-3}$ with eigenvectors $\mathbf{v}_1, \dots, \mathbf{v}_{n-3}$ and a triple eigenvalue $\lambda_{n-2} = \lambda_{n-1} =$

$\lambda_n = \lambda$ corresponding to a single eigenvector $\mathbf{v}$, the general solution of (2.5.8) is

$$\begin{aligned}\mathbf{u} = & C_1\mathbf{v}_1 e^{\lambda_1 t} + \cdots + C_{n-3}\mathbf{v}_{n-3}e^{\lambda_{n-3}t} \\ & + \left[C_{n-2}\mathbf{v} + C_{n-1}(\mathbf{w}_1 + \mathbf{v}t) + C_n(\mathbf{w}_3 + \mathbf{w}_2 t + \mathbf{v}t^2)\right]e^{\lambda t},\end{aligned} \tag{2.5.14}$$

where $C_1, \ldots, C_n$ are arbitrary constants and $\mathbf{w}_1$, $\mathbf{w}_2$, and $\mathbf{w}_3$ are, respectively, solutions of

$$\begin{cases}(\mathbf{A} - \lambda\mathbf{I})\mathbf{w}_1 = \mathbf{v} \\ (\mathbf{A} - \lambda\mathbf{I})\mathbf{w}_2 = 2\mathbf{v} \\ (\mathbf{A} - \lambda\mathbf{I})\mathbf{w}_3 = \mathbf{w}_2 .\end{cases} \tag{2.5.15}$$

From these examples, one can understand how to proceed in other cases.

REMARK 2.5.1 In general the eigenvalues are complex and therefore the solution is a linear combination of exponentials times sines and cosines (possibly times polynomials). ■

We conclude this section by considering the case in which (2.5.8) is derived from the single ordinary differential equation

$$\frac{d^n u}{dt^n} + a_{n-1}\frac{d^{n-1}u}{dt^{n-1}} + \cdots + a_1\frac{du}{dt} + a_0 u = 0 . \tag{2.5.16}$$

In this case, the solution is related to the roots of the characteristic polynomial

$$\lambda^n + a_{n-1}\lambda^{n-1} + a_1\lambda + a_0 = 0 \tag{2.5.17}$$

as follows:

1. Any distinct root of (2.5.17) corresponds to a solution of the form $Ce^{\lambda t}$ with C an arbitrary constant;
2. Any root of multiplicity r corresponds to a solution of the form $P_{r-1}(t)e^{\lambda t}$, where $P_{r-1}(t)$ is the general polynomial of degree $r - 1$.

For instance, if (2.5.17) with $n = 8$ has three distinct roots λ_1, λ_2, and λ_3 of multiplicity 1, a root λ_4 of multiplicity 2, and a root λ_5 of multiplicity 3, then the solution of (2.5.16) is

$$\begin{aligned}u(t) = & C_1 e^{\lambda_1 t} + C_2 e^{\lambda_2 t} + C_3 e^{\lambda_3 t} + (C_4 + C_5 t)e^{\lambda_4 t} \\ & + (C_6 + C_7 t + C_8 t^2)e^{\lambda_5 t} .\end{aligned} \tag{2.5.18}$$

2.6 Stability and Linearization

As already mentioned, it is possible to give a general procedure to search for analytic solutions only for some classes of linear differential equations and for very special nonlinear problems.

In most cases it is impossible to find analytic solutions. It is then desirable to have at least some knowledge of the qualitative behavior of the solution.

One may wonder about the following questions:

1. Is there a state such that, if it is occupied initially, it is mantained throughout the evolution of the system?
2. What happens if this state is not exactly occupied? Does the system remain "near" this state or not?
3. In the case of a positive answer, is it possible to write an analytic solution of the motion about this particular state?
4. What happens if there are small errors in the modelling equations?

The answers to the questions listed above are of relevant importance in the analysis of mathematical models. In fact, such answers can spread some light on the qualitative behavior predicted by the model. The present and the following section will give some criteria for carrying on this project.

2.6.1 Equilibrium configurations

The first question is readily solved by introducing the following definition.

DEFINITION 2.6.1 Equilibrium configurations
The states $\mathbf{u}_e$ such that

$$\mathbf{f}(\mathbf{u}_e) = \mathbf{0} \tag{2.6.1}$$

*are called **equilibrium configurations** of the autonomous system*

$$\frac{d\mathbf{u}}{dt} = \mathbf{f}(\mathbf{u}) . \tag{2.6.2}$$

In fact, if the following initial condition

$$\mathbf{u}(t = 0) = \mathbf{u}_e \tag{2.6.3}$$

is joined to (2.6.2), the initial-value problem (2.6.2–2.6.3) is solved by $\mathbf{u}(t) = \mathbf{u}_e$. Hence, to find the set of equilibrium solutions, it is necessary to solve a generally nonlinear system of algebraic equations.

Example 2.2 Equilibrium of the rotating pendulum
To find the relative equilibrium configuration (in the rotating frame) of the rotating pendulum dealt with in Section 2.2.2, one has to set the right-hand side of the evolution equation (2.2.15) equal to zero, obtaining

$$\begin{cases} u_2 = 0 \\ -\beta \sin u_1 (1 + \alpha \cos u_1) = 0 \,. \end{cases} \tag{2.6.4}$$

Solving (2.6.4) gives the configurations

$$\begin{cases} u_1 = 0\,, & u_2 = 0\,, \\ u_1 = \pi\,, & u_2 = 0\,, \\ u_1 = \cos^{-1}\dfrac{1}{\alpha}\,, & u_2 = 0\,, \end{cases} \tag{2.6.5}$$

where the third solution is possible only if $|\alpha| > 1$. □

2.6.2 The meaning of stable and unstable equilibrium

Assume that we now want to position the system in the equilibrium state. There are essentially two kinds of difficulties in accomplishing this project:

- Are we sure to position the system right in $\mathbf{u}_e$? For instance, in the previous example, our hand may tremble when placing the pendulum in the vertical position or there might be an imperfection in the localization of the centre of gravity (due for instance to inhomogeneity of the materials) or in the hinging mechanism.

- Are we sure that even if we succeed in positioning the system right in its equilibrium configuration, there will be no disturbance at all to perturb it? In the previous example, the angular velocity might not be exactly constant, there might be vibrations due to the presence of the engine, the hinge might tremble, or there might be a puff of air due, for instance, to our movements or even to our breathing, and so on.

The problem is, then, whether an initially perturbed system gradually returns to the equilibrium state, or at least remains next to it, or wanders away. These concepts are the object of the following definitions.

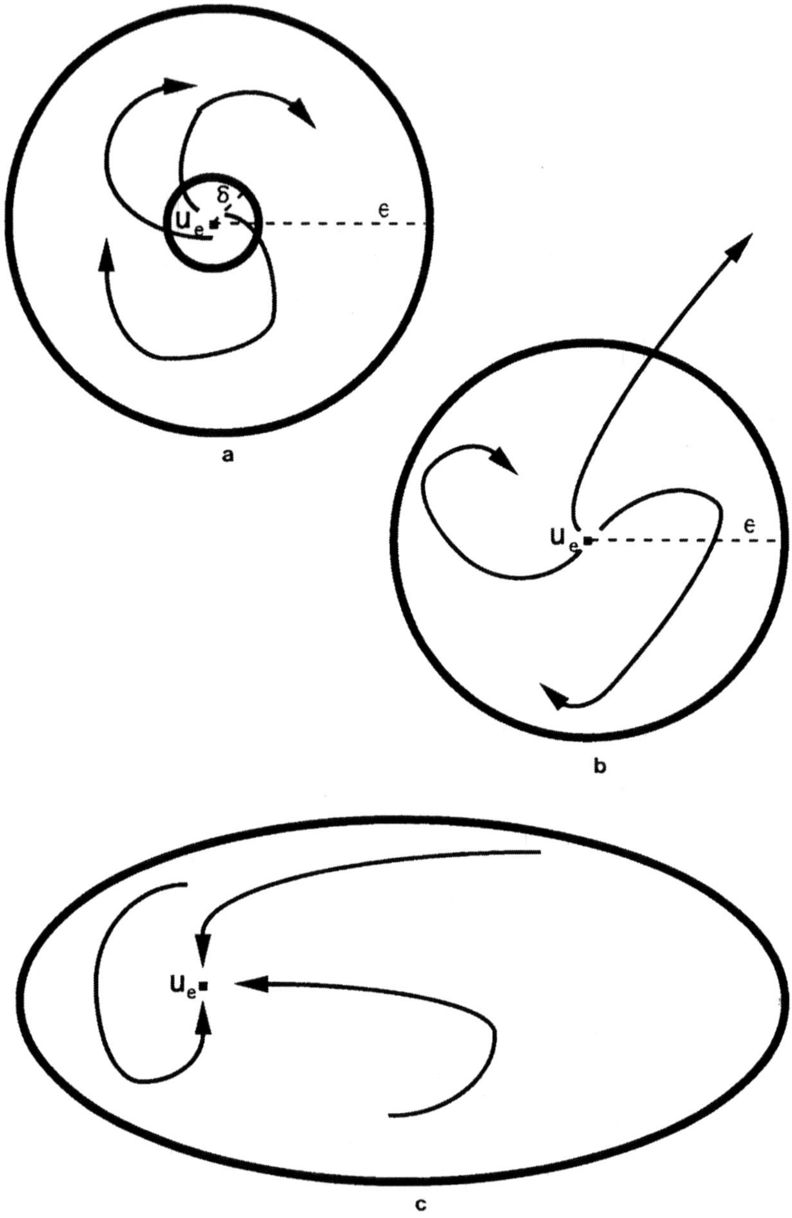

Figure 2.6.1 — Examples of (a) stable (b) unstable and (c) asymptotically stable equilibrium. In (c) $\mathcal{D}_e$ represents the basin of attraction. In (b) no matter how small we choose the neighborhood of $\mathbf{u}_e$, there always exists an initial condition $\mathbf{u}(0)$ such that $\mathbf{u}(t)$ sooner or later escapes from the ball of radius ε.

DEFINITION 2.6.2 Stable equilibrium
The equilibrium state $\mathbf{u}_e$ *is called* ***stable*** *if for any* $\varepsilon > 0$ *it is possible to find a* $\delta(\varepsilon) > 0$ *such that for any initial condition* $\mathbf{u}(0)$ *with*

$$\left\|\mathbf{u}(0) - \mathbf{u}_e\right\| < \delta(\varepsilon)\,,$$

the solution is such that

$$\left\|\mathbf{u}(t) - \mathbf{u}_e\right\| < \varepsilon \qquad \forall t \geq 0\,.$$

DEFINITION 2.6.3 Asymptotically stable equilibrium
A stable equilibrium state $\mathbf{u}_e$ *is also* ***asymptotically stable*** *if there is a neighborhood* $\mathcal{D}_e$ *of* $\mathbf{u}_e$ *such that if* $\mathbf{u}(0) \in \mathcal{D}_e$*, then*

$$\lim_{t \to +\infty} \mathbf{u}(t) = \mathbf{u}_e\,.$$

The largest possible $\mathcal{D}_e$ *is called the* ***basin of attraction*** *of* $\mathbf{u}_e$*.*

Of course, an equilibrium state that is not stable is called ***unstable***.

REMARK 2.6.1 The unstable (although equilibrium) configurations are impossible to be observed in nature. For this reason it is important to determine, among all equilibrium configurations, the stable ones. ■

REMARK 2.6.2 It is essential to realize that the definition of stability is local in nature. In fact, if we want the system to remain within a given tolerance near $\mathbf{u}_e$, the initial condition must be sufficiently close to $\mathbf{u}_e$. However, it is also essential to realize that an equilibrium position can be stable to small perturbations, but unstable to large ones. In this case, if the equilibrium position is asymptotically stable, the basin of attraction $\mathcal{D}_e$ is not the entire space, and only the solutions starting in $\mathcal{D}_e$ will tend towards $\mathbf{u}_e$. If this is the case, then $\mathbf{u}_e$ is said to be ***conditionally stable***, if not, then $\mathbf{u}_e$ is said to be ***globally stable***. ■

2.6.3 Linear stability

The simplest approach to establish the stability properties of an equilibrium configuration is to investigate what happens if the perturbation is very small. The stability condition we end up with is often referred to as the ***linear stability criterion***.

The starting point is the expansion of $\mathbf{f}$ in a Taylor series about the equilibrium state, which can be performed under a suitable regularity assumption on $\mathbf{f}$. In this way one can approximate $f_i(\mathbf{u})$, $i = 1, \dots, n$ with

$$f_i(\mathbf{u}) \cong f_i(\mathbf{u}_e) + (\mathbf{u} - \mathbf{u}_e) \cdot \nabla f_i(\mathbf{u}_e) + o\big(|\mathbf{u} - \mathbf{u}_e|\big) . \tag{2.6.6}$$

From the definition of equilibrium point $f_i(\mathbf{u}_e) = 0$ and therefore, for infinitesimal perturbations $\mathbf{v} = \mathbf{u} - \mathbf{u}_e$ about the equilibrium state, we can neglect the higher-order terms in the expansion and the differential system (2.6.2) can be approximated with its linearized form

$$\frac{d\mathbf{v}}{dt} = \mathbf{J}(\mathbf{u}_e)\mathbf{v} , \tag{2.6.7}$$

where $\mathbf{J}(\mathbf{u})$ is the Jacobian $n \times n$ matrix of the vector $\mathbf{f}$. The elements of such a matrix are

$$J_{ij} = \frac{\partial f_i}{\partial u_j}$$

so that the expression of $\mathbf{J}$ is

$$\mathbf{J} = \begin{pmatrix} \dfrac{\partial f_1}{\partial u_1} & \dfrac{\partial f_1}{\partial u_2} & \cdots & \dfrac{\partial f_1}{\partial u_n} \\ \dfrac{\partial f_2}{\partial u_1} & \dfrac{\partial f_2}{\partial u_2} & \cdots & \dfrac{\partial f_2}{\partial u_n} \\ \vdots & \vdots & \ddots & \vdots \\ \dfrac{\partial f_n}{\partial u_1} & \dfrac{\partial f_n}{\partial u_2} & \cdots & \dfrac{\partial f_n}{\partial u_n} \end{pmatrix} . \tag{2.6.8}$$

Equation (2.6.7) is a linear differential system with constant coefficients, which can be solved using the method described at the end of Section 2.5.

Recalling the results obtained there, one can write the solution as a sum of exponentials times cosine and sine functions (and possibly time polynomials). As $t \to \infty$, the dominating term is the exponential with the largest coefficient, which is the real part of an eigenvalue of the

Jacobian computed at equilibrium. Hence if the real part of all the eigenvalues is negative, then $\mathbf{v}$ tends to zero: The perturbation fades away. It is possible to relate these observations on the linearized system (2.6.7) to the behavior of the nonlinear system (2.6.2) via the following classic ***linearized stability criterion***.

THEOREM 2.6.1 ***(Linearized Stability Criterion)***
If $\mathbf{f}(\mathbf{u})$ *is twice continuously differentiable, denoting by* λ_i *the eigenvalues of the Jacobian matrix evaluated at the equilibrium state and by* $\Re e(\lambda_i)$ *the real part of* λ_i*, then:*

If $\forall i = 1, \ldots, n \;\; \Re e(\lambda_i) < 0$*, then* $\mathbf{u}_e$ *is asymptotically stable;*
If $\exists \hat{\imath}$ *such that* $\Re e(\lambda_{\hat{\imath}}) > 0$*, then* $\mathbf{u}_e$ *is unstable.* ■

REMARK 2.6.3 It is crucial to remark that this theorem guarantees the existence of a ***sufficiently small*** neighborhood $\mathcal{D}_e$ of $\mathbf{u}_e$ such that if $\mathbf{u}(0) \in \mathcal{D}_e$, then $\mathbf{u}(t)$ tends to $\mathbf{u}_e$, but does not give an algorithm for the actual computation of the basin of attraction $\mathcal{D}_e$. For this reason, this criterium is also named ***stability with respect to infinitesimal perturbations***. On the other hand, if $\mathbf{u}_e$ is linearly unstable, i.e., unstable with respect to infinitesimal perturbations, then it is also unstable to larger perturbations. However, nothing can be said, at the moment, about the long term behavior of the solution $\mathbf{u}(t)$, since the approximation (2.6.6) looses its validity as $\mathbf{u}(t)$ departs from $\mathbf{u}_e$. It may happen that $\mathbf{u}(t)$ goes to infinity, stays bounded, tends to a periodic orbit, or falls into another basin of attraction, tending toward a new stable configuration. ■

REMARK 2.6.4 If there is an $\hat{\imath}$ such that $\Re e(\lambda_{\hat{\imath}}) = 0$ and $\forall i \neq \hat{\imath}$, $\Re e(\lambda_i) < 0$, then the equilibrium configuration is called ***marginally stable*** (or neutrally stable). This is a "linear" concept. Nothing can be said in this case about the behavior of the solution of the nonlinear system. ■

Example 2.3 Linear stability for two-dimensional systems
Before going on, it is useful to specialize the form of the solution in the simple case where $n = 2$.

In this case one may have the following nondegenerated situations (distinct eigenvalues and $\det \mathbf{J} \neq 0$):

1. λ_1, λ_2 real
   Recalling (2.5.11), the solution of (2.6.7) can be written as

$$\mathbf{v}(t) = C_1 \mathbf{v}_1 e^{\lambda_1 t} + C_2 \mathbf{v}_2 e^{\lambda_2 t} \,,$$

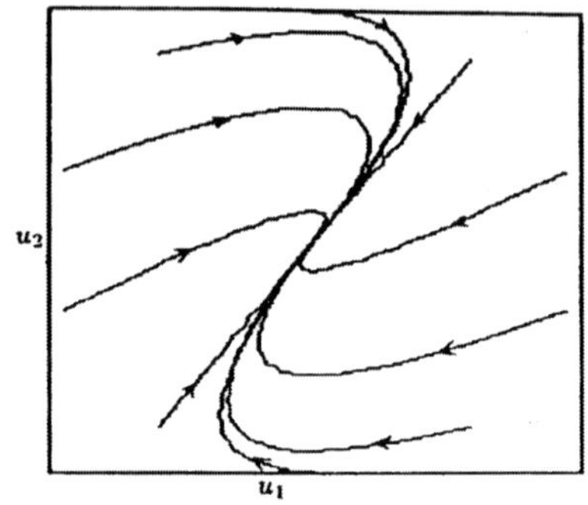

(a) Stable nodal point

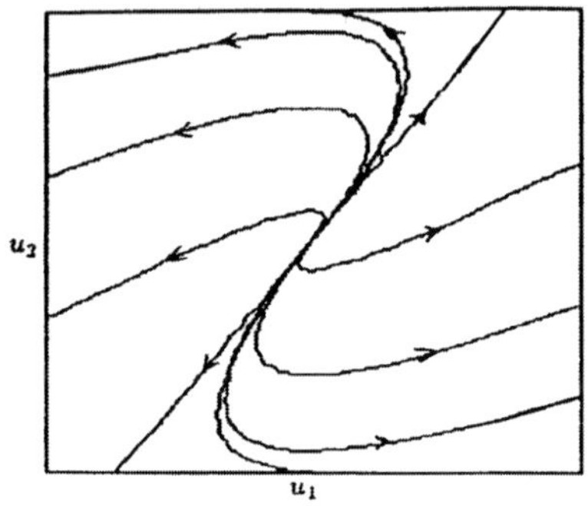

(b) Unstable nodal point

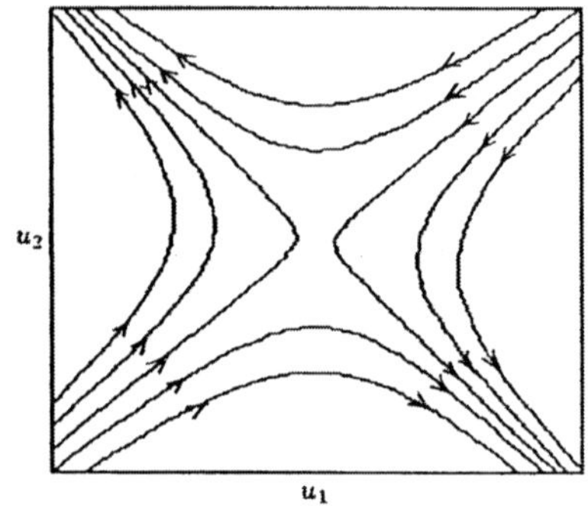

(c) Saddle point

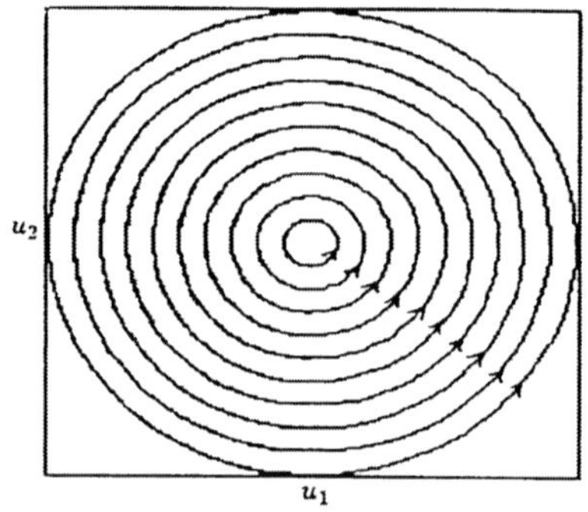

(d) Vortex point

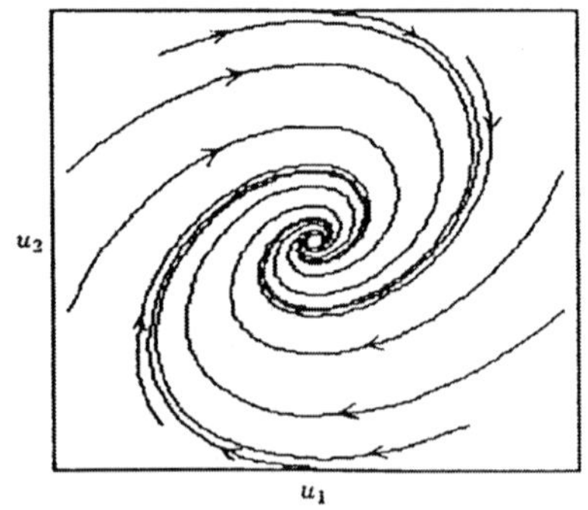

(e) Stable focal point

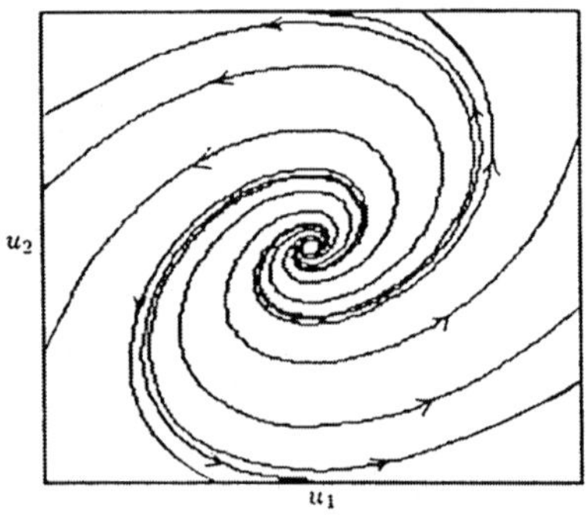

(f) Unstable focal point

Figure 2.6.2 — Orbits of two-dimensional linear systems in the phase plane.

where C_1 and C_2 are integration constants and $\mathbf{v}_1$ and $\mathbf{v}_2$ are the eigenvectors of $\mathbf{J}$.
If λ_1 and λ_2 are both negative, then

$$\lim_{t\to+\infty} \mathbf{v}(t) = \mathbf{0},$$

and this equilibrium point is called a ***stable node*** and is represented in the phase space in Fig. 2.6.2a.
If, instead, one of the eigenvalues is positive, then

$$\lim_{t\to+\infty} \mathbf{v}(t) = +\infty.$$

The equilibrium point is called an ***unstable node*** if both eigenvalues are positive (Fig. 2.6.2b) and a ***saddle point*** otherwise (Fig. 2.6.2c). In this latter case, two of the trajectories meet at the equilibrium point and two depart from it. They are therefore called ***stable and unstable manifold***.

2. $\lambda_1 = \lambda + i\omega$, $\lambda_2 = \lambda - i\omega$ complex conjugate
   In this case the solution of (2.6.7) can be written as

$$\mathbf{v}(t) = e^{\lambda t}\big[(C_1\mathbf{v}_1 + C_2\mathbf{v}_2)\cos\omega t + (C_2\mathbf{v}_1 - C_1\mathbf{v}_2)\sin\omega t\big],$$

   where $\mathbf{v}_1 + i\mathbf{v}_2$ is the eigenvector corresponding to $\lambda + i\omega$.
   If $\lambda = \Re e(\lambda_1) = \Re e(\lambda_2) < 0$, then

$$\lim_{t\to+\infty} \mathbf{v}(t) = 0$$

   and the equilibrium point is called a ***stable focus*** (Fig. 2.6.2e).
   If $\lambda = \Re e(\lambda_1) = \Re e(\lambda_2) > 0$, then

$$\lim_{t\to+\infty} \mathbf{v}(t) = +\infty$$

   and the equilibrium point is called an ***unstable focus*** (Fig. 2.6.2f). Finally, if $\lambda = \Re e(\lambda_1) = \Re e(\lambda_2) = 0$, then $\mathbf{v}(t)$ stays bounded and the equilibrium state is a ***vortex point*** (Fig. 2.6.2d).

Finally we remark that for $n = 2$ the eigenvalues of $\mathbf{J}$ are

$$\lambda = \frac{\operatorname{tr}\mathbf{J}}{2} \pm \sqrt{\frac{(\operatorname{tr}\mathbf{J})^2}{4} - \det\mathbf{J}}, \qquad (2.6.9)$$

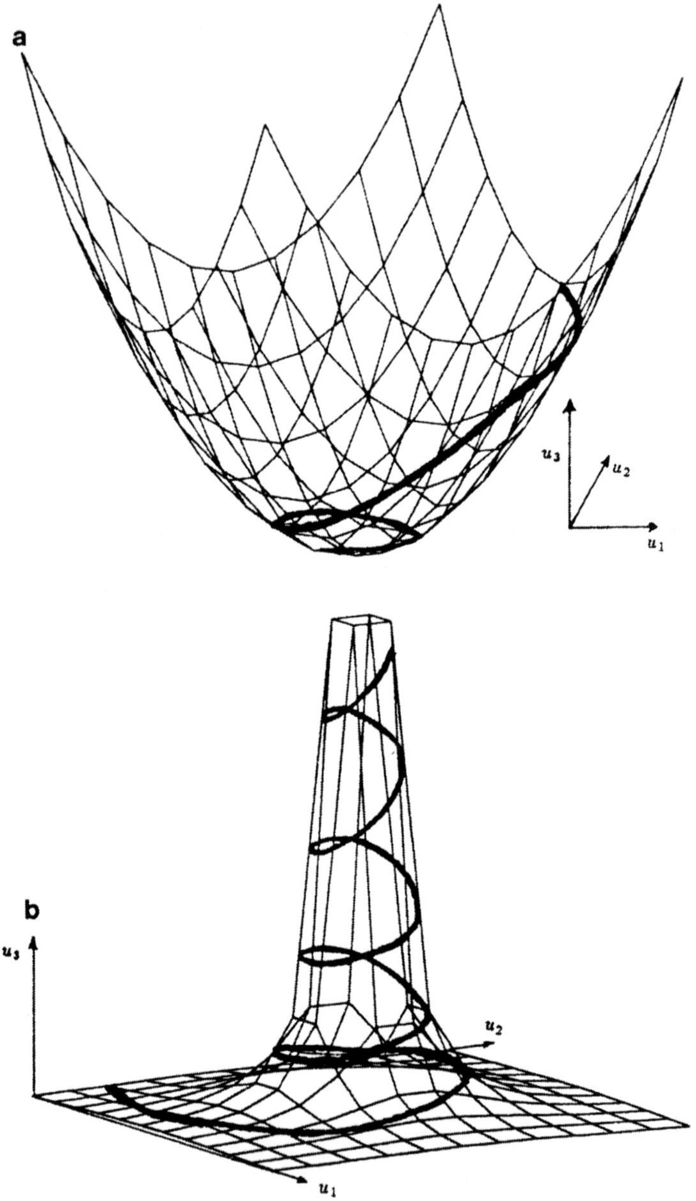

Figure 2.6.3 — Orbits of three-dimensional linear systems in the phase space when two eigenvalues (λ_1 and λ_2) are complex conjugate and one (λ_3) is real (a) $\Re_e(\lambda_1)\lambda_3 > 0$ (b) $\Re_e(\lambda_1)\lambda_3 < 0$. If $\lambda_3 > 0$ the trajectories are directed upward, otherwise they are directed downward.

where

$$\mathrm{tr}\,\mathbf{J} = \sum_{i=1}^{2} J_{ii} = J_{11} + J_{22}\,.$$

Therefore the eigenvalues have negative real part if and only if

$$\mathrm{tr}\;\mathbf{J}(\mathbf{u}_e) < 0 \qquad \text{and} \qquad \det \mathbf{J}(\mathbf{u}_e) > 0\,. \tag{2.6.10}$$

For $n \geq 3$ the classification of the equilibrium states is much more complex. Already for $n = 3$ there are 10 types of nondegenerate cases [ARN]. Some of them are plotted in Fig. 2.6.3. □

Example 2.4 ***Linear stability of Liénard's equation***
Consider the nonlinear circuit models introduced in Section 2.2.4 and in particular Liénard's equation (2.2.33) without forcing term

$$\frac{d^2u}{dt^2} + a(u)\frac{du}{dt} + b(u) = 0\,, \tag{2.6.11}$$

which can be written as a system of ordinary differential equations as

$$\begin{cases} \dfrac{du_1}{dt} = u_2 \\[2mm] \dfrac{du_2}{dt} = -a(u_1)u_2 - b(u_1)\,. \end{cases}$$

Its equilibrium configurations are given by $\mathbf{u}_e = (u_{1e}, u_{2e})$ with $u_{2e} = 0$ and u_{1e} such that $b(u_{1e}) = 0$.

The Jacobian is

$$\mathbf{J}(\mathbf{u}_e) = \begin{pmatrix} 0 & 1 \\ -\dfrac{da}{du_1}(u_1)u_2 - \dfrac{db}{du_1}(u_1) & -a(u_1) \end{pmatrix}(u_{1e}, 0)$$
$$= \begin{pmatrix} 0 & 1 \\ -\dfrac{db}{du_1}(u_{1e}) & -a(u_{1e}) \end{pmatrix}$$

and therefore

$$\mathrm{tr}\;\mathbf{J}(\mathbf{u}_e) = -a(u_{1e})\,, \qquad \det \mathbf{J}(\mathbf{u}_e) = \frac{db}{du_1}(u_{1e})\,.$$

Then, according to Eq.(2.6.10) if $a(u_{1e})$ and $db/du_1\,(u_{1e})$ are both positive, then the equilibrium solution $\mathbf{u}_e$ is asymptotically stable.

As an example, consider the modified Van der Pol equation (2.2.32) without forcing term

$$\frac{d^2u}{dt^2} - (\alpha + \beta u^2)\frac{du}{dt} + u = 0\,, \qquad (2.6.12)$$

which has a unique equilibrium for $u = du/dt = 0$, which, according to what was just stated, is stable if α is negative.

In more detail, the eigenvalues of the Jacobian

$$\mathbf{J}(\mathbf{u}_e) = \begin{pmatrix} 0 & 1 \\ -1 & \alpha \end{pmatrix}$$

are

$$\lambda = \frac{\alpha \pm \sqrt{\alpha^2 - 4}}{2}\,,$$

which are complex conjugate if $|\alpha| < 2$. One can then summarize the result in the Table 2.6.1.

α	Eigenvalues	Stability result
$\alpha \leq -2$	real and negative	stable node
$-2 < \alpha < 0$	complex with $\Re e(\lambda) < 0$	stable focus
$\alpha = 0$	purely imaginary ($\lambda = \pm i$)	vortex point
$0 < \alpha < 2$	complex with $\Re e(\lambda) > 0$	unstable focus
$\alpha \geq 2$	real and positive	unstable node

Table 2.6.1 — Classification of equilibrium points of (2.6.12).

□

2.6.4 Liapunov method

The linear stability criterium guarantees stability with respect to indefinitely small disturbances. To obtain this result one must solve the linear system (2.6.7), which is an approximation of the real model. In order to obtain more information on nonlinear stability criteria by using the right-hand side only of the ordinary differential equation, Liapunov suggested a method, which for this reason is usually called ***Liapunov direct method***. For those who are not familiar with stability theory, we can anticipate that the method consists in finding a function, called ***Liapunov function***, which essentially plays the role of a "generalized energy" for the system. If this "energy" decreases as the system evolves, i.e., if the system is "dissipative," then the system will tend to a stable configuration.

With this in mind we introduce the following concept.

DEFINITION 2.6.4 Liapunov function
Let $\mathbf{u}_e$ be an isolated equilibrium point of $d\mathbf{u}/dt = \mathbf{f}(\mathbf{u})$ in some open neighborhood $\mathcal{D}_e$ of $\mathbf{u}_e$. Suppose that there exists a function $V = V(\mathbf{u})$ which satisfies the following properties:

- *It is continuous in $\mathcal{D}_e$ and differentiable in $\mathcal{D}_e - \{\mathbf{u}_e\}$;*
- *It has a local minimum in $\mathbf{u}_e$, for instance*

$$V(\mathbf{u}_e) = 0\,, \qquad V(\mathbf{u}) > 0 \qquad \text{if} \ \ \mathbf{u} \in \mathcal{D}'_e - \{\mathbf{u}_e\}$$

  with $\mathcal{D}'_e$ open neighborhood of $\mathbf{u}_e$;
- *It is a nonincreasing function of time over any solution $\mathbf{u}(t)$ with initial condition $\mathbf{u}(0) = \mathbf{u}_0 \in \mathcal{D}_e - \{\mathbf{u}_e\}$ and therefore*

$$\frac{dV}{dt}\big(\mathbf{u}(t)\big) = \sum_{i=1}^{n} \frac{\partial V}{\partial u_i}\frac{du_i}{dt} = \sum_{i=1}^{n} \frac{\partial V}{\partial u_i} f_i\big(\mathbf{u}(t)\big) \le 0\,, \qquad \forall t \ge 0\,, \tag{2.6.13}$$

  *over every differentiable solution with initial condition $\mathbf{u}_0$. Then $V(\mathbf{u})$ is called a **Liapunov function**.*

THEOREM 2.6.2 ***(Liapunov Theorem)***
If there exists a Liapunov function V in a neighborhood $\mathcal{D}_e$ of the isolated equilibrium state $\mathbf{u}_e$, then $\mathbf{u}_e$ is stable. If, furthermore,

$$\frac{dV}{dt}\big(\mathbf{u}(t)\big) < 0\,,$$

then $\mathbf{u}_e$ is also asymptotically stable. ■

REMARK 2.6.5 The condition $dV/dt\big(\mathbf{u}(t)\big) < 0$ means that V decreases as the system evolves. This implies that the trajectory will

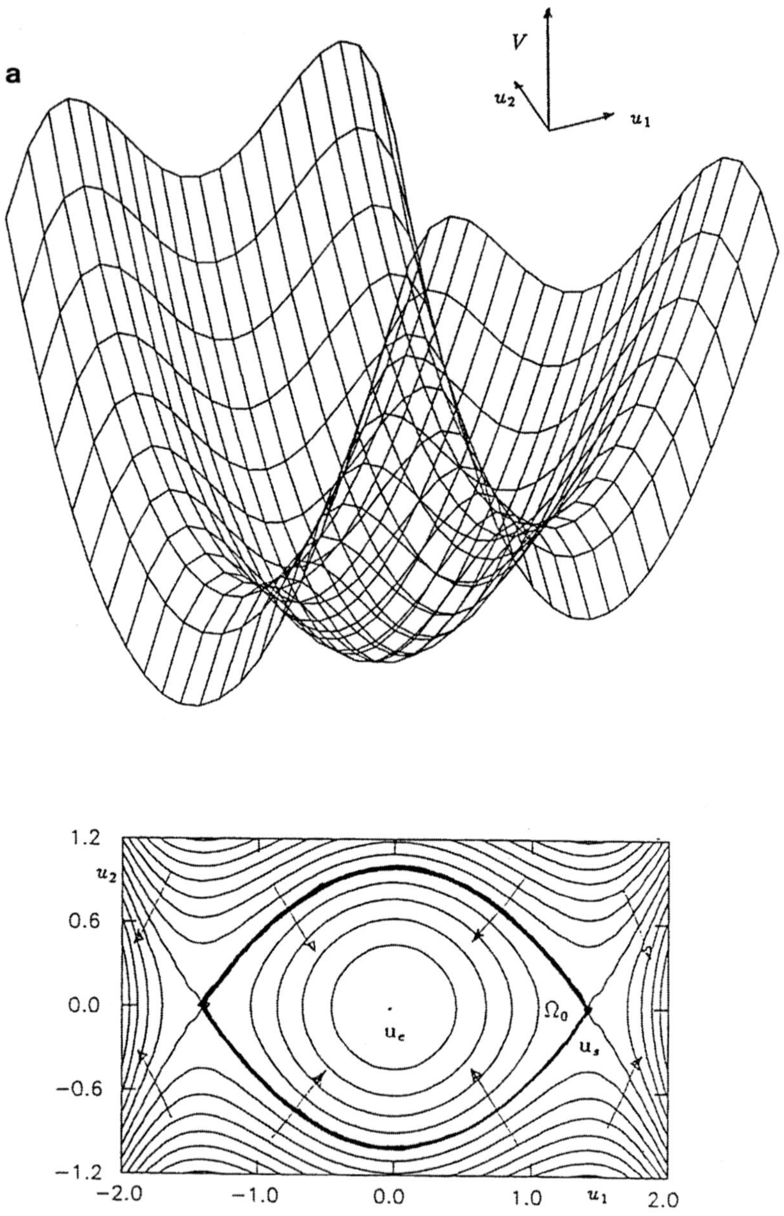
a
V
u_2
u_1
1.2
u_2
0.6
0.0
−0.6
−1.2
−2.0
−1.0
0.0
1.0
u_1
2.0
Ω_0
u_c
u_s

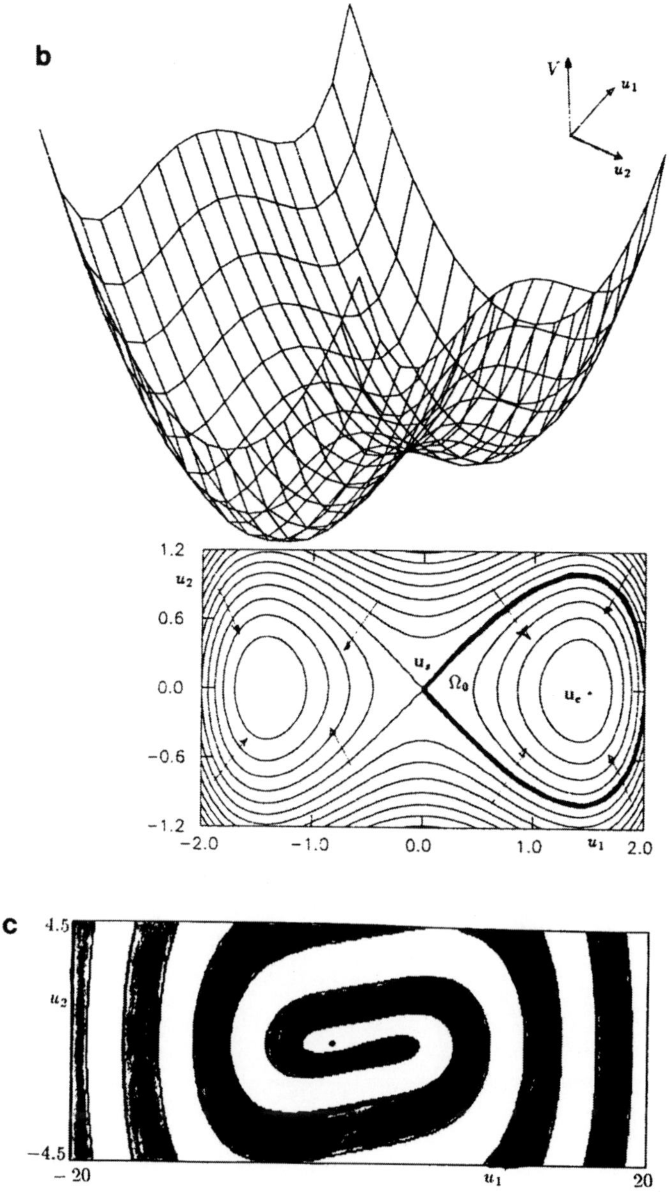

Figure 2.6.4 — Duffing's equation: Liapunov function and contour plot (the arrows indicate decreasing values of the contour lines) (a) for $\alpha > 0$ and $\beta < 0$ (b) for $\alpha < 0$ and $\beta > 0$. In (c) the basins of attraction of $\sqrt{-\alpha/\beta}$ (darker) and of $-\sqrt{-\alpha/\beta}$ (lighter) when $\alpha < 0$ and $\beta > 0$ are also shown. The black drop around $\sqrt{-\alpha/\beta}$ is the region Ω_0 also shown in (b).

remain in the region delimited by the level curve $V(\mathbf{u}) = V(\mathbf{u}_0)$. This observation helps to understand how the Liapunov function determines a region that certainly belongs to the basin of attraction. Consider, in fact, in the two-dimensional case, the level curve of the Liapunov function V = constant. Increasing the constant from $V(\mathbf{u}_e)$ on, the level curve may include say a saddle point $\mathbf{u}_s$ (see Fig. 2.6.4). Denote by Ω_0 the region enclosed by this level curve. Because of the previous observation, if $\mathbf{u}_0$ belongs to the region delimited by this level curve, then $V(\mathbf{u}_0) < V(\mathbf{u}_s)$ and, therefore, since $V\big(\mathbf{u}(t)\big) < V(\mathbf{u}_0)$, then $V\big(\mathbf{u}(t)\big) < V(\mathbf{u}_s)$. Therefore, $\mathbf{u}$ will always be within Ω_0 and will surely tend to $\mathbf{u}_e$. ■

Example 2.5 Stability for Van der Pol model
Consider again the modified Van der Pol equation (2.6.12) which can be written as

$$\begin{cases} \dfrac{du_1}{dt} = u_2 \\ \dfrac{du_2}{dt} = (\alpha + \beta u_1^2)u_2 - u_1 \,. \end{cases}$$

In order to study its stability, consider the function

$$V(u_1, u_2) = u_1^2 + u_2^2 \,,$$

which is a sort of energy. Of course, $V(u_1, u_2)$ is non-negative, vanishes only for $u_1 = u_2 = 0$, which is the equilibrium configuration of the Van der Pol equation, and is differentiable. We have to check what happens to its derivative along the trajectories

$$\begin{aligned} \frac{dV}{dt}\big(u_1(t), u_2(t)\big) &= \frac{\partial V}{\partial u_1}\big(u_1(t), u_2(t)\big)\frac{du_1}{dt} + \frac{\partial V}{\partial u_2}\big(u_1(t), u_2(t)\big)\frac{du_2}{dt} \\ &= 2u_1u_2 + 2u_2\big[(\alpha + \beta u_1^2)u_2 - u_1\big] \\ &= (\alpha + \beta u_1^2)u_2^2 \,, \end{aligned}$$

which is nonpositive if

$$\alpha + \beta u_1^2 \le 0 \,. \qquad (2.6.14)$$

In order to satisfy Eq.(2.6.14) in a neighborhood of $u_1 = u_2 = 0$, α must be nonpositive. If β is also nonpositive, then Eq.(2.6.14) is always satisfied and the equilibrium configuration is stable to any perturbation (globally stable). If, on the other hand, β is positive, Eq.(2.6.14) is satisfied only when

$$u_1^2 \le u_c^2 = \frac{|\alpha|}{\beta} \,.$$

Actually, since there cannot be solutions of the differential equation, apart from the equilibrium one, with either $u_1(t)$ or $u_2(t)$ vanishing, the equilibrium state is asymptotically stable for $\alpha < 0$ and also for $\alpha = 0$ provided that $\beta < 0$.

The level curve $V(\mathbf{u}) = V(u_c, 0)$ defines a disk Ω_0 in the (u_1, u_2)-plane, $u_1^2 + u_2^2 = |\alpha|/\beta$ in which $dV/dt < 0$. If $\mathbf{u}(t = 0) \in \Omega_0$, then $\mathbf{u}(t)$ is always in Ω_0 and therefore $u_1^2(t) \leq u_c^2$, which assures that dV/dt stays negative. Therefore Ω_0 is included in the basin of attraction. □

Liapunov Theorem 2.6.2 is partially inverted by the following theorem.

THEOREM 2.6.3 ***(Liapunov-Chetayev Theorem)***
Let $\mathbf{u}_e$ be an isolated equilibrium of $d\mathbf{u}/dt = \mathbf{f}(\mathbf{u})$ in some open neighborhood $\mathcal{D}_e$. If there exists a differentiable function $V = V(\mathbf{u})$ and an open set $\mathcal{D}_1$ such that

1. *$\mathbf{u}_e$ belongs to the border $\partial\mathcal{D}_1$ of $\mathcal{D}_1$*
2. *$V(\mathbf{u}) > 0$ in $\mathcal{D}_1$ and $V(\mathbf{u}) = 0$ on $\partial\mathcal{D}_1$*
3. *$\dfrac{dV}{dt} > 0$ in $\mathcal{D}_1$*

then $\mathbf{u}_e$ is unstable. ■

In the case of the models of discrete mechanics, it is often easy to construct a Liapunov function using the mechanical energy of the system. One can in fact prove the following theorem.

THEOREM 2.6.4 ***(Liapunov Function in Lagrangian Mechanics)***
For a Lagrangian system that is subject to conservative forces with potential energy U and possible dissipative forces, an equilibrium position $\mathbf{u}_e$ is stable if the potential energy U presents a local minimum in $\mathbf{u}_e$. ■

Example 2.6 Stability for Duffing's model
Consider the modified Duffing's equation, introduced in Section 2.2.4, without external forcing

$$\frac{d^2u}{dt^2} + a\frac{du}{dt} + \alpha u + \beta u^3 = 0\,. \tag{2.6.15}$$

If $a < 0$, then any equilibrium configuration is unstable, as proved in Example 2.5.

If $a > 0$, the term $a\,du/dt$ represents a viscous damping, while the term $\alpha u + \beta u^3$ is conservative with potential energy

$$U = \frac{\alpha}{2}u^2 + \frac{\beta}{4}u^4 .$$

Therefore the mechanical energy of the system has to decrease. We can then use Theorem 2.6.4 to study the stability properties of the equilibrium configurations

$$u_e = 0$$

$$u_e = \pm\sqrt{-\frac{\alpha}{\beta}} \qquad \text{if} \quad \alpha\beta < 0 .$$

From the sign of the second derivative of U we may understand when u_e is a minimum for U. We then obtain that

1. $u_e = 0$:
   - is unstable if $\alpha < 0$ (saddle);
   - is stable if $\alpha > 0$ (node if $0 < \alpha < \frac{a^2}{4}$, focus if $\alpha > \frac{a^2}{4}$).
2. $u_e = \pm\sqrt{-\frac{\alpha}{\beta}}$ (which exists only if $\alpha\beta < 0$):
   - are stable if $\alpha < 0$ (nodes if $-\frac{a^2}{8} < \alpha < 0$, foci if $\alpha < -\frac{a^2}{8}$);
   - are unstable if $\alpha > 0$ (nodes).

Furthermore if α and β are positive, then $u_e = 0$ is the only equilibrium configuration and is globally stable. If $\alpha > 0$ and $\beta < 0$, then $u_e = 0$ is certainly stable to perturbations with initial mechanical energy less than

$$U_{\max} = \frac{\alpha^2}{4|\beta|} ,$$

which is a subset of the basin of attraction (see Fig. 2.6.4).

Similarly, if $a < 0$ and $\beta > 0$, then $u_e = \pm\sqrt{-\alpha/\beta}$ is certainly stable if the initial mechanical energy is negative.

Consider the case $\alpha < -a^2/8$, $\beta > 0$. The spiral in Fig. 2.6.4c divides the phase space into two basins of attractions encircling each other. The states starting in the shaded region end up in $\sqrt{-\alpha/\beta}$, otherwise they end up in $-\sqrt{-\alpha/\beta}$.

In Fig. 2.6.4 both the basin of attractions (obtained by backward integration) and the regions, defined by the Liapunov theorem, are plotted. The reader is invited to study the linear stability of Duffing's equation and to determine the phase portrait. □

Unfortunately, the Liapunov method is not always successful. In fact, sometimes it is difficult to find a proper Liapunov function for a given problem, and often it may work only for certain ranges of the parameters.

2.6.5 Structural stability

Up to now, we have been discussing the stability of equilibria with respect to perturbations of the initial state. There is another stability idea involving the model itself called ***structural stability***, or ***total stability***, which is at least as important as the study of the stability with respect to perturbation of the equilibrium configuration. Unfortunately, it is more difficult to deal with.

In fact, there are several sources of perturbations that may affect the model itself. First, in conceiving the model, one is forced to restrict the agents influencing the motion. The other agents might be considered as perturbations of the model. In our mechanical example proposed in Section 2.2.2 friction in the hinge, viscous forces, etc., were all neglected. Despite of their smallness some of these agents will strongly affect the long term behavior of the pendulum. In fact, if dissipative forces are included in the model, then the system will ultimately reach a stable configuration. If, on the contrary, they are completely neglected, the systems orbits forever about it.

Furthermore, there might be some errors in computing experimentally the coefficients involved in $\mathbf{f}$ (say, gravitational acceleration, mass, length, spring constants, etc.).

Finally, there are unavoidable errors inherent in the derivation of the model itself. In fact, physical laws are often formulated in terms of a few (sometimes too few) macroscopic variables.

Some of these macroscopic variables (for instance, pressure, temperature, and so on) average the physics of countless microscopic particles. Hence the function $\mathbf{f}$ represents an average of a very complex dynamic system affected by random perturbation. Of course, it is unthinkable to examine the behavior of every single particle.

It is desirable, then, to have a model so good that small perturbations in $\mathbf{f}$ lead to solutions that are qualitatively the same as before.

This concept of stability, called ***structural stability***, is hard to verify and to obtain in practice. In fact, as already mentioned, adding even the smallest damping to our mechanical example changes the nature of the orbits from closed orbits around the stable state into trajectories that spiral toward it, so that the long term behavior is drastically different.

In the next section, we will show how a small error in the evaluation of a coefficient may change the character of an equilibrium state, say from stable to unstable, with a consequent change, which might be drastic or not, in the behavior of the system.

2.7 From Bifurcation to Chaos

Assume now that the differential system to be studied depends on a real parameter α

$$\frac{d\mathbf{u}}{dt} = \mathbf{f}(\mathbf{u}; \alpha) \tag{2.7.1}$$

and that $\mathbf{f}$ depends smoothly on $\mathbf{u}$ and α, which is usually the case in practical examples of mathematical models.

It is obvious that the solutions of (2.7.1) and therefore the localization of the equilibrium points depends on α. In particular, the condition

$$\mathbf{f}(\mathbf{u}; \alpha) = \mathbf{0} \tag{2.7.2}$$

implicitly defines a curve $\mathbf{u} = \mathbf{u}_e(\alpha)$ (or a set of curves) in the $(\mathbf{u}, \alpha)$-space, which represents the locus of equilibrium points. If it is not possible to write the solution of Eq.(2.7.2) explicitly, one can always trace this curve numerically [KEL].

2.7.1 Bifurcation diagrams

In this section, we will focus our attention on what happens when α varies, that is, how the equilibrium configurations and their stability properties change with α. Of course, the discussion is also valid when the system depends on more than one parameter, providing that all parameters but one are kept fixed and that the remaining one is allowed to vary acting on it from outside the system.

In order to visualize the dependence of the equilibrium points on α, it is useful to draw the curve $\mathbf{f}(\mathbf{u}, \alpha) = \mathbf{0}$ in the $(\mathbf{u}, \alpha)$-space. This, of course, can actually be done if $\mathbf{u}$ is at most two dimensional, obtaining a three-dimensional curve if $\mathbf{u} \in \mathbb{R}^2$ and a planar diagram if $u \in \mathbb{R}$.

If the system of differential equations can be written as a single ordinary differential equation in the state variable u, then the natural choice is to plot $u_e(\alpha)$. If this is not the case and $\mathbf{u} \in \mathbb{R}^n$, $n \geq 3$, then one can obtain partial pictures of the situation by illustrating graphically the

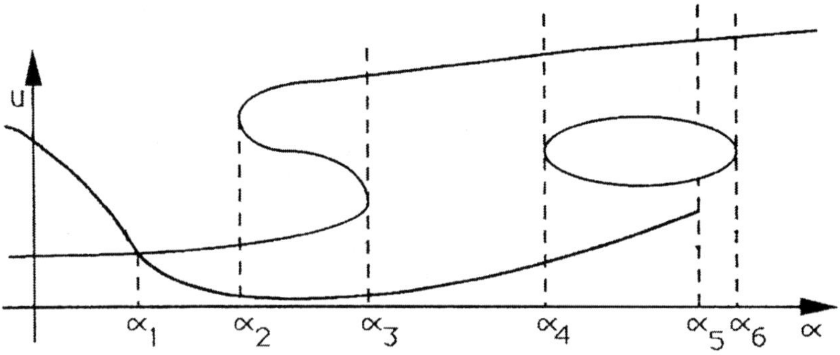

Figure 2.7.1 — Sample bifurcation diagram.

dependence of the components u_{ek} or of a norm of $\mathbf{u}_e$, such as

$$\|\mathbf{u}_e\| = \sqrt{\sum_{k=1}^{n} u_{ek}^2} \quad \text{or} \quad \|\mathbf{u}_e\| = \max_{1 \le k \le n} |u_{ek}|,$$

on α. The happiest choice is the one that allows us to capture the essential features of the stability properties of the system.

These diagrams are often called ***bifurcation diagrams*** (or branching diagrams or response diagrams). An example is given in Fig. 2.7.1. Typically, the curve $\mathbf{u}_e(\alpha)$ is made up of several pieces called ***branches***. They may intersect, be open, closed, or even end at certain values of α. Besides the location, the number of equilibrium solutions also may change with α. Referring to Fig. 2.7.1, one for instance has

- 1 solution if $\alpha = \alpha_1$ and $\alpha > \alpha_6$;
- 2 solutions if $\alpha < \alpha_1$, $\alpha_1 < \alpha < \alpha_2$, $\alpha_3 < \alpha < \alpha_4$, and $\alpha = \alpha_6$;
- 3 solutions if $\alpha = \alpha_2, \alpha_3, \alpha_4$ and $\alpha_5 < \alpha < \alpha_6$;
- 4 solutions if $\alpha_2 < \alpha < \alpha_3$ and $\alpha_4 < \alpha \le \alpha_5$.

The values $\alpha_1, \ldots, \alpha_m$, where the number of equilibrium solutions changes, are called ***branch points***.

DEFINITION 2.7.1 Bifurcation
We say that ***an equilibrium solution bifurcates from another*** *at* $\alpha = \alpha_b$ *if there are two distinct equilibrium solutions* $\hat{\mathbf{u}}_e(\alpha)$ *and* $\check{\mathbf{u}}_e(\alpha)$

continuous in α such that $\widehat{\mathbf{u}}_e(\alpha_b) = \check{\mathbf{u}}_e(\alpha_b)$. The common value

$$\{\alpha_b, \widehat{\mathbf{u}}_e(\alpha_b)\}$$

*is called a **bifurcation point**.*

Example 2.7 Bifurcation of Duffing's equation
Consider, for instance, the modified Duffing's equation (2.6.15)

$$\frac{d^2u}{dt^2} + a\frac{du}{dt} + \alpha u + \beta u^3 = 0\,.$$

Assume that β is a fixed positive constant, which can always be set equal to one by suitably rescaling the state variable (i.e., introducing $\widehat{u} = \sqrt{\beta}u$), and focus the attention on $u_e(\alpha)$.

From Example 2.6 recall that

- if $\alpha > 0$, then $u_e(\alpha) = 0$ is the only possible equilibrium configuration and is stable;
- if $\alpha < 0$, then $u_e(\alpha) = 0$ is unstable but there exist other two (symmetric) equilibrium configurations $u_e(\alpha) = \pm\sqrt{-\alpha}$, which are stable.

In Fig. 2.7.2 we have plotted the bifurcation diagram. In Fig. 2.7.2a we have just plotted the equilibrium curves $u_e(\alpha) = 0$ and $u_e(\alpha) = \pm\sqrt{-\alpha}$. In Fig. 2.7.2b we have used a classical convention to distinguish between stable and unstable equilibria, denoting them, respectively, with a heavy solid curve and a dashed one. The origin is a bifurcation point.

Finally, in Fig. 2.7.2c we have also indicated some typical trajectories, which spiral towards the stable configurations. It is to be observed that the character of these orbits dramatically changes as α changes sign, that is, as α crosses the bifurcation value.

The bifurcation diagram for the cases in which either a or β are negative constants are given in Fig 2.7.3. The reader is encouraged to derive them as an exercise.

The reader now has all the tools needed to look, again as an exercise, at the case in which α is considered fixed and β varies. □

It is also useful to see how these diagrams are related to the potential energy of the system. In fact, plotting the surface $U(u, \alpha; \beta)$ for different values of β, we realize that the stable branches are simply obtained by projecting in the (u, α)-plane the points of minimum of U (Fig. 2.7.4).

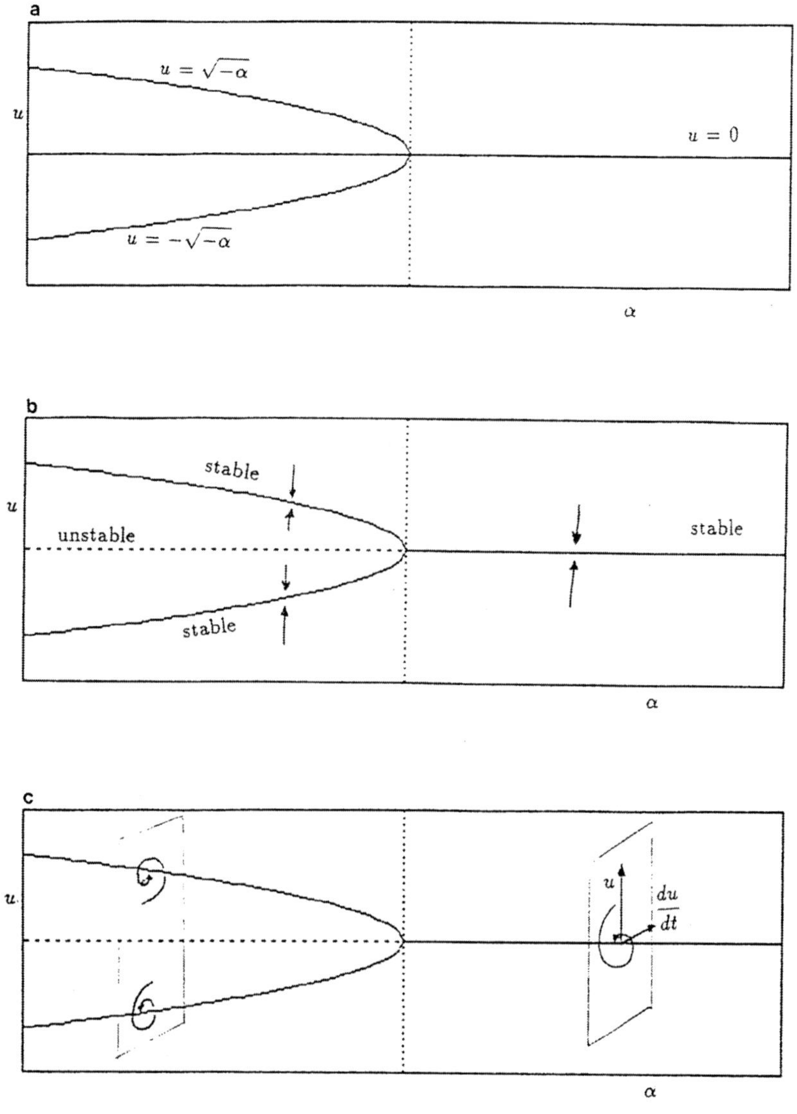

Figure 2.7.2 — Bifurcation diagram for the modified Duffing's equation without external forcing when a and β are positive.

(a) Lines represent the equilibrium configurations.

(b) Solid lines mean stability, dashed lines instability. Arrows indicate where the trajectories tend to.

(c) Sketch of phase diagrams as α varies.

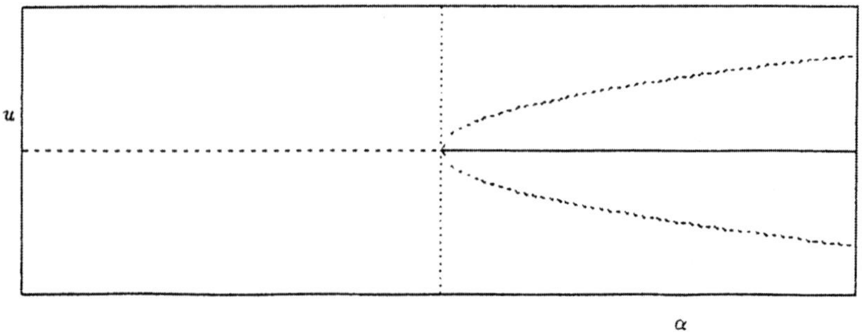

$a > 0 \qquad \beta < 0$

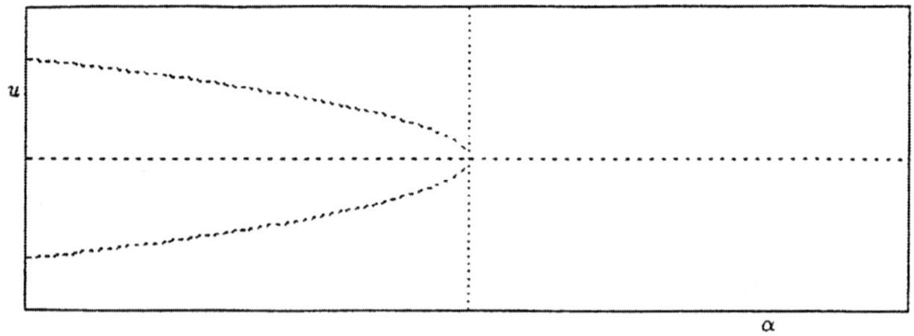

$a < 0 \qquad \beta > 0$

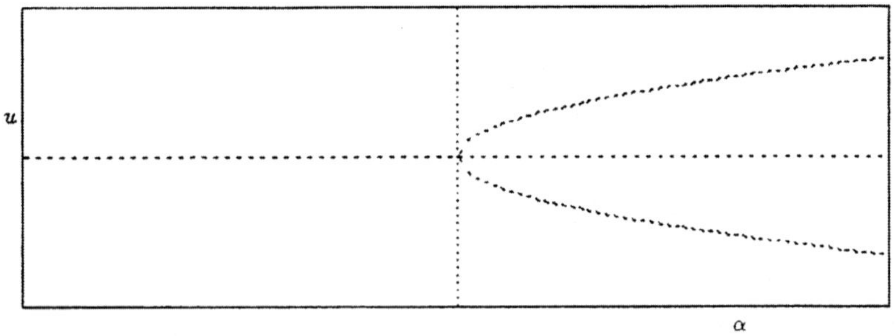

$a < 0 \qquad \beta < 0$

Figure 2.7.3 — Further bifurcation diagrams for the modified Duffing's equation without external forcing.

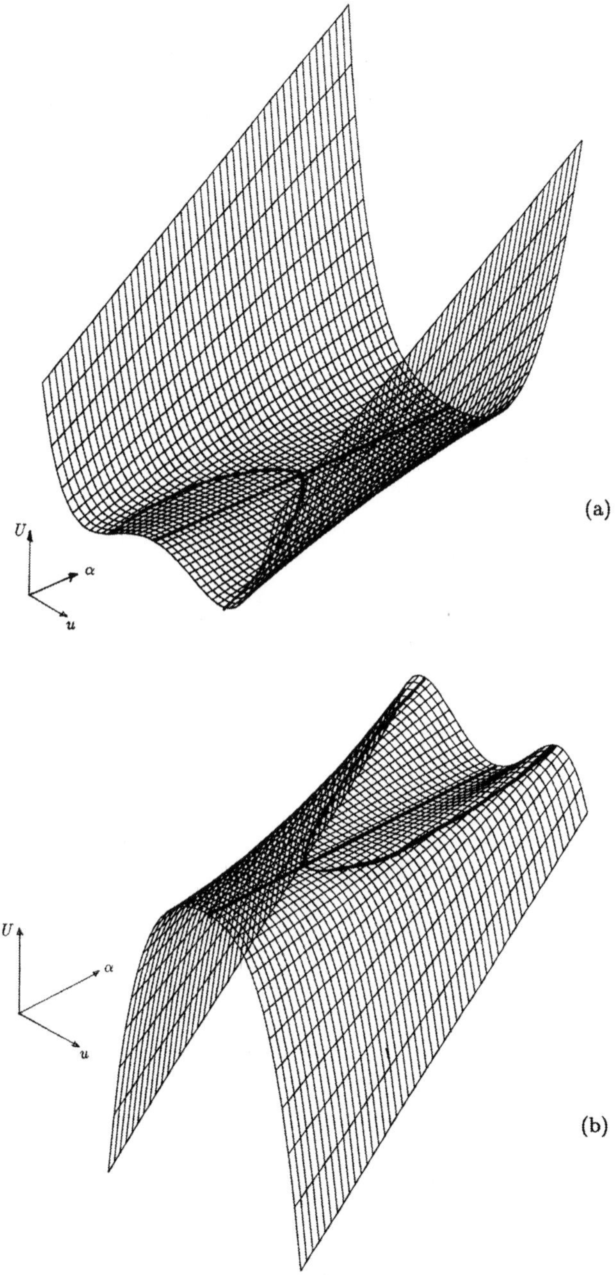

Figure 2.7.4 — Bifurcation diagrams and potential energy in the (a) supercritical and (b) subcritical case.

To clarify these concepts consider also the following example, which is a little bit more complex, but can help because the physics beyond it is conceptually very simple and the behavior of the system is more understandable.

Example 2.8 Bifurcation for rotating pendulum problem
Consider, for instance, the rotating pendulum model introduced in Section 2.2.2 with β fixed. We want, in fact, to focus our attention on what happens as α varies, say because we may influence the angular velocity Ω at which the pendulum is spinning or the rigidity k of the spring connecting the pendulum to the vertical rotation axis.

In Example 2.2 we found that the equilibrium configurations are

$$\begin{aligned} u_1 &= 0 \\ u_1 &= \pi \\ \cos u_1 &= \frac{1}{\alpha}, \qquad \text{if} \quad |\alpha| \geq 1 \,. \end{aligned}$$

Therefore, if $|\alpha| \leq 1$, there are two equilibrium points; if $|\alpha| > 1$, there are four equilibrium points. Hence $\alpha = \pm 1$ are branching points and also bifurcation points, since two distinct branches intersect there, as is evident from Fig. 2.7.5, where the bifurcation diagram in the (u_1, α)-plane with $u_1 \in [-\pi/2, 3\pi/2]$ is plotted.

What about the stability of these equilibrium configurations? Applying Theorem 2.6.4 the reader can easily verify that

$$\begin{aligned} u_1 &= 0 \quad \text{is stable} \quad \Longleftrightarrow \alpha \leq 1 \\ u_1 &= \pi \quad \text{is stable} \quad \Longleftrightarrow \alpha \leq -1 \\ \cos u_1 &= \frac{1}{\alpha} \quad \text{is stable} \quad \Longleftrightarrow \alpha \geq 1 \,. \end{aligned} \tag{2.7.3}$$

These conclusions are reported in Fig. 2.7.5.

Assume now that we may interact with the system and quasistatically change the value of α. For instance, with a screwdriver we could turn a screw that regulates either the rigidity of the spring or the angular velocity.

Referring to Figs. 2.7.5 and 2.7.6 one can describe the behavior of the system as follows. If the angular velocity is small, say $\alpha = -2$ or $\alpha = 0$, the potential energy U has a minimum in $u_1 = 0$ and therefore

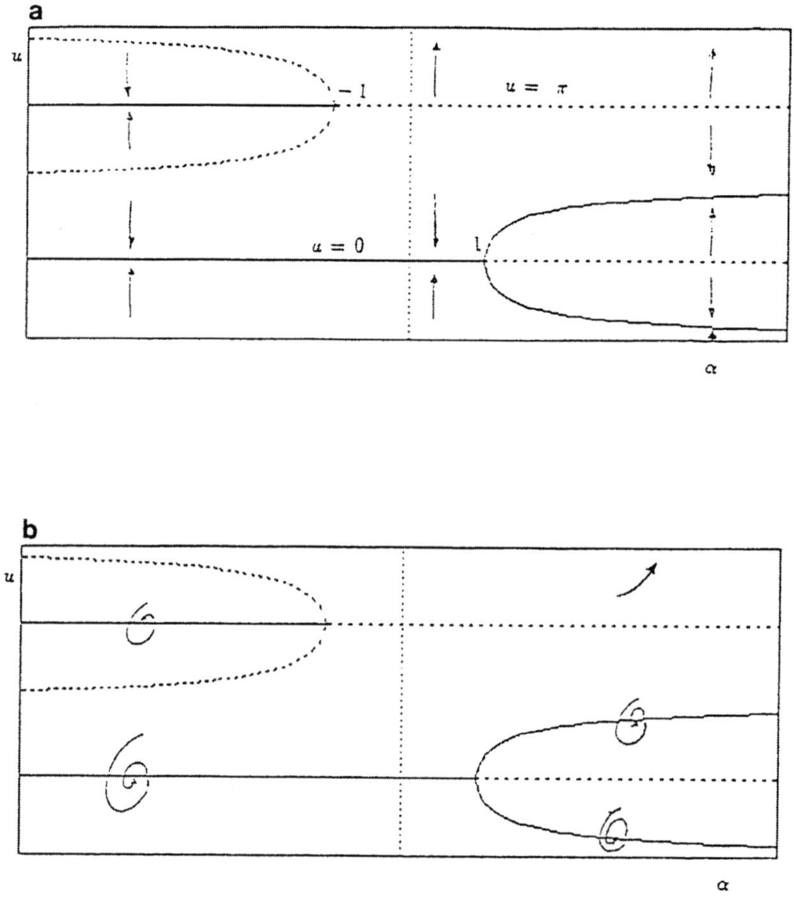

Figure 2.7.5 — Bifurcation diagrams for the rotating pendulum problem (Example 2.8). Solid lines indicate stable configurations and dashed lines unstable configurations.

(a) The arrows indicate where the solution tends for some particular values of α.

(b) Qualitative trajectories.

$u_1 = 0$ is stable. As soon as α crosses 1 (the bifurcation value), $u_1 = 0$ becomes a local maximum, whereas two minima appear nearby. Hence $u_1 = 0$ becomes unstable, while the two local minima, which become more and more pronounced as the angular velocity furtherly increases, represent the new stable equilibrium points.

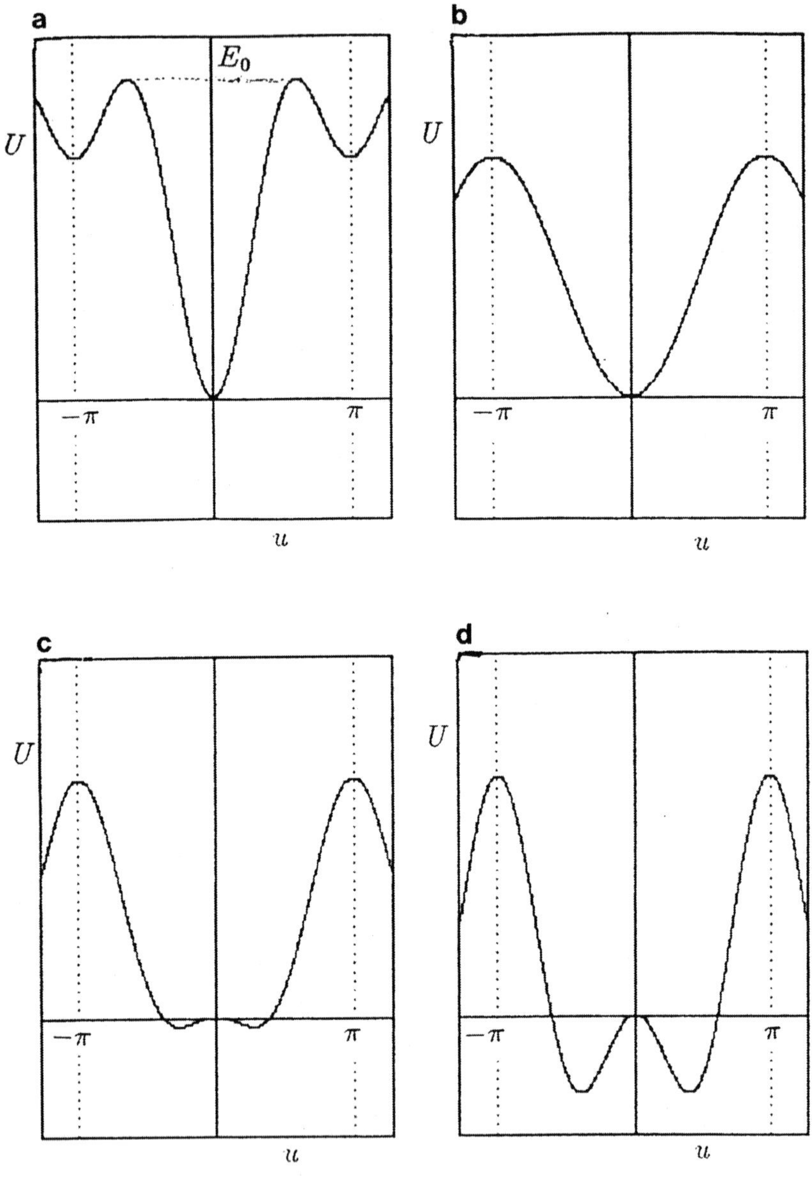

Figure 2.7.6 — Potential energy for Example 2.8 (rotating pendulum) for different values of α.
(a) $\alpha = -2$ (b) $\alpha = 0$ (c) $\alpha = 1.5$ (d) $\alpha = 3$.

This example may also help to understand the meaning of conditionally stable, unconditionally stable, and basin of attraction. In fact, if $\alpha = 0$ the solution of the ordinary differential equation will eventually tend to $u_1 = 0$ (modulo rotations of 2π) independently of its initial condition (unconditionally stable). If, instead, $\alpha = -2$ then the solution of (2.2.15) will tend to $u_1 = 0$ for sure only if its initial mechanical energy is less than the value E_0 shown in Fig. 2.7.6, otherwise it may fall in the other stable equilibrium configuration $u_1 = \pi$.
A similar argument can be given for each stable equilibrium point. □

We can now classify some types of bifurcations.

DEFINITION 2.7.2 Pitchfork and transcritical bifurcations
If one of the two branches intersecting at the bifurcation point $(\alpha_b, \mathbf{u}_b)$ is one sided (i.e. defined only for $\alpha \geq \alpha_b$ or for $\alpha \leq \alpha_b$), then the bifurcation is of **pitchfork** *type.*
If, instead, on both sides of the bifurcation point there are locally two solutions, then the bifurcation is called ***transcritical***.

Both bifurcation points in the previous example are of pitchfork type and they owe this name to the fork-like shape of the relative bifurcation diagram (see Fig. 2.7.5).

There is, however, a substantial difference between $\alpha = 1$, $u_1 = 0$ and $\alpha = -1$, $u_1 = \pi$. In fact, when $\alpha = 1$ is crossed, there is a smooth exchange of stability from $u_1 = 0$ to $\cos u_1 = 1/\alpha$ and *vice versa*.

On the contrary, the stability is locally lost at the bifurcation point $\alpha = -1$, i.e., there is no exchange of stability. This means that if the pendulum initially stays in the position $u_1 = 0$ for small angular velocity and then Ω is quasistatically increased, then, at a certain point (when α crosses 1), the pendulum will start oscillating about a new equilibrium position, say $u_1 = \cos^{-1}(1/\alpha)$. This event will not be dramatic since for $\alpha - 1$ small and positive, the new equilibrium position is approximatively given by $u_e \cong \sqrt{2(\alpha - 1)}$ and is therefore close to $u_1 = 0$.

If, instead, initially $u_1 = \pi$, then when a certain critical angular velocity, corresponding to $\alpha = -1$, is passed, the pendulum will suddenly drop from its upright position to oscillate wildly about the position $u_1 = 0$.

These two situations are, respectively, referred to as ***supercritical*** and ***subcritical bifurcation***.

The model (2.2.18)

$$\frac{du}{dt} = (\alpha - u)u\,, \tag{2.7.4}$$

which describes the evolution of a single species, and the following modified Duffing's equation

$$\frac{d^2u}{dt^2} + a\frac{du}{dt} - \alpha u + \beta u^2 + u^3 = 0\,, \qquad a > 0\,, \tag{2.7.5}$$

which may model a nonlinear circuit with a general cubic condenser, are, instead, examples of transcritical bifurcation as it is shown in Fig. 2.7.7. Again if a in Eq.(2.7.5) is negative, both branches are unstable.

The reader may find these bifurcation diagrams as an exercise.

Finally observe that in Fig. 2.7.7b there is a point P on one of the branches characterized by having $d\alpha/du = 0$ and $d^2\alpha/du^2 \neq 0$, which divides the branch into a stable part and an unstable part. This branching point, which is not a bifurcation point, is called a ***turning point.***

There is an analytic way to distinguish on an equilibrium curve among a regular point, a bifurcation point, and a turning point. In fact, in regular points the tangent to the curve exists and is not vertical. Since it is given by

$$\frac{d\mathbf{u}_e}{d\alpha} = -\mathbf{J}^{-1}\frac{\partial \mathbf{f}}{\partial \alpha}\,,$$

the Jacobian matrix $\mathbf{J}(\mathbf{u}_e, \alpha)$ must be nonsingular.

On the other hand in turning points the tangent to the equilibrium curve is vertical, while in bifurcation points it is not uniquely defined. Both cases correspond to the fact that the determinant of the Jacobian vanishes. At this point, to distinguish between the two cases, one has to compute $\partial \mathbf{f}/\partial \alpha(\mathbf{u}_e, \alpha)$. If the matrix obtained by attaching this vector to the Jacobian has maximum rank, then the branching point is a turning point, otherwise it is a bifurcation point. More properly, the following algebraic definitions can also be given.

DEFINITION 2.7.3 Turning point
The point $(\alpha_c, \mathbf{u}_c)$ of the equilibrium curve is a turning point if:

1. *The Jacobian matrix $\mathbf{J}$, computed at $(\mathbf{u}_c, \alpha_c)$, is singular;*
2. *The matrix*

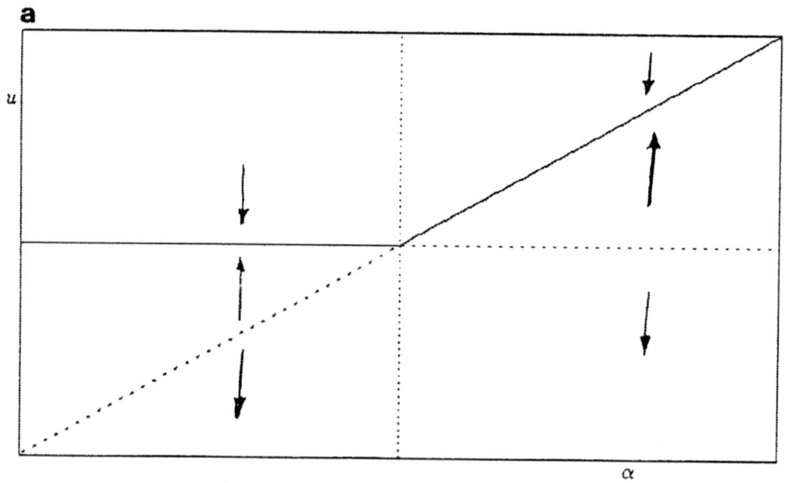

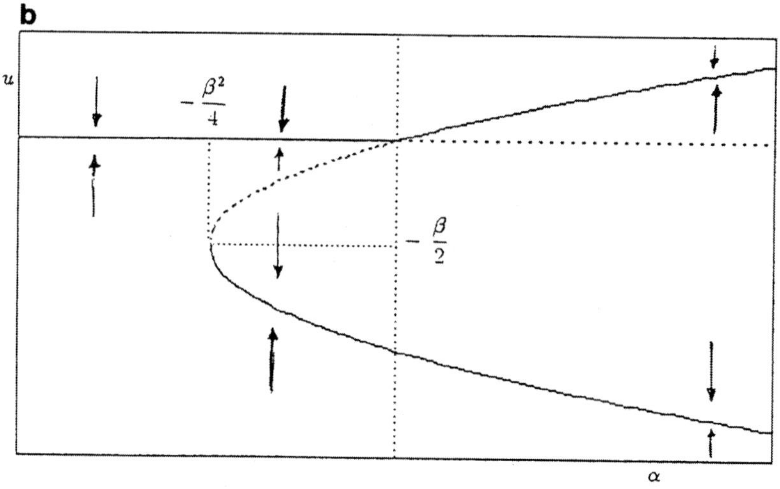

Figure 2.7.7 — Transcritical bifurcation.
(a) One species model (2.7.4).
(b) Generalized Duffing's equation (2.7.5) for a and β positive. If a is negative, both branches are unstable.

$$\begin{pmatrix} \dfrac{\partial f_1}{\partial u_1} & \dfrac{\partial f_1}{\partial u_2} & \cdots & \dfrac{\partial f_1}{\partial u_n} & \dfrac{\partial f_1}{\partial \alpha} \\ \dfrac{\partial f_2}{\partial u_1} & \dfrac{\partial f_2}{\partial u_2} & \cdots & \dfrac{\partial f_2}{\partial u_n} & \dfrac{\partial f_2}{\partial \alpha} \\ \vdots & \vdots & \ddots & \vdots & \vdots \\ \dfrac{\partial f_n}{\partial u_1} & \dfrac{\partial f_n}{\partial u_2} & \cdots & \dfrac{\partial f_n}{\partial u_n} & \dfrac{\partial f_n}{\partial \alpha} \end{pmatrix}, \tag{2.7.6}$$

computed in $(\mathbf{u}_c, \alpha_c)$, *is nonsingular;*

3. *There is a parametrization* $\widehat{\mathbf{u}}(s), \widehat{\alpha}(s)$ *with* $\widehat{\mathbf{u}}(0) = \mathbf{u}_c$, $\widehat{\alpha} = \alpha_c$ *and* $\dfrac{d^2\widehat{\alpha}}{ds^2}(0) \neq 0$.

DEFINITION 2.7.4 Simple bifurcation point
The point $(\alpha_c, \mathbf{u}_c)$ *of the equilibrium curve is a turning point if:*

1. *The Jacobian matrix* $\mathbf{J}(\mathbf{u}_c, \alpha_c)$ *is singular;*
2. *The matrix (2.7.6), still computed in* $(\mathbf{u}_c, \alpha_c)$, *is singular;*
3. *Exactly two branches intersect with two distinct tangents in* $(\alpha_c, \mathbf{u}_c)$.

2.7.2 Hysteresis

In the previous section we explained qualitatively the difference between supercritical and subcritical bifurcation. Consider now a subcritical bifurcation, which also presents a turning point. This situation is encountered, for instance, when dealing with the ordinary differential equation

$$\frac{d^2u}{dt^2} + \frac{du}{dt} - \alpha u - \beta u^3 + u^5 = 0\,, \qquad \beta > 0\,. \tag{2.7.7}$$

The bifurcation diagram given in Fig. 2.7.8 is characterized by a bifurcation point at $\alpha = \alpha_b = 0$ and two turning points at $\alpha = \alpha_t = -\beta^2/4$ with $u_t = \pm\sqrt{\beta/2}$.

For $\alpha < \alpha_t$, the only equilibrium configuration of (2.7.7) is $u = 0$. For $\alpha_t < \alpha < \alpha_b$, there are five equilibrium solutions, two of which are unstable, and for $\alpha > \alpha_b$, there are three equilibrium solutions, one of which is unstable.

If the system is initially in the trivial equilibrium configuration $u = 0$ with $\alpha < \alpha_t$ and α is quasistatically increased (β is again considered fixed), nothing happens in between α_t and α_b, but as soon as α becomes

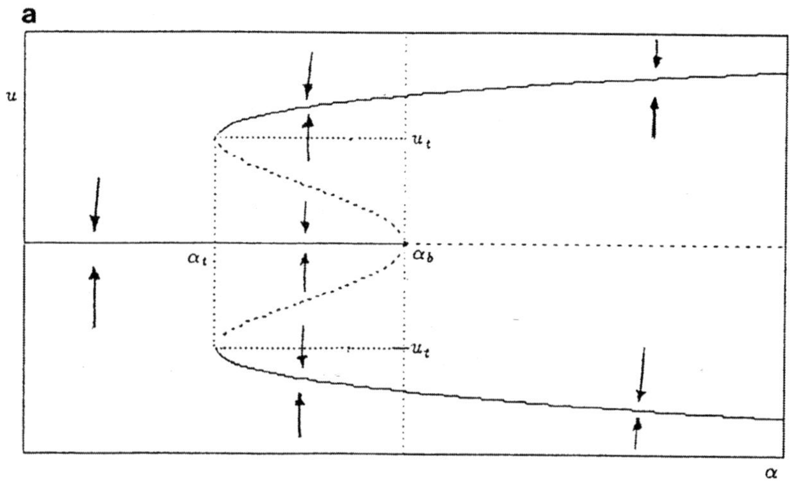

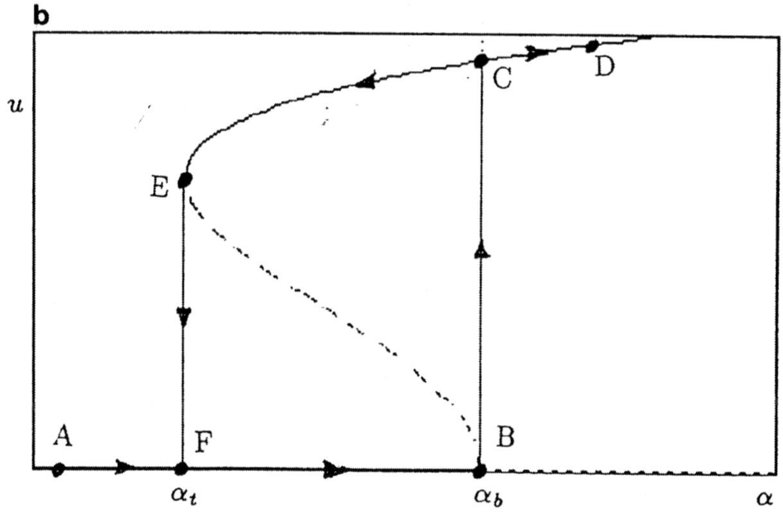

Figure 2.7.8 — Symmetric subcritical bifurcation with turning points. (a) Bifurcation diagram (b) Hysteresis loop.

larger than α_b the solution $u = 0$ looses stability (subcritically) and the system will tend toward one of the solutions

$$u_e = \pm\sqrt{\frac{\beta + \sqrt{\beta^2 + 4\alpha}}{2}}.$$

A further increase in α will only vary u_e, since u_e is a function of α, but the solution will continue to oscillate about a point on the bifurcating branch. The same happens as α decreases, till the turning point (α_t, u_t) is reached. If α is then furtherly decreased, the system will return to the only stable equilibrium configuration possible, the initial one. Hence, in the region $\alpha \in [\alpha_t, \alpha_b]$, the system exhibits hysteresis.

This phenomenon is not a mathematical lucubration, but occurs in many applications. Unfortunately, it is difficult to find an example so simple as to be understandable without introducing some basic knowledge in the specific field. We hope to convince and content the reader using the following home experiment. The reader is invited to perform the experiment.

Take a bicycle brake wire (or a metallic meter) and hold it up between your thumb and index as it is shown in Fig. 2.7.9. If the portion of wire over your hand is small, it will remain in its upright position, even if you strongly shake it (you are, say, in A of Fig. 2.7.8). Now raise the wire up, letting it slide very slowly between your fingers. At a certain point the wire will suddenly deflect to the right or to the left (it depends on the perturbation you have transmitted, although unwilling, to the wire) forming an angle with vertical. This means that you have just crossed the bifurcation point B and jumped to C. Mark this position. If you continue to raise the wire, you will only observe an increase in the angle formed by the tip of the wire and the tip of your fingers (you are in D). If you want and/or can, you might measure and record this angle and the related length of the wire.

Now, always very slowly, try to go back to the original position, observing that nothing happens when you cross the position you have marked before (the one corresponding to C). The angle slowly diminishes until, at a certain point (E), the wire suddenly snaps through to the upright position F. Mark also this position, which more or less corresponds to the value α_t. The hysteresis cycle is closed.

If you have recorded the data of this experiment you can use them to plot an approximate bifurcation diagram like the one shown in Fig. 2.7.8.

What happens if you take the wire in between the two marks? If you are, for instance, in the upright position and shake the wire just a little, it will stay there (you are still in the basin of attraction of $u = 0$), if you

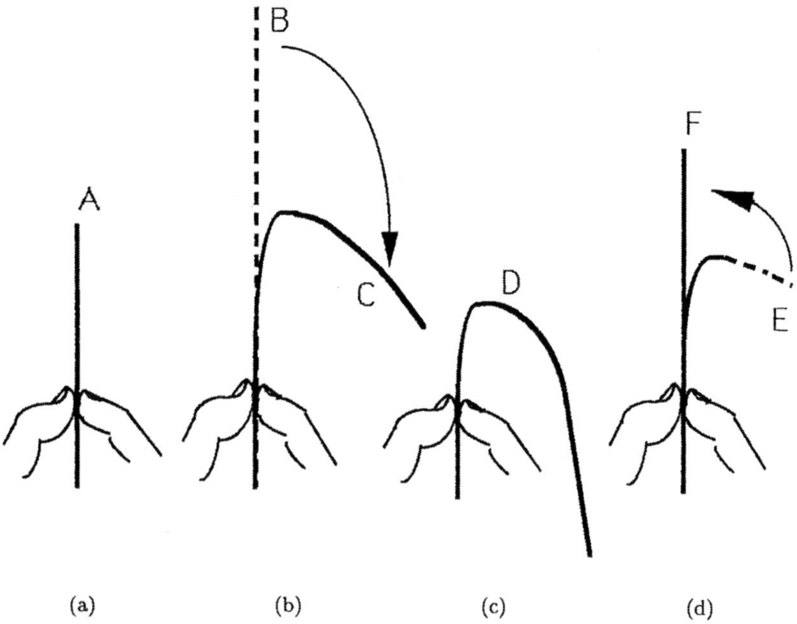

Figure 2.7.9 — Home experiment with a bicycle wire. Note that in (b) the wire is longer than in (d).

shake it strongly, the wire will drop to another equilibrium configuration. How the perturbation should be depends on where you are holding the wire and on the basin of attractions.

Finally, observe that if you are using an asymmetric metallic meter, then one of the two nontrivial equilibrium configurations (say right or left) is priviledged, due to the fact that the meter is concave toward the priviledged position. The subcritical bifurcation is then not symmetric. It is interesting and pedagogically instructive, in this case, to use the experimental data to draw a bifurcation diagram. In the dynamic case, the shifting landscape of the potential energy can help in understanding the dynamics of the hysteresis cycle just considered (Fig. 2.7.10).

2.7.3 Hysteresis cycles by cusp catastrophe

In this section we will consider briefly systems that depend on two parameters α and β, instead of one

$$\frac{d\mathbf{u}}{dt} = \mathbf{f}(\mathbf{u}; \alpha, \beta) \,. \tag{2.7.8}$$

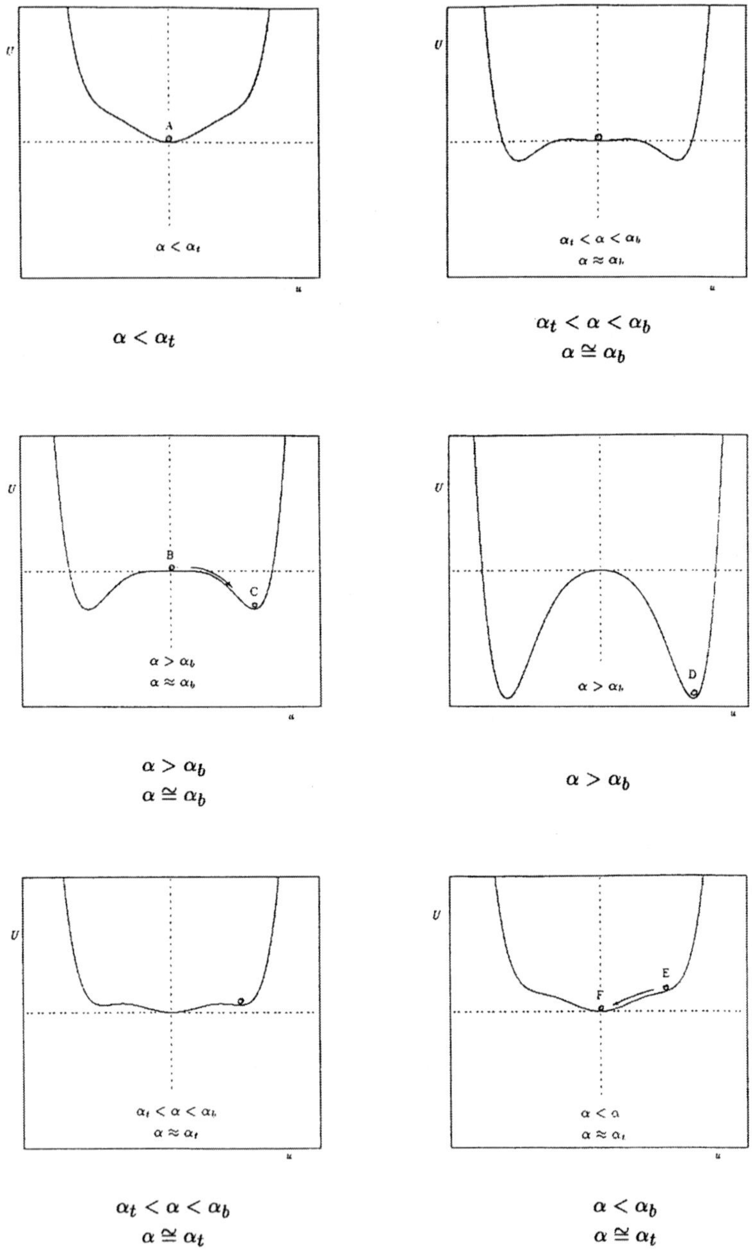

Figure 2.7.10 — Hysteresis representation for a mechanically conservative system through its potential energy.

It is assumed that the parameter β, considered fixed in the previous sections, may now change. One may also think of β as a perturbation of the model caused, for instance, by inaccuracies in measurements.

Usually a small change in β induces a correspondingly small change in the solution, at least for small times. However, in certain cases, the behavior of the solution can change so drastically that the phenomenon deserves to be qualified as ***catastrophic***.

Take, for instance, the following nonlinear circuit model

$$\frac{d^2u}{dt^2} + a\frac{du}{dt} - \alpha u + \beta u^3 + u^5 = 0\,, \qquad a > 0\,. \tag{2.7.9}$$

If we consider β fixed and draw the bifurcation diagram, we find a supercritical bifurcation if $\beta \geq 0$ and a subcritical one if $\beta < 0$. Hence as β changes the character of the bifurcation changes. Similarly for

$$\frac{d^2u}{dt^2} + a\frac{du}{dt} - \alpha u + \beta u^2 + u^3 = 0\,, \qquad a > 0\,, \tag{2.7.10}$$

we have a supercritical bifurcation when $\beta = 0$ and a transcritical one if $\beta \neq 0$.

The condition $\mathbf{f}(\mathbf{u}; \alpha, \beta) = \mathbf{0}$ implicitly defines in the $(\mathbf{u}, \alpha, \beta)$-space a surface $\mathbf{u}_e(\alpha, \beta)$ which represents the locus of equilibrium points. For instance, the surfaces corresponding to (2.7.9) and (2.7.10) are plotted in Fig. 2.7.11.

A more drastic difference occurs when we consider systems like

$$\frac{du}{dt} - \alpha u + u^3 + \beta = 0\,, \tag{2.7.11}$$

which for $\beta = 0$ represents the simplest example of supercritical bifurcation. For any given $\beta \neq 0$, the bifurcation disappears and the diagram looks like the one in Fig. 2.7.12, where we marked the turning points. This phenomenon is called ***simmetry breaking bifurcation***.

As previously stated, these turning points in one dimension are characterized by having both

$$f(u_t; \alpha_t, \beta) = 0 \qquad \text{and} \qquad \frac{\partial f}{\partial u}(u_t; \alpha_t, \beta) = 0\,,$$

and therefore the turning points are given by

$$\begin{cases} \alpha u - u^3 - \beta = 0 \\ \alpha - 3u^2 = 0\,, \end{cases}$$

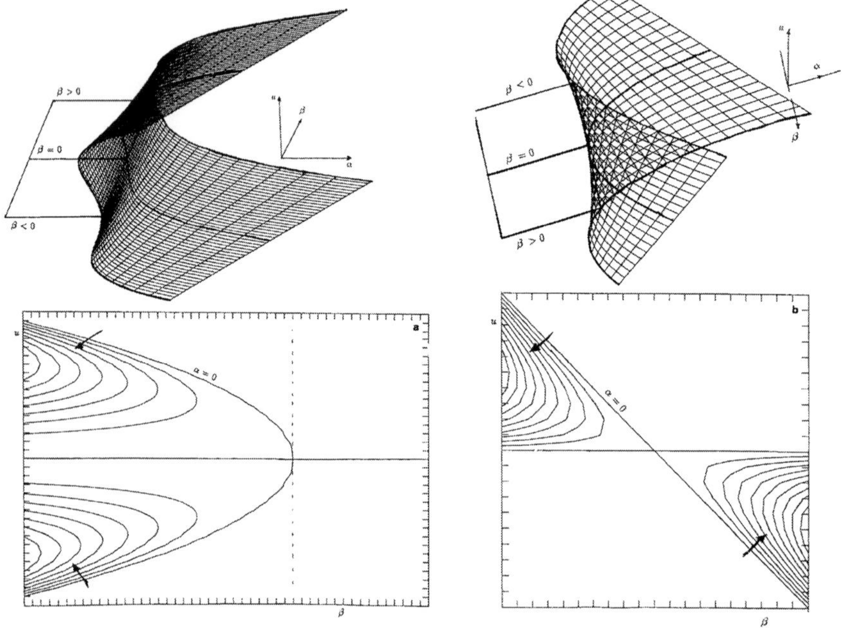

(a) (b)

Figure 2.7.11 — The equilibrium surface $\mathbf{u}_e(\alpha, \beta)$ and the contour plot (a) for equation (2.7.10) and (b) for equation (2.7.11).

which is solved by

$$u_t = \sqrt[3]{\frac{\beta}{2}} \tag{2.7.12a}$$

$$\alpha_t = 3\sqrt[3]{\frac{\beta^2}{4}}. \tag{2.7.12b}$$

In the (α, β)-plane the curve given by (2.7.12b) presents a cusp in $\alpha = \beta = 0$.

For this reason the phenomenon we are going to present in this section is called ***cusp catastrophe***.

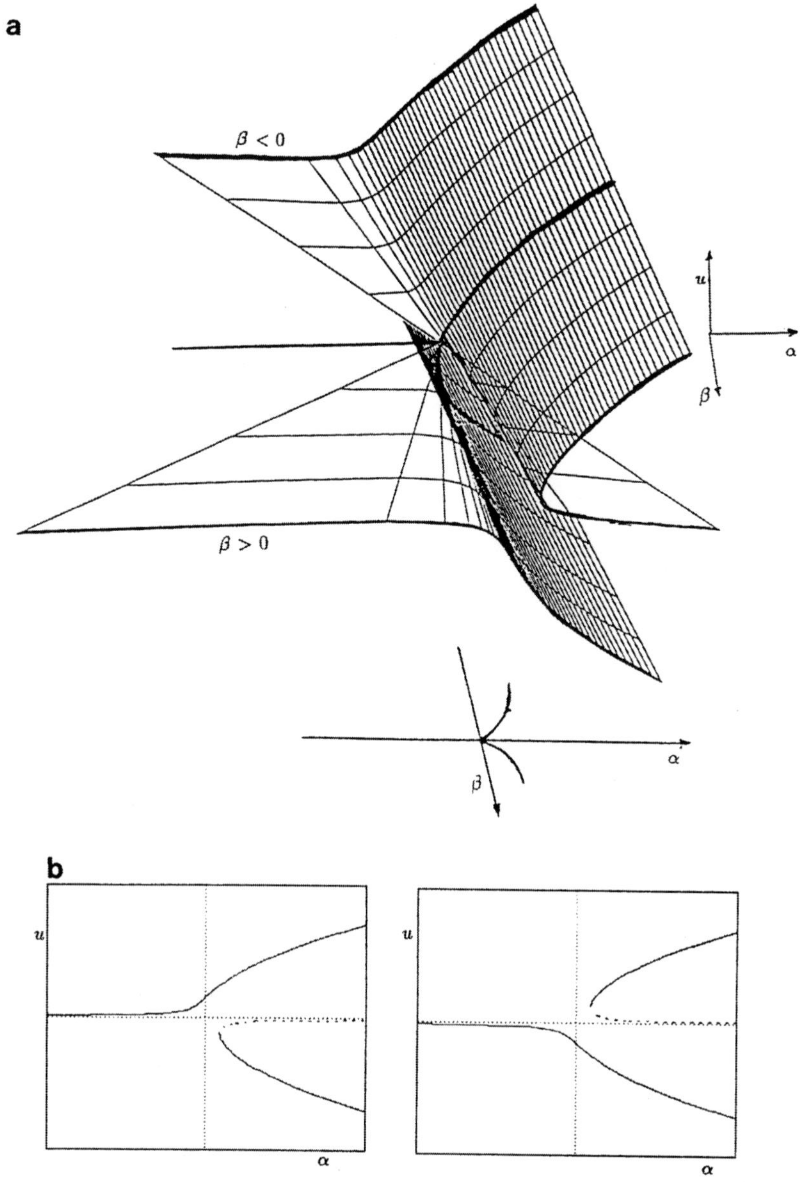

Figure 2.7.12 — Symmetry breaking for supercritical bifurcation
(a) Stable equilibrium surface $u_e(\alpha, \beta)$ with marked curves at $\beta = -1, 0, 1$ and projection in the (α, β)-plane of the values referring to the turning points (the cusp-like curve (2.7.13)).
(b) $u_e(\alpha)$ for $\beta = -1$ (left) and $\beta = 1$ (right).

REMARK 2.7.1 If β is fixed, there is no hysteresis loop while α varies. In fact even if the system is in a configuration belonging to the branch with a turning point and α decreases below the turning point, the system will jump to the other branch and will oscillate around it for ever, independently of how α is varied from then onwards. ■

REMARK 2.7.2 In the case of a subcritical bifurcation with turning points when $\beta = 0$, adding a β on the right-hand side of the differential equation has the only effect of deforming the hysteresis cycle. On the other hand, in the case of a transcritical bifurcation at $\beta = 0$, one has the formation of a hysteresis cycle or not according to the sign of β as it is evident in Fig. 2.7.13. ■

Coming back to the supercritical case, consider α fixed and examine the system as β varies. One has a bifurcation diagram such as those in Fig. 2.7.14a.

If $\alpha \leq 0$, then there is only one equilibrium point for any β, but if $\alpha > 0$, the situation is more complex. If $\beta < \beta_1$ or $\beta > \beta_2$, there is only one equilibrium state, and if $\beta \in [\beta_1, \beta_2]$, there are three possible equilibrium states, two of them are stable and one is unstable. In the sample model (2.7.11)

$$\beta_{1,2} = \pm \frac{2}{3\sqrt{3}} \alpha^{3/2} .$$

Therefore, in this case, the system will tend either to the upper or to the lower sheet of the pleated surface. It is then called ***bistable***.

As in the previous section, as β varies a hysteresis loop forms. This is evident in Fig. 2.7.14b.

As stated at the beginning of this section, in most of the cases when α and β vary, which may represent the slow environmental changes, the stable equilibrium configuration changes slightly. But if α and β move in the vicinity of some critical value, for instance a turning point, their variation can catapult the trajectory into another basin of attraction and therefore toward a new equilibrium point, which might be far away. This is particularly important in biological models, where, for instance, the succes of a given species lays in its ability to persist to random environmental changes. If, for instance, the environment characteristics change severely enough, this may cause a sudden jump to a configuration less favorable to life.

On the other hand, in Example 2.9 we will discuss a phenomenon in which chatastrophe supports life.

Till now, however, we have discussed a single equation considering u as state variable and α and β as constants under our control. However, in

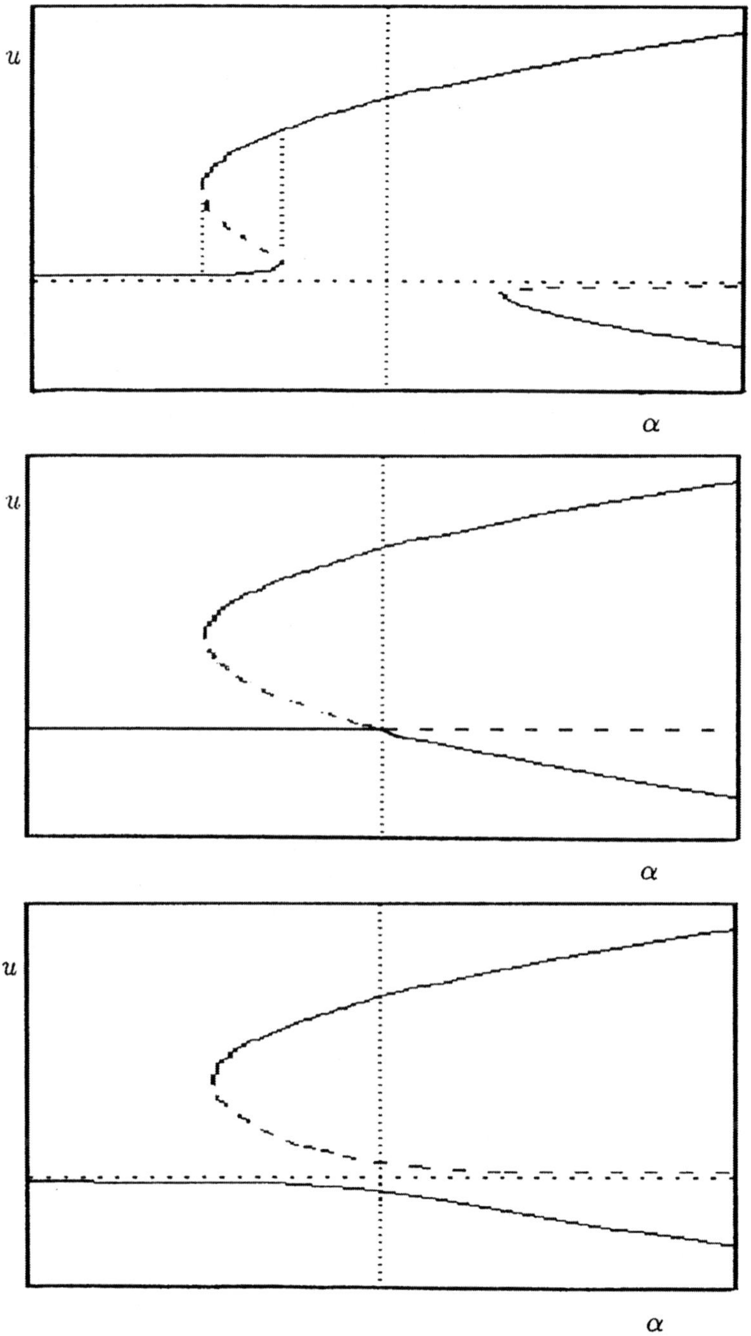

Figure 2.7.13 — Symmetry breaking for transcritical bifurcation.

some cases α and β can be considered, in some sense, as state variables themselves and there might be an automatic feedback that links the state variable with α and β via additional differential equations. In this way α and β determine the orbit $u(t)$, but as u varies it influences the values of α and β. Since trajectories usually move quickly toward their equilibrium points, these relationships prescribe the motion on the equilibrium surface, including the possible jumps, and may lead to the closure of the hysteresis cycle. An example is given in Fig. 2.7.14c, where

$$\frac{d\beta}{dt} = \varepsilon(u - u_c) \quad \text{with} \quad |u_c| < \sqrt{\frac{\alpha}{3}}$$

is attached to Eq.(2.7.11).

Example 2.9 Heart model

Some aspects of the dynamics of the heart can be modelled by schematizing the heart as a contracting and relaxing chamber with inlet and outlet valves. The relaxed state is called ***diastole***, the contracted state is called ***systole***.

An electrochemical stimulus causes the heart to contract, departing from its diastolic state. This contraction is slow at first, so that damaging backflows are avoided, and then a sudden push brings the heart to its systolic state. The way back to the relaxed state is similar, so that the cycle can be represented by Fig. 2.7.15.

Several researchers have tried to find a simple system of ordinary differential equations able to model this behavior. Zeeman [ZEE] proposed to choose the stimulus u_1 and the fiber length u_2 as state variables and then proposed the following model on the basis of known mathematical properies of the differential equations

$$\begin{cases} \dfrac{du_1}{dt} = \varepsilon(u_2 - \gamma)\,, \\[2mm] \dfrac{du_2}{dt} = \dfrac{1}{\varepsilon}\left(\alpha^2 u_2 - \dfrac{u_2^3}{3} - u_1\right), \end{cases} \qquad 0 < \alpha < \gamma\,. \qquad (2.7.13)$$

The parameter α represents the stretching due to pre-existent tension, which is provided by the blood pressure, and the parameter γ represents the diastole relaxed fiber length. Finally ε is a small parameter, which gives the effect of the two different contracting (and then relaxing) mechanisms in the heart beat.

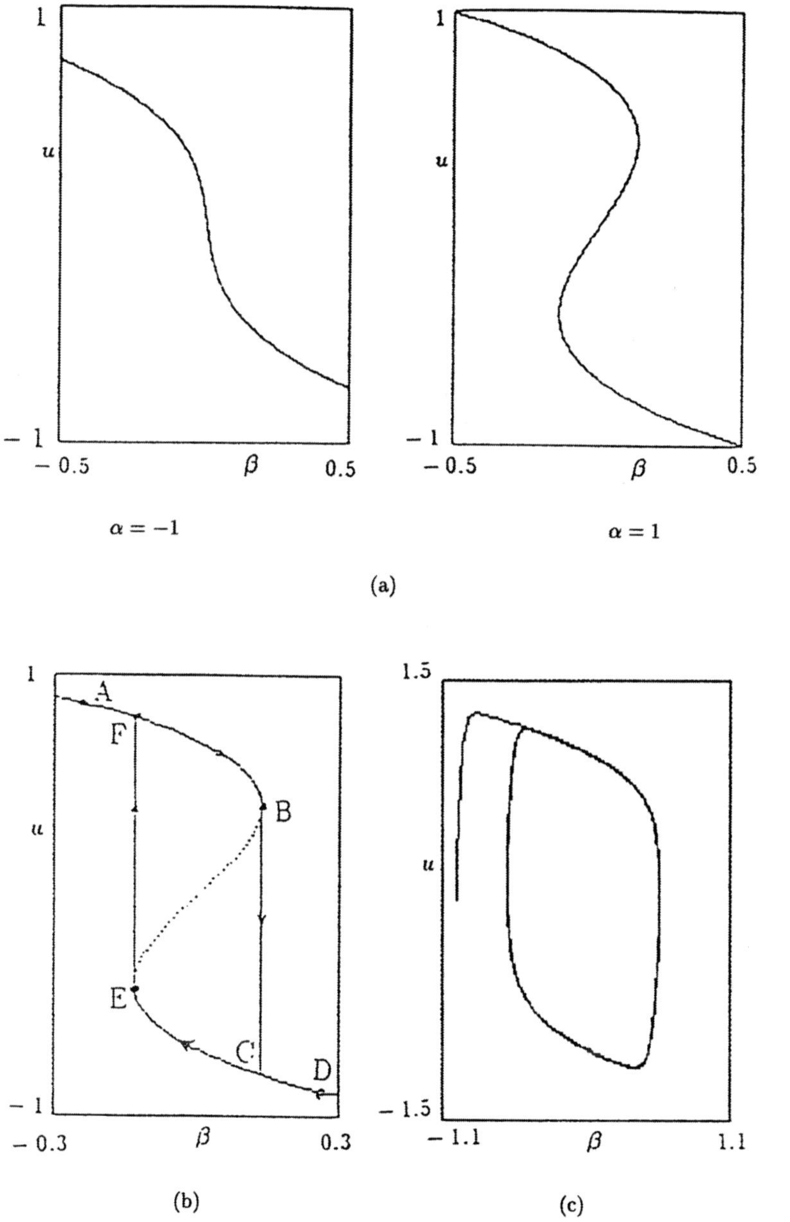

Figure 2.7.14 — Cusp catastrophe. (a) Bifurcation diagram for different α (b) hysteresis path (c) hysteresis loop for $\alpha = 1$ obtained integrating numerically Eq.(2.7.11) with $d\beta/dt = \varepsilon(u - u_e)$, $\varepsilon = u_e = 0.1$.

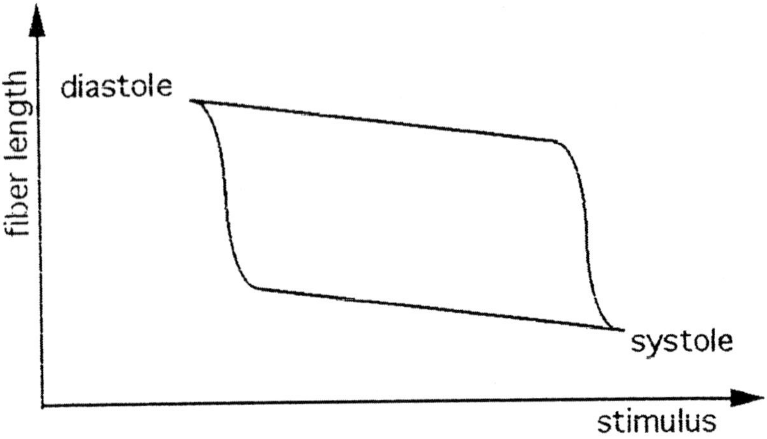

Figure 2.7.15 — Schematization of the heart pumping cycle.

The only equilibrium configuration of Eq.(2.7.13) is

$$\begin{cases} u_1 = \alpha^2\gamma - \dfrac{\gamma^3}{3} \\ u_2 = \gamma\,, \end{cases}$$

which is asymptotically stable if $0 < \alpha < \gamma$ as has been assumed, to assure that the diastolic state is the stable rest configuration. In fact, under this condition,

$$\mathbf{J}(\mathbf{u}_e) = \begin{pmatrix} 0 & \varepsilon \\ -\dfrac{1}{\varepsilon} & \dfrac{\alpha^2 - \gamma^2}{\varepsilon} \end{pmatrix}$$

has positive determinant and negative trace.

To understand Eq.(2.7.13) better, consider u_1 fixed so that we can more deeply analyze the second equation. Replacing u_1 with β we obtain an equation similar to Eq.(2.7.11), which was an example of the symmetry breaking of a supercritical bifurcation.

Referring to Figs. 2.7.12 and 2.7.16, we can now revisit the previous mathematical discussion in physiological terms. Assume that the stimulus goes from a minimum value β_1 at diastole to a maximum value β_2

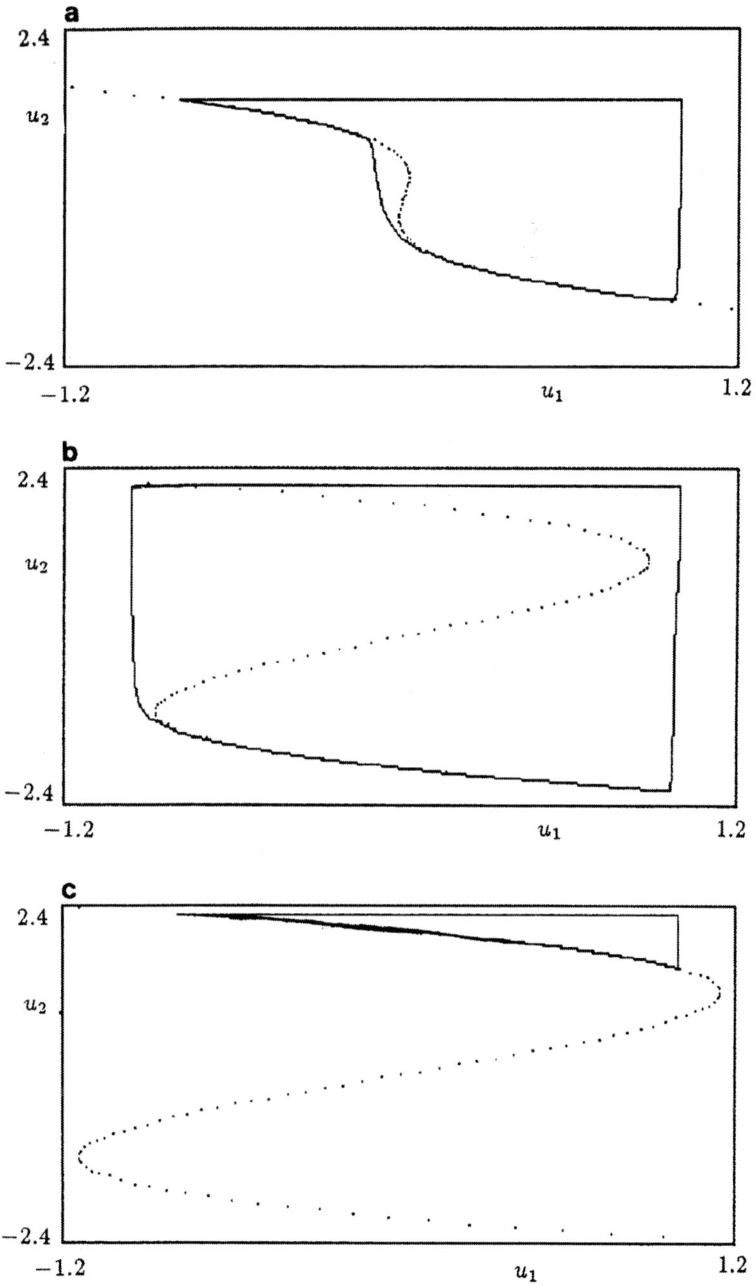

Figure 2.7.16 — Heart beat simulation at different tensions.
(a) $\alpha = 0.3$ (b) $\alpha = 1.1$ (c) $\alpha = 1.2$.

at systole. If tension α is low, the hysteresis cycle and therefore the heart beat is rather small (Fig. 2.7.16a). If the patient is scared or upset, the adrenalin causes a contraction of the arteries and an increase in the number of beats and in the blood pressure. Because of this, tension increases and the beat gets stronger (Fig. 2.7.16b). If a critical value is overcome, as may happen when a person with high blood pressure receives a sudden shock, then there is no beat at all (Fig. 2.7.16c), representing heart failure. In this situation the fiber moves uselessly in the neighborhood of the diastolic configuration. In order to recover, one has either to lower the tension of the patient or to administer a drug that temporarily increases the stimulus range.

Coming back to the full model (2.7.13), the cycle can be closed as follows. An external electrochemical stimulus forces u_1 to increase to a maximum value and therefore u_2 to contract slowly at first and then to jump to the lower sheet of the cusp surface. In the meantime, because of the first equation in (2.7.13), u_1 (the stimulus) decreases and therefore u_2 moves again slowly on the lower sheet of the cusp surface and then jumps back to the upper sheet finally reaching the diastole, the stable state. The external stimulus is then applied again and the cycle repeats itself. □

2.7.4 Limit cycles and Hopf bifurcation

In the previous section we approached the study of the stability of nonlinear systems and saw how an orbit repelled by an unstable point may tend to a new stable equilibrium state.

In many applications this is not the case. It may, in fact, happen that the solution approaches a periodic limit. We will say, in this case, that the solution tends to a cycle (see Fig. 2.7.17).

DEFINITION 2.7.5 Limit cycle
An orbit $\mathbf{u}(t)$ tends to a closed curve Γ called the ***limit cycle*** *if there exists a* ***period*** *$T > 0$ such that the sequence of points $\mathbf{u}(t+nT)$, $n \in \mathbb{N}$ tends to a point of Γ.*
A cycle is asymptotically stable if there exists a neighborhood $\mathcal{U}_\Gamma$ of Γ such that if $\mathbf{u}(0) \in \mathcal{U}_\Gamma$, then $\mathbf{u}(t)$ tends to Γ.

As we shall see, the relevance of limit cycles is basically related to the problem of how time-periodic behaviors may arise from the bifurcation of a steady state. This is essentially a two-dimensional problem. It is in fact impossible for a time-periodic solution to bifurcate from a steady one in

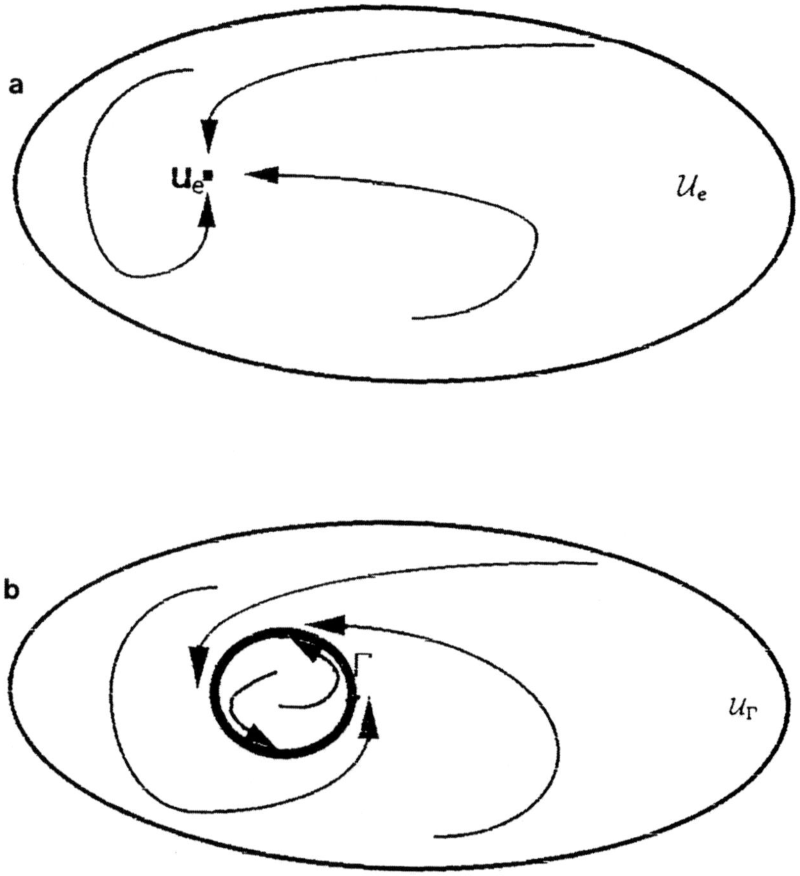

Figure 2.7.17 — (a) Asymptotic stability and (b) asympotically stable limit cycle.

one dimension. For this reason we consider this simpler case, referring to Iooss and Joseph [IaJ] for the treatment of higher-dimensional problems.

However, despite its importance in applications, even in this case, it is not easy to give general criteria for the existence and localization of limit cycles. For this reason, this subject still represents an active research field for both theoretical and applied scientists [STO], [YAN].

A first hint may be given by the famous Poincaré–Bendixson theorem.

THEOREM 2.7.1 ***(Poincaré–Bendixson Theorem)***
Consider the autonomous system

$$\frac{d\mathbf{u}}{dt} = \mathbf{f}(\mathbf{u}), \qquad \mathbf{u} \in \mathbb{R}^2, \tag{2.7.14}$$

and a domain $\mathcal{D}$ in the (u_1, u_2)-plane.

- *If $\mathcal{D}$ does not contain any stationary point and if no trajectory departs from $\mathcal{D}$, then $\mathcal{D}$ contains a limit cycle.*
- *If*

$$\nabla \cdot \mathbf{f} = \frac{\partial f_1}{\partial u_1} + \frac{\partial f_2}{\partial u_2}$$

is continuous and does not change sign in $\mathcal{D}$, then no limit cycle can exist in $\mathcal{D}$. ■

The main difficulty in applying this theorem is in finding a suitable domain $\mathcal{D}$. It is clear that the nonexistence criterium is very easy to check. It is a bit more difficult to control the existence hypothesis, the meaning of which is shown in Fig. 2.7.18.

REMARK 2.7.3 The Poincaré–Bendixson theorem only works in two dimensions. This is essentially due to the fact that orbits cannot intersect and therefore no trajectory can escape from the domain inside the cycle. In more dimensions, escape is possible and this allows much richer dynamics, such as trajectories on tori or even chaos. ■

Stronger theorems can be obtained for special classes of differential equations. For instance, for Liénard's equation

$$\frac{d^2u}{dt^2} + a(u)\frac{du}{dt} + b(u) = 0, \tag{2.7.15}$$

which was introduced in Section 2.2.4, the following theorem can be proved.

THEOREM 2.7.2 ***(Limit Cycle for Liénard's Equation)***
If a and b are continuous and continuously differentiable functions and if:

1. $u\,b(u) > 0, \qquad \forall u \neq 0;$

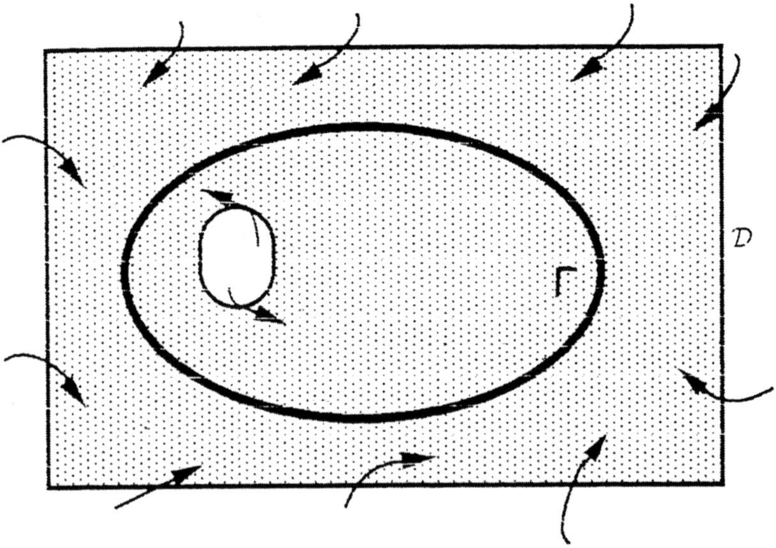

Figure 2.7.18 — Poincaré–Bendixson theorem. Γ is a limit cycle and all orbits enter the domain $\mathcal{D}$.

2. $a(u) < 0 \iff u \in (-\alpha, \beta)$, *with* $\alpha, \beta > 0$;

3. $a(u)$ *is not integrable at* $\pm\infty$;

then every nontrivial solution of Liénard's equation (2.7.15) is either a limit cycle or tends toward it.
If, in addition:

4. $a(u)$ *is an even function;*

5. $b(u)$ *is an odd function with* $b(u) > 0\,, \quad \forall u > 0$,

then Liénard's equation (2.7.15) has a unique stable limit cycle. ∎

Similarly for the more general differential equation

$$\frac{d^2u}{dt^2} + a\left(u, \frac{du}{dt}\right)\frac{du}{dt} + b(u) = 0\,, \tag{2.7.16}$$

we have the following theorem.

THEOREM 2.7.3 ***(Levinson–Smith Theorem)***
If:

1. $ub(u) > 0 \qquad \forall u \neq 0;$
2. $b(u)$ *is not integrable at* $+\infty$;
3. $a(0,0) < 0$;
4. $\exists\, u_0, M > 0$, *such that*

$$a(u,v) \geq \begin{cases} 0\,, & \text{if } |u| > u_0 \\ -M\,, & \text{if } |u| \leq u_0; \end{cases}$$

5. $\exists\, u_1 > u_0$, *such that*

$$\int_{u_0}^{u_1} a\big(u, v(u)\big)\, du \geq 10 M u_0\,,$$

where $v(u)$ *is an arbitrary positive and monotonically decreasing function of* u,

then Eq.(2.7.16) has at least one limit cycle. ∎

Consider again, now, systems with a parameter

$$\frac{d\mathbf{u}}{dt} = \mathbf{f}(\mathbf{u};\alpha)$$

and assume that there is an equilibrium state $\mathbf{u}_e$, which is stable for $\alpha < \alpha_b$ and unstable otherwise. It may happen that for $\alpha > \alpha_b$ the solution tends to a time-periodic orbit. The system then undergoes a bifurcation in which trajectories will not spiral toward a new stable equilibrium point, but will tend to a limit cycle (see Fig. 2.7.19). This is called ***Hopf bifurcation*** and to recognize it, the following theorem is helpful.

THEOREM 2.7.4 ***(Hopf Theorem)***
Let $\mathbf{u}_e(\alpha) \in \mathbb{R}^n$ *be asymptotically stable for* $\alpha < \alpha_b$ *and unstable for* $\alpha > \alpha_b$. *If at criticality the Jacobian* $\mathbf{J}$ *of* $\mathbf{f}$ *has a simple pair of purely imaginary eigenvalues denoted by* λ, *that is*

$$\Re e\big(\lambda(\alpha = \alpha_b)\big) = 0 \text{ and } \Im m\big(\lambda(\alpha = \alpha_b)\big) \neq 0\,,$$

and all the other eigenvalues have a negative real part and furthermore

$$\frac{d}{d\alpha}\Re e\big(\lambda(\alpha = \alpha_b)\big) > 0\,, \tag{2.7.17}$$

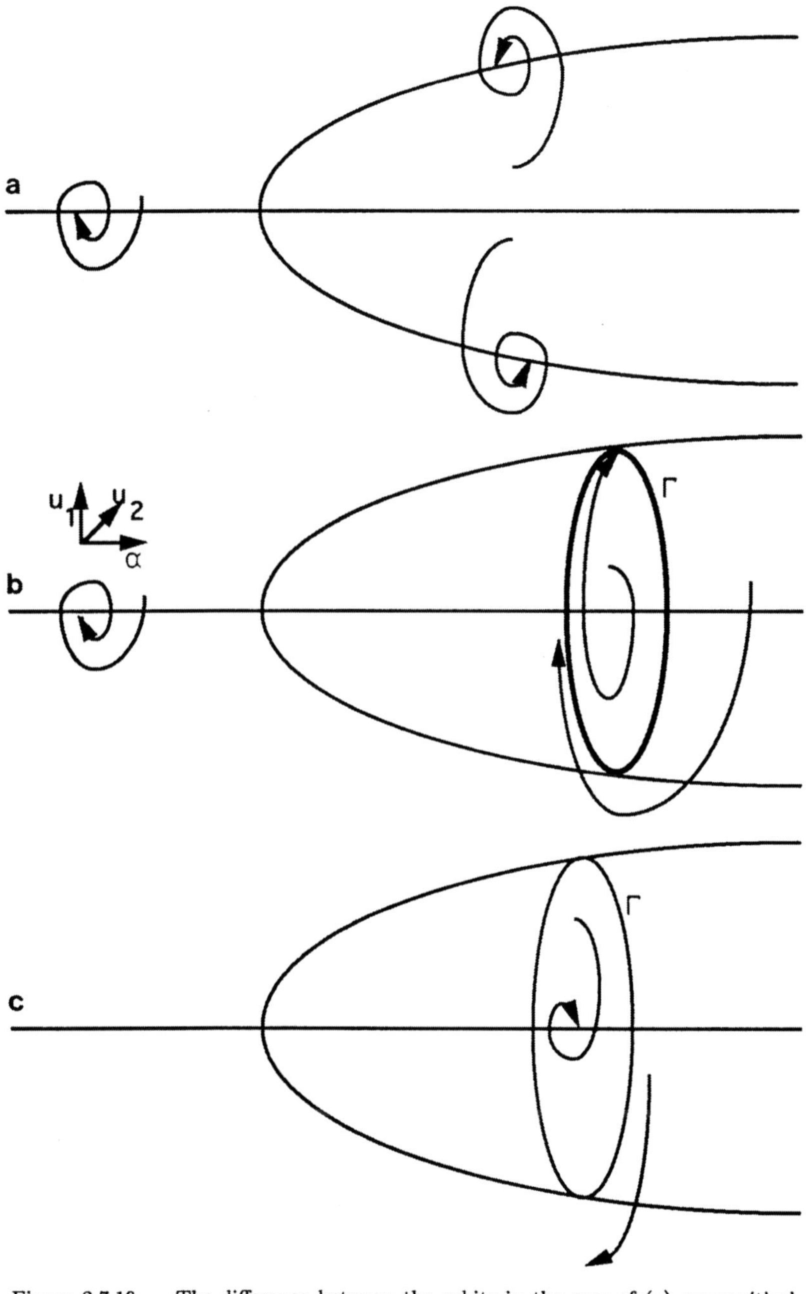

Figure 2.7.19 — The difference between the orbits in the case of (a) supercritical bifurcation, (b) Hopf bifurcation with stable limit cycle, and (c) Hopf bifurcation with unstable limit cycle.

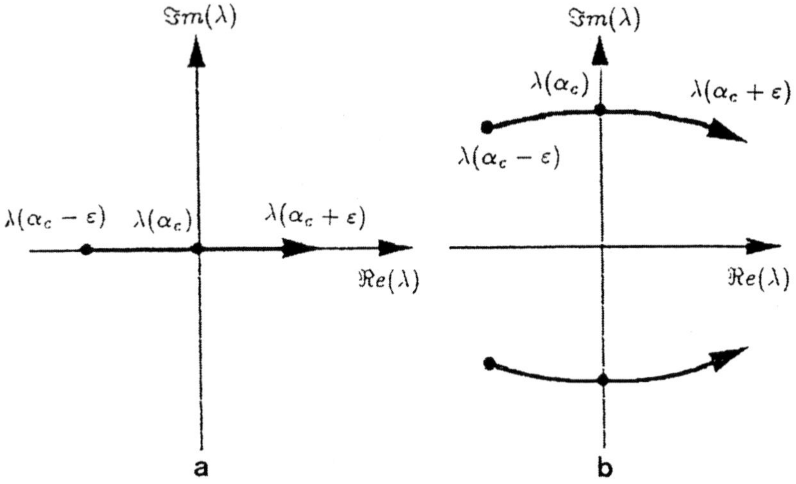

Figure 2.7.20 — Behavior of the critical eigenvalues as α increases from stable values to unstable ones in the case of (a) stationary bifurcation and (b) Hopf bifurcation.

then for α sufficiently near α_b, there exists a limit cycle with initial period

$$T = \frac{2\pi}{\Im m\big(\lambda(\alpha = \alpha_b)\big)}\,.$$

If $\mathbf{u}(\alpha = \alpha_b)$ *is locally asymptotically stable, then the limit cycle is stable.* ∎

We recall that, in the case of stationary bifurcation, by Definition 2.7.4, one of the eigenvalues vanishes at criticality (as implied by the singularity of the Jacobian). Instead, in the case of Hopf bifurcation, because of the transversality condition (2.7.17), the eigenvalues λ of $\mathbf{J}$ cross the imaginary axis from left to right away from the origin as α, increasing from stable values to unstable ones, crosses criticality (see Fig. 2.7.20).

Generally, the Hopf theorem does not tell whether a stable (supercritical) limit cycle exists for $\alpha > \alpha_b$ or an unstable (subcritical) limit cycle exists for $\alpha < \alpha_b$, inside of which all orbits spiral towards $\mathbf{u}_e$ and outside of which orbits diverge, as shown in Fig. 2.7.19.

In order to know whether the limit cycle is stable or unstable, we must prove that $\mathbf{u}_e$ is locally asymptotically stable at criticality. This can be done using the Liapunov method. In fact, linear stability criteria are useless because at criticality the eigenvalues are purely imaginary, that is $\mathbf{u}_e(\alpha = \alpha_b)$ is neutrally stable. This precludes any conclusion on the character of the nonlinear system (as observed in Remark 2.6.4). This second part is, however, certainly not trivial.

The following examples may help the reader to understand limit cycles and Hopf bifurcation.

Example 2.10 Limit cycle for the Van der Pol equation
As an example, consider the modified Van der Pol equation introduced in Section 2.2.4

$$\frac{d^2u}{dt^2} - (\alpha + \beta u^2)\frac{du}{dt} + u = 0\,, \tag{2.7.18}$$

with $\beta < 0$.

From the physical point of view, we see that if $\alpha > 0$ the energy of the system is dissipated for u large enough, but some energy is pumped into the system for $|u| < \sqrt{\alpha/|\beta|}$. It is this balance between the damping for large u and the destabilizing excitation for small u that gives rise to self-sustained oscillations. It is easy to check that this equation satisfies the hypothesis of Theorem 2.7.2, therefore a stable limit cycle exists for $\alpha > 0$ and a supercritical Hopf bifurcation occurs at $\alpha = 0$. This can be also obtained using Theorem 2.7.4. In fact, in Example 2.4 we showed that $u = 0$ is the only equilibrium state that is asymptotically stable for $\alpha < 0$ and unstable for $\alpha > 0$ and that the eigenvalues related to the linear problem are

$$\lambda = \frac{\alpha \pm \sqrt{\alpha^2 - 4}}{2}\,.$$

Therefore, near the origin

$$\Re e(\lambda) = \frac{\alpha}{2} \quad \text{and} \quad \frac{d}{d\alpha}\,\Re e\big(\lambda(\alpha)\big) = \frac{1}{2} > 0\,.$$

Furthermore, at criticality

$$\Im m\big(\lambda(\alpha = 0)\big) = \pm i \neq 0\,.$$

From the Hopf theorem, then, there exists a limit cycle. To establish if the limit cycle is subcritical (unstable) or supercritical (stable), one has to verify that $u = 0$ is locally asymptotically stable for $\alpha = 0$. But

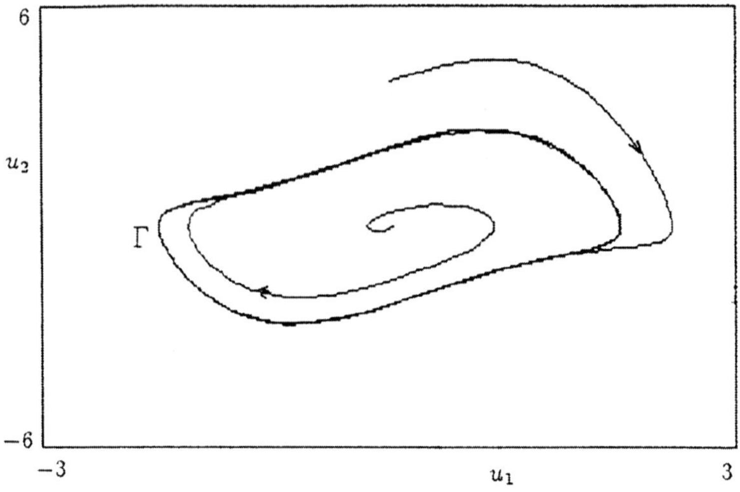

Figure 2.7.21 — Integration of the Van der Pol equation.

this has been proved and remarked upon at the end of Example 2.6. In Fig. 2.7.21, Eq.(2.7.18) is integrated showing the trajectories towards the limit cycle.

Observe that if $\beta > 0$, then again the Hopf theorem holds, but one does not have that $u = 0$ is locally asymptotically stable for $\alpha = 0$. In fact, in this case the limit cycle exists but is unstable. This can be readily proved by introducing the new variables

$$\widetilde{\alpha} = -\alpha\,, \quad \widetilde{\beta} = -\beta\,, \quad \widetilde{t} = -t\,. \tag{2.7.19}$$

In this way, we again end up with Eq.(2.7.18) with $\widetilde{\beta} < 0$. Hence there is a stable limit cycle for $\widetilde{\alpha} > 0$. But since time, $\widetilde{t}$, is reversed, what is found stable is actually unstable for the original time t. Therefore, we actually found that there is an unstable limit cycle for $\alpha < 0$.

This time reversing technique is actually used to search for unstable limit cycles. □

Example 2.11 Hopf bifurcation for May's model
Consider a predator-prey system evolving according to May's model, Eqs.(2.2.20, 2.2.21)

$$\frac{du_1}{dt} = B_1 u_1 \left(1 - \frac{u_1}{M_1}\right) - \frac{H_1 H_M u_1 u_2}{H_M + H_1 u_1}$$
$$\frac{du_2}{dt} = B_2 u_2 \left(1 - \frac{u_2}{M_2 u_1}\right), \tag{2.7.20}$$

with B_1, B_2, M_1, M_2, H_1, and H_M positive constants.

Besides the extinction equilibrium configuration given by

$$u_{1e} = M_1 , \quad u_{2e} = 0 , \tag{2.7.21}$$

there is another equilibrium configuration given by

$$u_{2e} = M_2 u_{1e} \tag{2.7.22}$$

and by the positive root of

$$B_1 \left(1 - \frac{u_1}{M_1}\right) = \frac{H_1 H_M M_2 u_1}{H_M + H_1 u_1}, \tag{2.7.23}$$

which is of

$$B_1 H_1 u_1^2 + (B_1 H_M - B_1 M_1 H_1 + H_1 H_M M_1 M_2) u_1 - B_1 M_1 H_M = 0 .$$

The Jacobian matrix of the linearized system related to Eq.(2.7.20) is

$$\mathbf{J} = \begin{pmatrix} J_{11} & J_{12} \\ J_{21} & J_{22} \end{pmatrix}, \tag{2.7.24}$$

where

$$J_{11} = B_1 \left(1 - 2\frac{u_1}{M_1}\right) - \frac{H_1 H_M^2 u_2}{(H_M + H_1 u_1)^2},$$
$$J_{12} = -\frac{H_1 H_M u_1}{H_M + H_1 u_1},$$
$$J_{21} = \frac{B_2 u_2^2}{M_2 u_1^2},$$
$$J_{22} = B_2 \left(1 - 2\frac{u_2}{M_2 u_1}\right).$$

By evaluating the trace and the determinant of (2.7.24) in the extinction equilibrium (2.7.21) it is easy to verify that this state is unstable.

Evaluation of (2.7.24) in (2.7.22, 2.7.23) yields

$$\mathbf{J}(\mathbf{u} = \mathbf{u}_e) = \begin{pmatrix} B_e & -\dfrac{H_1 H_M u_{1e}}{H_M + H_1 u_{1e}} \\ B_2 M_2 & -B_2 \end{pmatrix}, \tag{2.7.25}$$

where

$$B_e = B_1 \left(1 - 2\frac{u_{1e}}{M_1}\right) - \frac{H_1 H_M^2 M_2 u_{1e}}{(H_M + H_1 u_{1e})^2},$$

which, using (2.7.23), can be written as

$$B_e = -\frac{B_1}{M_1} u_{1e} + \frac{H_1^2 H_M M_2 u_{1e}^2}{(H_M + H_1 u_{1e})^2}.$$

The determinant of (2.7.25) is

$$\det \mathbf{J} = \frac{B_1}{M_1} u_{1e} + \frac{H_1 H_M^2 M_2 u_{1e}^2}{(H_M + H_1 u_{1e})^2}$$

and therefore is always positive, while

$$\operatorname{tr} \mathbf{J} = B_e - B_2$$

changes sign as B_2 crosses B_e.

The eigenvalues $\lambda_i(B_2)$ of $\mathbf{J}$ are complex for B_2 sufficiently near B_e, with

$$\Re e\big(\lambda_i(B_2)\big) = \frac{1}{2} \operatorname{tr} \mathbf{J} = \frac{B_e - B_2}{2}.$$

Therefore, the equilibrium state (2.7.22, 2.7.23) is asymptotically stable if $B_2 > B_e$ and unstable if $B_2 < B_e$. Of course, instability can occur only if B_e is positive. That is, when

$$\alpha = \frac{B_1}{H_1 M_1 M_2} < \frac{1}{4}$$

$$\gamma = \frac{H_1 M_1}{H_M} \in \left(\frac{1 - \alpha - \sqrt{1 - 4\alpha}}{\alpha}, \frac{1 - \alpha + \sqrt{1 - 4\alpha}}{\alpha}\right).$$

In this example it is more difficult to establish whether the equilibrium configuration is asymptotically stable or not at criticality $B_e = B_2$. In spite of this, since

$$\Re e\big(\lambda_i(B_e)\big) = 0, \quad \Im m\big(\lambda_i(B_e)\big) \neq 0,$$

and the real part of the eigenvalues increases as B_2 decreases, that is, as we move from the stable to the unstable region, Hopf bifurcation Theorem 2.7.4 assures the existence of a limit cycle. Numerically it can be verified, as expected, that the limit cycle is stable.

From the biological point of view this means that there is a periodic evolution in the number of predators and prey. In fact, when there is abundant food the number of predators increases, but this drastically reduces the amount of available prey. The predators, then, start starving to death. Due to the smaller number of predators, the number of prey starts increasing again and the cycle repeats itself forever, as shown in Fig. 2.7.22. □

As was done for stationary bifurcation, it is possible to illustrate Hopf bifurcation graphically. A stable limit cycle is usually indicated with a series of small solid circles, while an unstable limit cycle is indicated with a series of small open circles, as indicated in Fig. 2.7.23.

The position of the circles usually indicates the minimum and the maximum value of the oscillations.

In the supercritical case, see Fig. 2.7.23b, an exchange of stability occurrs at α_c between the stationary solution and the time-periodic one. The amplitude of the oscillations continuously increases from zero as α departs from α_c. This kind of Hopf bifurcation is called ***soft loss of stability*** or ***soft generation of limit cycle***.

Similar to what we saw in Section 2.7.2, and as we shall see in the following example, it may occur that the subcritical periodic branch turns back at $\alpha = \alpha_t$ gaining stability, as shown in Fig. 2.7.23c. Then, as in Section 2.7.2, as α crosses α_c, the system suddenly undergoes large amplitude periodic oscillations and as α returns below α_t (not α_c) the system will again tend toward the stationary solution. A kind of hysteresis cycle sets up. This kind of Hopf bifurcation is called ***hard loss of stability*** or ***hard generation of limit cycle***.

Between α_t and α_c, the stable configuration, the unstable configuration, and the stable limit cycle coexist. In two dimensions the phase diagram may look like that presented in Fig. 2.7.23c. Observations similar to those made in Section 2.7.2 can be reproposed here with the obvious changes. We remark here that if the control parameter α depends on another variable, or is varied from outside the system so that it can go back and forth between a value less than α_t and a value greater than α_c, then the system exhibits a series of sudden ***bursts***, that is, periods of strong oscillation, usually of small duration, to periods of relative quiescence. Such a switch between active and inactive states is of interest in chemistry, biology, and physiology. Here we will discuss the case of

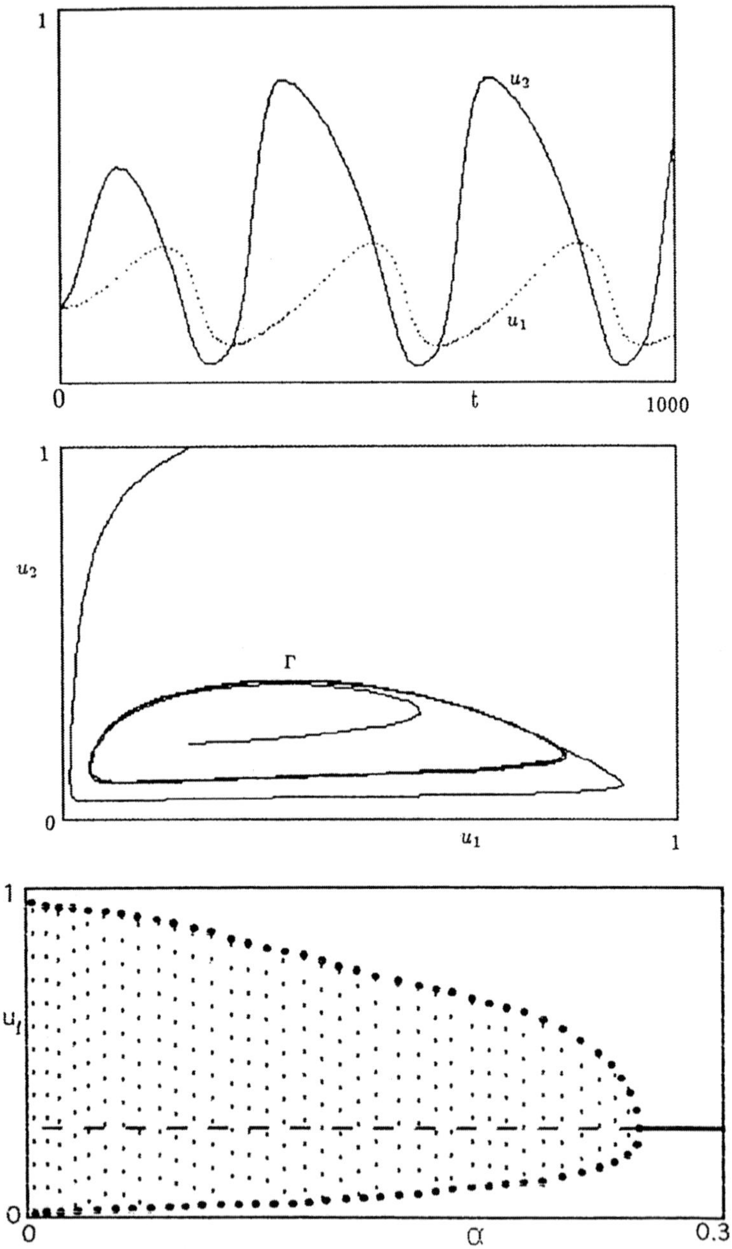

Figure 2.7.22 — Simulation of May's predator-prey model. Temporal evolution and limit cycle in the phase plane for $\alpha = 0.1$. The Hopf bifurcation diagram is also shown.

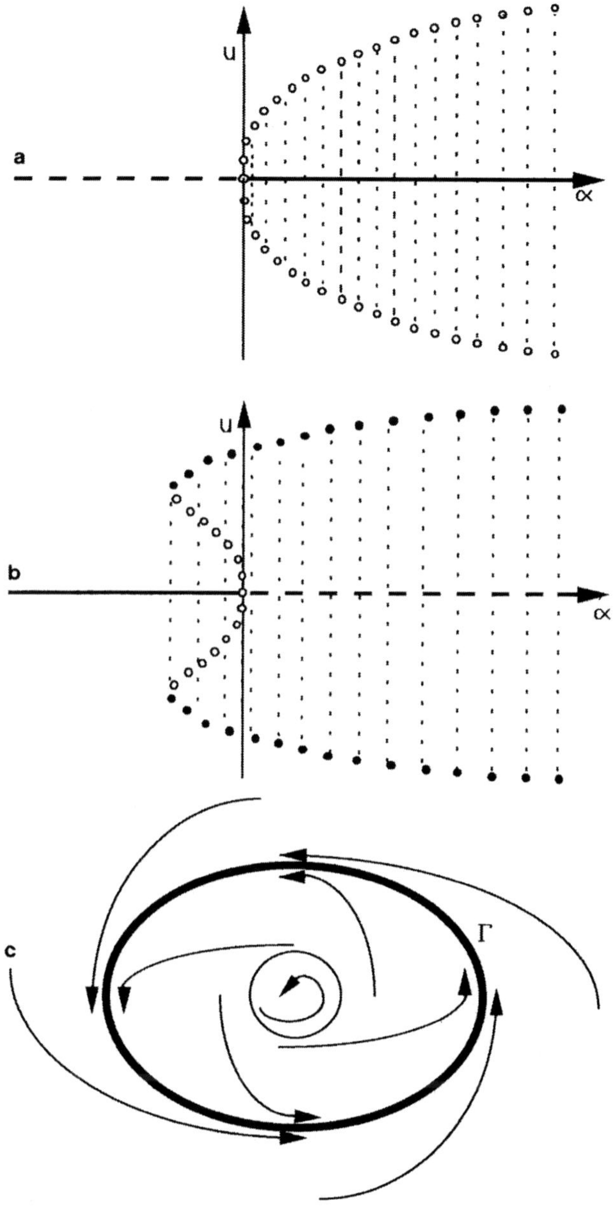

Figure 2.7.23 — Hopf bifurcation diagrams. Solid circles represent stable limit cycles and open circles represent unstable limit cycles. (a) Subcritical bifurcation, (b) subcritical bifurcation with turning points, (c) phase diagram related to case (b) for $\alpha_t < \alpha < \alpha_b$.

the propagation of an electric impulse down a nerve axon introduced in Section 2.2.5.

Example 2.12 Hopf bifurcation for the FitzHugh nerve impulse model
Consider the FitzHugh model introduced in Section 2.2.5

$$\begin{cases} \dfrac{du_1}{dt} = \dfrac{1}{\varepsilon}\left(u_1 - \dfrac{u_1^3}{3} - u_2 + \alpha\right) \\ \dfrac{du_2}{dt} = \varepsilon\,(u_1 - \beta u_2 + \gamma)\,, \end{cases} \tag{2.7.26}$$

where

$$\beta \in (0,1)\,, \quad \gamma \in \left(1 - \frac{2}{3}\beta, 1\right)\,, \text{ and } \varepsilon \tag{2.7.27}$$

are fixed parameters, with ε small with respect to unity. We want to study the behavior of the system as α varies. We recall that α models the applied current.

The equilibrium states of (2.7.26) are given by

$$\begin{cases} u_2 = \dfrac{u_1 + \gamma}{\beta} \\ \alpha = \dfrac{u_1^3}{3} + \dfrac{1-\beta}{\beta}u_1 + \dfrac{\gamma}{\beta} \end{cases} \tag{2.7.28}$$

and the Jacobian is

$$\mathbf{J} = \begin{pmatrix} \dfrac{1-u_1^2}{\varepsilon} & -\dfrac{1}{\varepsilon} \\ \varepsilon & -\beta\varepsilon \end{pmatrix}. \tag{2.7.29}$$

Then we have

$$\det \mathbf{J} = 1 - \beta + \beta u_1^2\,,$$

which is always positive because of (2.7.27) and

$$\operatorname{tr} \mathbf{J} = \frac{1-u_1^2}{\varepsilon} - \beta\varepsilon\,,$$

which is negative if and only if $u_1^2 > 1 - \beta\varepsilon^2$. Therefore, only in this case is the equilibrium configuration (2.7.28) asymptotically stable. If $u_1^2 < 1 - \beta\varepsilon^2$, then it is unstable.

The eigenvalues of $\mathbf{J}$ at criticality, i.e., when $u_1 = \pm\sqrt{1-\beta\varepsilon^2}$ reduce to

$$\lambda = \pm i\sqrt{\det \mathbf{J}}\,. \tag{2.7.30}$$

Furthermore, according to (2.6.9), $\Re e(\lambda) = -\mathrm{tr}\,\mathbf{J}/2$ near criticality and therefore

$$\frac{d\Re e(\lambda)}{d\alpha} = \frac{\beta u_1}{\varepsilon \det \mathbf{J}} \neq 0\,. \tag{2.7.31}$$

Thus, the points $u_1 = \pm\sqrt{1-\beta\varepsilon^2}$ on the curve (2.7.28) are Hopf bifurcation points.

Furthermore, it is possible to see that both these points are subcritical with hard generation of limit cycles. A typical bifurcation diagram is given in Fig. 2.7.24b. This is typical and essential for the onset of nerve impulses. In fact when a stimulating current α passes a critical level (the Hopf bifurcation point), a strong oscillatory motion immediately sets up reaching its maximum amplitude. Similarly as α decreases below the critical value, the oscillations will suddenly stop. In Fig. 2.7.24a, Eq.(2.7.26) is integrated; bursts set in or smooth out as α_c or α_t are passed by.

Finally we want to remark that the fact that the two Hopf bifurcation points are connected by the same time-varying branch is not uncommon.

□

2.7.5 Attractors and chaos

In the previous sections we have encountered systems of ordinary differential equations that possess solutions that either spiral towards a stable point or toward a limit cycle. This means that as t goes to infinity, the solution rests in a stable configuration or oscillates periodically. Furthermore, unless for very peculiar cases, if two orbits start sufficiently near, they end up in the same configuration.

Sometimes, this situation does not occur. The oscillations are irregular, appear aperiodic, and two orbits that start next to each other, no matter how near, depart exponentially from each other. In spite of this, orbits seem to end all in the same region. Think of the motion of a particle in a turbulent liquid. With some luck we can say at most that the particle will end up, after a given time, in a certain region.

Without going into detail, it should be clear what is usually meant by attractor. A stable point or cycle is an attractor. The fuzzy region described by an infinite set of points, which represents the asymptotic behavior of a chaotic system, is called ***a strange attractor***.

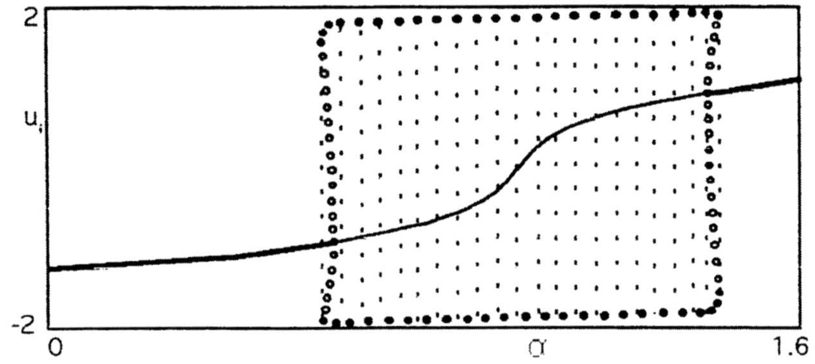

(a)

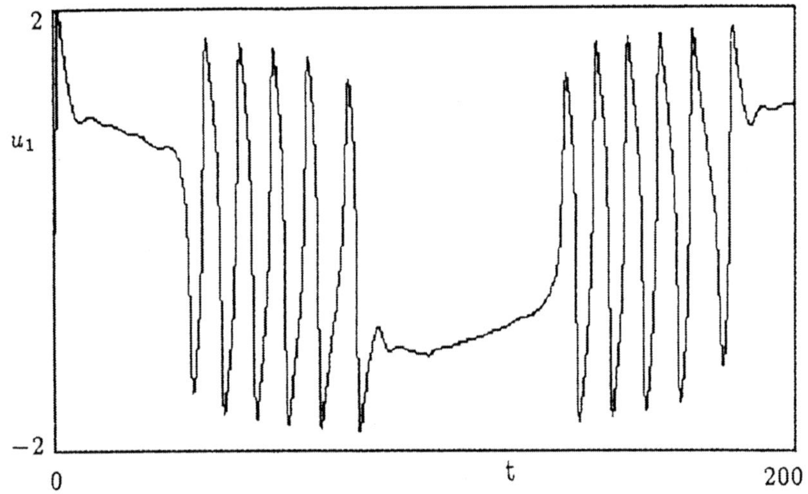

(b)

Figure 2.7.24 — (a) Bursts for FitzHugh nerve impulse propagation model with $\alpha = \frac{5}{18} + \frac{11}{36}\left|1 - \frac{t}{100}\right|$, $\beta = 0.8$, $\gamma = 0.9$ and (b) bifurcation diagram for $\beta = 0.8$ and $\gamma = 0.9$.

The sensitivity of the solution from the initial data has many practical consequences. Consider, for instance, a sample experiment of a chaotic system. Since there are always small deviations in setting the system in the initial state or perturbations while the experiment goes on, the system is less and less predictable as time goes by.

In a similar way, if we are trying to simulate the system with a numerical code, then the predicted trajectory will depend crucially on the method, on the precision, and even on the computer used. In spite of this, the general figure composed by the trajectory is essentially independent of these physical and mechanical perturbations.

This unpredictability is why it is impossible to have a long term weather forecast, but only some general predictions.

The integration of a system of ordinary differential equation exhibiting a strange attractor might be considered a problem that is never accurate enough and never finished. The result may be interpreted as coming from a nondeterministic problem. But, pay attention! The resulting chaotic motion and unpredictability is, instead, governed by deterministic equations with well-defined initial conditions and coefficients and is not affected by any stochastic perturbations.

The most famous example of a system of ordinary differential equations exhibiting a strange attractor is the Lorentz model

$$\begin{cases} \dfrac{du_1}{dt} = -\alpha(u_1 - u_2) \\ \dfrac{du_2}{dt} = \beta u_1 - u_2 - u_1 u_3 \\ \dfrac{du_3}{dt} = -\gamma u_3 + u_1 u_2 \,, \end{cases} \tag{2.7.32}$$

where α, β, and γ are real positive parameters. Equation (2.7.32) was originally derived while studying the Bénard problem. This problem consists in the study of the stability properties and of the onset of convective and turbulent motions in a two-dimensional fluid heated from below and cooled from above. The variables of the resulting partial differential equation are expanded into an infinite number of modes, three of which are considered nonvanishing. Roughly speaking, u_1 measures the rate of convective overturning and u_2 and u_3 measure, respectively, the horizontal and the vertical temperature variation. The parameters α and β are called, respectively, the Prandtl and Rayleigh numbers, while γ is a measure of the region under consideration. A complete analysis of this model is given by Sparrow [SPW]. The result of its integration in the phase space is given in Fig. 2.7.25.

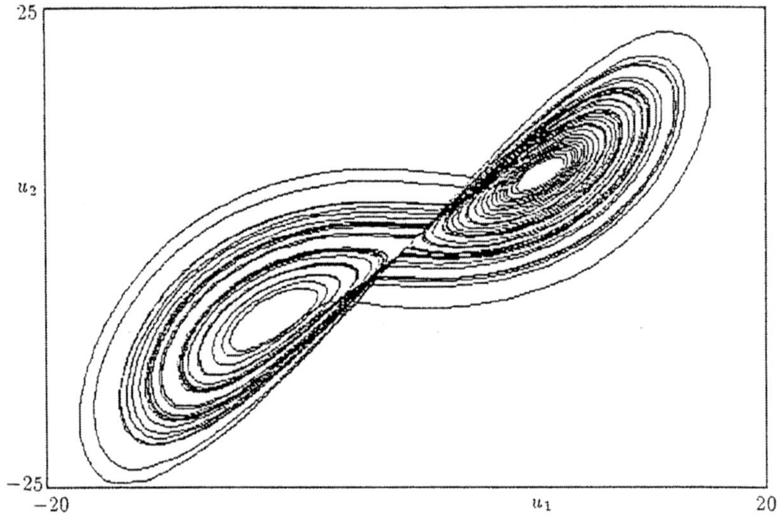

Figure 2.7.25 — Numerical integration of Lorentz model.

Other examples are the forced modified Duffing's and Van der Pol equations (2.2.31) and (2.2.32).

Another useful tool for describing the attractors is the ***stroboscopic map***. This map is obtained by observing the system periodically and marking with a point its position in the phase space. It is similar to obtaining a stroboscopic picture by flashing a light.

In Eqs. (2.2.31) and (2.2.32), the time interval between two successive flashes can be taken as the period of the forcing term and a time of ten periods is waited at the beginning to eliminate the effects of the initial conditions.

Figure 2.7.26 assembles one thousand stroboscopic points generated by a specific trajectory. Apparently, the points indicate the formations of lines nested in a complex structure. One may argue that perhaps these fuzzy lines are filled in as time goes to infinity (in fact, as already mentioned, a numerical calculation of a strange attractor can never be considered as completed). This impression is false. In fact an enlargement of any of these lines reveals that they are made up of sublines organized in substructures, and so on.

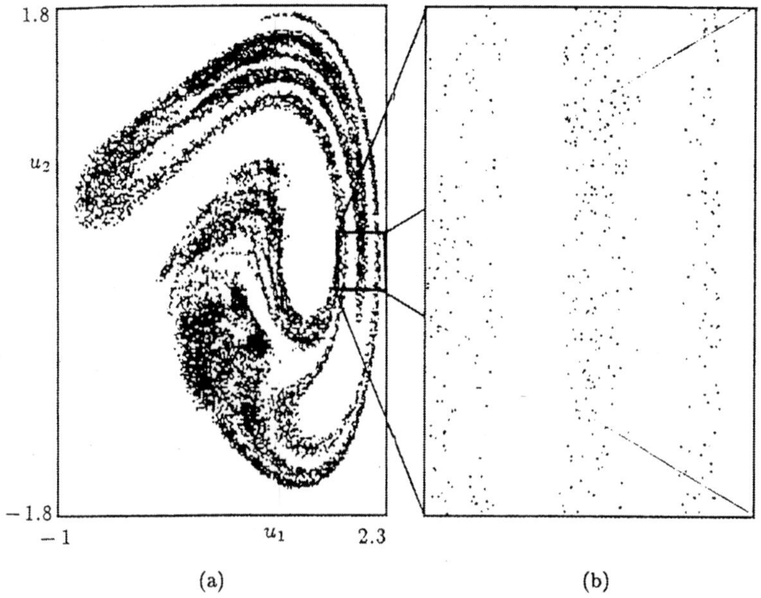

Figure 2.7.26 — Stroboscopic map for the forced Duffing's equation (a) and its enlargement (b).

It is natural to ask at this point how chaos develops from a stationary solution as the coefficients of the ordinary differential equation vary and, if possible, how to determine for the parameters a range in which chaotic behavior has to be expected. Essentially, the problem is to determine how a turbulent motion develops from a laminar flow and for which value of the parameter involved in the ordinary differential equations the laminar flow breaks up to become turbulent.

Unfortunately these problems are not well settled. What is sure is that bifurcation plays a fundamental role. Several routes to chaos may be conceived and they are all more or less supported by both theoretical and experimental evidence. So chaos may develop through repeated sequences of Hopf bifurcations, through what is called a torus bifurcation, through a sequence of period doubling bifurcations, so that the period quickly approaches infinity, or through the appearence of aperiodic vibrations during a periodic oscillation of the system.

Several other hypotheses have been proposed and they, perhaps, do not contradict each other. What is sure is that chaos is initiated by bifurcation and develops through several other bifurcations.

2.8 Numerical Methods for Initial-Value Problems

After having obtained all possible information from a qualitative study of the mathematical problem, one can then develop the numerical simulation of the system, or better of the conjectured mathematical model.

We will not enter here into details that are more proper for a numerical analysis textbook, but we feel like giving some advice to the reader on how to approach, choose, develop, and control numerical schemes.

There are essentially two features regarding the solution of differential equations that cannot be represented exactly on a computer. In fact, one usually wants to know the solution in a finite time interval, a continuum set that the computer cannot handle, and one has to evaluate derivatives, which are obtained by a limit process that cannot be simulated numerically.

As a basis of any numerical scheme, one has, then, to introduce a time step $h = \Delta t$ to discretize the interval in very small parts and an approximation formula for the derivatives. In this way, the original differential equation is replaced by a difference equation. It is clear that since we replaced an infinite process with a finite one, an error is introduced as an unavoidable toll to be paid. This error is called the ***truncation error***.

DEFINITION 2.8.1 Truncation error

*The **truncation error** T_{err} is the amount by which the solution of the differential equation*

$$\frac{d\mathbf{u}}{dt} = \mathbf{f}(t, \mathbf{u}) \tag{2.8.1}$$

fails to satisfy the approximate equation. More precisely, it is the norm of the difference between the solution to the differential equation and the numerical solution divided by the time step used in the numerical scheme.

Furthermore, an obvious and essential requirement for the numerical scheme is that it must be able to simulate better and better the original ordinary differential equation by suitably refining the step size, that is, the discretized equation should tend to the ordinary differential equation to which it is related as h goes to zero.

DEFINITION 2.8.2 Consistency of an approximation method
An approximation method is called ***consistent*** *with the differential equation if the truncation error goes to zero as the time step goes to zero.*

There is, however, another important requirement a numerical scheme has to satisfy, which is a bit more technical but can be easily explained by recalling the concepts of equilibrium and stability introduced in Section 2.6. There we saw that formally a system can occupy indefinitely an unstable state, but as soon as it is perturbed it wanders away. A similar thing happens for numerical methods. Here the unavoidable perturbations are the truncation and round-off errors and they should not be amplified without bound as time goes by. One then has to verify that the applied numerical scheme is stable.

DEFINITION 2.8.3 Absolute stability
An approximation method is called ***absolutely stable*** *in a point ah of the complex plane if the sequence $\{u_n\}$ generated by the method applied to the model equation*

$$\frac{du}{dt} = au \tag{2.8.2}$$

with time step $\Delta t = h$, is bounded.

The ***stability region*** *is the set of points $ah \in \mathbb{C}$ for which the method is absolutely stable.*

A method is called ***A-stable*** *if the stability region includes the whole negative half plane $\Re e(ah) \leq 0$.*

Though the stability region is defined using the trivial model equation (2.8.2), it gives a fairly complete picture of the behavior of the numerical method.

How can one use this information? Even if we have a nonlinear model we can consider a linearized form of it

$$\frac{d\mathbf{u}}{dt} = \mathbf{Au} \tag{2.8.3}$$

and compute the eigenvalues $\lambda_1, \ldots, \lambda_n$ of $\mathbf{A}$, the so-called spectrum of the linearized model. If $h\lambda_1, \ldots, h\lambda_n$ all belong to the stability region of a certain numerical method, then the numerical errors are not amplified as the integration is continued and therefore we can consider it for simulating the original nonlinear model. If, on the other hand, there are some eigenvalues outside the stability region, then we must be aware that the numerical errors grow exponentially in time with an exponent proportional to the distance between the stability region and the eigenvalue times the time step. The method has to be stopped as this error gets larger than the desired tolerance.

Besides the accuracy and stability requirements, there are other qualities that characterize a good numerical method: It should be efficient, economically convenient, adaptive (in the sense that the time step can be easily changed during numerical integration), versatile (so that the same computer program can then be adapted for several similar problems without much effort), and should have reasonable storage requirements.

Before writing his own codes, the reader should also know that there are mathematical libraries, for instance [IMS, NAG], which include routines adequate for any nonstiff ordinary differential equation problem. We encourage the reader to use the software already prepared by specialists, at least as a starting point. However, after that, since these routines are written to work in general cases, the reader can choose to write his own numerical code so that he can throughly exploit all the characteristics and properties he knows about the model with which he is dealing.

Whenever possible, before starting the actual simulation, several tests should be performed, not necessarily in this order:

- Control the stability properties of the numerical method keeping in mind the spectrum related to the model.
- Test the code on problems with a known analytic solution.
- Check the accuracy and convergence on decreasing h, and eventually set h in agreement with the desired accuracy. The standard way to determine whether the scheme is implemented correctly is to run the code for several (decreasing) values of h. The resulting a posteriori estimate should agree with the theoretical a priori estimate when h is not too large.
- Control equilibrium configurations and their stability.
- Compare the evolution about stable states against known analytic results.
- Set parameters to values that give rise to some special property of the solution, an analytic (or quasianalytic) solution or that reduce the problem to one already studied by other authors.
- Check limit cases (small and long-time behavior, etc.).

- If the final system of ordinary differential equations that is being integrated can be written in a different equivalent form, compare the results obtained by integrating the two systems.
- Compare the results obtained using different numerical methods (for instance, a routine belonging to an available computer library).

In conclusion, the flow chart of Table 2.8.1 could represent a suggested, but sometimes ideal, procedure.

The following sections are written in the form of a *vade mecum* for the user. We try to summarize the essential features of some well-known methods pointing out their accuracy and stability properties and weighing their pros and cons.

Before dealing with the numerical methods, it is useful recalling that the initial-value problem

$$\begin{cases} \dfrac{d\mathbf{u}}{dt} = \mathbf{f}(t, \mathbf{u}) \\ \mathbf{u}(t_0) = \mathbf{u}_0 \end{cases}$$

can be written in integral form as

$$\mathbf{u}(t) = \mathbf{u}_0 + \int_{t_0}^{t} \mathbf{f}\big(s, \mathbf{u}(s)\big)\, ds\,, \tag{2.8.4}$$

or, by partitioning the interval $[t_0, t]$ in

$$[t_0, t] = [t_0, t_1] \cup [t_1, t_2] \cup \cdots \cup [t_{n-1}, t_n]\,,$$

with $t_n = t$, as

$$\mathbf{u}(t) = \mathbf{u}_0 + \sum_{i=0}^{n-1} \int_{t_i}^{t_{i+1}} \mathbf{f}\big(s, \mathbf{u}(s)\big)\, ds\,. \tag{2.8.5}$$

2.8.1 Forward Euler method

The simplest way to approximate the time derivative is by setting

$$\frac{d\mathbf{u}}{dt}(t_i) \cong \frac{\mathbf{u}_{i+1} - \mathbf{u}_i}{h}\,, \tag{2.8.6}$$

where $\mathbf{u}_i = \mathbf{u}(t_i)$ and $h = \Delta t = t_{i+1} - t_i$ is the time step.

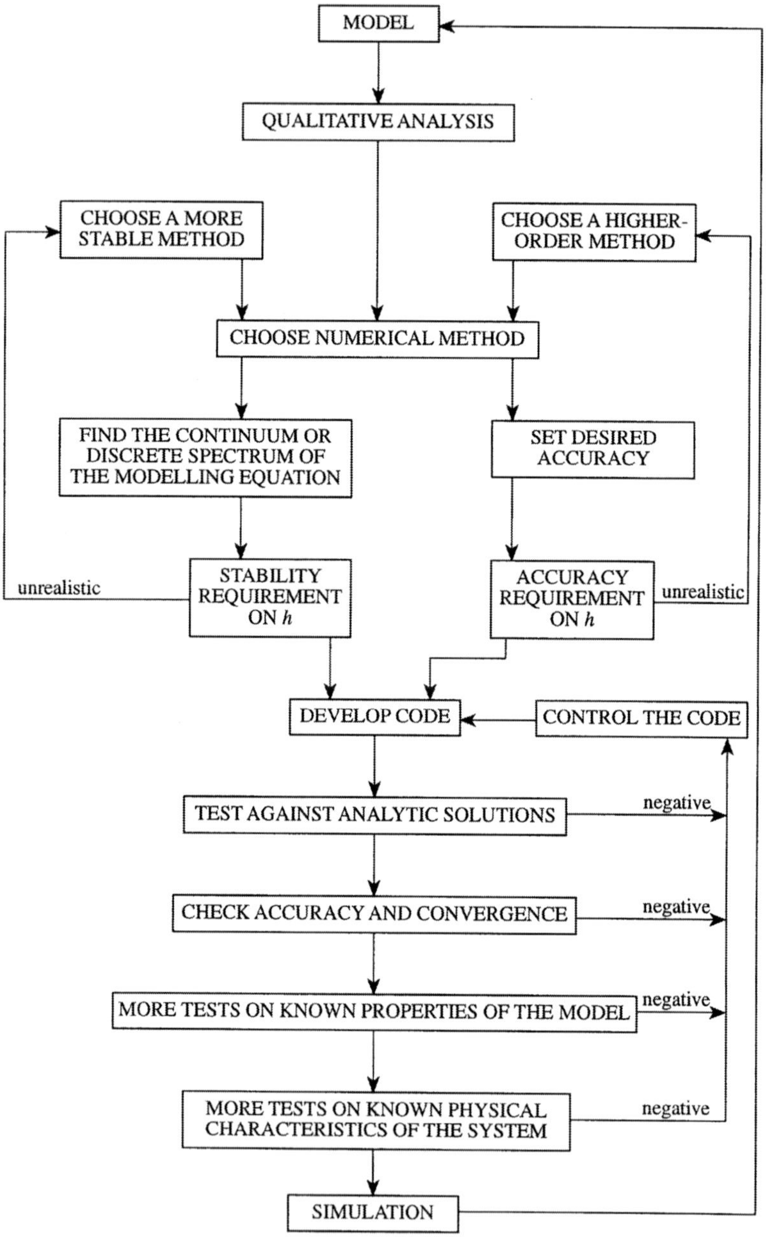

Table 2.8.1 — Testing the model.

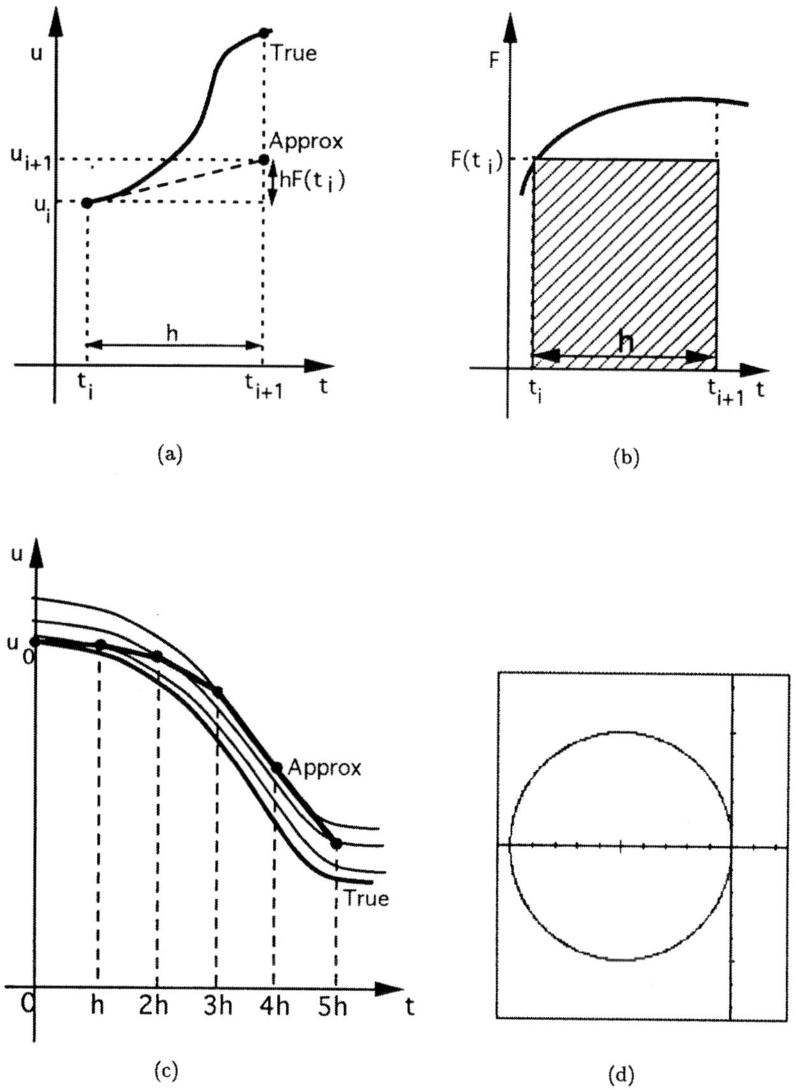

Figure 2.8.1 — Forward Euler method. (a) Differential and (b) integral representation. (c) Sketch of how the numerical method approximates the solution. The family of curves is a set of exact solutions obtained by starting from different initial conditions. (d) Absolute stability region.

In the forward Euler method the right-hand side of the differential equation (2.8.1) is then evaluated at time t_i, which corresponds somewhat to following the direction of the tangent to the solution in t_i, in fact,

$$\mathbf{f}(t_i, \mathbf{u}_i) = \frac{d\mathbf{u}}{dt}(t_i),$$

see Fig. 2.8.1a, or to approximating each integral in (2.8.5) with the area of the rectangle with heigth $\mathbf{f}(t_i, \mathbf{u}_i)$.

We then have

$$\mathbf{u}_{i+1} = \mathbf{u}_i + h\mathbf{f}(t_i, \mathbf{u}_i). \tag{2.8.7}$$

For each component the truncation error is

$$\begin{aligned} T_{\text{err}} &= \frac{u_{i+1} - u_i}{h} - f(t_i, u_i) \cong \frac{du}{dt}(t_i) + \frac{h}{2!}\frac{d^2u}{dt^2}(\xi) - \frac{du}{dt}(t_i) \\ &= \frac{h}{2}\frac{d^2u}{dt^2}(\xi), \quad \text{with} \quad \xi \in (t_i, t_{i+1}). \end{aligned} \tag{2.8.8}$$

For simplicity of notation we have omitted to write the subscript k. Troughout this chapter the truncation error is intended per component. Of course, the point ξ in which the second derivative in (2.8.8) is evaluated changes from component to component. According to (2.8.8), the forward Euler method is then correct in power of h only up to the first power and for this reason it is called a ***first-order method***.

Although this is the simplest conceivable method, it is usually not accurate enough and should be avoided whenever possible.

The stability region is the set of complex ah such that

$$|\lambda| = |1 + ah| \leq 1 \tag{2.8.9}$$

as plotted in Fig. 2.8.1d. It implies, for instance, that the method is unstable if the linearized version of the ordinary differential equation has an imaginary eigenvalue.

2.8.2 Backward Euler method

The backward Euler method differs from the previous one in the evaluation of $\mathbf{f}$ in $(t_{i+1}, \mathbf{u}_{i+1})$, so that the value at the $(i+1)$-th step

$$\mathbf{u}_{i+1} = \mathbf{u}_i + h\mathbf{f}(t_{i+1}, \mathbf{u}_{i+1}) \tag{2.8.10}$$

is not explicitly given in terms of known quantities. The method is then called ***implicit***.

DEFINITION 2.8.4 Explicit and implicit methods
*A method is called **explicit** if the value $\mathbf{u}_{i+1}$ is given in term of previously computed values $\mathbf{u}_j$, $j \leq i$. Otherwise, it is called **implicit**.*

Generally speaking, explicit methods are more suitable if one wants to focus on the details of the transient behavior, since these methods usually need a smaller time step, but, of course, this also implies a longer computer time. Implicit methods are, instead, more suitable for quickly reaching the steady state.

Coming back to the backward Euler method, its accuracy is the same as the forward Euler method, but the stability region is much larger, as usually happens for implicit methods. In fact, the stability region is given by

$$|\lambda| = \left|\frac{1}{1-ah}\right| \leq 1 \Longleftrightarrow |ah-1| \geq 1\,, \tag{2.8.11}$$

which is the region outside the unit circle with center at $ah = 1$ plotted in Fig. 2.8.2c. Since it includes the whole negative half plane, the method is A-stable and therefore the restrictions on the time step are only related to accuracy requirements.

This method corresponds to following the direction of the tangent in the station $i+1$ from $\mathbf{u}_i$, or to approximating each integral in (2.8.5) by the area of the rectangle of height $\mathbf{f}(t_{i+1}, \mathbf{u}_{i+1})$.

If we are dealing with a system of linear differential equations, then we can explicitly determine the value $\mathbf{u}_{i+1}$. In this case, in fact, the discretized form of (2.8.3)

$$\mathbf{u}_{i+1} = \mathbf{u}_i + h\mathbf{A}(t_{i+1})\mathbf{u}_{i+1}$$

can be rewritten as

$$\mathbf{u}_{i+1} = \left[\mathbf{I} - h\mathbf{A}(t_{i+1})\right]^{-1}\mathbf{u}_i\,, \tag{2.8.12}$$

where $\mathbf{I}$ is the identity matrix and $\mathbf{I} - h\mathbf{A}(t_{i+1})$ is certainly nonsingular for suitably small h. For the inversion of large matrices, methods like the conjugate gradient, among others, are suggested [HaY].

If on the other hand $\mathbf{f}(t, \mathbf{u})$ is nonlinear, the value $\mathbf{u}_{i+1}$ can be approximated by a forward Euler method, so that the modified scheme can

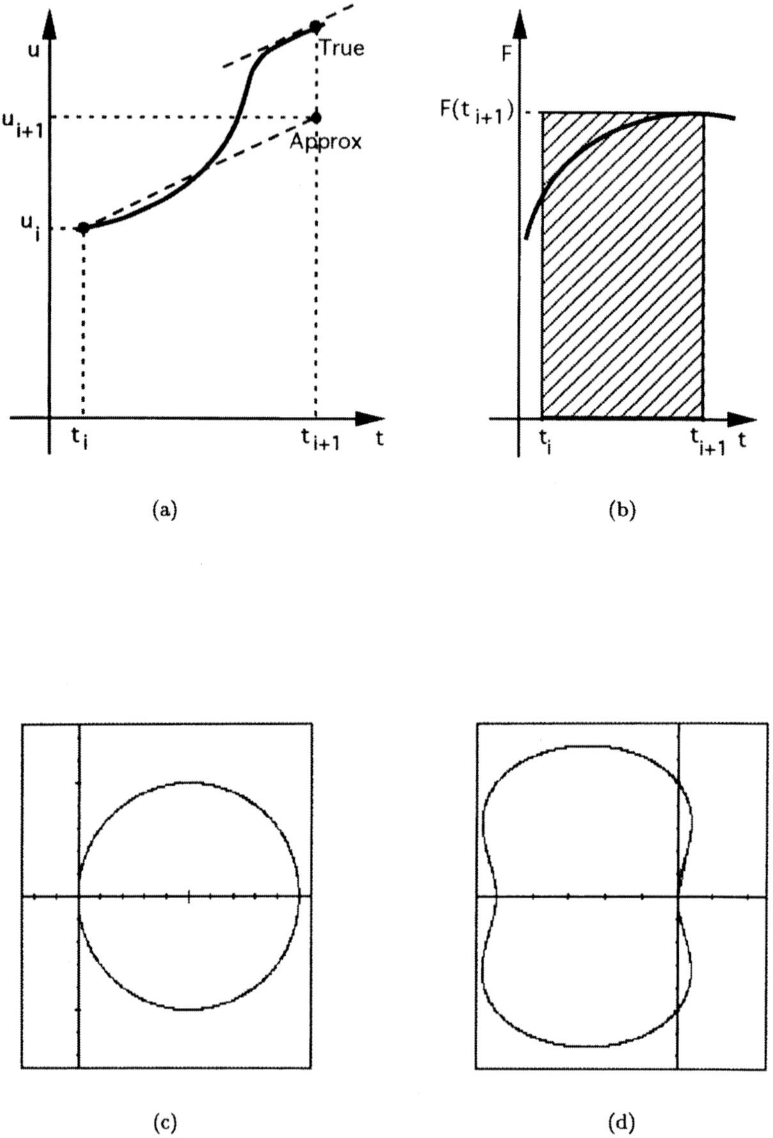

Figure 2.8.2 — Backward Euler method. (a) Differential and (b) integral representation and absolute stability region for (c) the fully implicit method and (d) for its modification (2.8.13).

be written in vector form as

$$\mathbf{u}_{i+1}^{*} = \mathbf{u}_i + h\mathbf{f}(t_i, \mathbf{u}_i) \tag{2.8.13a}$$

$$\mathbf{u}_{i+1} = \mathbf{u}_i + h\mathbf{f}(t_{i+1}, \mathbf{u}_{i+1}^{*}) \,. \tag{2.8.13b}$$

Equation (2.8.13a) is termed the ***predictor***, since it gives a first prediction of the value of $\mathbf{u}$ at $t = t_{i+1}$. Equation (2.8.13b) is termed ***corrector***, since it corrects the value obtained by the predictor. The pair of equations (2.8.13) form the simplest example of ***predictor-corrector*** scheme.

Though the accuracy of (2.8.13) does not change, the region of stability is reduced, as evident in Fig. 2.8.2d, since it is now given by

$$|\lambda| = \left|1 + ah + (ah)^2\right| \leq 1 \,. \tag{2.8.14}$$

Of course, one could apply the correction several times

$$\mathbf{u}_{i+1}^{*} = \mathbf{u}_i + h\mathbf{f}(t_i, \mathbf{u}_i)$$

$$p \text{ times} \begin{cases} \mathbf{u}_{i+1} = \mathbf{u}_i + h\mathbf{f}(t_{i+1}, \mathbf{u}_{i+1}^{*}) \\ \mathbf{u}_{i+1}^{*} = \mathbf{u}_{i+1} \,. \end{cases}$$

This does not change the order of the method but decreases the constant in front of the second derivative of $\mathbf{u}$ in (2.8.8). This process is called correcting to convergence. Note that the convergence is not toward the solution of (2.8.1) but toward the solution of its discretized version (2.8.10). If more iterations are needed, then perhaps the step size is too large.

2.8.3 Crank–Nicolson method

The Crank–Nicolson method consists in averaging the values of $\mathbf{f}$ in t_i and t_{i+1} so that it can be written as

$$\mathbf{u}_{i+1} = \mathbf{u}_i + \frac{h}{2}\left[\mathbf{f}(t_i, \mathbf{u}_i) + \mathbf{f}(t_{i+1}, \mathbf{u}_{i+1})\right] . \tag{2.8.15}$$

This corresponds to proceeding for half the time step along the value of the derivative at t_i and for the remaining part along the derivative at t_{i+1}, or equivalently to evaluating the integral in (2.8.4) by the trapezoidal rule.

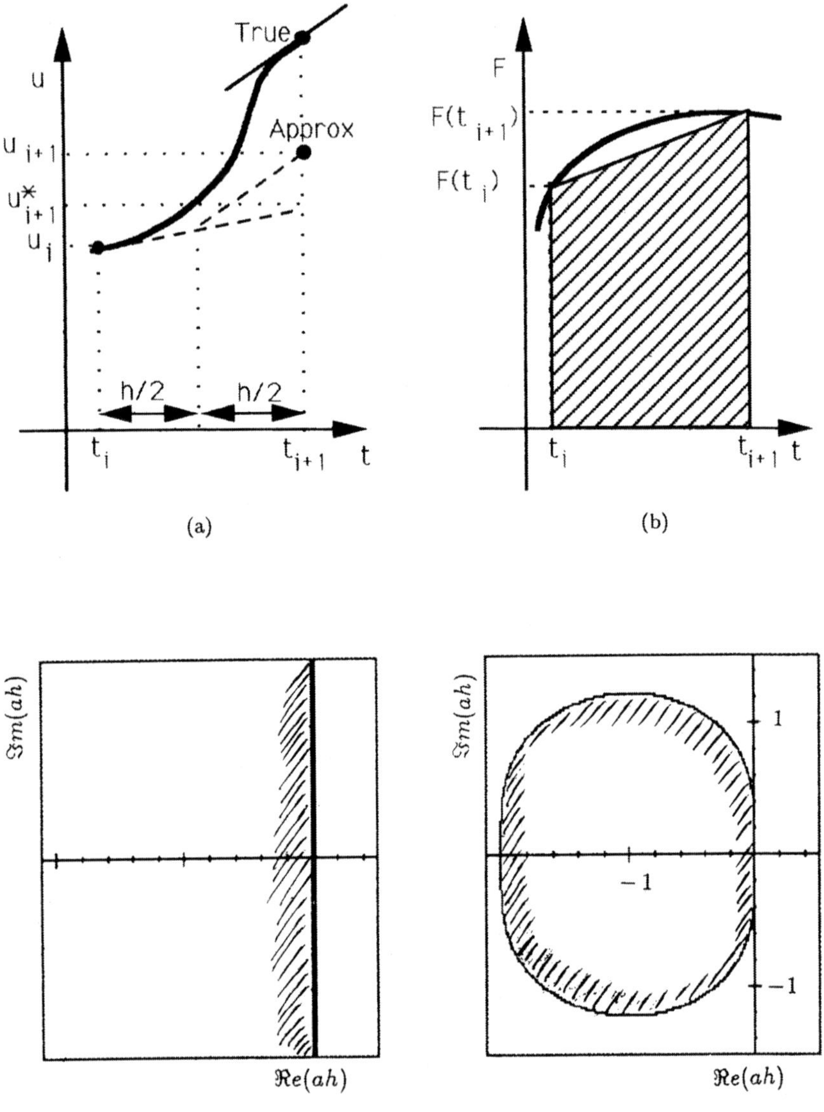

Figure 2.8.3 — Crank–Nicolson method. (a) Differential and (b) integral representation and absolute stability region for (c) the fully implicit method and (d) its modification (2.8.19).

The method is second order since for each component the truncation error is

$$T_{\mathrm{err}} = -\frac{h^2}{12}\frac{d^3u}{dt^3}(\xi)\,, \qquad \xi \in [t_i, t_{i+1}]\,. \tag{2.8.16}$$

The stability region is the whole negative half plane

$$|\lambda| = \left|\frac{2-ah}{2+ah}\right| \le 1 \Longleftrightarrow \Re e(ah) \le 0\,, \tag{2.8.17}$$

and therefore the method is A-stable.

The same remarks made for the backward Euler method also hold for this implicit method. In particular, if the system of differential equations is linear, then $\mathbf{u}_{i+1}$ can be written explicitly as

$$\mathbf{u}_{i+1} = \left[\mathbf{I} - \frac{h}{2}\mathbf{A}(t_{i+1})\right]^{-1}\left[\mathbf{I} + \frac{h}{2}\mathbf{A}(t_i)\right]\mathbf{u}_i\,. \tag{2.8.18}$$

If, on the other hand, $\mathbf{f}(t, \mathbf{u})$ is nonlinear, then forward Euler method

$$\mathbf{u}^*_{i+1} = \mathbf{u}_i + h\mathbf{f}(t_i, \mathbf{u}_i) \tag{2.8.19a}$$

can be used as a predictor to evaluate $\mathbf{u}_{i+1}$ as

$$\mathbf{u}_{i+1} = \mathbf{u}_i + \frac{h}{2}\left[\mathbf{f}(t_i, \mathbf{u}_i) + \mathbf{f}(t_{i+1}, \mathbf{u}^*_{i+1})\right]. \tag{2.8.19b}$$

This method is called also second-order Runge–Kutta method.

The stability region of this modified method is described by the complex values of ah such that

$$|\lambda| = \left|1 + ah + \frac{(ah)^2}{2}\right| \le 1 \tag{2.8.20}$$

and is represented in Fig. 2.8.3d.

2.8.4 Runge–Kutta methods

The second-order method (2.8.19) can be rewritten in the form

$$\mathbf{u}_{i+1} = \mathbf{u}_i + \frac{h}{2}(\mathbf{K}_1 + \mathbf{K}_2) \tag{2.8.21a}$$

with

$$\begin{aligned} \mathbf{K}_1 &= \mathbf{f}(t_i, \mathbf{u}_i) \\ \mathbf{K}_2 &= \mathbf{f}(t_i + h, \mathbf{u}_i + h\mathbf{K}_1)\,, \end{aligned} \tag{2.8.21b}$$

which is a special form of the methods obtained by linearly combining the values

$$\mathbf{K}_p = \mathbf{f}\big(t_i + \alpha_p h, \mathbf{u}_i + h(\beta_{p1}\mathbf{K}_1 + \cdots + \beta_{p,p-1}\mathbf{K}_{p-1})\big)\,, \quad p = 1, \ldots, r$$

of the function at different stations. By suitably choosing the coefficients α_p and β_{pq}, it is possible to obtain the so-called ***Runge–Kutta methods*** of order r. For instance, when $r = 3$ one obtains the third-order method

$$\mathbf{u}_{i+1} = \mathbf{u}_i + \frac{h}{6}(\mathbf{K}_1 + 4\mathbf{K}_2 + \mathbf{K}_3)\,, \tag{2.8.22a}$$

with

$$\begin{aligned} \mathbf{K}_1 &= \mathbf{f}(t_i, \mathbf{u}_i) \\ \mathbf{K}_2 &= \mathbf{f}\left(t_i + \frac{h}{2}, \mathbf{u}_i + \frac{h}{2}\mathbf{K}_1\right) \\ \mathbf{K}_3 &= \mathbf{f}\big(t_i + h, \mathbf{u}_i + h(2\mathbf{K}_2 - \mathbf{K}_1)\big)\,. \end{aligned} \tag{2.8.22b}$$

For $r = 4$, one obtains the fourth-order method

$$\mathbf{u}_{i+1} = \mathbf{u}_i + \frac{h}{6}(\mathbf{K}_1 + 2\mathbf{K}_2 + 2\mathbf{K}_3 + \mathbf{K}_4)\,, \tag{2.8.23a}$$

with

$$\begin{aligned} \mathbf{K}_1 &= \mathbf{f}(t_i, \mathbf{u}_i) \\ \mathbf{K}_2 &= \mathbf{f}\left(t_i + \frac{h}{2}, \mathbf{u}_i + \frac{h}{2}\mathbf{K}_1\right) \\ \mathbf{K}_3 &= \mathbf{f}\left(t_i + \frac{h}{2}, \mathbf{u}_i + \frac{h}{2}\mathbf{K}_2\right) \\ \mathbf{K}_4 &= \mathbf{f}(t_i + h, \mathbf{u}_i + h\mathbf{K}_3)\,. \end{aligned} \tag{2.8.23b}$$

Both the third- and the fourth-order methods are closely related to Simpson's rule applied to the integrals in (2.8.5). (See Fig. 2.8.4a.)

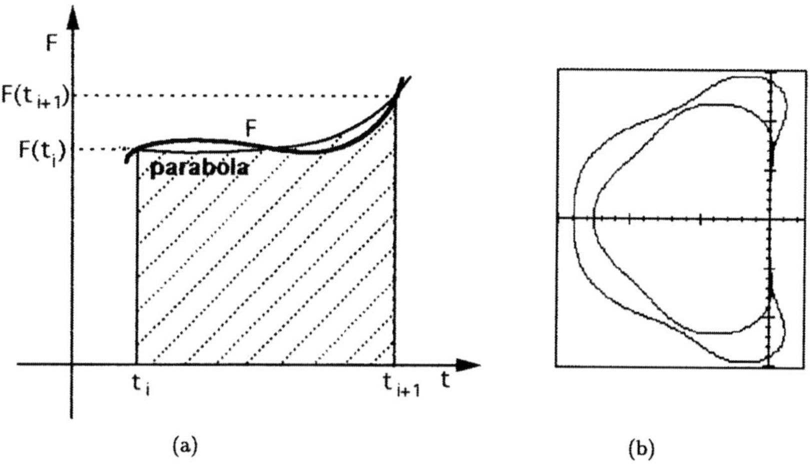

Figure 2.8.4 — (a) Integral representation and (b) stability region of Runge–Kutta methods of third and fourth order. (The one related to the second-order method (2.8.21) is given in Fig. 2.8.3d).

The stability regions of the third- and fourth-order methods, given, respectively, by

$$
\begin{aligned}
|\lambda| &= \left|1 + ah + \frac{(ah)^2}{2} + \frac{(ah)^3}{6}\right| \le 1 \\
|\lambda| &= \left|1 + ah + \frac{(ah)^2}{2} + \frac{(ah)^3}{6} + \frac{(ah)^4}{24}\right| \le 1 ,
\end{aligned}
\tag{2.8.24}
$$

include part of the imaginary axis (precisely, respectively, the intervals $[-\sqrt{3}, \sqrt{3}]$ and $[-2\sqrt{2}, 2\sqrt{2}]$) and intersect the real axis in the points -2.51 and -2.79, respectively.

These methods are very popular for their adaptability and versatility and work quite well for nonstiff problems. The disadvantage is that several evaluations of **f** have to be performed per time step and therefore if the computation of **f** is very heavy, this method may demand an excessive amount of labor per step and become inconvenient.

Third- and fourth-order Runge–Kutta methods (2.8.22) and (2.8.23) are recomended in dealing with problems for which the spectrum is not available, since they include both a part of the imaginary axis and a part of the negative real axis (in spite of being explicit).

With respect to the multistep methods that will be dealt with in Sections 2.8.6 and 2.8.7, Runge–Kutta methods (as the methods dealt with in Sections 2.8.1–2.8.2) have the advantage of being self-starting and adaptive, in the sense that the time step can be changed at any moment according to an estimate of the local error. Unfortunately, it is not trivial to get this estimate, as it is for the predictor-corrector methods (as shown in Section 2.8.7). In this case, one can either integrate again using a method of the same order but halving the time step, or integrate again with a higher-order scheme. The time step has to be decreased if the difference between the two values (over h) is larger than a specified maximum tolerance and can be increased if this difference is smaller than the minimum required tolerance. Of course, this checking is time consuming.

A well-known method for accomplishing this project with the least effort possible is the so-called Runge–Kutta–Fehlberg scheme, which we recall here

$$
\begin{aligned}
\mathbf{u}_{i+1} &= \mathbf{u}_i + \frac{h}{20520}(2375\mathbf{K}_1 + 11264\mathbf{K}_3 + 10985\mathbf{K}_4 - 4104\mathbf{K}_5) \\
\mathbf{u}^*_{i+1} &= \mathbf{u}_i + \frac{h}{282150}(33440\mathbf{K}_1 + 146432\mathbf{K}_3 + 142805\mathbf{K}_4 \\
&\qquad - 50787\mathbf{K}_5 + 10260\mathbf{K}_6)\,.
\end{aligned}
\tag{2.8.25a}
$$

with

$$
\begin{aligned}
\mathbf{K}_1 &= \mathbf{f}(t_i, \mathbf{u}_i) \\
\mathbf{K}_2 &= \mathbf{f}\left(t_i + \frac{h}{4}, \mathbf{u}_i + \frac{h}{4}\mathbf{K}_1\right) \\
\mathbf{K}_3 &= \mathbf{f}\left(t_i + \frac{3}{8}h, \mathbf{u}_i + \frac{3}{32}h(\mathbf{K}_1 + 3\mathbf{K}_2)\right) \\
\mathbf{K}_4 &= \mathbf{f}\left(t_i + \frac{12}{13}h, \mathbf{u}_i + \frac{h}{2197}(1932\mathbf{K}_1 - 7200\mathbf{K}_2 + 7296\mathbf{K}_3)\right) \\
\mathbf{K}_5 &= \mathbf{f}\left(t_i + h, \mathbf{u}_i + \frac{h}{4104}(8341\mathbf{K}_1 - 32832\mathbf{K}_2 \right. \\
&\qquad \left. + 29440\mathbf{K}_3 - 845\mathbf{K}_4)\right)
\end{aligned}
$$

$$\mathbf{K}_6 = \mathbf{f}\bigg(t_i + \frac{h}{2}, \mathbf{u}_i + \frac{h}{20520}(-6080\mathbf{K}_1 + 41040\mathbf{K}_2 - 28352\mathbf{K}_3 + 9295\mathbf{K}_4 - 5643\mathbf{K}_5)\bigg). \qquad (2.8.25b)$$

2.8.5 Leap-frog method

The leap-frog method corresponds to approximating the derivative of $\mathbf{u}$ by

$$\frac{d\mathbf{u}}{dt}(t_i) \cong \frac{\mathbf{u}(t_{i+1}) - \mathbf{u}(t_{i-1})}{2h} \qquad (2.8.26)$$

and evaluating the right-hand side in the midpoint, so that one has

$$\mathbf{u}_{i+1} = \mathbf{u}_{i-1} + 2h\mathbf{f}(t_i, \mathbf{u}_i), \qquad (2.8.27)$$

which corresponds to following, from the point $(t_{i-1}, \mathbf{u}_{i-1})$, the direction of the tangent in the i-th station, or to using the midpoint rule to evaluate the integral in (2.8.4).

The truncation error is

$$T_{\text{err}} = \frac{h^2}{6}\frac{d^3u}{dt}(\xi), \qquad \xi \in [t_{i-1}, t_{i+1}], \qquad (2.8.28)$$

and therefore the method is of second-order. However, its stability region is very small, since the two roots of

$$ah = \frac{1}{2}\left(\lambda - \frac{1}{\lambda}\right) \qquad (2.8.29)$$

both have modulus less that 1 only if ah is imaginary with $|ah| \leq 1$, as is evident from Fig. 2.8.5. Therefore, the model can be used exclusively for problems with imaginary eigenvalues.

This method is the simplest one that uses the values of $\mathbf{u}$ at previous times. This implies an improvement in accuracy but needs a method to start up the computation. In order to mantain the accuracy, the starting procedure must be of the same order, for instance one may choose the second-order Runge–Kutta method (2.8.13).

Besides, the leap-frog method is a time-reversible or nondissipative method. However, since it is stable only on a segment of the complex (ah)-plane for the model problem, extra care is needed in practical situations. Furthermore, the leap-frog method is subject to oscillations of period $2h$, which arise from an extraneous solution to the difference

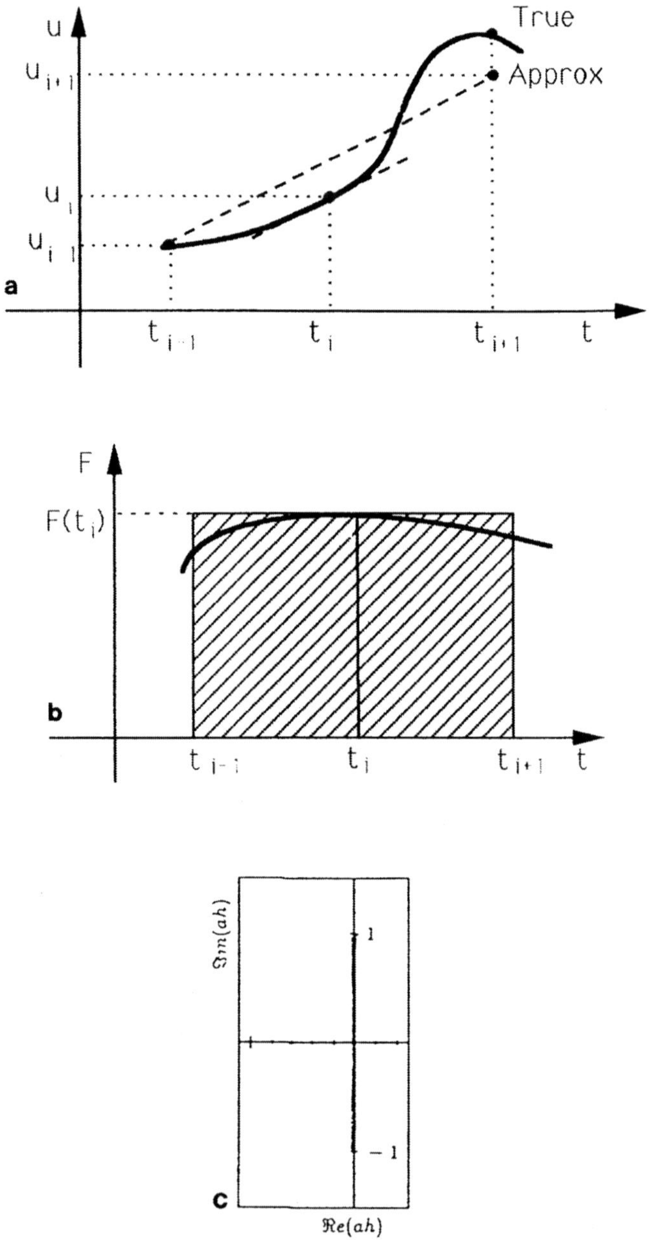

Figure 2.8.5 — Leap-frog method. (a) Differential and (b) integral representation and (c) absolute stability region.

equation. However, these oscillations can be controlled by every so often averaging the solution at two consecutive steps.

2.8.6 Adams–Bashforth methods

All the previous methods (but the leap-frog method) are one-step methods, in the sense that they use information about the solution $\mathbf{u}_i$ in a single point $t = t_i$ to get the next approximation $\mathbf{u}_{i+1}$ at $t = t_{i+1}$. On the other hand, the leap-frog method profits from what happened at a previous step $t = t_{i-1}$. Multistep methods generalize this approach using more information on what happened in the past. In fact, referring to Eq.(2.8.4), one can replace $\mathbf{f}(t, \mathbf{u})$ with a polynomial $\mathbf{P}_m(t)$ of order m, which interpolates the data obtained at $m+1$ previous steps $t_{i-m}, t_{i-m+1}, \ldots, t_{i-1}, t_i$, so that one can estimate

$$\mathbf{u}_{i+1} = \mathbf{u}_i + h \int_{t_i}^{t_{i+i}} \mathbf{P}_m(t)\, dt\,. \tag{2.8.30}$$

One then obtains

$$\begin{aligned}
m = 1 &\Longrightarrow \mathbf{u}_{i+1} = \mathbf{u}_i + \frac{h}{2}\left[3\mathbf{f}(t_i, \mathbf{u}_i) - \mathbf{f}(t_{i-1}, \mathbf{u}_{i-1})\right] \\
m = 2 &\Longrightarrow \mathbf{u}_{i+1} = \mathbf{u}_i + \frac{h}{12}\left[23\mathbf{f}(t_i, \mathbf{u}_i) - 16\mathbf{f}(t_{i-1}, \mathbf{u}_{i-1})\right. \\
&\qquad\qquad \left. + 5\mathbf{f}(t_{i-2}, \mathbf{u}_{i-2})\right] \\
m = 3 &\Longrightarrow \mathbf{u}_{i+1} = \mathbf{u}_i + \frac{h}{24}\left[55\mathbf{f}(t_i, \mathbf{u}_i) - 59\mathbf{f}(t_{i-1}, \mathbf{u}_{i-1})\right. \\
&\qquad\qquad \left. + 37\mathbf{f}(t_{i-2}, \mathbf{u}_{i-2}) - 9\mathbf{f}(t_{i-3}, \mathbf{u}_{i-3})\right].
\end{aligned} \tag{2.8.31}$$

The truncation errors are, respectively,

$$\begin{aligned}
m = 1 &\Longrightarrow T_{\text{err}} = \frac{5}{12} h^2 \frac{d^3 u}{dt^3}(\xi)\,, \qquad \xi \in [t_{i-1}, t_{i+1}] \\
m = 2 &\Longrightarrow T_{\text{err}} = \frac{3}{8} h^3 \frac{d^4 u}{dt^4}(\xi)\,, \qquad \xi \in [t_{i-2}, t_{i+1}] \\
m = 3 &\Longrightarrow T_{\text{err}} = \frac{251}{720} h^4 \frac{d^5 u}{dt^5}(\xi)\,, \qquad \xi \in [t_{i-3}, t_{i+1}]\,,
\end{aligned} \tag{2.8.32}$$

and therefore the methods in (2.8.31) are, respectively, of second, third and fourth order.

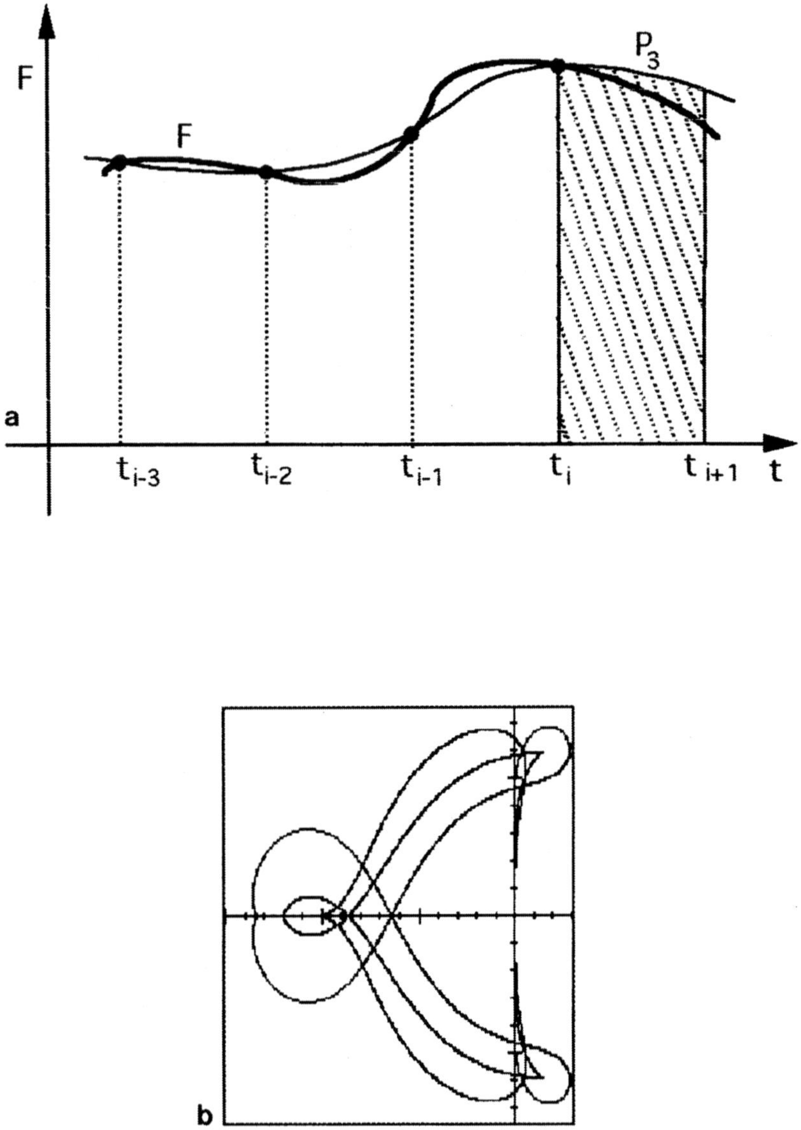

Figure 2.8.6 — Adams–Bashforth methods. (a) Integral representation with interpolation points heavily marked. (b) Absolute stability regions for the second- ($m = 1$), third- ($m = 2$), and fourth- ($m = 3$) order methods.

The stability regions are those enclosed, respectively, by the curves

$$
\begin{aligned}
m = 1 &\Longrightarrow ah = 2\,\frac{\lambda^2 - \lambda}{3\lambda - 1} \\
m = 2 &\Longrightarrow ah = 12\,\frac{\lambda^3 - \lambda^2}{23\lambda^2 - 16\lambda + 5} \\
m = 3 &\Longrightarrow ah = 24\,\frac{\lambda^4 - \lambda^3}{55\lambda^3 - 59\lambda^2 + 37\lambda - 9},
\end{aligned}
\tag{2.8.33}
$$

with $|\lambda| = 1$. Fig. 2.8.6b shows that they get smaller as the order of the method increases. For instance the intersections with the real axis are, respectively, -1, $-6/11$, and $-3/10$. The first, which is forward Euler method, and the second-order Adams–Bashforth method do not include any part of the imaginary axis, whereas the third-order method includes the imaginary interval $[-0.723, 0.723]$ and the fourth-order method includes the interval $[-0.430, 0.430]$. Actually, though the second-order method does not include any part of the imaginary axis, it is very close to it and so the method is only weakly unstable with respect to problems with imaginary eigenvalues. For this reason it can still be used in this case but not for long temporal integrations.

One could obtain in this way methods of any order, which, of course, become more and more complex, but always less complex than the corresponding Runge–Kutta method of the same order. In fact, in (2.8.31) both $\mathbf{u}_{i-j}$ and $\mathbf{f}(t_{i-j}, \mathbf{u}_{i-j})$, $j \geq 1$ are already known from the previous steps. So, if they are stored, which requires just a bit more computer memory, we only need one more function evaluation. In contrast, the fourth-order Runge–Kutta method needs four function evaluations per step. Hence multistep methods are convenient if the evaluation of $\mathbf{f}$ is complicated.

The disadvantage of this class of methods is that they need a starting procedure for the very first steps, which must be of the same order in order to keep the truncation error always of the same size. For this purpose, Runge–Kutta methods are often used. Furthermore, it is difficult to adapt the step size during integration, since one uses information from several previous steps at given temporal distances. If one needs to do that and still wants to use multistep methods, then in the transition between one time step and another one must use either a suitable interpolation formula on the previous nodes or a different adaptive method of the same order (say, a Runge–Kutta method). It is essential again that the Runge–Kutta method be of the same order, otherwise error propagation may render the solution useless, regardless of the precision of the multistep method being used. Finally, the stability regions of an

Adams–Bashforth method is much smaller than the one related to the corresponding Runge–Kutta method of the same order.

2.8.7 Adams–Moulton methods

Consider now the same procedure used for Adams–Bashforth methods but replace $\mathbf{f}(t, \mathbf{u})$ with the polynomial of degree m that interpolates the data obtained at

$$t = t_{i-m+1}, t_{i-m+2}, \ldots, t_i, t_{i+1}$$

(u_{i+1} is still unknown). The integration over the interval $[t_i, t_{i+1}]$ yields an implicit method, usually called the ***Adams–Moulton method.***

According to the number of nodes used, one has

$$\begin{aligned}
m = 1 &\Longrightarrow \mathbf{u}_{i+1} = \mathbf{u}_i + \frac{h}{2}\left[\mathbf{f}(t_{i+1}, \mathbf{u}_{i+1}) + \mathbf{f}(t_i, \mathbf{u}_i)\right] \\
m = 2 &\Longrightarrow \mathbf{u}_{i+1} = \mathbf{u}_i + \frac{h}{12}\left[5\mathbf{f}(t_{i+1}, \mathbf{u}_{i+1}) + 8\mathbf{f}(t_i, \mathbf{u}_i) - \mathbf{f}(t_{i-1}, \mathbf{u}_{i-1})\right] \\
m = 3 &\Longrightarrow \mathbf{u}_{i+1} = \mathbf{u}_i + \frac{h}{24}\left[9\mathbf{f}(t_{i+1}, \mathbf{u}_{i+1}) + 19\mathbf{f}(t_i, \mathbf{u}_i)\right. \\
&\qquad \left. - 5\mathbf{f}(t_{i-1}, \mathbf{u}_{i-1}) + \mathbf{f}(t_{i-2}, \mathbf{u}_{i-2})\right].
\end{aligned} \tag{2.8.34}$$

Observe that the formula for $m = 1$ is just the Crank–Nicolson method (2.8.15).

The truncation errors are, respectively,

$$\begin{aligned}
m = 2 &\Longrightarrow T = -\frac{1}{24} h^3 \frac{d^4u}{dt^4}(\xi), \qquad \xi \in [t_{i-1}, t_{i+1}], \\
m = 3 &\Longrightarrow T = -\frac{19}{720} h^4 \frac{d^5u}{dt^5}(\xi), \qquad \xi \in [t_{i-2}, t_{i+1}].
\end{aligned} \tag{2.8.35}$$

Observe that the truncation errors (2.8.35) relative to these implicit methods are much smaller, though of the same order, than those (2.8.24) related to the corresponding explicit methods. This is due to the fact that Adams–Bashforth methods are obtained by integrating the interpolating polynomial outside its range of interpolation, while in Adams–Moulton methods the integration of the interpolating polynomial is over the last subinterval of interpolation, which is a more accurate procedure.

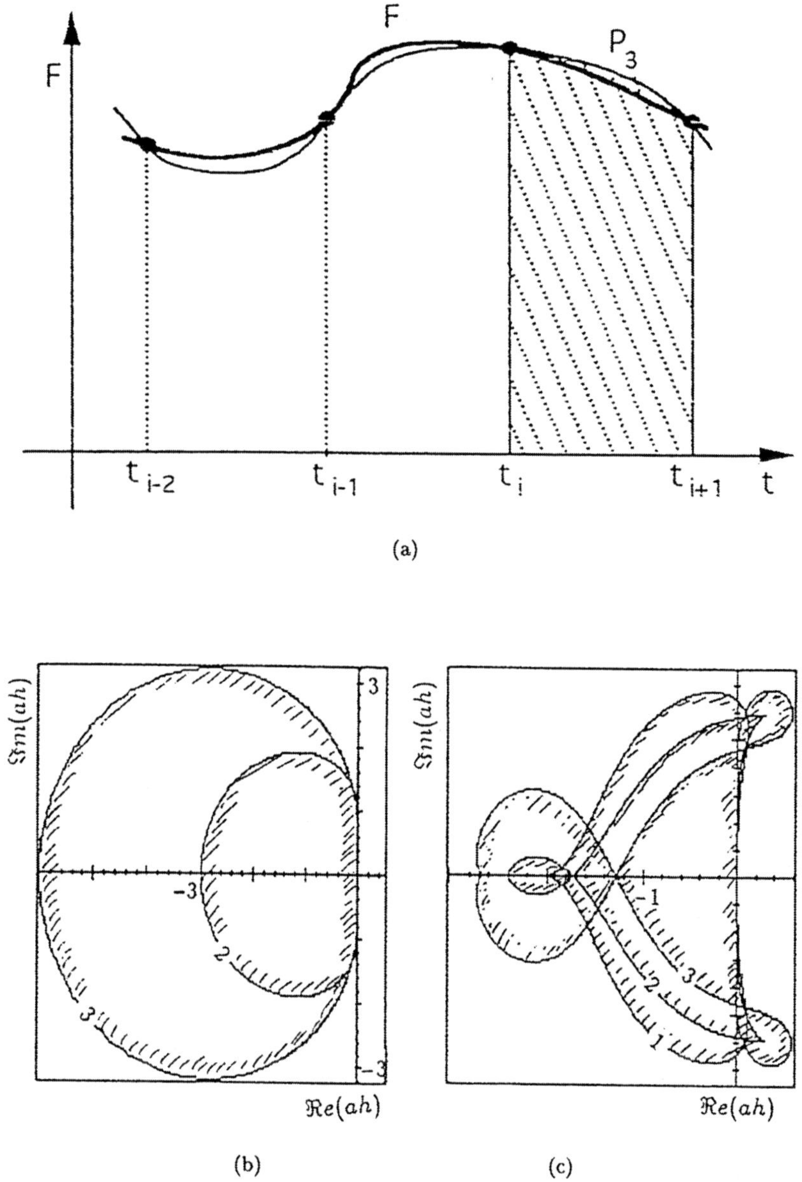

Figure 2.8.7 — (a) Integral representation and (b) absolute stability region of Adams–Moulton methods of third and fourth order. That related to the second-order method is given in Fig. 2.8.3c. (c) Absolute stability region for Adams predictor-corrector schemes.

The stability regions are delimited by the curves

$$
\begin{aligned}
m = 2 &\Longrightarrow ah = 12\frac{\lambda^2 - \lambda}{5\lambda^2 + 8\lambda - 1}, \\
m = 3 &\Longrightarrow ah = 24\frac{\lambda^3 - \lambda^2}{9\lambda^3 + 19\lambda^2 - 5\lambda + 1},
\end{aligned} \tag{2.8.36}
$$

with $|\lambda| = 1$. Figure 2.8.7b shows that they are much larger than the regions obtained via explicit methods, but do not cross the imaginary axis. Furthermore, they decrease in size as the order increases.

As for all implicit methods, if the system is linear $\mathbf{f} = \mathbf{A}\mathbf{u}$, then one can explicitly write

$$
\begin{aligned}
m = 3 &\Longrightarrow \mathbf{u}_{i+1} = \left[\mathbf{I} - \frac{5}{12}h\mathbf{A}(t_{i+1})\right]^{-1}\left[\mathbf{u}_i + \frac{h}{12}\mathbf{A}(t_i)(8\mathbf{u}_i - \mathbf{u}_{i-1})\right], \\
m = 4 &\Longrightarrow \mathbf{u}_{i+1} = \left[\mathbf{I} - \frac{9}{24}h\mathbf{A}(t_{i+1})\right]^{-1} \\
&\qquad \times \left[\mathbf{u}_i + \frac{h}{24}\mathbf{A}(t_i)(19\mathbf{u}_i - 5\mathbf{u}_{i-1} + \mathbf{u}_{i-2})\right].
\end{aligned} \tag{2.8.37}
$$

If, on the other hand, $\mathbf{f}$ is nonlinear, the Adams–Bashforth methods of the same order can be used as predictors. So, for instance, the fourth-order Adams predictor-corrector method can be written as

$$
\begin{aligned}
\mathbf{u}^*_{i+1} &= \mathbf{u}_i + \frac{h}{24}\left[55\mathbf{f}(t_i, \mathbf{u}_i) - 59\mathbf{f}(t_{i-1}, \mathbf{u}_{i-1})\right. \\
&\quad \left. + 37\mathbf{f}(t_{i-2}, \mathbf{u}_{i-2}) - 9\mathbf{f}(t_{i-3}, \mathbf{u}_{i-3})\right], \\
\mathbf{u}_{i+1} &= \mathbf{u}_i + \frac{h}{24}\left[9\mathbf{f}(t_{i+1}, \mathbf{u}^*_{i+1}) + 19\mathbf{f}(t_i, \mathbf{u}_i)\right. \\
&\quad \left. - 5\mathbf{f}(t_{i-1}, \mathbf{u}_{i-1}) + \mathbf{f}(t_{i-2}, \mathbf{u}_{i-2})\right].
\end{aligned} \tag{2.8.38}
$$

Usually, iterating the correction is not worthwhile, since it improves the final result very little.

The predictor-corrector schemes inherit all advantage and disadvantage of all multistep methods. For instance, an accurate starting procedure is needed (say, a fourth-order Runge–Kutta method) making it more complex to implement, but, on the other hand, it only needs two function evaluations per step (the others have been already computed and opportunely stored).

Another advantage of the predictor-corrector scheme is that the error committed during the integration between t_i and t_{i+1} can be readily estimated as

$$\mathbf{u}_{i+1} - \widehat{\mathbf{u}}(t_{i+1}) \cong \begin{cases} \frac{1}{6}(\mathbf{u}_{i+1} - \mathbf{u}^*_{i+1})\,, & \text{if } m = 2 \\ \frac{1}{10}(\mathbf{u}_{i+1} - \mathbf{u}^*_{i+1})\,, & \text{if } m = 3 \\ \frac{19}{270}(\mathbf{u}_{i+1} - \mathbf{u}^*_{i+1})\,, & \text{if } m = 4\,, \end{cases} \tag{2.8.39}$$

where $\widehat{\mathbf{u}}(t)$ is the solution of the system of ordinary differential equations with initial condition $\widehat{\mathbf{u}}(t_i) = \mathbf{u}_i$, $\mathbf{u}^*_{i+1}$ is the predicted value, and $\mathbf{u}_{i+1}$ is the value after correction at $t = t_{i+1}$.

The stability regions in Fig. 2.8.7c are delimited by the curves

$$\begin{aligned} m = 1 &\Longrightarrow \lambda^2 - \left[1 + ah + \frac{3}{4}(ah)^2\right]\lambda + \frac{(ah)^2}{4} = 0\,, \\ m = 2 &\Longrightarrow \lambda^3 - \left[1 + \frac{13}{12}ah + \frac{115}{144}(ah)^2\right]\lambda^2 + \frac{ah}{12}\left[1 + \frac{20}{3}ah\right]\lambda \\ &\qquad - \frac{25}{144}(ah)^2 = 0\,, \\ m = 3 &\Longrightarrow \lambda^4 - \left[1 + \frac{7}{6}ah + \frac{495}{576}(ah)^2\right]\lambda^3 + \frac{ah}{24}\left[5 + \frac{531}{24}ah\right]\lambda^2 \\ &\qquad - \frac{ah}{24}\left[1 + \frac{111}{8}ah\right]\lambda + \frac{81}{576}(ah)^2 = 0\,, \end{aligned} \tag{2.8.40}$$

with $|\lambda| = 1$, which cross the real axis at -2, $-12/5$, and $-24/9$ and the imaginary axis at ± 1.29, ± 1.2, and ± 0.97, respectively.

Another method to find the solution to the implicit method (2.8.34) even in the nonlinear case is by Newton's iterative method, which is written (the suffix refers to the iteration)

$$\begin{aligned} \mathbf{u}^{n+1}_{i+1} = \mathbf{u}^n_{i+1} - &\left[\mathbf{I} - \frac{h}{2}\mathbf{J}(t_{i+1}, \mathbf{u}^n_{i+1})\right]^{-1} \\ &\times \left[\mathbf{u}^n_{i+1} - \frac{h}{2}\mathbf{f}(t_{i+1}, \mathbf{u}^n_{i+1}) - \frac{h}{2}\mathbf{f}(t_i, \mathbf{u}_i)\right]\,, \end{aligned} \tag{2.8.41}$$

when $m = 1$ (i.e., Crank–Nicolson method)

$$\mathbf{u}_{i+1}^{n+1} = \mathbf{u}_{i+1}^{n} - \left[\mathbf{I} - \frac{5h}{12}\mathbf{J}(t_{i+1}, \mathbf{u}_{i+1}^{n})\right]^{-1} \left[\mathbf{u}_{i+1}^{n} - \frac{5h}{12}\mathbf{f}(t_{i+1}, \mathbf{u}_{i+1}^{n}) - \frac{8h}{12}\mathbf{f}(t_i, \mathbf{u}_i) + \frac{h}{12}\mathbf{f}(t_{i-1}, \mathbf{u}_{i-1})\right], \tag{2.8.42}$$

when $m = 2$ and

$$\mathbf{u}_{i+1}^{n+1} = \mathbf{u}_{i+1}^{n} - \left[\mathbf{I} - \frac{9h}{24}\mathbf{J}(t_{i+1}, \mathbf{u}_{i+1}^{n})\right]^{-1} \left[\mathbf{u}_{i+1}^{n} - \frac{9h}{24}\mathbf{f}(t_{i+1}, \mathbf{u}_{i+1}^{n}) - \frac{19h}{24}\mathbf{f}(t_i, \mathbf{u}_i) + \frac{5h}{24}\mathbf{f}(t_{i-1}, \mathbf{u}_{i-1}) - \frac{h}{24}\mathbf{f}(t_{i-2}, \mathbf{u}_{i-2})\right], \tag{2.8.43}$$

when $m = 3$. In all these formulas $\mathbf{J}$ represents the Jacobian matrix.

Actually the inverse matrix is not actually computed exactly, since this is inefficient, but in some sense approximated. In this way the method converges more slowly, but is more efficient [HIN]. These methods, as all the fully implicit methods already presented, are suitable for stiff problems.

2.9 Scientific Programs

In Section 2.8 we gave some guidelines on how to choose and control a numerical method before passing to massive computation and simulation. In this section we give some guidelines to those who are not familiar with scientific programming on how to actually write a computer program. We then describe in detail some sample computer programs stored in the diskette accompaining this book, which will be useful for the simulation and quantitative analysis of discrete models. We tried to write them in the simplest way possible using the basilar Basic commands and avoiding more advanced techniques, which might compact the routines or speed up the computation. After some practice, the reader can improve the routines introducing his own adjustments.

2.9.1 Dimensionless variables

As a very first step it is often recomended to write the problem in dimensionless form, i.e., to identify a characteristic time scale τ and n reference measures U_i of the state variables and to introduce the dimensionless time $\widetilde{t} = t/\tau$ and state variables $\widetilde{u}_i = u_i/U_i$. In this way, the initial-value problem can be rewritten as

$$\begin{cases} \dfrac{d\widetilde{\mathbf{u}}}{d\widetilde{t}} = \widetilde{\mathbf{f}}(\widetilde{t}, \widetilde{\mathbf{u}}; \widetilde{\mathbf{r}}) \\ \widetilde{\mathbf{u}}(0) = \widetilde{\mathbf{u}}_0 \,, \end{cases}$$

where $\widetilde{u}_{i0} = u_{i0}/U_i$ and $\widetilde{f}_i(\widetilde{t}, \widetilde{\mathbf{u}}; \widetilde{\mathbf{r}}) = f_i(\tau\widetilde{t}, U_1\widetilde{u}_1, \ldots, U_n\widetilde{u}_n; \mathbf{r})\tau/U_i$.

REMARK 2.9.1 The vector $\widetilde{\mathbf{r}}$ has a smaller dimension than $\mathbf{r}$, as will soon be evident in Example 2.13 and this is one of the advantages that justifies the introduction of the dimensionless quantities. ■

It is better to use as τ and U_i fixed and easily measurable quantities or to let them stem out of the problem itself.

Example 2.13 Population dynamics
To clarify the concepts introduced above consider the initial-value problem related to the one-population model (2.2.18)

$$\begin{cases} \dfrac{dN}{dt} = B_0 N \left(1 - \dfrac{N}{M}\right) \\ N(0) = N_0 \,, \end{cases} \tag{2.9.1}$$

with B_0 and M constant.

The initial-value problem, and therefore the solution, depends on three constants B_0, M, and N_0, in particular $\mathbf{r} = \{B_0, M\}$.

Introducing now the dimensionless quantities

$$\widetilde{t} = \frac{t}{\tau} \qquad \text{and} \qquad \widetilde{N} = \frac{N}{\mathcal{N}} \,, \tag{2.9.2}$$

with τ and $\mathcal{N}$ still to be specified, Eq.(2.9.1) can be rewritten

$$\begin{cases} \dfrac{d\widetilde{N}}{d\widetilde{t}} = B_0 \tau \widetilde{N} \left(1 - \dfrac{\mathcal{N}}{M}\widetilde{N}\right) \\ \widetilde{N}(0) = \dfrac{N_0}{\mathcal{N}} \,. \end{cases} \tag{2.9.3}$$

Choosing

$$\tau = 1/B_0 \qquad \text{and} \qquad \mathcal{N} = M \,, \tag{2.9.4}$$

the initial-value problem becomes

$$\begin{cases} \dfrac{d\widetilde{N}}{d\widetilde{t}} = \widetilde{N}\left(1 - \widetilde{N}\right) \\ \widetilde{N}(0) = \widetilde{N}_0 \,. \end{cases} \tag{2.9.5}$$

The ordinary differential equation in (2.9.5) does not depend on any parameter and the initial-value problem, and therefore the solution, only depends on the initial value $\widetilde{N}_0 = N_0/M$. A complete study of the model is obtained by integrating (2.9.5) as $\widetilde{N}_0 \in \mathbb{R}_+$, while a complete study of the original problem in dimensional form would be obtained by integrating (2.9.1) varying $(N_0, M, B_0) \in \mathbb{R}^3_+$, which is uselessly lengthy investigation. □

REMARK 2.9.2 One may look at the choice (2.9.4) as focusing on what happens as the initial condition changes. As an alternative, one might choose $\mathcal{N} = N_0$ so that the dimensionless form of Eq.(2.9.1) becomes

$$\begin{cases} \dfrac{d\widetilde{N}}{d\widetilde{t}} = \widetilde{N}\left(1 - \dfrac{\widetilde{N}}{\widetilde{M}}\right) \\ \widetilde{N}(0) = 1 \,, \end{cases} \tag{2.9.6}$$

where $\widetilde{M} = M/N_0$. This corresponds to fixing the initial condition and focusing on what happens as the parameters change.

Though philosophically different, the two choices are mathematically equivalent, since a trivial rescaling of the solution of (2.9.6) gives the solution of (2.9.5) or of the dimensional problem (2.9.1). We mention this to point out that it is not important how the problem is written in dimensionless form. Among all possible choices it is convenient to use the one in which τ and U_i can be easily measured (possibly once and for all). ■

Writing the problem in dimensionless form is not compulsory, but recomended for the reasons stated above. Of course, one may prefer to keep the physical meaning of the dependent and independent variables, and it is convenient to do that only if some specific tests have to be performed. Essentially the choice depends on the domain of the parameter to be spanned.

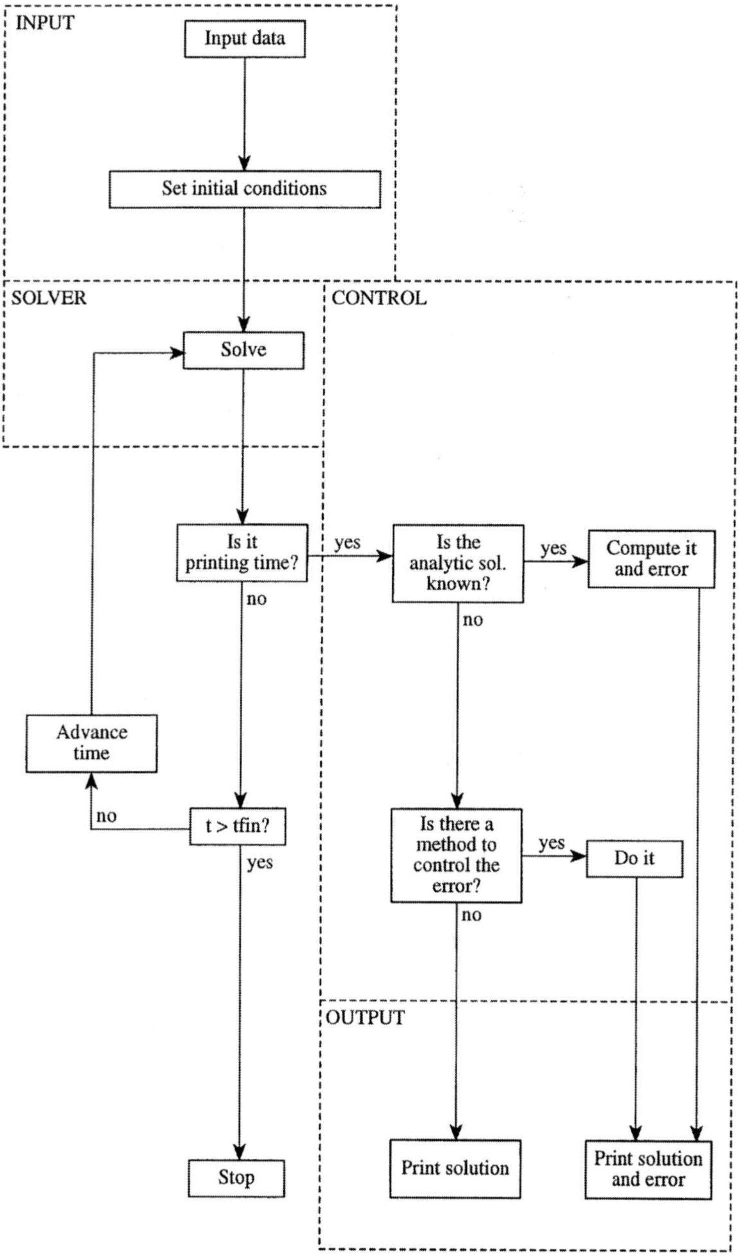

Table 2.9.1 — Flow chart.

2.9.2 The flow chart

As a second step one has to determine the logic line of the program. It is difficult to give general guidelines for performing this step since everyone has his own way of thinking. We can state that the program may be divided into an input part, an Ordinary Differential Equation (ODE) solver, an output part, and, if possible, a control part, as shown in Table 2.9.1.

Sometimes it is convenient to sketch first a rough flow chart similar to that shown in Table 2.9.1 and then one or more detailed charts to develop in detail each and every block. This second part can only be done after having collected all information on the model and on the method to use. Therefore, in order to be specific, we will focus on the following simple system.

Example 2.14 Coupled oscillator

Consider the initial-value problem related to the coupled oscillator described in Fig. 2.9.1

$$\begin{cases} m_1 \dfrac{d^2 x_1}{dt^2} = -k_1(x_1)x_1 + k_2(x_2 - x_1)(x_2 - x_1) \\ m_2 \dfrac{d^2 x_2}{dt^2} = -k_2(x_2 - x_1)(x_2 - x_1) \\ x_1(0) = x_{10} \\ \dfrac{dx_1}{dt}(0) = v_{10} \\ x_2(0) = x_{20} \\ \dfrac{dx_2}{dt}(0) = v_{20}\,, \end{cases}$$

which can be written as

$$\begin{cases} \dfrac{du_1}{dt} = u_2 \\ \dfrac{du_2}{dt} = c_1 u_1 + c_2 u_3 \\ \dfrac{du_3}{dt} = u_4 \\ \dfrac{du_4}{dt} = c_3 u_1 + c_4 u_3\,, \end{cases} \tag{2.9.7a}$$

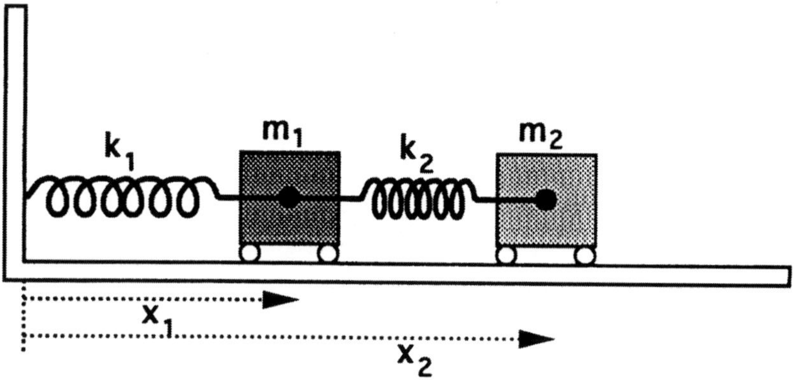

Figure 2.9.1 — Coupled oscillator.

with

$$\begin{cases} u_1(0) = x_{10} \\ u_2(0) = v_{10} \\ u_3(0) = x_{20} \\ u_4(0) = v_{20}\,, \end{cases} \tag{2.9.7b}$$

where

$$\mathbf{u} = \left\{ x_1, \frac{dx_1}{dt}, x_2, \frac{dx_2}{dt} \right\},$$

and

$$c_1 = -\frac{k_1 + k_2}{m_1}, \qquad c_2 = \frac{k_2}{m_1}, \qquad c_3 = \frac{k_2}{m_2}, \qquad c_4 = -\frac{k_2}{m_2}$$

are functions of u_1 and u_3.

If k_1 and k_2 are constants, we can write the analytic solution of (2.9.7) explicitly

$$\begin{cases} u_1 = A_1 \cos \Omega_1 t + A_2 \cos \Omega_2 t + A_3 \sin \Omega_1 t + A_4 \sin \Omega_2 t \\ u_2 = -A_1 \Omega_1 \sin \Omega_1 t - A_2 \Omega_2 \sin \Omega_2 t + A_3 \Omega_1 \cos \Omega_1 t + A_4 \Omega_2 \cos \Omega_2 t \\ u_3 = B_1 \cos \Omega_1 t + B_2 \cos \Omega_2 t + B_3 \sin \Omega_1 t + B_4 \sin \Omega_2 t \\ u_4 = -B_1 \Omega_1 \sin \Omega_1 t - B_2 \Omega_2 \sin \Omega_2 t + B_3 \Omega_1 \cos \Omega_1 t + B_4 \Omega_2 \cos \Omega_2 t\,, \end{cases} \tag{2.9.8}$$

where

$$\Omega_{1,2} = \sqrt{\frac{-(c_1+c_4) \pm \sqrt{(c_1-c_4)^2+4c_2c_3}}{2}}$$

and

$$A_1 = \frac{c_1x_{10}+c_2x_{20}+\Omega_2^2x_{10}}{\Omega_2^2-\Omega_1^2}$$

$$A_2 = -\frac{c_1x_{10}+c_2x_{20}+\Omega_1^2x_{10}}{\Omega_2^2-\Omega_1^2}$$

$$A_3 = \frac{c_1v_{10}+c_2v_{20}+\Omega_2^2v_{10}}{\Omega_1(\Omega_2^2-\Omega_1^2)}$$

$$A_4 = -\frac{c_1v_{10}+c_2v_{20}+\Omega_1^2v_{10}}{\Omega_2(\Omega_2^2-\Omega_1^2)}$$

$$B_1 = -\frac{(c_1+\Omega_1^2)A_1}{c_2}$$

$$B_2 = -\frac{(c_1+\Omega_2^2)A_2}{c_2}$$

$$B_3 = -\frac{(c_1+\Omega_1^2)A_3}{c_2}$$

$$B_4 = -\frac{(c_1+\Omega_2^2)A_4}{c_2}\,.$$

The spectrum of the ordinary differential equation in (2.9.7) (which is already linear) has two couples of imaginary eigenvalues corresponding to the two frequencies Ω_1 and Ω_2. □

To fully understand the programs we strongly recomend that the reader do the Exercises referring to Chapter 2. These exercises will slowly bring the reader to the understanding of the main features and differences between the methods presented here.

2.9.3 RK4: A fourth-order Runge–Kutta program

In this section we will describe line-by-line the fourth-order Runge–Kutta program RK4 stored in the diskette accompaining this book and printed at the end of this section. For simplicity the same structure will be held for the other programs presented in this chapter.

The input part is a list of input commands and options that set the type of output, the initial conditions, and the parameters of the ordinary differential equation.

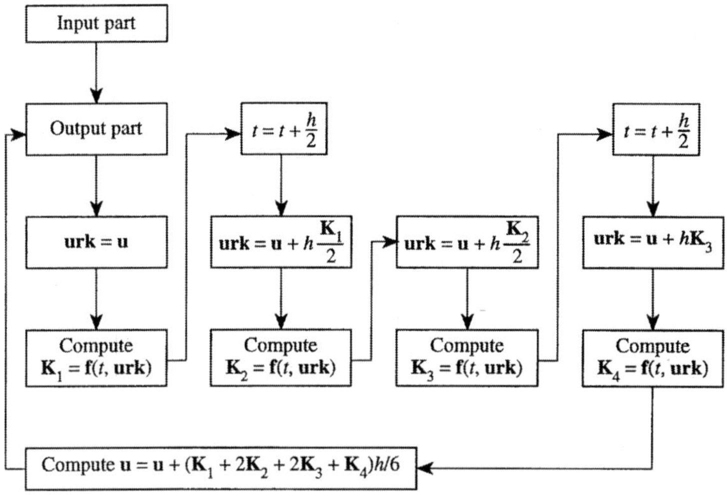

Table 2.9.2 — Runge–Kutta solver.

After initializing time and state variables to their initial values, the output part governs, according to what has been set in the input part, the type of output. If the analytic solution is known and the print-out option is chosen, then the error committed is also computed and printed. When the final time has passed by, then the program passes, if this is the case, to the next initial-value problem (related to different initial conditions) and then asks if the same thing has to be repeated with a different time step. This is useful, for instance, when checking the accuracy or the stability of the method.

The Runge–Kutta solver represented in Table 2.9.2 is made up of a series of DO-loops that establishes the values at which to compute the right-hand side of the equation. Finally, Eq.(2.8.23a) is evaluated and the program goes back to the output part.

We preferred to place at the end of the program the integration routine and the subroutine to compute the right-hand side of the differential equation (respectively, at lines 2000–2990 and 3000–3200), so that they can be easily identified, modified, and/or extracted.

We now intend to describe line-by-line the computer program related to the flow charts just described.

In lines 10–39 we indicated the variables used with an indication of their meaning. In using this program for other applications these variables should not be used as parameters of the differential equations or of its analytic solution. The variables used in lines 40-49 are instead used as parameters of the differential equations. These variables can be redefined and used.

Line 200 defines the dimension of the arrays and line 240 clears the screen.

Lines 250-780 are input lines. Among them lines 510-650 define the parameters of the differential equations, which are also used for the analytic solution. These lines can be deleted or should be changed in using this program for other applications.

Line 800 defines the coefficients used by the Runge–Kutta method. We return here if at the end of the integration a different time step is desired.

Line 803 prints $h\lambda$, which should be less than $2\sqrt{2}$ for stability. To be deleted in other applications.

Line 805 clears the screen from the input lines.

A DO-loop which refers to the integration of the different initial-value problems begins at line 810 and ends at line 2320.

Line 820 initializes time and line 830 initializes a printing flag. Then lines 870-890 set the initial values of the state variable.

If one wants the values of the state variable and the error committed (`out$=''yes''`), then line 900 jumps the lines 940-1030, which gives the commands relative to the plotting area and sends the program to line 1070 where the analytic solution is computed (if available, lines 1070-1110). Then it jumps again to line 1260, where it is decided whether or not to print, according to the chosen print-out interval (`prout`). Lines 1270-1360 are the printing commands. Line 1320 computes the error made and line 1350 stops computation to give some time to read the print-outs (if desired). Finally line 1370 updates the flag and line 1380 determines if the final time is passed by.

On the other hand, if one wants to plot the solution then line 940 sets the screen type, lines 950-980 print the values at the extrema of the plotting area, line 990 locates the position for printing the next input line (see line 2330). Then, line 1000 defines the plotting area and line 1010 sets the end points of the plotting area. Finally lines 1020 or 1030 sets the location of the beginning pixel. If the reader access to a better screen, the quality of the graph can be improved by suitably changing the values given in lines 940-1000.

After computing the analytic solution at lines 1080-1110 (if available), line 1170 or line 1200 connects the solution at the previous step with the updated value. If the analytic solution is known line 1180 or 1210 dots it. Then the printing commands are jumped and the program determines if the final time is passed by.

We then enter the core of the program.

The right-hand side of the differential equation $\mathbf{f}(t, \mathbf{u})$ is given in the subroutine on lines 3000-3200. Referring then to Eq.(2.8.23b), lines 2030-2060 compute $\mathbf{K}_1 = \mathbf{f}(t_i, \mathbf{u}_i)$ and line 2090 stores it in k1. Line 2070 increases t of half the time step and lines 2080-2110 set the value $\mathbf{u} + h\mathbf{K}_1/2$ at which to compute $\mathbf{f}$ again (line 2120). This is $\mathbf{K}_2$ and is stored in k2 at line 2140. In the same way lines 2130-2160 set the value $\mathbf{u} + h\mathbf{K}_2/2$ at which to compute $\mathbf{f}$ again (line 2170). Finally, line 2180 increases time of half the time step and lines 2190-2220 set the value $\mathbf{u} + h\mathbf{K}_3$ at which to compute $\mathbf{f}$ (line 2230). The new value of $\mathbf{u}$ is then computed in line 2260 and the program goes back to the output commands.

We note that in order to use this program for other applications it is enough to introduce the function $\mathbf{f}$ at lines 3070-3100 and, if known, the analytic solution at lines 1070-1110.

RK4.BAS

```
1 REM ************************************************************
2 REM *      Solution of a system of differential equations      *
3 REM *           using fourth-order Runge-Kutta method          *
4 REM ************************************************************
5 REM
10 REM ***********************************************************
12 REM *                      Variables used                      *
13 REM ***********************************************************
14 REM ndim = number of first-order ODE's (<20)
15 REM ngraf = number of plots referring to different IVP's (<10)
16 REM u(ndim) = state variable
17 REM uan(ndim) = analytic solution
18 REM u0(ngraf,ndim)= initial condition of the state variable
19 REM f(ndim) = right-hand side of differential equations
20 REM usug(ndim) = suggested values for initial data
21 REM t = time
22 REM h = time step (h2=h/2, h6=h/6)
23 REM tfin = final time of integration
24 REM prout = print-out interval
25 REM out$ = flag for print-out option
26 REM prstop$ = flag for inserting a pause after print-outs
27 REM iprint = flag used for switching the printing procedure
28 REM an$ = flag indicating if the analytic solution is available
```

```
29 REM phase$ = flag related to the type of graph to plot
30 REM kgraf, kgraf1, kgraf2 = indicate the components to plot
31 REM ptmin, ptmax, pumin, pumax = extrema of graph
32 REM urk(ndim), k1-k4(ndim) = variables used in Runge-Kutta
40 REM ***********************************************************
41 REM *               Next variables can be reused              *
42 REM ***********************************************************
43 REM m1, m2 = masses
44 REM kk1, kk2 = rigidity constants
45 REM c1-c4 = constant coefficients of the example
46 REM a1-a4, b1-b4, om1, om2, dom = variables used for the
   analytic solution
200 DIM u(20), uan(20), u0(10, 20), urk(20), f(20), usug(10),
   k1(20), k2(20), k3(20), k4(20)
210 REM **********************************************************
220 REM *                      Input lines                       *
230 REM **********************************************************
240 CLS 0
250 PRINT "****************************************************"
260 PRINT "* Solution of a system of differential equations    *"
270 PRINT "* using fourth-order Runge-Kutta method             *"
280 PRINT "****************************************************"
290 PRINT ""
300 PRINT "When more that one quantity is requested, separate
   with a comma"
310 PRINT "Values in square brackets are the suggested values,
   not default"
320 INPUT "Number of differential equations [4]"; ndim
330 INPUT "Time step and final time [.1,100]"; h, tfin
340 PRINT "Number of superimposed plots with different initial
   data [1]"
350 INPUT " or print-out of the result of a single IVP [0]"; ngraf
360 out$ = "no"
370 IF ngraf = 0 THEN ngraf = 1:  out$ = "yes"
380 IF out$ = "yes" THEN INPUT "Print-out interval [.5]"; prout
390 IF out$ = "yes" THEN INPUT "Do you want to stop at every
   print-out [y]"; prstop$
400 usug(1) = 1:  usug(2) = 2:  usug(3) = 0:  usug(4) = 0
410 FOR igraf = 1 TO ngraf
420 PRINT "Initial data for plot n."; igraf
430 FOR i = 1 TO ndim
440 PRINT "Initial value u0 of component n."; i;
450 IF ndim = 4 THEN PRINT " ["; usug(i); "]";
460 INPUT u0(igraf, i)
470 NEXT i
480 NEXT igraf
490 INPUT "Masses of the test coupled oscillator [1,.01 or .1,1]";
   m1,m2
```

```
500 INPUT "Spring rigidity of the test coupled oscillator [1,.01
    or .1,1]"; kk1, kk2
510 c1 = -(kk1 + kk2) / m1
520 c2 = kk2 / m1
530 c3 = kk2 / m2
540 c4 = -kk2 / m2
550 om1 = SQR((-(c1 + c4) + SQR((c1 - c4) ^2 + 4 * c2 * c3)) / 2)
560 om2 = SQR((-(c1 + c4) - SQR((c1 - c4) ^2 + 4 * c2 * c3)) / 2)
570 dom = om2 ^2 - om1 ^2
580 a1 = (c1 * u0(1, 1) + c2 * u0(1, 3) + u0(1, 1) * om2 ^2) / dom
590 a2 = -(c1 * u0(1, 1) + c2 * u0(1, 3) + u0(1, 1) * om1 ^2)
    / dom
600 a3 = (c1 * u0(1, 2)+c2 * u0(1, 4)+u0(1, 2)* om2^2)
    / (dom * om1)
610 a4 = -(c1*u0(1, 2)+c2 * u0(1, 4)+u0(1, 2)* om1^2)
    / (dom * om2)
620 b1 = -(c1 + om1^2) * a1 / c2
630 b2 = -(c1 + om2^2) * a2 / c2
640 b3 = -(c1 + om1^2) * a3 / c2
650 b4 = -(c1 + om2^2) * a4 / c2
660 an$ = "n"
670 IF ngraf < 2 THEN INPUT "Is the analytical solution known
    [y]"; an$
680 IF out$ = "yes" THEN GOTO 800
690 INPUT "Do you want a phase diagram [y] or the temporal diagram
    [n]"; phase$
700 IF phase$ = "y" THEN INPUT "On which components do you want
    to focus? (Indicate two of them 1,2,3,...) [1,2]",
    kgraf1, kgraf2
710 IF phase$ = "n" THEN INPUT "Which component do you want to
    plot? (Indicate the component 1,2,3,...) [1]", kgraf
720 ptmin = 0: ptmax = tfin
730 IF phase$ = "y" THEN INPUT "Minimum and maximum abscissa of
    the graph [-3,3]"; ptmin, ptmax
740 IF phase$ = "y" THEN INPUT "Minimum and maximum ordinate of
    the graph [-2,2 or -4,4]"; pumin, pumax
750 IF phase$ = "n" THEN INPUT "Minimum and maximum ordinate of
    the graph [-3,3]"; pumin, pumax
760 REM ***********************************************************
770 REM *                    End of input lines                    *
780 REM ***********************************************************
800 h2 = h / 2:  h6 = h / 6
803 IF om1 * om2 * m1 * kk1 * m2 * kk2 <> 0 THEN PRINT
    "|h*lambda| ="; h*om1; h*om2, "Press RETURN": INPUT scratch
805 CLS 0
810 FOR igraf = 1 TO ngraf
820 t = 0
830 iprint = 0
```

```
840 REM **********************************************************
850 REM *                  Set initial data                       *
860 REM **********************************************************
870 FOR i = 1 TO ndim
880 u(i) = u0(igraf, i)
890 NEXT i
900 IF out$ = "yes" THEN GOTO 1070
910 REM **********************************************************
920 REM *                Set up plotting area                     *
930 REM **********************************************************
940 KEY OFF: SCREEN 2
950 LOCATE 14, 10:  PRINT ptmin
960 LOCATE 14, 70:  PRINT ptmax
970 LOCATE 13, 7:  PRINT pumin
980 LOCATE 2, 5:  PRINT pumax
990 LOCATE 15, 20
1000 VIEW (80, 10)-(560, 100), , 1
1010 WINDOW (ptmin, pumin)-(ptmax, pumax)
1020 IF phase$ = "n" THEN PSET (0, u0(igraf, kgraf))
1030 IF phase$ = "y" THEN PSET (u0(igraf, kgraf1), u0(igraf,
   kgraf2))
1040 REM *********************************************************
1050 REM *     If known, insert here the analytic solution       *
1060 REM *********************************************************
1070 IF an$ = "n" THEN GOTO 1120
1080 uan(1) = a1 * COS(om1 * t) + a2 * COS(om2 * t) +
   a3 * SIN(om1 * t) + a4 * SIN(om2 * t)
1090 uan(2) = -a1 * om1 * SIN(om1 * t) - a2 * om2 * SIN(om2 * t) +
   a3 * om1 * COS(om1 * t) + a4 * om2 * COS(om2 * t)
1100 uan(3) = b1 * COS(om1 * t) + b2 * COS(om2 * t) +
   b3 * SIN(om1 * t) + b4 * SIN(om2 * t)
1110 uan(4) = -b1 * om1 * SIN(om1 * t) - b2 * om2 * SIN(om2 * t) +
   b3 * om1 * COS(om1 * t) + b4 * om2 * COS(om2 * t)
1120 IF out$ = "yes" THEN GOTO 1260
1130 REM *********************************************************
1140 REM *                 Plotting commands                     *
1150 REM *********************************************************
1160 IF phase$ = "y" THEN GOTO 1200
1170 LINE -(t, u(kgraf))
1180 IF an$ = "y" THEN PSET (t, uan(kgraf)):  PSET (t, u(kgraf))
1190 GOTO 1380
1200 LINE -(u(kgraf1), u(kgraf2))
1210 IF an$ = "y" THEN PSET (uan(kgraf1), uan(kgraf2)):  PSET
   (u(kgraf1), u(kgraf2))
1220 GOTO 1380
1230 REM *********************************************************
1240 REM *                 Print-out commands                    *
1250 REM *********************************************************
```

```
1260 IF t < prout * iprint - h/10 AND t <> 0 THEN GOTO 1380
1270 PRINT TAB(15); "t="; t
1280 FOR i = 1 TO ndim
1290 PRINT TAB(10); "u("; i; ")="; u(i),
1300 IF an$ = "n" THEN GOTO 1340
1310 IF uan(i) = 0 THEN uan(i) = .0000001
1320 er = 1 - u(i) / uan(i)
1330 PRINT uan(i), "error="; er
1340 NEXT i
1350 IF prstop$ = "y" THEN INPUT scratch
1360 PRINT
1370 iprint = iprint + 1
1380 IF t > tfin - h/10 THEN GOTO 2320
2000 REM ***********************************************************
2010 REM *             Fourth-order Runge-Kutta routine            *
2020 REM ***********************************************************
2030 FOR i = 1 TO ndim
2040 urk(i) = u(i)
2050 NEXT i
2060 GOSUB 3000
2070 t = t + h2
2080 FOR i = 1 TO ndim
2090 k1(i) = f(i)
2100 urk(i) = u(i) + h2 * k1(i)
2110 NEXT i
2120 GOSUB 3000
2130 FOR i = 1 TO ndim
2140 k2(i) = f(i)
2150 urk(i) = u(i) + h2 * k2(i)
2160 NEXT i
2170 GOSUB 3000
2180 t = t + h2
2190 FOR i = 1 TO ndim
2200 k3(i) = f(i)
2210 urk(i) = u(i) + h * k3(i)
2220 NEXT i
2230 GOSUB 3000
2240 FOR i = 1 TO ndim
2250 k4(i) = f(i)
2260 u(i) = u(i) + h6 * (k1(i) + 2 * (k2(i) + k3(i)) + k4(i))
2270 NEXT i
2280 GOTO 1070
2290 REM ***********************************************************
2300 REM *                 End Runge-Kutta routine                 *
2310 REM ***********************************************************
2320 NEXT igraf
2330 INPUT "Different time step?  (0 if not)", h
2340 IF h <> 0 THEN GOTO 800
```

```
2350 END
3000 REM *******************************************************
3010 REM *          Subroutine for r.h.s. of equations         *
3020 REM *            insert here the function F(t,u)          *
3030 REM *              without using the variables            *
3040 REM *            indicated in lines 10-40, since          *
3050 REM *          they are already used in the program       *
3060 REM *******************************************************
3070 f(1) = urk(2)
3080 f(2) = c1 * urk(1) + c2 * urk(3)
3090 f(3) = urk(4)
3100 f(4) = c3 * urk(1) + c4 * urk(3)
3190 RETURN
3200 END
```

2.9.4 RKF: A Runge–Kutta–Fehlberg program

The program RKF has the same structure of RK4 with minor changes in lines 1-2000. The major changes are, of course, in the ODE solver and in the addition of a control part. Lines 2025, 2070, 2125, 2180, 2235, and 2280 set the times and lines 2040, 2100, 2150, 2210, 2250, and 2310 set the values of $\mathbf{u}$ at which to compute $\mathbf{f}(t, \mathbf{u})$ (lines 2060, 2120, 2170, 2230, 2270, and 2330), i.e., the quantities $\mathbf{K}_i$, $i = 1, \ldots, 6$ in Eq.(2.8.25b).

The main difference consists in the computation of two estimates of $\mathbf{u}$ by a fourth- and a fifth-order method and in an error control, which permits an automatic adaptation of the time step.

More precisely, the two values of $\mathbf{u}$ given in Eq.(2.8.25a) are computed in lines 2360 and 2365. The difference between these two values over the time step is computed in line 2367 and the maximum of the difference over the components is stored in er in line 2369. If this error is too large with respect to the desired accuracy, then the time step is halved in line 2410 and the integration is repeated. If the error is not too large then the time and the value of the state variable are updated (lines 2420-2450). If however the error is too small then on line 2460 the time step is doubled.

Several runs can now be done observing how the time step changes as the system evolves and how it depends on the required accuracy. Changes in time step become more dramatic (and therefore RKF more useful) when systems with behaviors like the Van der Pol equation are studied. For this reason we also stored in the diskette the program VANDRPOL, which performs its simulation. Once more, the reader is referred to the Exercises to Chapter 2.

RKF.BAS

```
1 REM ************************************************************
2 REM *      Solution of a system of differential equations      *
3 REM *            using Runge-Kutta-Fehlberg routine             *
4 REM ************************************************************
5 REM
10 REM ***********************************************************
12 REM *                     Variables used                      *
13 REM ***********************************************************
14 REM ndim = number of first-order ODE's (ndim < 20)
15 REM ngraf = number of plots referring to different IVP's
16 REM u(ndim) = state variable
17 REM uan(ndim) = analytic solution
18 REM u0(ngraf,ndim)= initial condition of the state variable
19 REM uu(ndim), ustar(ndim) = approximations of u
20 REM f(ndim) = right-hand side of differential equations
21 REM usug(ndim) = suggested values for initial data
22 REM t = time
23 REM h = time step (h2=h/2, h6=h/6)
24 REM tfin = final time of integration
25 REM tol = tolerance
26 REM dif, er = differences between the outputs of the two
   schemes
27 REM prout = print-out interval
28 REM out$ = flag for print-out option
29 REM prstop$ = flag for inserting a pause after print-outs
30 REM iprint = flag used for switching the printing procedure
31 REM an$ = flag indicating if the analytic solution is available
32 REM phase$ = flag related to the type of graph to plot
33 REM kgraf, kgraf1, kgraf2 = indicate the components to plot
34 REM ptmin, ptmax, pumin, pumax = extrema of graph
35 REM urk(ndim), k1-k6(ndim) = variables used in Runge-Kutta
40 REM ***********************************************************
41 REM *             Next variables can be reused                *
42 REM ***********************************************************
43 REM m1, m2 = masses
44 REM kk1, kk2 = rigidity constants
45 REM c1-c4 = constant coefficients of the example
46 REM a1-a4, b1-b4, om1, om2, dom = variables used for the
   analytic solution
200 DIM u(20), uan(20), u0(10, 20), urk(20), uu(20), ustar(20)
205 DIM f(20), usug(10), k1(20), k2(20), k3(20), k4(20), k5(20),
   k6(20)
210 REM **********************************************************
220 REM *                     Input lines                        *
230 REM **********************************************************
240 CLS 0
```

```
250 PRINT "*****************************************************"
260 PRINT "*   Solution of a system of differential equations   *"
270 PRINT "*          using Runge-Kutta-Fehlberg routine        *"
280 PRINT "*****************************************************"
290 PRINT ""
300 PRINT "When more that one quantity is requested, separate
    with a comma"
310 PRINT "Values in square brackets are the suggested values,
    not default"
320 INPUT "Number of differential equations [4]"; ndim
330 INPUT "Initial time step and final time [1,100]"; h, tfin
333 INPUT "Desired tolerance [.001]"; tol
340 PRINT "Number of superimposed plots with different initial
    data [1]"
350 INPUT " or print-out of the result of a single IVP [0]"; ngraf
360 out$ = "no"
370 IF ngraf = 0 THEN ngraf = 1:  out$ = "yes"
380 IF out$ = "yes" THEN INPUT "Print-out interval [.5]"; prout
390 IF out$ = "yes" THEN INPUT "Do you want to stop at every
    print-out [y]"; prstop$
400 usug(1) = 1:  usug(2) = 2:  usug(3) = 0:  usug(4) = 0
410 FOR igraf = 1 TO ngraf
420 PRINT "Initial data for plot n."; igraf
430 FOR i = 1 TO ndim
440 PRINT "Initial value u0 of component n."; i;
450 IF ndim = 4 THEN PRINT " ["; usug(i); "]";
460 INPUT u0(igraf, i)
470 NEXT i
480 NEXT igraf
490 INPUT "Masses of the test coupled oscillator [1,.01 or .1,1]";
    m1,m2
500 INPUT "Spring rigidity of the test coupled oscillator [1,.01
    or .1,1]"; kk1, kk2
510 c1 = -(kk1 + kk2) / m1
520 c2 = kk2 / m1
530 c3 = kk2 / m2
540 c4 = -kk2 / m2
550 om1 = SQR((-(c1 + c4) + SQR((c1 - c4)^2 + 4 * c2 * c3)) / 2)
560 om2 = SQR((-(c1 + c4) - SQR((c1 - c4)^2 + 4 * c2 * c3)) / 2)
570 dom = om2 ^2 - om1 ^2
580 a1 = (c1 * u0(1, 1) + c2 * u0(1, 3) + u0(1, 1) * om2^2) / dom
590 a2 = -(c1 * u0(1, 1) + c2 * u0(1, 3) + u0(1, 1) * om1^2) / dom
600 a3 = (c1*u0(1, 2) + c2*u0(1, 4) + u0(1, 2)*om2^2) / (dom*om1)
610 a4 = -(c1*u0(1, 2)+ c2*u0(1, 4)+ u0(1, 2)*om1^2) / (dom*om2)
620 b1 = -(c1 + om1^2) * a1 / c2
630 b2 = -(c1 + om2^2) * a2 / c2
640 b3 = -(c1 + om1^2) * a3 / c2
650 b4 = -(c1 + om2^2) * a4 / c2
```

```
660 an$ = "n"
670 IF ngraf < 2 THEN INPUT "Is the analytical solution known
    [y]"; an$
680 IF out$ = "yes" THEN GOTO 800
690 INPUT "Do you want a phase diagram [y] or the temporal
    diagram [n]"; phase$
700 IF phase$ = "y" THEN INPUT "On which components do you want
    to focus? (Indicate two of them 1,2,3,...)  [1,2]",
    kgraf1, kgraf2
710 IF phase$ = "n" THEN INPUT "Which component do you want to
    plot? (Indicate the component 1,2,3,...)  [1]", kgraf
720 ptmin = 0:  ptmax = tfin
730 IF phase$ = "y" THEN INPUT "Minimum and maximum abscissa of
    the graph [-3,3]"; ptmin, ptmax
740 IF phase$ = "y" THEN INPUT "Minimum and maximum ordinate of
    the graph [-2,2 or -4,4]"; pumin, pumax
750 IF phase$ = "n" THEN INPUT "Minimum and maximum ordinate of
    the graph [-3,3]"; pumin, pumax
760 REM ************************************************************
770 REM *                  End of input lines                      *
780 REM ************************************************************
800 CLS 0
810 FOR igraf = 1 TO ngraf
820 t = 0
830 iprint = 0
840 REM ************************************************************
850 REM *                  Set initial data                        *
860 REM ************************************************************
870 FOR i = 1 TO ndim
880 u(i) = u0(igraf, i)
890 NEXT i
900 IF out$ = "yes" THEN GOTO 1070
910 REM ************************************************************
920 REM *                 Set up plotting area                     *
930 REM ************************************************************
940 KEY OFF: SCREEN 2
950 LOCATE 14, 10:  PRINT ptmin
960 LOCATE 14, 70:  PRINT ptmax
970 LOCATE 13, 7:  PRINT pumin
980 LOCATE 2, 5:  PRINT pumax
990 LOCATE 15, 20
1000 VIEW (80, 10)-(560, 100), , 1
1010 WINDOW (ptmin, pumin)-(ptmax, pumax)
1020 IF phase$ = "n" THEN PSET (0, u0(igraf, kgraf))
1030 IF phase$ = "y" THEN PSET (u0(igraf, kgraf1), u0(igraf,
   kgraf2))
1040 REM ************************************************************
1050 REM *      If known, insert here the analytic solution        *
```

```
1060 REM ************************************************************
1070 IF an$ = "n" THEN GOTO 1120
1080 uan(1) = a1 * COS(om1 * t) + a2 * COS(om2 * t) +
   a3 * SIN(om1 * t) + a4 * SIN(om2 * t)
1090 uan(2) = -a1 * om1 * SIN(om1 * t) - a2 * om2 * SIN(om2 * t) +
   a3 * om1 * COS(om1 * t) + a4 * om2 * COS(om2 * t)
1100 uan(3) = b1 * COS(om1 * t) + b2 * COS(om2 * t) +
   b3 * SIN(om1 * t) + b4 * SIN(om2 * t)
1110 uan(4) = -b1 * om1 * SIN(om1 * t) - b2 * om2 * SIN(om2 * t) +
   b3 * om1 * COS(om1 * t) + b4 * om2 * COS(om2 * t)
1120 IF out$ = "yes" THEN GOTO 1260
1130 REM ************************************************************
1140 REM *                    Plotting commands                      *
1150 REM ************************************************************
1155 LOCATE 15, 35:  PRINT "h ="; h
1160 IF phase$ = "y" THEN GOTO 1200
1170 LINE -(t, u(kgraf))
1180 IF an$ = "y" THEN PSET (t, uan(kgraf)):  PSET (t, u(kgraf))
1190 GOTO 1380
1200 LINE -(u(kgraf1), u(kgraf2))
1210 IF an$ = "y" THEN PSET (uan(kgraf1), uan(kgraf2)):  PSET
   (u(kgraf1), u(kgraf2))
1220 GOTO 1380
1230 REM ************************************************************
1240 REM *                    Print-out commands                     *
1250 REM ************************************************************
1260 IF t < prout * iprint - h/10 AND t <> 0 THEN GOTO 1380
1270 PRINT TAB(15); "t="; t
1280 FOR i = 1 TO ndim
1290 PRINT TAB(10); "u("; i; ")="; u(i),
1300 IF an$ = "n" THEN GOTO 1340
1310 IF uan(i) = 0 THEN uan(i) = .0000001
1320 er = 1 - u(i) / uan(i)
1330 PRINT uan(i), "error="; er
1340 NEXT i
1350 IF prstop$ = "y" THEN INPUT scratch
1360 PRINT
1370 iprint = iprint + 1
1380 IF t > tfin - h/10 THEN GOTO 2960
2000 REM ************************************************************
2010 REM *               Runge-Kutta-Fehlberg routine                *
2020 REM ************************************************************
2025 ti = t
2030 FOR i = 1 TO ndim
2040 urk(i) = u(i)
2050 NEXT i
2060 GOSUB 3000
2070 t = ti + h / 4
```

```
2080 FOR i = 1 TO ndim
2090 k1(i) = f(i)
2100 urk(i) = u(i) + h / 4 * k1(i)
2110 NEXT i
2120 GOSUB 3000
2125 t = ti + 3 * h / 8
2130 FOR i = 1 TO ndim
2140 k2(i) = f(i)
2150 urk(i) = u(i) + 3 * h / 32 * (k1(i) + 3 * k2(i))
2160 NEXT i
2170 GOSUB 3000
2180 t = ti + 12 / 13 * h
2190 FOR i = 1 TO ndim
2200 k3(i) = f(i)
2210 urk(i) = u(i)+h / 2197*(1932*k1(i) - 7200*k2(i) + 7296*k3(i))
2220 NEXT i
2230 GOSUB 3000
2235 t = ti + h
2237 FOR i = 1 TO ndim
2240 k4(i) = f(i)
2250 urk(i) = u(i) + h / 4104 * (8341 * k1(i) - 32832 * k2(i) +
   29440 * k3(i) - 845 * k4(i))
2260 NEXT i
2270 GOSUB 3000
2280 t = ti + h / 2
2290 FOR i = 1 TO ndim
2300 k5(i) = f(i)
2310 urk(i) = u(i) + h / 20520 * (-6080 * k1(i) + 41040 * k2(i) -
   28352 * k3(i) + 9295 * k4(i) - 5643 * k5(i))
2320 NEXT i
2330 GOSUB 3000
2335 er = 0
2340 FOR i = 1 TO ndim
2350 k6(i) = f(i)
2360 ustar(i) = u(i) + h / 282150 * (33440 * k1(i) + 146432 *
   k3(i) + 142805 * k4(i) - 50787 * k5(i) + 10260 * k6(i))
2365 uu(i) = u(i) + h / 20520 * (2375 * k1(i) + 11264 * k3(i) +
   10985 * k4(i) - 4104 * k5(i))
2367 dif = ABS(uu(i) - ustar(i)) / h
2369 IF dif > er THEN er = dif
2370 NEXT i
2380 REM *******************************************************
2390 REM *  Set new time step if er is too large or too small   *
2400 REM *******************************************************
2410 IF er > tol THEN h = h / 2:  GOTO 2030
2420 t = ti + h
2430 FOR i = 1 TO ndim
2440 u(i) = uu(i)
```

```
2450 NEXT i
2460 IF er < tol / 100 THEN h = 2 * h
2920 REM *******************************************************
2930 REM *          End of Runge-Kutta-Fehlberg routine        *
2940 REM *******************************************************
2950 GOTO 1070
2960 NEXT igraf
2970 INPUT "Different initial conditions?  (y or n)", scr$
2980 IF scr$ = "y" THEN GOTO 400
2990 END
3000 REM *******************************************************
3010 REM *           Subroutine for r.h.s. of equations        *
3020 REM *            insert here the function f(t,u)          *
3030 REM *              without using the variables            *
3040 REM *            indicated in lines 10-40, since          *
3050 REM *         they are already used in the program        *
3060 REM *******************************************************
3070 f(1) = urk(2)
3080 f(2) = c1 * urk(1) + c2 * urk(3)
3090 f(3) = urk(4)
3100 f(4) = c3 * urk(1) + c4 * urk(3)
3190 RETURN
3200 END
```

2.9.5 AB4: A fourth-order Adams–Bashforth program

The structure of the program AB4 is again the same as the previous one but for the fact that the output part is now a subroutine moved to lines 2800-2990. It should, however, be easy for the reader to identify the different parts of the program and the meaning of each command following the description given in Section 2.9.3.

The only important difference remains, of course, in the core of the program where the fourth-order Runge–Kutta method is used only for the first three steps (lines 2000-2380). The value of **f** at the different time steps $t = 0, h, 2h, 3h$ is respectively stored in `f3`, `f2`, `f1`, and `f` (lines 2070-2110 and 2390-2430). The Adams–Bashforth method is then introduced at lines 2500-2590. After having advanced time (line 2630) and returned from the output routine (line 2640), the values of **f** at previous times are rearranged (lines 2680-2720) and the Adams–Bashforth routine is repeated till the final time of integration is reached.

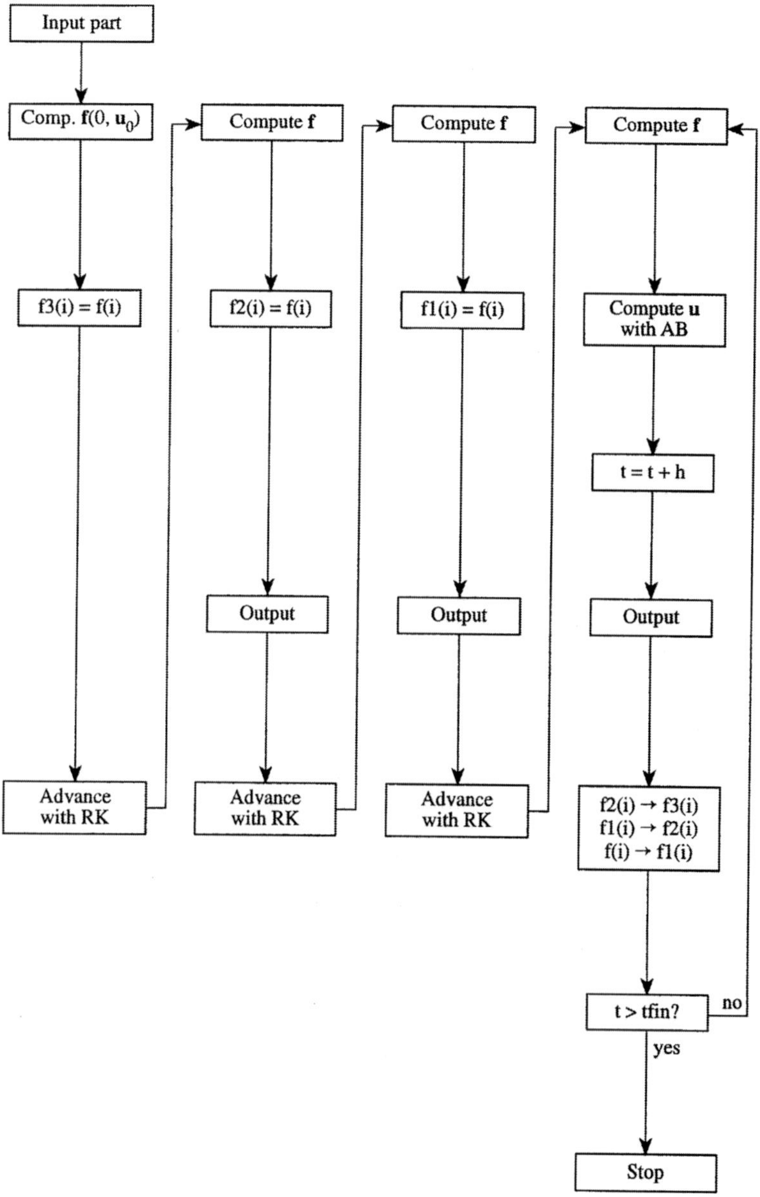

Table 2.9.3 — Adams–Bashforth flow chart.

AB4.BAS

```
1 REM *************************************************************
2 REM *      Solution of a system of differential equations       *
3 REM *         using fourth-order Adams-Bashforth method         *
4 REM *************************************************************
5 REM
10 REM ************************************************************
12 REM *                     Variables used                       *
13 REM ************************************************************
14 REM ndim = number of first-order ODE's (<20)
15 REM ngraf = number of plots referring to different IVP's (<10)
16 REM u(ndim) = state variable
17 REM uan(ndim) = analytic solution
18 REM u0(ngraf,ndim)= initial condition of the state variable
19 REM f(ndim) = right-hand side of differential equations
20 REM f1-f3(ndim) = right-hand side at previous times
21 REM usug(ndim) = suggested values for initial data
22 REM t = time
23 REM h = time step (h2=h/2, h6=h/6, h24=h/24)
24 REM tfin = final time of integration
25 REM prout = print-out interval
26 REM out$ = flag for print-out option
27 REM prstop$ = flag for inserting a pause after print-outs
28 REM iprint = flag used for switching the printing procedure
29 REM an$ = flag indicating if the analytic solution is available
30 REM phase$ = flag related to the type of graph to plot
31 REM kgraf, kgraf1, kgraf2 = indicate the components to plot
32 REM ptmin, ptmax, pumin, pumax = extrema of graph
33 REM usub(ndim) = variables used for evaluating r.h.s.
34 REM k1-k4(ndim) = variables used for Runge-Kutta
40 REM ************************************************************
41 REM *              Next variables can be reused                *
42 REM ************************************************************
43 REM m1, m2 = masses
44 REM kk1, kk2 = rigidity constants
45 REM c1-c4 = constant coefficients of the example
46 REM a1-a4, b1-b4, om1, om2, dom = variables used for the
    analytic solution
200 DIM u(20), uan(20), u0(10, 20), usub(20), f(20), f1(20),
    f2(20), f3(20)
205 DIM usug(10), k1(20), k2(20), k3(20), k4(20)
210 REM ***********************************************************
220 REM *                      Input lines                        *
230 REM ***********************************************************
240 CLS 0
250 PRINT "*****************************************************"
260 PRINT "*   Solution of a system of differential equations  *"
```

```
270 PRINT "*       with fourth-order Adams-Bashforth method      *"
280 PRINT "*****************************************************"
290 PRINT ""
300 PRINT "When more than one quantity is requested, separate
   with a comma"
310 PRINT "Values in square brackets are the suggested values,
   not default"
320 INPUT "Number of differential equations [4]"; ndim
330 INPUT "Time step and final time [.05,100]"; h, tfin
340 PRINT "Number of superimposed plots with different initial
   data [1]"
350 INPUT " or print-out of the result of a single IVP [0]"; ngraf
360 out$ = "no"
370 IF ngraf = 0 THEN ngraf = 1:  out$ = "yes"
380 IF out$ = "yes" THEN INPUT "Print-out interval [.5]"; prout
390 IF out$ = "yes" THEN INPUT "Do you want to stop at every
   print-out [y]"; prstop$
400 usug(1) = 1:  usug(2) = 2:  usug(3) = 0:  usug(4) = 0
410 FOR igraf = 1 TO ngraf
420 PRINT "Initial data for plot n."; igraf
430 FOR i = 1 TO ndim
440 PRINT "Initial value u0 of component n."; i;
450 IF ndim = 4 THEN PRINT " ["; usug(i); "]";
460 INPUT u0(igraf, i)
470 NEXT i
480 NEXT igraf
490 INPUT "Masses of the test coupled oscillator [1,.01 or .1,1]";
   m1,m2
500 INPUT "Spring rigidity of the test coupled oscillator [1,.01
   or .1,1]"; kk1, kk2
510 c1 = -(kk1 + kk2) / m1
520 c2 = kk2 / m1
530 c3 = kk2 / m2
540 c4 = -kk2 / m2
550 om1 = SQR((-(c1 + c4) + SQR((c1 - c4)^2 + 4*c2*c3)) / 2)
560 om2 = SQR((-(c1 + c4) - SQR((c1 - c4) ^2 + 4 * c2 * c3)) / 2)
570 dom = om2 ^2 - om1 ^2
580 a1 = (c1 * u0(1, 1) + c2 * u0(1, 3) + u0(1, 1) * om2 ^2) / dom
590 a2 = -(c1 * u0(1, 1) + c2 * u0(1, 3)+u0(1, 1) * om1 ^2) / dom
600 a3 = (c1*u0(1, 2)+ c2*u0(1, 4)+ u0(1, 2)*om2^2) / (dom*om1)
610 a4 = -(c1*u0(1, 2)+ c2*u0(1, 4)+ u0(1, 2)*om1^2) / (dom*om2)
620 b1 = -(c1 + om1^2) * a1 / c2
630 b2 = -(c1 + om2^2) * a2 / c2
640 b3 = -(c1 + om1^2) * a3 / c2
650 b4 = -(c1 + om2^2) * a4 / c2
660 an$ = "n"
670 IF ngraf < 2 THEN INPUT "Is the analytical solution known
   [y]"; an$
```

```
680 IF out$ = "yes" THEN GOTO 800
690 INPUT "Do you want a phase diagram [y] or the temporal
    diagram [n]"; phase$
700 IF phase$ = "y" THEN INPUT "On which components do you want
    to focus? (Indicate two of them 1,2,3,...)  [1,2]",
    kgraf1, kgraf2
710 IF phase$ = "n" THEN INPUT "Which component do you want to
    plot? (Indicate the component 1,2,3,...)  [1]", kgraf
720 ptmin = 0:  ptmax = tfin
730 IF phase$ = "y" THEN INPUT "Minimum and maximum abscissa of
    the graph [-3,3]"; ptmin, ptmax
740 IF phase$ = "y" THEN INPUT "Minimum and maximum ordinate of
    the graph [-2,2 or -4,4]"; pumin, pumax
750 IF phase$ = "n" THEN INPUT "Minimum and maximum ordinate of
    the graph [-3,3]"; pumin, pumax
760 REM ***********************************************************
770 REM *                   End of input lines                     *
780 REM ***********************************************************
800 h24 = h / 24:  h2 = h / 2:  h6 = h / 6
803 IF om1 * om2 * m1 * kk1 * m2 * kk2 <> 0 THEN PRINT "
    |h*lambda| ="; h * om1; h * om2, "Press RETURN": INPUT scratch
805 CLS 0
810 FOR igraf = 1 TO ngraf
820 t = 0
830 iprint = 0
840 REM ***********************************************************
850 REM *                   Set initial data                       *
860 REM ***********************************************************
870 FOR i = 1 TO ndim
880 u(i) = u0(igraf, i)
890 NEXT i
900 IF out$ = "yes" THEN GOSUB 2800:  GOTO 2000
910 REM ***********************************************************
920 REM *                 Set up plotting area                     *
930 REM ***********************************************************
940 KEY OFF: SCREEN 2
950 LOCATE 14, 10:  PRINT ptmin
960 LOCATE 14, 70:  PRINT ptmax
970 LOCATE 13, 7:  PRINT pumin
980 LOCATE 2, 5:  PRINT pumax
990 LOCATE 15, 20
1000 VIEW (80, 10)-(560, 100), , 1
1010 WINDOW (ptmin, pumin)-(ptmax, pumax)
1020 IF phase$ = "n" THEN PSET (0, u0(igraf, kgraf))
1030 IF phase$ = "y" THEN PSET (u0(igraf, kgraf1),
    u0(igraf, kgraf2))
2000 REM ***********************************************************
2010 REM *  Fourth-order Runge-Kutta routine to starting values   *
```

```
2020 REM ***********************************************************
2040 FOR i = 1 TO ndim
2050 usub(i) = u(i)
2060 NEXT i
2070 GOSUB 3000
2080 FOR i = 1 TO ndim
2090 f3(i) = f(i)
2110 NEXT i
2150 FOR irk = 1 TO 3
2160 GOSUB 3000
2170 t = t + h2
2180 FOR i = 1 TO ndim
2190 k1(i) = f(i)
2200 usub(i) = u(i) + h2 * k1(i)
2210 NEXT i
2220 GOSUB 3000
2230 FOR i = 1 TO ndim
2240 k2(i) = f(i)
2250 usub(i) = u(i) + h2 * k2(i)
2260 NEXT i
2270 GOSUB 3000
2280 t = t + h2
2290 FOR i = 1 TO ndim
2300 k3(i) = f(i)
2310 usub(i) = u(i) + h * k3(i)
2320 NEXT i
2330 GOSUB 3000
2340 FOR i = 1 TO ndim
2350 k4(i) = f(i)
2360 u(i) = u(i) + h6 * (k1(i) + 2 * (k2(i) + k3(i)) + k4(i))
2370 usub(i) = u(i)
2380 NEXT i
2390 GOSUB 3000
2400 FOR i = 1 TO ndim
2410 IF irk = 1 THEN f2(i) = f(i)
2420 IF irk = 2 THEN f1(i) = f(i)
2430 NEXT i
2440 GOSUB 2800
2450 NEXT irk
2460 REM ***********************************************************
2470 REM *                   End Runge-Kutta routine                 *
2480 REM ***********************************************************
2500 REM ***********************************************************
2510 REM *           Fourth-order Adams-Bashforth routine            *
2520 REM ***********************************************************
2530 FOR i = 1 TO ndim
2540 usub(i) = u(i)
2550 NEXT i
```

```
2560 GOSUB 3000
2570 FOR i = 1 TO ndim
2580 u(i) = u(i) + h24*(55*f(i) - 59*f1(i) + 37*f2(i) - 9*f3(i))
2590 NEXT i
2600 REM *********************************************************
2610 REM *               End Adams-Bashforth routine             *
2620 REM *********************************************************
2630 t = t + h
2640 GOSUB 2800
2650 REM *********************************************************
2660 REM *                Updating previous values               *
2670 REM *********************************************************
2680 FOR i = 1 TO ndim
2690 f3(i) = f2(i)
2700 f2(i) = f1(i)
2710 f1(i) = f(i)
2720 NEXT i
2730 IF t < tfin - h/10 THEN GOTO 2530
2740 NEXT igraf
2750 INPUT "Different time step?  (0 if not)", h
2760 IF h <> 0 THEN GOTO 800
2770 END
2800 REM *********************************************************
2801 REM *                    Output subroutine                  *
2802 REM *********************************************************
2805 REM *********************************************************
2806 REM *      If known, insert here the analytic solution      *
2807 REM *********************************************************
2810 IF an$ = "n" THEN GOTO 2835
2815 uan(1) = a1 * COS(om1 * t) + a2 * COS(om2 * t) +
   a3 * SIN(om1 * t) + a4 * SIN(om2 * t)
2820 uan(2) = -a1 * om1 * SIN(om1 * t) - a2 * om2 * SIN(om2 * t) +
   a3 * om1 * COS(om1 * t) + a4 * om2 * COS(om2 * t)
2825 uan(3) = b1 * COS(om1 * t) + b2 * COS(om2 * t) +
   b3 * SIN(om1 * t) + b4 * SIN(om2 * t)
2830 uan(4) = -b1 * om1 * SIN(om1 * t) - b2 * om2 * SIN(om2 * t) +
   b3 * om1 * COS(om1 * t) + b4 * om2 * COS(om2 * t)
2835 IF out$ = "yes" THEN GOTO 2880
2837 REM *********************************************************
2838 REM *                   Plotting commands                   *
2839 REM *********************************************************
2840 IF phase$ = "y" THEN GOTO 2860
2845 LINE -(t, u(kgraf))
2850 IF an$ = "y" THEN PSET (t, uan(kgraf)):  PSET (t, u(kgraf))
2855 GOTO 2990
2860 LINE -(u(kgraf1), u(kgraf2))
2865 IF an$ = "y" THEN PSET (uan(kgraf1), uan(kgraf2)):  PSET
   (u(kgraf1), u(kgraf2))
```

```
2870 GOTO 2990
2875 REM ********************************************************
2876 REM *                   Print-out commands                 *
2877 REM ********************************************************
2880 IF t <= prout * iprint - h/10 AND t <> 0 THEN GOTO 2990
2885 PRINT TAB(15); "t="; t
2890 FOR i = 1 TO ndim
2900 PRINT TAB(10); "u("; i; ")="; u(i),
2910 IF an$ = "n" THEN GOTO 2950
2920 IF uan(i) = 0 THEN uan(i) = .0000001
2930 er = 1 - u(i) / uan(i)
2940 PRINT uan(i), "error="; er
2950 NEXT i
2960 IF prstop$ = "y" THEN INPUT scratch
2970 PRINT
2980 iprint = iprint + 1
2990 RETURN
2995 END
3000 REM ********************************************************
3010 REM *           Subroutine for r.h.s. of equations         *
3020 REM *            insert here the function f(t,u)           *
3030 REM *              without using the variables             *
3040 REM *            indicated in lines 10-40, since           *
3050 REM *          they are already used in the program        *
3060 REM ********************************************************
3070 f(1) = usub(2)
3080 f(2) = c1 * usub(1) + c2 * usub(3)
3090 f(3) = usub(4)
3100 f(4) = c3 * usub(1) + c4 * usub(3)
3190 RETURN
3200 END
```

2.9.6 APC4: A fourth-order predictor-corrector program

The predictor-corrector program APC4 differs from AB4 in the fact that the value of $\mathbf{u}$ at $t = t_{i+1}$ is first predicted by the Adams–Bashforth method (line 2490), then after having advanced time (line 2510) and having stored $\mathbf{f}(t_i, \mathbf{u}_i)$ in `f0` (line 2530), the prediction is used to evaluate the correction $\mathbf{f}(t_{i+1}, \mathbf{u}_{i+1})$ (line 2560). This value is needed to evaluate Eq.(2.8.38) (lines 2570-2582). (See Table 2.9.4.)

Finally, the error estimate (2.8.39) is used to adapt the time step with the same procedure explained in Section 2.9.4 for RKF. If the time step is adapted, the program goes back to the Runge–Kutta part to find the new starting values.

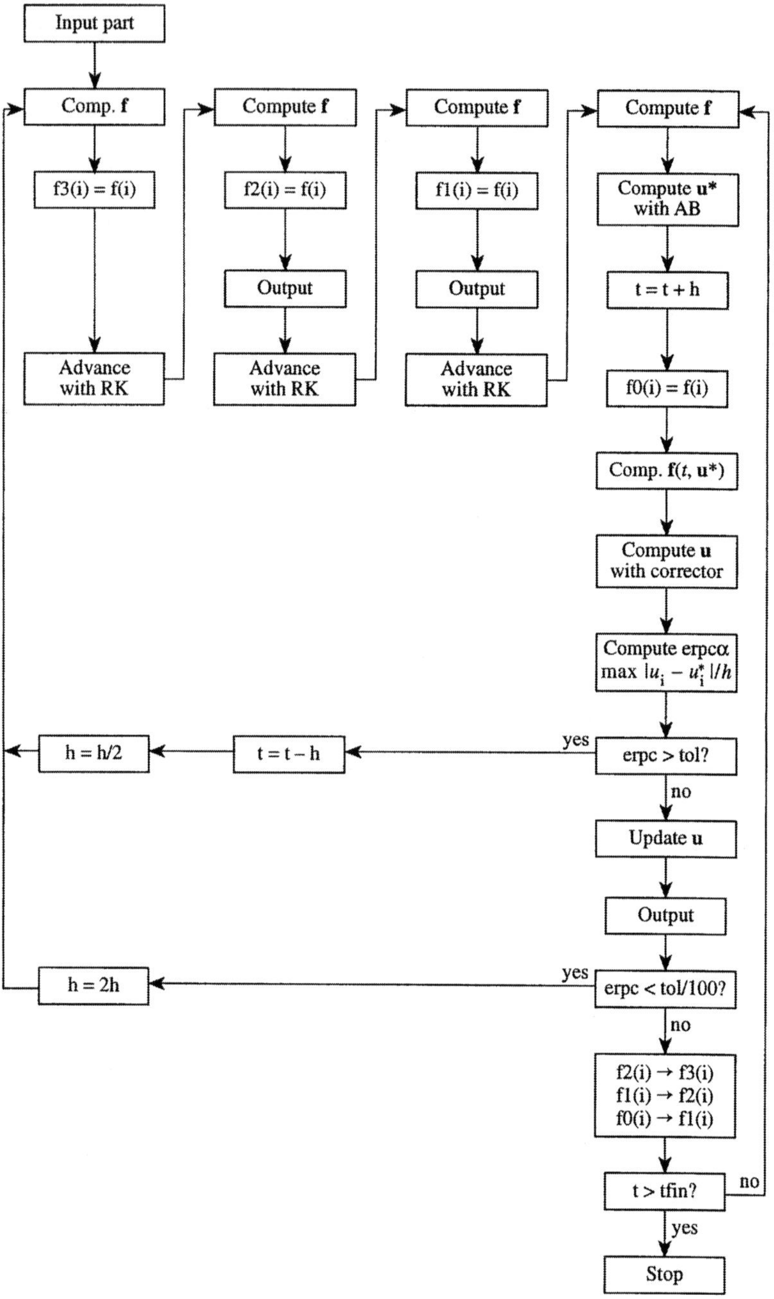

Table 2.9.4 — Predictor-corrector flow chart.

APC4.BAS

```
1 REM ************************************************************
2 REM *      Solution of a system of differential equations      *
3 REM *       using fourth-order predictor-corrector method      *
4 REM ************************************************************
5 REM
10 REM ***********************************************************
12 REM *                     Variables used                      *
13 REM ***********************************************************
14 REM ndim = number of first-order ODE's (<20)
15 REM ngraf = number of plots referring to different IVP's (<10)
16 REM u(ndim) = state variable
17 REM upred(ndim) = predicted value for u
18 REM ucorr(ndim) = corrected value for u
19 REM uan(ndim) = analytic solution
20 REM u0(ngraf,ndim) = initial condition of the state variable
21 REM f(ndim) = right-hand side of differential equations
22 REM f0-f3(ndim) = right-hand side at previous times
23 REM usug(ndim) = suggested values for initial data
24 REM t = time
25 REM h = time step (h2=h/2, h6=h/6, h24=h/24)
26 REM tfin = final time of integration
27 REM tol = tolerance
28 REM dif, er = differences between the outputs of the two
   schemes
29 REM prout = print-out interval
30 REM out$ = flag for print-out option
31 REM prstop$ = flag for inserting a pause after print-outs
32 REM iprint = flag used for switching the printing procedure
33 REM an$ = flag indicating if the analytic solution is available
34 REM phase$ = flag related to the type of graph to plot
35 REM kgraf, kgraf1, kgraf2 = indicate the components to plot
36 REM ptmin, ptmax, pumin, pumax = extrema of graph
37 REM usub(ndim) = variables used for evaluating r.h.s.
38 REM k1-k4(ndim) = variables used for Runge-Kutta
40 REM ***********************************************************
41 REM *              Next variables can be reused               *
42 REM ***********************************************************
43 REM m1, m2 = masses
44 REM kk1, kk2 = rigidity constants
45 REM c1-c4 = constant coefficients of the example
46 REM a1-a4, b1-b4, om1, om2, dom = variables used for the
   analytic solution
200 DIM u(20), uan(20), u0(10, 20), usub(20), upred(20),
   ucorr(20), usug(10)
205 DIM f(20), f0(20), f1(20), f2(20), f3(20), k1(20), k2(20),
   k3(20), k4(20)
```

```
210 REM *********************************************************
220 REM *                      Input lines                      *
230 REM *********************************************************
240 CLS 0
250 PRINT "*****************************************************"
260 PRINT "*   Solution of a system of differential equations   *"
270 PRINT "*    with fourth-order predictor-corrector method    *"
280 PRINT "*****************************************************"
290 PRINT ""
300 PRINT "When more than one quantity is requested, separate
   with a comma"
310 PRINT "Values in square brackets are the suggested values,
   not default"
320 INPUT "Number of differential equations [4]"; ndim
330 INPUT "Time step and final time [.1,100]"; h, tfin
335 INPUT "Required tolerance [.001]"; tol
340 PRINT "Number of superimposed plots with different initial
   data [1]"
350 INPUT " or print-out of the result of a single IVP [0]"; ngraf
360 out$ = "no"
370 IF ngraf = 0 THEN ngraf = 1:  out$ = "yes"
380 IF out$ = "yes" THEN INPUT "Print-out interval [.5]"; prout
390 IF out$ = "yes" THEN INPUT "Do you want to stop at every
   print-out [y]"; prstop$
400 usug(1) = 1:  usug(2) = 2:  usug(3) = 0:  usug(4) = 0
410 FOR igraf = 1 TO ngraf
420 PRINT "Initial data for plot n."; igraf
430 FOR i = 1 TO ndim
440 PRINT "Initial value u0 of component n."; i;
450 IF ndim = 4 THEN PRINT " ["; usug(i); "]";
460 INPUT u0(igraf, i)
470 NEXT i
480 NEXT igraf
490 INPUT "Masses of the test coupled oscillator [1,.01 or .1,1]";
   m1,m2
500 INPUT "Spring rigidity of the test coupled oscillator [1,.01
   or .1,1]"; kk1, kk2
510 c1 = -(kk1 + kk2) / m1
520 c2 = kk2 / m1
530 c3 = kk2 / m2
540 c4 = -kk2 / m2
550 om1 = SQR((-(c1 + c4) + SQR((c1 - c4)^2 + 4 * c2 * c3)) / 2)
560 om2 = SQR((-(c1 + c4) - SQR((c1 - c4)^2 + 4 * c2 * c3)) / 2)
570 dom = om2^2 - om1^2
580 a1 = (c1 * u0(1, 1) + c2 * u0(1, 3) + u0(1, 1) * om2^2) / dom
590 a2 = -(c1 * u0(1, 1) + c2 * u0(1, 3) + u0(1, 1) * om1^2) / dom
600 a3 = (c1*u0(1, 2) + c2*u0(1, 4) + u0(1, 2)*om2^2) / (dom*om1)
610 a4 = -(c1*u0(1, 2)+ c2*u0(1, 4)+ u0(1, 2)*om1^2) / (dom*om2)
```

```
620 b1 = -(c1 + om1 ^2) * a1 / c2
630 b2 = -(c1 + om2 ^2) * a2 / c2
640 b3 = -(c1 + om1 ^2) * a3 / c2
650 b4 = -(c1 + om2 ^2) * a4 / c2
660 an$ = "n"
670 IF ngraf < 2 THEN INPUT "Is the analytical solution known
    [y]"; an$
680 IF out$ = "yes" THEN GOTO 800
690 INPUT "Do you want a phase diagram [y] or the temporal
    diagram [n]"; phase$
700 IF phase$ = "y" THEN INPUT "On which components do you want
    to focus?  (Indicate two of them 1,2,3,...)  [1,2]",
    kgraf1, kgraf2
710 IF phase$ = "n" THEN INPUT "Which component do you want to
    plot?  (Indicate the component 1,2,3,...)  [1]", kgraf
720 ptmin = 0:  ptmax = tfin
730 IF phase$ = "y" THEN INPUT "Minimum and maximum abscissa of
    the graph [-3,3]"; ptmin, ptmax
740 IF phase$ = "y" THEN INPUT "Minimum and maximum ordinate of
    the graph [-2,2 or -4,4]"; pumin, pumax
750 IF phase$ = "n" THEN INPUT "Minimum and maximum ordinate of
    the graph [-3,3]"; pumin, pumax
760 REM ***********************************************************
770 REM *                    End of input lines                   *
780 REM ***********************************************************
800 CLS 0
810 FOR igraf = 1 TO ngraf
820 t = 0
830 iprint = 0
840 REM ***********************************************************
850 REM *                    Set initial data                     *
860 REM ***********************************************************
870 FOR i = 1 TO ndim
880 u(i) = u0(igraf, i)
890 NEXT i
900 IF out$ = "yes" THEN GOSUB 2800:  GOTO 2000
910 REM ***********************************************************
920 REM *                   Initialize graphics                   *
930 REM ***********************************************************
940 KEY OFF: SCREEN 2
950 LOCATE 14, 10:  PRINT ptmin
960 LOCATE 14, 70:  PRINT ptmax
970 LOCATE 13, 7:  PRINT pumin
980 LOCATE 2, 5:  PRINT pumax
990 LOCATE 15, 20
1000 VIEW (80, 10)-(560, 100), , 1
1010 WINDOW (ptmin, pumin)-(ptmax, pumax)
1020 IF phase$ = "n" THEN PSET (0, u0(igraf, kgraf))
```

```
1030 IF phase$ = "y" THEN PSET (u0(igraf, kgraf1), u0(igraf,
   kgraf2))
2000 REM *******************************************************
2010 REM *     Fourth-order Runge-Kutta routine to compute      *
2011 REM *                   starting values                    *
2020 REM *******************************************************
2025 h24 = h / 24:  h2 = h / 2:  h6 = h / 6
2030 FOR i = 1 TO ndim
2040 usub(i) = u(i)
2050 NEXT i
2060 GOSUB 3000
2070 FOR i = 1 TO ndim
2080 f3(i) = f(i)
2090 NEXT i
2100 FOR irk = 1 TO 3
2110 GOSUB 3000
2120 t = t + h2
2130 FOR i = 1 TO ndim
2140 k1(i) = f(i)
2150 usub(i) = u(i) + h2 * k1(i)
2160 NEXT i
2170 GOSUB 3000
2180 FOR i = 1 TO ndim
2190 k2(i) = f(i)
2200 usub(i) = u(i) + h2 * k2(i)
2210 NEXT i
2220 GOSUB 3000
2230 t = t + h2
2240 FOR i = 1 TO ndim
2250 k3(i) = f(i)
2260 usub(i) = u(i) + h * k3(i)
2270 NEXT i
2280 GOSUB 3000
2290 FOR i = 1 TO ndim
2300 k4(i) = f(i)
2310 u(i) = u(i) + h6 * (k1(i) + 2 * (k2(i) + k3(i)) + k4(i))
2320 usub(i) = u(i)
2330 NEXT i
2340 GOSUB 3000
2350 FOR i = 1 TO ndim
2360 IF irk = 1 THEN f2(i) = f(i)
2370 IF irk = 2 THEN f1(i) = f(i)
2380 NEXT i
2390 GOSUB 2800
2400 NEXT irk
2410 REM *******************************************************
2420 REM *               End Runge-Kutta routine                *
2430 REM *******************************************************
```

```
2434 REM ***********************************************************
2435 REM *        Fourth-order predictor-corrector routine       *
2436 REM ***********************************************************
2440 FOR i = 1 TO ndim
2450 usub(i) = u(i)
2460 NEXT i
2470 GOSUB 3000
2474 REM ***********************************************************
2475 REM *                    Predictor part                     *
2476 REM ***********************************************************
2480 FOR i = 1 TO ndim
2490 upred(i) = u(i)+ h24*(55*f(i) - 59 * f1(i)+ 37*f2(i)
   - 9*f3(i))
2500 NEXT i
2504 REM ***********************************************************
2505 REM *                    Corrector part                     *
2506 REM ***********************************************************
2510 t = t + h
2520 FOR i = 1 TO ndim
2530 f0(i) = f(i)
2540 usub(i) = upred(i)
2550 NEXT i
2560 GOSUB 3000
2570 FOR i = 1 TO ndim
2580 ucorr(i) = u(i) + h24 * (9 * f(i) + 19 * f0(i) - 5 * f1(i)
   + f2(i))
2582 NEXT i
2584 REM ***********************************************************
2585 REM *             End predictor-corrector routine           *
2586 REM ***********************************************************
2590 erpc = 0
2600 FOR i = 1 TO ndim
2610 dif = ABS(ucorr(i) - upred(i)) * 19 / (270 * h)
2620 IF dif > erpc THEN erpc = dif
2630 NEXT i
2634 REM ***********************************************************
2635 REM *   Set new time step if er is too large or too small   *
2636 REM ***********************************************************
2640 IF erpc > tol THEN t = t - h:  h = h / 2:  GOTO 2000
2650 FOR i = 1 TO ndim
2660 u(i) = ucorr(i)
2670 NEXT i
2680 GOSUB 2800
2690 IF erpc < tol / 100 THEN h = 2 * h:  GOTO 2000
2694 REM ***********************************************************
2695 REM *               Updating previous values                *
2696 REM ***********************************************************
2700 FOR i = 1 TO ndim
```

```
2710 f3(i) = f2(i)
2720 f2(i) = f1(i)
2730 f1(i) = f0(i)
2740 NEXT i
2750 IF t < tfin - h/10 THEN GOTO 2440
2760 NEXT igraf
2770 INPUT "Different initial conditions? (y or n)"; scr$
2780 IF scr$ = "y" THEN GOTO 400
2790 END
2800 REM ***********************************************************
2801 REM *                    Output subroutine                    *
2802 REM ***********************************************************
2805 REM ***********************************************************
2806 REM *      If known, insert here the analytic solution       *
2807 REM ***********************************************************
2810 IF an$ = "n" THEN GOTO 2835
2815 uan(1) = a1 * COS(om1 * t) + a2 * COS(om2 * t) +
   a3 * SIN(om1 * t) + a4 * SIN(om2 * t)
2820 uan(2) = -a1 * om1 * SIN(om1 * t) - a2 * om2 * SIN(om2 * t) +
   a3 * om1 * COS(om1 * t) + a4 * om2 * COS(om2 * t)
2825 uan(3) = b1 * COS(om1 * t) + b2 * COS(om2 * t) +
   b3 * SIN(om1 * t) + b4 * SIN(om2 * t)
2830 uan(4) = -b1 * om1 * SIN(om1 * t) - b2 * om2 * SIN(om2 * t) +
   b3 * om1 * COS(om1 * t) + b4 * om2 * COS(om2 * t)
2835 IF out$ = "yes" THEN GOTO 2880
2837 REM ***********************************************************
2838 REM *                    Plotting commands                    *
2839 REM ***********************************************************
2840 LOCATE 15, 35:  PRINT "h ="; h
2842 IF phase$ = "y" THEN GOTO 2860
2845 LINE -(t, u(kgraf))
2850 IF an$ = "y" THEN PSET (t, uan(kgraf)):  PSET (t, u(kgraf))
2855 GOTO 2990
2860 LINE -(u(kgraf1), u(kgraf2))
2865 IF an$ = "y" THEN PSET (uan(kgraf1), uan(kgraf2)):  PSET
   (u(kgraf1), u(kgraf2))
2870 GOTO 2990
2875 REM ***********************************************************
2876 REM *                   Print-out commands                    *
2877 REM ***********************************************************
2880 IF t <= prout * iprint - h/10 AND t <> 0 THEN GOTO 2990
2885 PRINT TAB(15); "t="; t
2890 FOR i = 1 TO ndim
2900 PRINT TAB(10); "u("; i; ")="; u(i),
2910 IF an$ = "n" THEN GOTO 2950
2920 IF uan(i) = 0 THEN uan(i) = .0000001
2930 eran = 1 - u(i) / uan(i)
2940 PRINT uan(i), "error="; eran
```

```
2950 NEXT i
2960 IF prstop$ = "y" THEN INPUT scratch
2970 PRINT
2980 iprint = iprint + 1
2990 RETURN
2995 END
3000 REM *****************************************************
3010 REM *          Subroutine for r.h.s. of equations       *
3020 REM *           insert here the function f(t,u)         *
3030 REM *             without using the variables           *
3040 REM *           indicated in lines 10-40, since         *
3050 REM *         they are already used in the program      *
3060 REM *****************************************************
3070 f(1) = usub(2)
3080 f(2) = c1 * usub(1) + c2 * usub(3)
3090 f(3) = usub(4)
3100 f(4) = c3 * usub(1) + c4 * usub(3)
3190 RETURN
3200 END
```

Problems for Chapter 2

Problem 2.1

Consider two rods OA and AB hinged to each other in A so that the system moves in a vertical plane. The rod OA is hinged in O, while the other is linked through a spring (hinged in B) at a horizontal slide at a distance h over the origin.

1. Find the equilibrium configurations and discuss their stability.
2. Linearize the model around the stable configurations and compute the analytic solution.
3. Integrate numerically the nonlinear system performing the tests suggested in Section 2.8.

Note to Problem 2.1 *The reader is referred to the examples of Section 2.2.2 and to the contents of Sections 2.6 and 2.8.*

Problem 2.2

Discuss the stability of the equilibrium configurations of Eqs.(2.7.4), (2.7.5), (2.7.7), and (2.7.8) studying the related bifurcation diagrams.

Problem 2.3

Derive the bifurcation diagrams for Eqs.(2.7.9) and (2.7.10) at fixed α and varying β. Perform the same calculations for fixed β and varying α.

Problem 2.4

Consider the program RK4 and its application to the analysis of the dynamics of the model in Example 2.14. The reader is invited to deal with the following items:

1. Understand the dynamics described by the model by first running the program with one of the suggested values ($k_1 = m_1 = 0.1$, $k_2 = m_2 = 1$ or $k_1 = m_1 = 1$, $k_2 = m_2 = 0.01$) and then decreasing either k_2 or m_2 or both.
2. Increase h and print the values of the state variable. Examine the dependence of the error on the time step and check that for h small it goes as h^4. Plot the solution and observe how accuracy is lost.
3. The spectrum of the differential operator, includes already in the program, has two imaginary eigenvalues corresponding to the two

frequencies. Theory predicts instability as soon as the time step overcomes the value $2\sqrt{2}$ over the maximum of the eigenvalues. This occurs for $k_1 = m_1 = 0.1$, $k_2 = m_2 = 1$ at $h \cong 0.819$ and for $k_1 = m_1 = 1$, $k_2 = m_2 = 0.01$ at $h \cong 2.69$. The following runs are suggested

$$
\begin{array}{cc}
h = 0.81,\ 0.82 & h = 2.68,\ 2.7 \\
\texttt{tfin} = 500 & \texttt{tfin} = 500 \\
m_1 = k_1 = 0.1 & m_1 = k_1 = 1 \\
m_2 = k_2 = 1 & m_2 = k_2 = 0.01 \\
u_1 \in [-20, 20] & u_1 \in [-20, 20] .
\end{array}
$$

The reader can then slowly increase h to observe how the time needed for the instability to build up depends on the distance between the eigenvalue and the contour of the stability region plotted in Fig. 2.8.4b. Observe also that accuracy is lost before instability sets in.

4. Integrate a linear system of ordinary differential equations with real negative eigenvalues, checking for the stability criterium.

5. In Fig. 2.2.1 the quantity $k(x)x$ is plotted versus x. Find a model of spring rigidity satisfying the behavior depicted in Fig. 2.2.1 and study how the quantitative results change when this nonconstant rigidity is considered.

6. Construct programs that use the second- and third-order Runge–Kutta methods. We here explicitly write down the lines relating to the second-order method to substitute for lines 2070-2270

```
t = t + h
FOR i=1 TO ndim
k1(i)=f(i)
urk(i)=u(i)+h*k1(i)
NEXT i
GOSUB 3000
FOR i=1 TO ndim
k2(i)=f(i)
u(i)=u(i)+h2*(k1(i)+k2(i))
NEXT i
```

Do the same for the third-order method. The same exercises proposed in items 2–5 can now be repeated. Furthermore, compare the outputs obtained using the same time step for the different order routines and the time step needed for each program to obtain a desired accuracy. Observe that the second-order Runge–Kutta method is unstable for problems with imaginary eigenvalues, but the stability region is very near the imaginary axis for small time steps (see Fig. 2.8.3d). For this reason the method can still be used for a not too long period of integration. The following runs are suggested:

(`h,tfin`) =(0.1,1000),(0.1,6000),(0.2,1000),(0.3,1000),(0.4,300),
(0.5,200)
with $m_1 = k_1 = 1$, $m_2 = k_2 = 0.01$, $u_1 \in (-30, 30)$.

Problem 2.5
Repeat the previous exercise for the program AB4. In addition, compare the time step h required by RK4 and by AB4 to reach a given accuracy.

Problem 2.6
Referring to Examples 2.5 and 2.10, run the program VANDRPOL using increasing values of $\alpha = -\beta$ starting from $\alpha = 1$, so that the nonlinear effects become more and more important. Observe how the time step changes during integration. For $\alpha = -\beta > 5$ the value of the ordinate of the diagram has to be adjusted say to $u_2 \in [-10, 10]$ for $\alpha = 5$ and to $u_2 \in [-15, 15]$ for $\alpha = 10$. Better temporal diagrams are obtained for `tfin`= 50.

Referring to Section 2.7.5, add a periodic forcing term.

Problem 2.7
Integrate Van der Pol equation with APC4 after a proper choice of the initial conditions and parameters that characterize the model.

Problem 2.8
Consider some of the nonlinear differential systems introduced in this chapter, choose the ODE solver, and compare the theoretical results given in the relative examples with the simulation. Interesting simulations are:

1. The pendulum problem dealt with in Section 2.2.2. Does the singularity disturb?
2. The particle dynamics problem solved analytically in Example 2.1. Observe the difficulties that arise, for instance, when $\alpha = -2$ in jumping over the singularities.
3. The heart model introduced in Example 2.9. Add a triggering mechanism like the one in Fig. 2.7.11.
4. The biological models by Hodgkin and Huxley and by Fitzhugh introduced in Section 2.2.5 and developed also in Example 2.12.

Compare the time required to complete the integration of an initial-value problem related to one of the previous models at a given accuracy using RK4, AB4, RKF, and APC4.

Chapter 3

CONTINUOUS MODELS

3.1 Introduction

Modelling in continuum physics is a fundamental chapter in the history of mathematical physics and, in general, in the history of sciences. In fact, the development of mathematical physics in past times has been essentially based on the formulation of the models of continuum systems: Fluid dynamics, elasticity, continuum statistical theories, etc. These models are nowadays the fundamental background of the mathematical models of natural, applied, and technological sciences.

As we have already seen in Chapter 1, ***continuous dynamic models*** are characterized by the fact that the state variable $\mathbf{u}$ depends on time and space variables. In general this dependence is continuous and one can write

$$\mathbf{u} = \mathbf{u}(t, \mathbf{x}) : \quad [0, T] \times \mathcal{D} \longmapsto \mathbb{R}^n , \tag{3.1.1}$$

where $\mathcal{D} \subseteq \mathbb{R}^3$ is the domain of the space variable and $\mathbf{u}$ is an n-dimensional vector.

If the state variable is a scalar and the space variable is in dimension one, then one simply has

$$u = u(t, x) : \quad [0, T] \times [a, b] \longmapsto \mathbb{R} . \tag{3.1.2}$$

In static continuous models the dependence on time drops and the state variable is

$$\mathbf{u} = \mathbf{u}(\mathbf{x}) : \quad \mathcal{D} \longmapsto \mathbb{R}^n . \tag{3.1.3}$$

Generally, continuous models are formulated in terms of partial differential equations. This is not, however, a general rule and other types of equations may be encountered: Finite differences, integral equations, integrodifferential equations, and so on. Some examples will be given here and in the Chapter 4.

Examples of continuous dynamic models were presented in Chapter 1, see Examples 1.1, 1.2, 1.4, and 1.6. The mathematical modelling of the physical systems dealt with in Chapter 1 will be revisited in this chapter within a more general framework. However, it has to be clear that the study of continuous models is a huge subject that actually would deserve several volumes. The present chapter therefore gives only some general guidelines about the modelling procedure, and applies them to some sample (however, not always simple) systems.

Keeping in mind all the modelling aspects, the reader can then choose the continuous system he wants to study in detail. To understand it fully, to master the operative methods, and to keep updated on the subject will require his best research effort.

This chapter is organized in a fashion similar to Chapter 2. The first part deals with mathematical modelling methods and model classifications; the second part with the formulation of mathematical problems and with the methods for the analysis of problems.

Mathematical modelling of continuous systems follows the derivation lines described in Chapter 2. The first step consists in selecting the state variable, then suitable equilibrium and conservation equations can be used in order to obtain the evolution equation for the state variable. One needs an equation (either equilibrium or conservation) for each component of the state variable.

The modelling procedure requires, as for discrete models, the use of phenomenological constitutive equations suitable, as we shall see, for describing the physical behavior of the matter that constitutes the physical system.

Mathematical problems that refer to the analysis of continuous dynamic models are generally initial-boundary-value problems. Namely, the evolution equation to be solved needs additional information on the initial and boundary conditions, that is, the values of the state variable at the initial time and at the boundary of the space variables. How to give these conditions is not as simple as it is for discrete systems.

Some mathematical methods will then be proposed in this chapter in order to provide the reader with effective methods for obtaining quantitative results for a large variety of nonlinear problems. Finally, as in Chapter 2, some computer programs will be given as starting point for exercises.

Having completed this preliminary excursus, the contents of this chapter can now be reported and discussed in detail.

The second section deals with the general procedure of mathematical modelling. The procedure is then applied to deducing a simple model describing the small vibrations of an elastic string in Section 3.3, to the study of continuum systems in Section 3.4, to the derivation of the Maxwell's model of electromagnetism in Section 3.5, and to the study of a biological system in Section 3.6.

The seventh section then provides a classification of models based on the nature of the phenomena modelled as well as on the structure of the equation. This section also deals with the analysis of the fundamental qualitative difference between hyperbolic, parabolic, and elliptic models.

Section 3.8, similar to the corresponding section of Chapter 2, deals with the formulation of mathematical problems, with some concise indications on the correct formulation and the well-posedness of problems.

The sections that follow are devoted to mathematical methods to obtain quantitative results with particular attention to nonlinear problems. In detail, Section 3.9 deals with finite difference methods, while Section 3.10 deals with collocation methods. Section 3.11 provides a brief presentation of domain decomposition methods. Finally, the last section deals with some applications and proposes some scientific programs related to the solution of specific problems.

It needs to be understood that although the reader will find, in this chapter, a large variety of information on effective methods, the content is not and does not claim to be complete. Nevertheless, several bibliographical indications will be given in order to cover the field of mathematical methods, which will not be specifically dealt with in this chapter.

As we have mentioned, the cultural background of this chapter is the classical mathematical physics [TaS, RIC] and the methods of classical analysis applied to the solution of problems in mathematical physics. Useful fundamental books are, among others, the classical volume by Courant and Hilbert [CaH] for the analysis of linear problems and that by Lions and Magenes [LaM] for several interesting nonlinear initial-boundary-value problems.

A deeper analysis of continuum mechanics in its general foundation is given by Truesdell [TRU] and Gurtin [GUT]. It is, however, easier to look directly at books dealing with specific fields in continuum mechanics, such as Chorin and Marsden [CaM] for fluid dynamics (especially inviscid) and Joseph and Renardy [JaR] for two-fluid dynamics, Astarita and Marrucci [AaM] for non-Newtonian models, and Gurtin [GUR] for solid mechanics. As far as heat conduction models are concerned, an easy review of the models available in the literature is given by Joseph and Preziosi [JaP]. A simple introduction to electromagnetic models can be found in Cottingham and Greenwood [CaG] and Müller [MUL].

At present, the literature on mathematical models in biology is vast and it can even be difficult for an applied mathematician who is taking his first steps in the field to orient himself in the jungle of the large number of articles published in this area.

We can indicate, among several others, the pioneer collection of articles, edited by Rosen [ROS] on the foundations of mathematical biology. These articles cover a broad range of topics and constitute a fundamental source of information in the field. Additional useful readings are the books by Lin and Segel [LaS] and by Segel [SEG], which deal with several methodological aspects of mathematical modelling in the natural sciences and of the interplay between mathematics and the analysis of mathematical models. A complete discussion of modelling in biological sciences can be found in the book by Murray [MUR], which covers a very broad range of interests.

An easy introduction to the fundamentals of partial differential equations is given by Duchateau and Zachmann [DaZ], while a more advanced treatment with attention to applications is given by Courant and Hilbert [CaH] and by Jeffreys and Jeffreys [JaJ].

As far as numerical methods are concerned Duchateau and Zachmann [DaZ] and Sewell [SEW] provide an easy introduction to finite difference methods. A deeper approach can be found in Strikwerda [STI] and Forsythe and Wasow [FaW]. Collocation and spectral methods are instead studied in Gottlieb and Orszag [GaO] and in Canuto, et al. [CaA], where a particular attention is paid to their application to fluid-dynamics models. In the same book an introduction to domain decomposition methods can also be found. The bibliography on this last subject is, however, expanding quickly, therefore the reader should look at articles published in specific journals. A recent collection is the proceedings edited by Chan [CHA]. Among other articles, we suggest for their simplicity and/or their review attitude those by Quarteroni and coauthors [GaQ], [GQS], and [QUA].

There are finally several books dealing with the simulation of specific models. For instance, the volumes by Hirsh [HIR] lean to the study of fluid-dynamic problems, while Anderson, Tannehill, and Pletcher [ATP] also deals with heat transfer problems.

3.2 Mathematical Modelling

This section deals with some methodological aspects of the mathematical modelling of continuous systems. In the next sections, some modelling procedures are proposed and then applied to the study of a number of systems chosen as an example.

As usual, the derivation of a mathematical model suitable for describing a certain physical system requires a preliminary phenomenological observation both of the physical system and of its connection with the outer environment. Then the modelling procedure can follow these steps:

Step 1 Select the state variable.

Step 2 Model the interaction between the actual system and the outer environment.

Step 3 Define the equilibrium, conservation, and/or balance equations, etc., which involve the state variable.

Step 4 Model the material behavior of the system by means of phenomenological models, which will be called ***constitutive laws***.

Step 5 Derive, using the equations dealt with in Steps 3 and 4, a mathematical model that consists of a suitable set of equations, which describes the time and space evolution of the state variable.

Then, once the mathematical model has been formulated, the usual procedure for its analysis can be organized: Formulation of proper initial and boundary conditions, qualitative analysis, solution of mathematical problems, and validation of the model.

Although the scheme does not include all possibilities and the examples given in the next sections will not cover all conceivable types of modelling procedures, the reader still will receive from their treatment a quantity of information sufficient to achieve a general understanding of the methodology.

As mentioned in Step 3, to derive a model several lines can be followed:

- Observe the quantities that set a system in equilibrium (in a material reference frame) and derive of the linked equilibrium equations;
- Observe the quantities that are preserved during the evolution of the system and derive of the linked conservation equations;
- Observe the causes that influence certain overall quantities and derive of the linked balance equations;
- Simule the system by using simpler physical models and/or well-known mathematical equations that possess properties similar to those that characterize the system to be modelled and simulated;
- Model the microscopic interactions in order to define the statistical behavior of complex systems and the equations that govern the evolution of some macroscopic (or ensemble) quantities (e.g., statistical mechanics, population modelling, rarefied gas-dynamics, fluid-solid mixtures, bubbly liquids, porous media, etc.).

Some examples of models and of the related modelling techniques will be given.

In Section 3.3 a simple model for the vibration of an elastic string will serve as an example of equilibrum equations. In Section 3.4 an introduction to continuous mechanics will be given as an example of conservation and balance equations. In Section 3.5 Maxwell's model of electromagnetism will be dealt with as an example of a system whose modelling equations may have different origin. Finally Section 3.6 will deal with the direct simulation of a biological model, namely the propagation of impulses along a nerve fiber. This is a system so complex that only direct simulation by known mathematical equations or the use of well-known simple physical systems seems practical, at the moment.

We will not give here any example of statistical modelling, since their treatment needs deep concepts of probability theory. Most models in physics have this origin (e.g., statistical mechanics models, Boltzmann equation, models in two-phase fluid-dynamics, etc.). Due to this feature, an introduction to statistical models will be presented in Chapter 4.

3.3 Modelling the Vibration of an Elastic String

As an example of modelling procedure by means of equilibrium equations, consider, with reference to Fig. 3.3.1, an elastic string fixed at both its extrema. The modelling procedure for the dynamics of the system can be based upon the following assumptions:

- The string is held in A and B by a strong tension directed along the string. This implies that gravity can be neglected and, therefore, that the string takes at equilibrium an "essentially" straight configuration, identified by the unit vector $\mathbf{i}$.
- The string is subjected to small displacements from the equilibrium configuration.
- The motion of the string is localized on a plane, identified by the unit vectors $\mathbf{i}$ and $\mathbf{j}$ and every point of the string moves along $\mathbf{j}$, i.e., perpendicularly to the equilibrium state.
- The independent variables are the time t and the location x along the string.
- The state variable is the perpendicular displacement

$$u = u(t, x) : \quad [0, T] \times [0, \ell] \longmapsto \mathbb{R} . \tag{3.3.1}$$

- A fluid-dynamic drag is exerted on the string by the outer environment. The related phenomenological model states that the force acting upon the string element Δx is

$$\mathbf{F} = -h \left(\left| \frac{\partial u}{\partial t} \right| \right) \frac{\partial u}{\partial t} \Delta x \, \mathbf{j} , \tag{3.3.2}$$

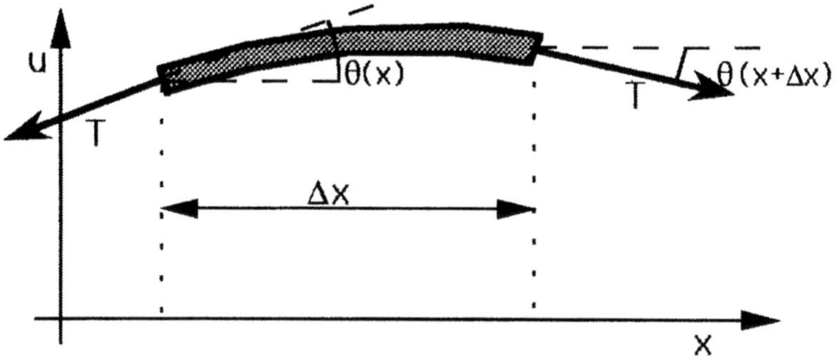

Figure 3.3.1 — Representation of a vibrating string.

where h is a positive function of the velocity modulus.

- A constant internal tension $T(x) = T_0$ can be assumed in the framework of linear elasticity.

Referring then to Fig. 3.3.1, if θ is the slope of the string, then the component of the tension along $\mathbf{j}$ acting over the element Δx of the string is

$$\Delta T = T_0 \sin\theta(x + \Delta x) - T_0 \sin\theta(x) . \tag{3.3.3}$$

Since

$$\sin\theta \cong \frac{\Delta u}{\sqrt{\Delta u^2 + \Delta x^2}} \cong \frac{\Delta u}{\Delta x} \cong \frac{\partial u}{\partial x} , \tag{3.3.4}$$

for small deformations, we can obtain the approximation

$$\Delta T \cong T_0 \left[\frac{\partial u}{\partial x}(x + \Delta x) - \frac{\partial u}{\partial x}(x) \right] \cong T_0 \Delta x \frac{\partial^2 u}{\partial x^2}(x) . \tag{3.3.5}$$

In conclusion, the element Δx is subject to the following actions:

- $T_0 \dfrac{\partial^2 u}{\partial x^2} \Delta x$ is the action of the internal tension;
- $-\rho \dfrac{\partial^2 u}{\partial t^2} \Delta x$ is the inertial force;
- $-h \left(\left| \dfrac{\partial u}{\partial t} \right| \right) \dfrac{\partial u}{\partial t} \Delta x$ is the fluid-dynamic drag.

In the material reference frame, these forces set the system in equilibrium (Newton's mechanical model or, more properly, d'Alembert's principle). This yields the following continuous dynamic model

$$\frac{\partial^2 u}{\partial t^2} = c^2 \frac{\partial^2 u}{\partial x^2} - \frac{1}{\rho} h \left(\left| \frac{\partial u}{\partial t} \right| \right) \frac{\partial u}{\partial t}, \tag{3.3.6}$$

where

$$c = \sqrt{\frac{T_0}{\rho}}. \tag{3.3.7}$$

REMARK 3.3.1 The model was derived under the assumption of small deformations of the string. Consequently the validity of the model is limited to the case of small vibrations. If the model is used to describe large deformations, then unreliable descriptions of the physical reality would follow. ■

3.4 Mathematical Models of Continuum Mechanics

As a more complex model that uses conservation and balance equations, we will consider the motion of a continuous body. We point out that the first simplification in the derivation of this mathematical model is just the ***continuum assumption***. This is certainly an approximation, since intermolecular distances are always positive quantities. On the other hand, when these distances are sufficiently small with respect to the characteristic dimensions of the body, the continuum hypothesis becomes reasonable and can be regarded as an acceptable simplification. If this hypothesis is no longer applicable, e.g., for a rarefied gas, then mathematical models have to be derived at a molecular scale.

Consider, with reference to Fig. 3.4.1, a continuous system which at the initial time occupies the region $\mathcal{B}_0$. Referring to a fixed reference frame, a particle of the system can be identified by the position $\mathbf{X}$ occupied at the initial time. At time t the continuous system has deformed to a new configuration $\mathcal{B}_t$. In particular the marked particle

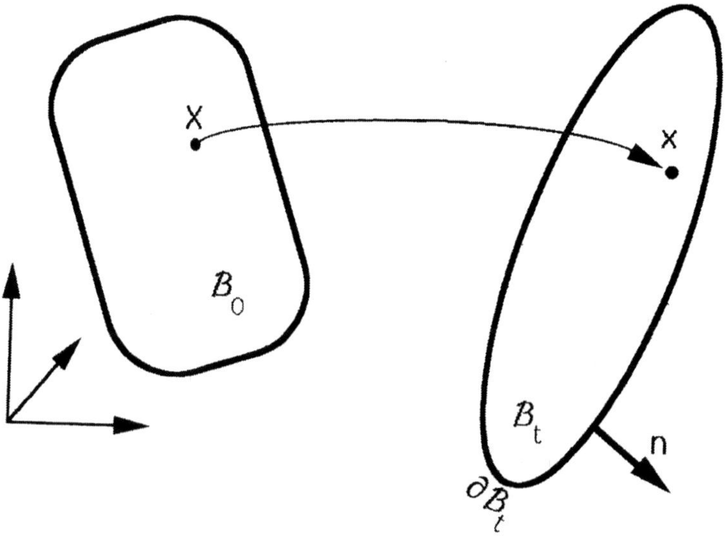

Figure 3.4.1 — Representation of a continuous system.

X has moved to a new position **x**. The motion can then be described by giving $\mathbf{x} = \mathbf{x}(\mathbf{X}, t)$. The velocity of the particle **X** is then defined by

$$\mathbf{v} = \frac{d\mathbf{x}}{dt} = \frac{\partial}{\partial t}\mathbf{x}(\mathbf{X}, t) .$$

The evolution of the continuous system can be described using two different points of view, which are strongly related to the choice of independent variables.

DEFINITION 3.4.1 Lagrangian and Eulerian coordinates
Choosing **X** *and* t *as independent variables means referring to the initial configuration and following the vicissitudes of the particle* **X** *throughout its motion. One has then chosen a so-called* ***Lagrangian*** *point of view. Choosing* **x** *and* t *as independent variables means referring to the present configuration and looking at what happens in the point* **x** *as the system evolves. One has then chosen a so-called* ***Eulerian*** *point of view.*

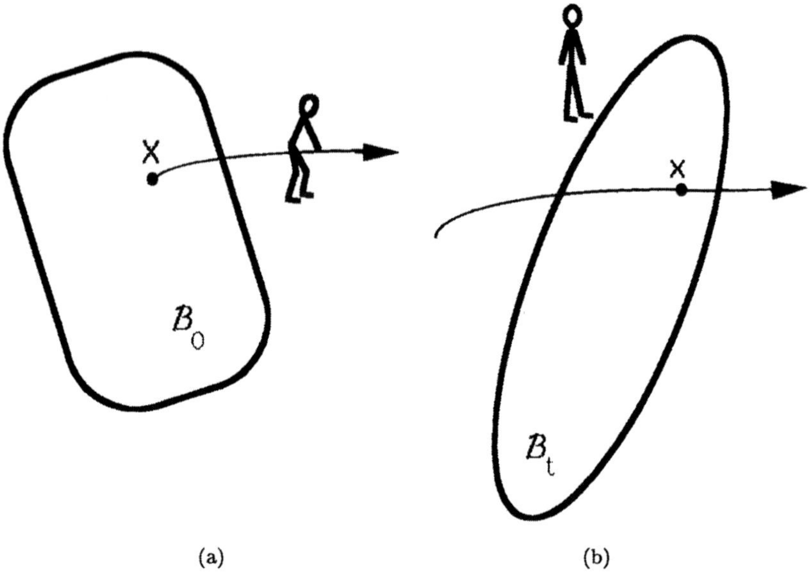

Figure 3.4.2 — (a) Lagrangian and (b) Eulerian points of view.

To describe the evolution of a quantity $f(t, \mathbf{x})$, one has to compute the ***material time derivative***

$$\frac{d}{dt} f(t, \mathbf{x}) = \frac{\partial}{\partial t} f(t, \mathbf{x}) + \mathbf{v}(t, \mathbf{x}) \cdot \nabla f(t, \mathbf{x}) \,, \tag{3.4.1}$$

where

$$\nabla f = \frac{\partial f}{\partial x} \mathbf{e}_1 + \frac{\partial f}{\partial y} \mathbf{e}_2 + \frac{\partial f}{\partial z} \mathbf{e}_3$$

denotes the gradient of f.

3.4.1 General procedure

Let $p = p(t, \mathbf{x})$ be a certain scalar physical quantity, i.e., a scalar macroscopic observable referred to the unit volume. We want to study the evolution of the global quantity

$$P(t) = \int_{\mathcal{B}_t} p(t, \mathbf{x}) \, dV \,, \tag{3.4.2}$$

obtained by integrating p over the material volume $\mathcal{B}_t$.

The temporal variation of P can be due to a "flux" φ of something through the boundary $\partial\mathcal{B}_t$ and to a "source" term, so that one has

$$\frac{dP}{dt} = \frac{d}{dt}\int_{\mathcal{B}_t} p(t,\mathbf{x})\,dV = \int_{\partial\mathcal{B}_t} \varphi(t,\mathbf{x};\mathbf{n})d\sigma + \int_{\mathcal{B}_t} s(t,\mathbf{x})\,dV\,, \quad (3.4.3)$$

where $\mathbf{n}$ is the outward normal to the boundary $\partial\mathcal{B}_t$ of $\mathcal{B}_t$.

Recalling some basic theorems it is possible to deduce from the integral form of the balance equation (3.4.3) a more useful differential form. The first theorem that will be used to accomplish this project is the so-called ***transport theorem***, which deals with the computation of the time derivative of an integral over a material volume. The second theorem, usually called ***Cauchy's fundamental theorem of continuum mechanics***, gives an important characterization of the "flux" term. Finally, the third theorem is the classical ***Gauss divergence theorem***, which relates an integral over a closed surface with an integral over the enclosed volume.

THEOREM 3.4.1 *(Transport Theorem)*
If p is a continuously differentiable function and $\mathcal{B}_t$ is a closed and connected material volume, then

$$\frac{d}{dt}\int_{\mathcal{B}_t} p(t,\mathbf{x})\,dV = \int_{\mathcal{B}_t}\left(\frac{dp}{dt} + p\nabla\cdot\mathbf{v}\right)dV\,. \quad (3.4.4)$$

■

THEOREM 3.4.2 *(Cauchy's Theorem of Continuum Mechanics)*
If $p(t,\mathbf{x})$, $\varphi(t,\mathbf{x};\mathbf{n})$, $s(t,\mathbf{x})$, and $\mathbf{v}(t,\mathbf{x})$ in Eqs.(3.4.3) and (3.4.4) are bounded and regular, then φ is linear in $\mathbf{n}$

$$\varphi(t,\mathbf{x};\mathbf{n}) = \mathbf{\Psi}(t,\mathbf{x})\cdot\mathbf{n}\,, \quad (3.4.5)$$

where $\mathbf{\Psi}$ is the vector with components $\Psi_i(t,\mathbf{x}) = \varphi(t,\mathbf{x};\mathbf{e}_i)$. ■

THEOREM 3.4.3 *(Gauss Divergence Theorem)*
Let $\mathcal{B}_t$ be a bounded region with smooth boundary surface $\partial\mathcal{B}_t$ with unit outward normal $\mathbf{n}$. If $\mathbf{\Psi}$ is continuous in $\mathcal{B}_t$ and piecewise differentiable in $\mathcal{B}_t - \partial\mathcal{B}_t$, then

$$\int_{\partial\mathcal{B}_t} \mathbf{\Psi}\cdot\mathbf{n}\,d\sigma = \int_{\mathcal{B}_t} \nabla\cdot\mathbf{\Psi}\,dV\,,$$

where

$$\nabla \cdot \boldsymbol{\Psi} = \sum_{i=1}^{3} \frac{\partial \Psi_i}{\partial x_i} = \frac{\partial \Psi_x}{\partial x} + \frac{\partial \Psi_y}{\partial y} + \frac{\partial \Psi_z}{\partial z}$$

is the divergence of $\boldsymbol{\Psi}$*.* ■

Applying these theorems to Eq.(3.4.3) yields

$$\frac{dp}{dt} + p\nabla \cdot \mathbf{v} = \nabla \cdot \boldsymbol{\Psi} + s\,, \tag{3.4.6}$$

or using (3.4.1)

$$\frac{\partial p}{\partial t} + \nabla \cdot (p\mathbf{v}) = \nabla \cdot \boldsymbol{\Psi} + s\,. \tag{3.4.7}$$

The same line can be followed for vector quantities $\mathbf{p}$. In this case, the vector integral form

$$\frac{d}{dt}\int_{\mathcal{B}_t} \mathbf{p}(t,\mathbf{x})\,dV = \int_{\partial\mathcal{B}_t} \boldsymbol{\Phi}(t,\mathbf{x};\mathbf{n})d\sigma + \int_{\mathcal{B}_t} \mathbf{s}(t,\mathbf{x})\,dV\,. \tag{3.4.8}$$

leads to the differential form

$$\frac{d\mathbf{p}}{dt} + \mathbf{p}\nabla \cdot \mathbf{v} = \nabla \cdot \boldsymbol{\Psi} + \mathbf{s}\,, \tag{3.4.9}$$

where $\boldsymbol{\Psi}$ is now a tensor of rank two with components

$$\Psi_{ij}(t,\mathbf{x}) = \mathbf{e}_i \cdot \boldsymbol{\Phi}(t,\mathbf{x};\mathbf{e}_j)$$

and

$$\nabla \cdot \boldsymbol{\Psi} = \sum_{i=1}^{3}\left(\sum_{j=1}^{3} \frac{\partial \Psi_{ij}}{\partial x_j}\right)\mathbf{e}_i\,.$$

3.4.2 Conservation of mass

The mechanical evolution of a continuum body is described when we succeed in determining the evolution of its mass density

$$\rho(t,\mathbf{x}) = \frac{dm}{dV}$$

and of its the velocity field $\mathbf{v}(t, \mathbf{x})$.

What do we know? First, we know that if $\mathcal{B}_t$ is a material volume (and there is no creation or destruction of mass, as is usually the case), then the mass

$$M(t) = \int_{\mathcal{B}_t} \rho(t, \mathbf{x}) \, dV \tag{3.4.10}$$

contained in $\mathcal{B}_t$ has to be preserved ***(conservation of mass)***, that is,

$$\frac{dM}{dt} = 0 \, .$$

Referring then to Eq.(3.4.3), $\varphi = s = 0$, therefore Eq.(3.4.7) reduces to

$$\frac{\partial \rho}{\partial t} + \nabla \cdot (\rho \mathbf{v}) = 0 \, . \tag{3.4.11}$$

REMARK 3.4.1 If we know that the material is incompressible, as it is essentially for all fluids and solids, then we have that $\rho = \rho_0$ is constant, which leads to the ***incompressibility*** condition

$$\nabla \cdot \mathbf{v} = 0 \, . \tag{3.4.12}$$

■

3.4.3 Momentum balance

The second thing we know is that the forces acting on the body will change its momentum

$$\mathbf{Q}(t) = \int_{\mathcal{B}_t} \rho(t, \mathbf{x}) \mathbf{v}(t, \mathbf{x}) \, dV$$

and its angular momentum

$$\mathbf{K}_0(t) = \int_{\mathcal{B}_t} (\mathbf{P}(t, \mathbf{x}) - \mathbf{0}) \times \rho(t, \mathbf{x}) \mathbf{v}(t, \mathbf{x}) \, dV \, .$$

A basic assumption in continuum mechanics is that the forces are distributed in the body and can be distinguished in ***body forces*** $\rho \mathbf{b}$, which act on the body from outside, e.g., the gravitational force, and in contact forces $\mathbf{t}$, which act on the boundary $\partial \mathcal{B}_t$ of the body. These are molecular forces that particles in contact with $\partial \mathcal{B}_t$ exert on the body.

REMARK 3.4.2 The previous modelling assumptions, though seemingly natural, are restrictive. In fact, among other things, we may observe the following:

- Internal gravitational forces are not considered;
- As far as contact forces are concerned, it is assumed that the force acting on the surface element $d\sigma$ is $\mathbf{t}\,d\sigma$ (***simple continuum assumption***); a better approximation would be to include a momentum $\mathbf{m}\,d\sigma$ acting on the surface element (***polar continuum***);
- Because of the small radius of action of molecular forces, it is assumed that the force applied to the surface element referred to $\mathbf{x}$ is linked only to the particles immediately adjacent to $\mathbf{x}$, so that one can assume that $\mathbf{t}$ depends only on the point $\mathbf{x}$ and on the normal to the surface element in $\mathbf{x}$, that is $\mathbf{t} = \mathbf{t}(t, \mathbf{x}; \mathbf{n})$. ■

In this case, in Eq.(3.4.8) $\mathbf{p} = \rho\mathbf{v}$, $\boldsymbol{\Phi} = \mathbf{t}$, and $\mathbf{s} = \rho\mathbf{b}$. Thus from (3.4.9) we have

$$\frac{d}{dt}(\rho\mathbf{v}) + \rho\mathbf{v}\nabla\cdot\mathbf{v} = \rho\mathbf{b} + \nabla\cdot\mathbf{T}\,, \tag{3.4.13}$$

where $\mathbf{T}$ is the so-called ***stress tensor.***

Using the conservation of mass (3.4.11) and Eq.(3.4.1), Eq.(3.4.13) can be reduced to

$$\rho\left(\frac{\partial\mathbf{v}}{\partial t} + \mathbf{v}\cdot\nabla\mathbf{v}\right) = \rho\mathbf{b} + \nabla\cdot\mathbf{T}\,, \tag{3.4.14}$$

where

$$\mathbf{v}\cdot\nabla\mathbf{v} = \sum_{i=1}^{3} v_i\frac{\partial\mathbf{v}}{\partial x_i} = v_x\frac{\partial\mathbf{v}}{\partial x} + v_y\frac{\partial\mathbf{v}}{\partial y} + v_z\frac{\partial\mathbf{v}}{\partial z}\,.$$

By the balance of angular momentum it can be proved that $\mathbf{T}$ is a symmetric tensor. This is a direct consequence of the simple continuum hypothesis. It does not hold in polar continuum mechanics.

3.4.4 Energy balance

If one is also interested in describing thermodynamic phenomena, then one has to introduce another state variable capable of describing the energy of the system.

First, we can distinguish between two types of energies, a mechanical one

$$\mathcal{T}(t) = \int_{\mathcal{B}_t} \frac{1}{2}\rho(t,\mathbf{x})v^2(t,\mathbf{x})\, dV$$

linked to the macroscopic motion of the body, and an internal one

$$\mathcal{E}(t) = \int_{\mathcal{B}_t} \rho(t,\mathbf{x})\epsilon(t,\mathbf{x})\, dV$$

linked to the microscopic motion of the molecules of the body. In principle, and this is in fact the case, we cannot discount the possibility that part of the mechanical energy of the body is transferred to mechanical energy of its molecules (the internal energy) and *vice versa*. One has then to study the temporal variation of their sum. Certainly, the power generated by the forces acting on the body

$$\Pi_m(t) = \int_{\mathcal{B}_t} \rho\mathbf{b}\cdot\mathbf{v}\, dV + \int_{\partial\mathcal{B}_t} \mathbf{t}\cdot\mathbf{v}\, d\sigma$$

will play a crucial role, but the energy of the system can also change because of conduction through the boundary and radiation; so we have also to consider the term

$$\Pi_q(t) = \int_{\mathcal{B}_t} \rho r\, dV + \int_{\partial\mathcal{B}_t} h\, d\sigma\ .$$

Hence the balance of total energy

$$\frac{d}{dt}(\mathcal{T}+\mathcal{E}) = \Pi_m + \Pi_q$$

yields

$$\frac{d}{dt}\int_{\mathcal{B}_t} \rho\left(\frac{v^2}{2}+\epsilon\right) dV = \int_{\partial\mathcal{B}_t} (\mathbf{t}\cdot\mathbf{v}+h)\, d\sigma + \int_{\mathcal{B}_t} \rho(\mathbf{b}\cdot\mathbf{v}+r)\, dV\ , \quad (3.4.15)$$

which is sometimes referred to as ***first principle of thermodynamics***.

Computing the mechanical energy term from the balance of momentum (3.4.13), Eq.(3.4.15) reduces to

$$\frac{d}{dt}\int_{\mathcal{B}_t} \rho\epsilon\, dV = \int_{\mathcal{B}_t} \mathbf{T}:\mathbf{D} + \int_{\partial\mathcal{B}_t} h\, d\sigma + \int_{\mathcal{B}_t} \rho r\, dV\ , \qquad (3.4.16)$$

where the first term on the right-hand side of the equality sign is defined as

$$\mathbf{T}:\mathbf{D}=\sum_{i,j=1}^{3} T_{ij}D_{ij}$$

and

$$D_{ij}=\frac{1}{2}\left(\frac{\partial v_i}{\partial x_j}+\frac{\partial v_j}{\partial x_i}\right)$$

are the components of the ***rate of strain tensor*** **D**.

The usual procedure yields the following balance of internal energy

$$\rho\left(\frac{\partial \epsilon}{\partial t}+\mathbf{v}\cdot\nabla\epsilon\right)=\mathbf{T}:\mathbf{D}-\nabla\cdot\mathbf{q}+\rho r\,, \tag{3.4.17}$$

where $\mathbf{q}$ is the heat flux related to h through $h=-\mathbf{q}\cdot\mathbf{n}$, (recall Theorem 3.4.2), r is the energy supply, and $\mathbf{T}:\mathbf{D}$ represents the increase of internal energy by friction.

3.4.5 Constitutive laws

The system of balance equations

$$\begin{cases} \frac{\partial \rho}{\partial t}+\nabla\cdot(\rho\mathbf{v})=0 \\ \rho\left(\frac{\partial \mathbf{v}}{\partial t}+\mathbf{v}\cdot\nabla\mathbf{v}\right)=\rho\mathbf{b}+\nabla\cdot\mathbf{T}\,, \qquad \text{with } \mathbf{T}=\mathbf{T}^T\,, \end{cases} \tag{3.4.18}$$

which is in charge of describing the mechanical evolution of a continuum body cannot actually fulfill its scope in this form, since it consists of only four equations in the ten unknowns ρ, $\mathbf{v}$, and T_{ij} with $i\le j$. The introduction of the energy balance would even worsen the situation since it introduces four more unknowns, ϵ and $\mathbf{q}$, and only one equation.

This is, however, not surprising since Eq.(3.4.18), at the moment, can describe the motion of a fluid as well as that of an elastic body or of a plastic material.

Of course, all these materials will respond very differently to the same strain, for instance a pinch. The difference lies in how they respond to the deformations. Considering Eq.(3.4.18) we have then to express the dependence of the stress tensor $\mathbf{T}$ on the state variables ρ and $\mathbf{v}$, the so-called ***constitutive laws***. When this is accomplished the system has the same number of equations and unknowns. We have then ***closed*** the system and can start considering the model for simulation.

Of course, there is a large variety of materials and an even larger variety of constitutive laws (sometimes there are several laws modelling the same material under different conditions). A first selection can be operated by taking into account the general features of the constitutive laws, but the final choice can only be operated via the validation procedure.

If one is dealing with a process with thermodynamic effects, then the temperature θ has to be included as a state variable and the energy balance should also be considered. So the system is closed when the functional relation

$$\begin{aligned} \mathbf{T} &= \mathcal{T}(\rho, \mathbf{v}, \theta) \\ \epsilon &= \mathcal{E}(\rho, \mathbf{v}, \theta) \\ \mathbf{q} &= \mathcal{Q}(\rho, \mathbf{v}, \theta) \end{aligned} \tag{3.4.19}$$

are given.

The constitutive equations (3.4.19) must, however, satisfy the second principle of thermodynamics, which acts on their form as a restriction to be satisfied for any thermodynamic process. However, this topic is too specialized and we will simply give some examples of constitutive equations in the pure mechanical case and in the case of a rigid-body undergoing thermodynamic processes.

Perfect fluid

This material can only exert stresses normal to the surface element and not shear stresses, hence

$$\mathbf{t}(t, \mathbf{x}; \mathbf{n}) = -p(t, \mathbf{x})\mathbf{n} \quad \Longrightarrow \quad \mathbf{T} = -p(t, \mathbf{x})\mathbf{I}\,, \tag{3.4.20}$$

where p is the so-called ***pressure*** and $\mathbf{I}$ is the identity matrix. Equation (3.4.20) is, however, still not enough to close the system. In fact, as stated in Eq.(3.4.19), we have still to link p to the state variables ρ and $\mathbf{v}$. If in (3.4.20) the dependence on $(t, \mathbf{x})$ is through the density, i.e., $p(t, \mathbf{x}) = p(\rho(t, \mathbf{x}))$, then the fluid is called ***isentropic***. Many (not too rarefied) gases can often be viewed as isentropic with

$$p(\rho) = \alpha \rho^\gamma \quad \text{with} \quad \gamma \geq 1\,. \tag{3.4.21}$$

The related modelling equations are the ***compressible Euler equations***

$$\begin{cases} \dfrac{\partial \rho}{\partial t} + \nabla \cdot (\rho \mathbf{v}) = 0 \\ \rho \left(\dfrac{\partial \mathbf{v}}{\partial t} + \mathbf{v} \cdot \nabla \mathbf{v} \right) = -p'(\rho) \nabla \rho + \rho \mathbf{b}\,. \end{cases} \tag{3.4.22}$$

In some situations the behavior of low viscous fluids such as water can also be described by (3.4.20), but in this case one can impose the incompressibility condition getting the ***incompressible Euler equations***

$$\begin{cases} \rho_0 \left(\frac{\partial \mathbf{v}}{\partial t} + \mathbf{v} \cdot \nabla \mathbf{v} \right) = -\nabla p + \rho_0 \mathbf{b} \\ \nabla \cdot \mathbf{v} = 0 \, , \end{cases} \tag{3.4.23}$$

where the pressure p now plays the role of a Lagrangian parameter to satisfy the incompressibility constraint. Both Euler models allow, as we shall see, the propagation and formation of shock waves.

Viscous incompressible fluid

This material behaves like a perfect fluid in static conditions, while in dynamic conditions it can exert shear stresses proportional to the rate of strain tensor $\mathbf{D}$

$$\mathbf{T} = -p\mathbf{I} + 2\mu \mathbf{D} \, , \tag{3.4.24}$$

where $\mu > 0$ is the viscosity of the fluid.

This constitutive equation describes well the behavior of viscous fluids such as oil, but is also used for low viscous fluids as water in situations in which its viscosity cannot be neglected. The related model

$$\begin{cases} \rho_0 \left(\frac{\partial \mathbf{v}}{\partial t} + \mathbf{v} \cdot \nabla \mathbf{v} \right) = -\nabla p + \mu \nabla^2 \mathbf{v} + \rho_0 \mathbf{b} \\ \nabla \cdot \mathbf{v} = 0 \end{cases} \tag{3.4.25}$$

is called the ***incompressible Navier–Stokes model.*** In Eq.(3.4.25) ∇^2 is the Laplacian operator defined by

$$\nabla^2 \mathbf{v} = \sum_{i=1}^{3} \frac{\partial^2 \mathbf{v}}{\partial x_i^2} = \frac{\partial^2 \mathbf{v}}{\partial x^2} + \frac{\partial^2 \mathbf{v}}{\partial y^2} + \frac{\partial^2 \mathbf{v}}{\partial z^2} \, .$$

This model smooths out any discontinuity and therefore is not able to support shock waves.

Viscoelastic fluid

This material has the property of remembering what happened in the past. Many different constitutive equations possessing this characteristic can be proposed. A popular integrodifferential model is the ***linear***

viscoelastic model

$$\mathbf{T}(t,\mathbf{x}) = -p(t,\mathbf{x})\mathbf{I} + 2\int_0^{+\infty} G(s)\mathbf{D}(t-s,\mathbf{x})\,ds\,, \qquad (3.4.26)$$

where G is a positive fast descreasing function of time, say a combination of exponentials with negative exponents, called relaxation kernel.

In the steady case, the fluid behaves as a viscous one with viscosity

$$\mu = \int_0^{+\infty} G(s)\,ds\,,$$

but in unsteady problems, the stress at time t depends on the past deformations. The resulting integrodifferential model

$$\begin{cases} \rho_0\left(\dfrac{\partial \mathbf{v}}{\partial t} + \mathbf{v}\cdot\nabla\mathbf{v}\right) = -\nabla p + \displaystyle\int_0^{+\infty} G(s)\nabla^2\mathbf{v}(t-s)\,ds + \rho_0\mathbf{b} \\ \nabla\cdot\mathbf{v} = 0 \end{cases} \qquad (3.4.27)$$

describes well the behavior of polymeric fluids like plastic melts.

Other popular constitutive equations for viscoelastic fluids are the so-called ***convective Maxwell's models***, which add to the incompressible condition and to the general balance law (3.4.14) another set of partial differential equations

$$\tau\Bigg[\frac{\partial \mathbf{T}}{\partial t} + \mathbf{v}\cdot\nabla\mathbf{T} + \alpha(\mathbf{T}\nabla\mathbf{v} + \nabla\mathbf{v}^T\mathbf{T}) + (\alpha-1)(\nabla\mathbf{v}\mathbf{T} + \mathbf{T}\nabla\mathbf{v}^T)\Bigg] + \mathbf{T} = -p\mathbf{I} + 2\mu\mathbf{D}\,, \qquad (3.4.28)$$

with $\alpha \in [0,1]$ that relates the components of the stress tensor to the components of the rate of strain tensor (τ is a small characteristic time called relaxation time). For $\tau = 0$ we again have the viscous model. Though these models have some characteristics in common with the linear viscoelastic model, they differ on several aspects [AaM]. The decision on which model works best depends only on how well it describes the phenomenon at hand. For instance, assume that we want to simulate the effects of a rod steadily turned into a fluid using the model (3.4.28). If we set $\alpha = 0$, then the fluid would climb on the rod. If, instead, $\alpha = 1$, then the fluid would dip near the rod. This experiment would suggest that $\alpha = 0$ works better, but there are other situations in which different values of α should be considered to explain other phenomena or to avoid unreliable theoretical predictions (e.g., strange instabilities). What we can actually say is that neither (3.4.27) nor (3.4.28) can be taken for granted and work in every situation. However, among the models recalled here, the model that seems to work well in most cases is (3.4.28) with $\alpha = 0$, the so-called ***upper convected Maxwell's model.*** All viscoelastic models presented here allow the propagation of shock waves.

Hyperelastic solid

The stress of this material is described by introducing a function W of the Cauchy–Green strain tensor

$$\mathbf{B} = \mathbf{F}\mathbf{F}^T \quad \text{with} \qquad F_{ij} = \frac{\partial x_i}{\partial X_j},$$

called the ***specific energy of deformation***. In the general case the relation is not trivial for the unskilled reader. However, if the material is incompressible, then one has that W depends on $\mathbf{B}$ only through the traces of $\mathbf{B}$ and of its inverse $\mathbf{B}^{-1}$, that is $W = W(I_+, I_-)$ and

$$\mathbf{T} = -p\mathbf{I} + 2\rho_0 \left(\frac{\partial W}{\partial I_+}\mathbf{B} - \frac{\partial W}{\partial I_-}\mathbf{B}^{-1} \right). \tag{3.4.29}$$

This model describes well the behavior of elastic materials like rubber also to large deformations, but it is rather complex, even in the oversimplified case in which W is simply a linear function of I_+, the so-called ***neo-Hookeian material***,

$$W = \frac{K}{2}(I_+ - 3), \tag{3.4.30}$$

or equivalently

$$\mathbf{T} = -p\mathbf{I} + \rho_0 K\mathbf{B}. \tag{3.4.31}$$

Linear elasticity

An easier model can be deduced in the case of small deformations. If, in fact, the displacement $\mathbf{u}(t, \mathbf{X}) = \mathbf{x}(t, \mathbf{X}) - \mathbf{X}$ and its derivatives are assumed to be small (which also implies that it is not necessary to distinguish between derivatives versus $\mathbf{x}$ and versus $\mathbf{X}$), then it is possible to prove that

$$\mathbf{B} \cong \mathbf{I} + \nabla\mathbf{u} + \nabla\mathbf{u}^T.$$

Therefore, the constitutive law (3.4.31) simplifies to

$$\mathbf{T} = -p\mathbf{I} + \rho_0 K\big(\nabla\mathbf{u} + \nabla\mathbf{u}^T\big)\,. \tag{3.4.32}$$

Substituting (3.4.32) back into the balance law (3.4.14) and taking into account the incompressibility condition yields the equation of motion

$$\begin{cases} \rho_0 \dfrac{\partial^2 \mathbf{u}}{\partial t^2} = -\nabla p + \rho_0 K \nabla^2 \mathbf{u} \\ \nabla \cdot \mathbf{u} = 0\,, \end{cases} \tag{3.4.33}$$

which well describes the small deformations of elastic materials. One can recognize that (3.4.33) is a generalization to more dimensions of the string model (3.3.6) in the absence of external forces.

All the preceding examples are concerned with stress-strain constitutive relations. In order to include thermodynamics, more complex constitutive relations should be introduced.

Of course, this complicates things, therefore we will only consider the case of a rigid material with no radiation, since, in this case, one has only to give the constitutive equations relating the specific internal energy ϵ and the heat flux $\mathbf{q}$ to the temperature θ and use the energy equation

$$\rho_0 \left(\frac{\partial \epsilon}{\partial t} + \mathbf{v}\cdot\nabla\epsilon\right) = -\nabla\cdot\mathbf{q}\,. \tag{3.4.34}$$

If ϵ is only a function of the present value of the temperature, $\epsilon = \epsilon(\theta)$, then by the chain rule

$$\frac{d\epsilon}{dt} = c_v \frac{d\theta}{dt}\,, \tag{3.4.35}$$

where $c_v = d\epsilon/d\theta$ is the specific heat at constant volume, a practically constant quantity. The relation between the heat flux and the temperature are more complex.

Fourier's model

The heat flux is proportional to the gradient of temperature and travels from warmer parts of the body to colder ones

$$\mathbf{q} = -\mathbf{K}\nabla\theta\,, \tag{3.4.36}$$

where $\mathbf{K}$ is the ***thermal conductivity tensor***. The heat equation is then

$$\rho_0 c_v \frac{\partial \theta}{\partial t} = \nabla \cdot \big(\mathbf{K}(t, \mathbf{x}; \theta) \nabla \theta\big) , \tag{3.4.37}$$

which in the hysotropic and homogeneous case $\mathbf{K} = K\mathbf{I}$ with K positive constant, can be rewritten

$$\frac{\partial \theta}{\partial t} = \kappa \nabla^2 \theta , \tag{3.4.38}$$

where $\kappa = K/\rho_0 c_v$ is the ***thermal diffusivity***.

Cattaneo's model

In this model a "thermal inertia" is taken into account and the following constitutive law is proposed for the heat flux

$$\tau \frac{\partial \mathbf{q}}{\partial t} + \mathbf{q} = -K \nabla \theta , \qquad K > 0 . \tag{3.4.39}$$

The heat equation is then

$$\tau \frac{\partial^2 \theta}{\partial t^2} + \frac{\partial \theta}{\partial t} = \kappa \nabla^2 \theta . \tag{3.4.40}$$

Gurtin and Pipkin's model

Similar to the linear viscoelastic model, in this model it is assumed that both the internal energy and the heat flux are influenced by the past temperature evolution. Then, for instance, Eq.(3.4.35) is no longer valid. A popular model characterized by this property is that proposed by Gurtin and Pipkin [GaP]

$$\epsilon(t, \mathbf{x}) = b + \gamma \theta + \int_0^{+\infty} G_\epsilon(s) \theta(t - s, \mathbf{x})\, ds , \tag{3.4.41}$$

$$\mathbf{q}(t, \mathbf{x}) = -\int_0^{+\infty} G_q(s) \nabla \theta(t - s, \mathbf{x})\, ds , \tag{3.4.42}$$

where G_ϵ and G_q are non-negative fast decreasing (to zero) functions.

The heat equation then becomes

$$\rho_0\left[\gamma\frac{\partial\theta}{\partial t}+\int_0^{+\infty}G_\epsilon(s)\frac{\partial\theta}{\partial t}(t-s,\mathbf{x})\,ds\right]$$
$$=\int_0^{+\infty}G_q(s)\nabla^2\theta(t-s,\mathbf{x})\,ds\,. \tag{3.4.43}$$

If $G_\epsilon(s)=0$ and $G_q=Ke^{-s/\tau}/\tau$, it is possible to prove that the integrodifferential model can be reduced to Cattaneo's model (3.4.40) with $\gamma=c_v$.

3.5 Mathematical Models of Electromagnetism

Consider now the problem of defining a mathematical model suitable for describing the evolution of a set of charged particles moving in a vacuum environment.

Observation of physical reality leads to the following conclusions:

- Two electric charges at rest attract each other with a force proportional to their charges and inversely proportional to the square of the distance (Coulomb's law).
- This interaction may be described through the introduction of an electric field $\mathbf{E}$ defined, in each point, as the force felt in that point by a very small test charge divided by its charge q.
- Charges, when moving, may be subjected to forces perpendicular to their velocities. If this is the case, in that point there is a magnetic field defined by

$$\mathbf{F}=q\mathbf{v}\times\mathbf{B}\,.$$

Then, globally, a charge moving in an electromagnetic field feels a Lorentz force $q(\mathbf{E}+\mathbf{v}\times\mathbf{B})$.
- When charges are distributed, it is convenient to introduce the charge density ρ. The total charge in a material volume $\mathcal{B}_t$

$$Q=\int_{\mathcal{B}_t}\rho\,dV$$

is preserved during the evolution of the system. Therefore, proceeding as for the conservation of mass in Section 3.4.2, we end up with

$$\frac{\partial\rho}{\partial t}+\nabla\cdot\mathbf{j}=0\,, \tag{3.5.1}$$

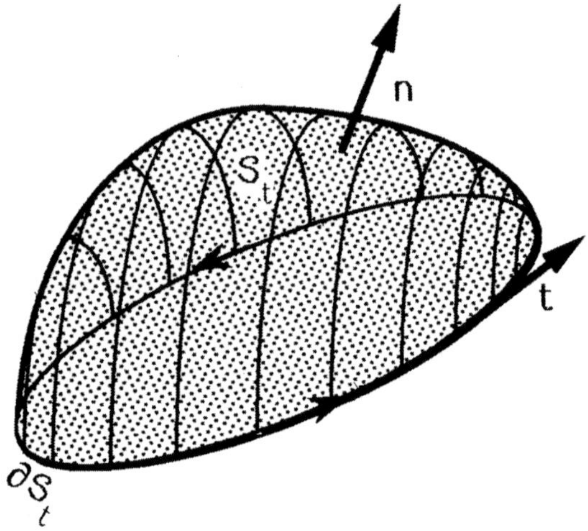

Figure 3.5.1 — Surface geometry.

where $\mathbf{j} = \rho\mathbf{v}$ is the current density.

- Electromagnetic currents are generators of magnetic fields. Ampere's law gives the value of the magnetic field generated by a stationary current

$$\mathbf{B}(\mathbf{y}) = -\frac{\mu_0}{4\pi}\int_{\mathbb{R}^3} \mathbf{j}(\mathbf{x}) \times \nabla \frac{1}{|\mathbf{y}-\mathbf{x}|}\, dV\,, \qquad (3.5.2)$$

where μ_0 is the so-called permeability constant.

In order to proceed we will need the following classical theorems.

THEOREM 3.5.1 *(Stokes' Theorem)*
If S_t is an open regular surface with border ∂S_t and $\mathbf{u}$ is a vector field continuous with its first derivatives in S_t, then

$$\int_{\partial S_t} \mathbf{u}\cdot\mathbf{t}\, d\ell = \int_{S_t} \nabla\times\mathbf{u}\cdot\mathbf{n}\, d\sigma\,,$$

where $\mathbf{t}$ is the tangent to ∂S_t, $\mathbf{n}$ is the normal to S_t oriented as shown in Fig. 3.5.1, $\nabla\times$ is the curl operator

$$(\nabla\times\mathbf{u})_i = \varepsilon_{ijk}\frac{\partial u_k}{\partial x_j}\,,$$

and ε_{ijk} is the Levi-Civita symbol. ■

THEOREM 3.5.2 ***(Transport Theorem for Surfaces)***
If $\mathbf{u}$ *is a vector field continuous with its first derivatives on a regular material surface* $\mathcal{S}_t$*, then*

$$\frac{d}{dt}\int_{\mathcal{S}_t} \mathbf{u}\cdot\mathbf{n}\,d\sigma = \int_{\mathcal{S}_t}\left(\frac{d\mathbf{u}}{dt} + \mathbf{u}\nabla\cdot\mathbf{v} - \mathbf{u}\cdot\nabla\mathbf{v}\right)\cdot\mathbf{n}\,d\sigma$$

$$= \int_{\mathcal{S}_t}\left[\frac{\partial\mathbf{u}}{\partial t} + \nabla\times(\mathbf{u}\times\mathbf{v}) + \mathbf{v}\nabla\cdot\mathbf{u}\right]\cdot\mathbf{n}\,d\sigma\,.$$

■

3.5.1 Maxwell's equations

We will now deal briefly with the derivation of a set of equations called ***Maxwell's equations*** governing the evolution of $\mathbf{E}$ and $\mathbf{B}$.

The first equation derives from Coulomb's law. It can be, in fact, proved that the flux of electric field through a closed surface $\partial\mathcal{B}_t$ is proportional to the total charge Q_{in} contained in $\mathcal{B}_t$

$$\int_{\partial\mathcal{B}_t} \mathbf{E}\cdot\mathbf{n}\,d\sigma = \frac{1}{\varepsilon_0}\int_{\mathcal{B}_t}\rho\,dV = Q_{in}\,,$$

where ε_0 is the so-called dielectric constant or, using the divergence theorem 3.4.3

$$\nabla\cdot\mathbf{E} = \frac{\rho}{\varepsilon_0}\,. \tag{3.5.3}$$

Taking the divergence of (3.5.2) and observing that

$$\nabla\cdot\left(\nabla\frac{1}{|\mathbf{y}-\mathbf{x}|}\times\mathbf{j}(\mathbf{x})\right) = \mathbf{j}(\mathbf{x})\cdot\nabla\times\left(\nabla\frac{1}{|\mathbf{y}-\mathbf{x}|}\right) = 0\,,$$

where derivatives are with respect to $\mathbf{y}$, we obtain that

$$\nabla\cdot\mathbf{B} = 0\,, \tag{3.5.4}$$

so that the flux of a magnetic field generated by a stationary electric current is null.

Taking instead the curl of (3.5.2) and observing that in the stationary case

$$\nabla\times\left(\mathbf{j}\times\nabla\frac{1}{|\mathbf{y}-\mathbf{x}|}\right) = \nabla^2\frac{1}{|\mathbf{y}-\mathbf{x}|}\mathbf{j}\,,$$

we obtain

$$\nabla \times \mathbf{B} = \mu_0 \mathbf{j}\,, \tag{3.5.5}$$

where we used the fact that

$$\nabla^2 \frac{1}{|\mathbf{y}-\mathbf{x}|} = -4\pi\delta(\mathbf{y}-\mathbf{x})\,,$$

where δ is the Dirac delta.

REMARK 3.5.1 In the unsteady case, Eq.(3.5.5) is inconsistent with (3.5.1). In fact, taking the divergence of (3.5.5) and recalling that

$$\nabla \cdot (\nabla \times \mathbf{u}) = 0\,, \qquad \forall \mathbf{u}$$

leads to $\nabla \cdot \mathbf{j} = 0$, which is in contrast with (3.5.1). ∎

To solve the contradiction pointed out in the previous remark, Maxwell successfully argued that a term $\mathbf{a}$ should have been added to the left-hand side of Eq.(3.5.5). Taking now the divergence of (3.5.5) and using (3.5.1) and (3.5.3) leads then to

$$0 = \mu_0 \nabla \cdot \mathbf{j} + \nabla \cdot \mathbf{a} = -\mu_0 \frac{\partial \rho}{\partial t} + \nabla \cdot \mathbf{a} = -\varepsilon_0 \mu_0 \frac{\partial}{\partial t}(\nabla \cdot \mathbf{E}) + \nabla \cdot \mathbf{a}\,,$$

that is,

$$\mathbf{a} = \varepsilon_0 \mu_0 \frac{\partial \mathbf{E}}{\partial t}\,. \tag{3.5.6}$$

Equation (3.5.2) takes then the generalized form

$$\nabla \times \mathbf{B} = \mu_0 \mathbf{j} + \varepsilon_0 \mu_0 \frac{\partial \mathbf{E}}{\partial t}\,. \tag{3.5.7}$$

REMARK 3.5.2 The added term (3.5.6), called displacement current by Maxwell, is often small, since $\mu_0\varepsilon_0$ is the inverse of the square of the speed of light, but it is essential for explaining phenomena like the propagation of electromagnetic waves. ∎

Finally, Faraday observed that a current was induced to flow in a closed circuit when immersed in a time-varying magnetic field or, conversely, when a circuit was moved in a fixed magnetic field. It is not difficult to understand that in this last case a Lorentz-type force is exerted on the free electrons. These observations lead to the conclusion that the variation of magnetic flux through a surface induces an electromotive force along its boundary

$$\frac{d}{dt}\int_{S_t} \mathbf{B}\cdot\mathbf{n}\,d\sigma = -\int_{\partial S_t} (\mathbf{E}+\mathbf{v}\times\mathbf{B})\cdot\mathbf{t}\,d\ell\,. \tag{3.5.8}$$

Recalling the Transport Theorem 3.5.2, Eq.(3.5.8) can be rewritten as

$$\int_{S_t}\left[\frac{\partial \mathbf{B}}{\partial t} + \nabla\times(\mathbf{B}\times\mathbf{v}) + \mathbf{v}\nabla\cdot\mathbf{B}\right]\cdot\mathbf{n}\,d\sigma$$

$$= -\int_{\partial S_t} \mathbf{E}\cdot\mathbf{t}\,d\ell - \int_{\partial S_t} \mathbf{v}\times\mathbf{B}\cdot\mathbf{t}\,d\ell\,.$$

Using Eq.(3.5.4) and applying Stokes' theorem to the right-hand side yields the differential form of Faraday's law

$$\frac{\partial \mathbf{B}}{\partial t} = -\nabla\times\mathbf{E}\,. \tag{3.5.9}$$

The system of equations (3.5.3), (3.5.4), (3.5.7), and (3.5.9) is the so-called Maxwell's model of electromagnetism. The quantities ρ and $\mathbf{j}$ are related by the continuity equation (3.5.1).

As in Section 3.4, Maxwell's model is still not sufficient to give the evolution of $\mathbf{E}$, $\mathbf{B}$, ρ, and $\mathbf{j}$. Constitutive equations are needed. For instance, in the case of charged particles moving in a vacuum environment, we need a set of equations that describe the motion of the particles, as shown in Example 3.2.

Moreover, these equations are not independent. In fact, taking the divergence of (3.5.7) and (3.5.9) yields (3.5.1) and (3.5.4). In particular, taking the divergence of Faraday's law (3.5.9), one has

$$0 = \nabla\cdot\nabla\times\mathbf{E} = -\nabla\cdot\frac{\partial \mathbf{B}}{\partial t} = -\frac{\partial \nabla\cdot\mathbf{B}}{\partial t}\,,$$

that is, $\nabla\cdot\mathbf{B}$ does not depend on time. This justifies the fact that $\nabla\cdot\mathbf{B} = 0$, which was proved in the stationary case, is also valid in the nonstationary case.

REMARK 3.5.3 Historically, the route to deriving Maxwell's model was not as easy as shown here. Some of the equations where thought in integral form, some in differential form, some as simplified versions, then refined. ∎

Example 3.1 Electromagnetic waves
As an application, consider Maxwell's equations in a space empty of everything but for the presence of an electromagnetic field. Maxwell's model reduces then to

$$\begin{cases} \nabla \cdot \mathbf{E} = \nabla \cdot \mathbf{B} = 0 \\ \nabla \times \mathbf{B} = \varepsilon_0 \mu_0 \dfrac{\partial \mathbf{E}}{\partial t} \\ \nabla \times \mathbf{E} = -\dfrac{\partial \mathbf{B}}{\partial t} \, . \end{cases}$$

If we eliminate $\mathbf{B}$, then we obtain

$$\varepsilon_0 \mu_0 \frac{\partial^2 \mathbf{E}}{\partial t^2} = \nabla^2 \mathbf{E} \, .$$

Similarly we can obtain

$$\frac{\partial^2 \mathbf{B}}{\partial t^2} = \frac{1}{\varepsilon_0 \mu_0} \nabla^2 \mathbf{B} \, .$$ □

Example 3.2 Plasma dynamics
A plasma is an ensemble of mostly charged (but also uncharged) particles. We will consider the case of a null total charge. An example is the ionosphere around the earth. The ionization is caused by absorption of ultraviolet radiation. In this case, the electrons are much lighter than the positive ions and so, if the frequency of the electric field is high enough, then they are the only ones to move. This is the so-called ***one-fluid plasma hypothesis***.

Deducing the model describing the behavior of a plasma is not easy, since Maxwell's model has to be joined to the equations of motion of the single particles. Without going into details that involve concepts to be introduced in Chapter 4, we will simply say that if $\mathcal{N}$ is the total number per unit volume of particles with charge q and mass m and $\mathbf{v}$ is the mean velocity of the particles, then $\mathcal{N}q$ is the charge density, $\mathcal{N}m$ is the mass density, $\mathcal{N}q\mathbf{v}$ is the electric current, and $\mathcal{N}m\mathbf{v}$ is the momentum.

Then the following fluid-dynamic model can be proposed

$$\nabla \times \mathbf{E} = -\frac{\partial \mathbf{B}}{\partial t}$$

$$\nabla \times \mathbf{B} = \mu_0 \mathcal{N} q \mathbf{v} + \varepsilon_0 \mu_0 \frac{\partial \mathbf{E}}{\partial t}$$

$$\frac{\partial}{\partial t}(\mathcal{N} q) + \nabla \cdot (\mathcal{N} q \mathbf{v}) = 0$$

$$\frac{d}{dt}(\mathcal{N} m \mathbf{v}) = \mathcal{N} q(\mathbf{E} + \mathbf{v} \times \mathbf{B}),$$

where the last equation is nothing other than Newton's model with a Lorentz force.

Among other simplifications, the reader can easily recognize that gravity and mutual repulsion forces have been neglected and that neither viscous nor pressure terms appear in Newton's equation (***cold plasma hypothesis***). If pressure is taken under consideration, then one has to give the constitutive relation linking p to $\mathcal{N}$, for instance the adiabatic one $p = \alpha(\mathcal{N}_0 + \mathcal{N})^\gamma$. □

3.5.2 Constitutive laws

Things become much more complicated when dealing with more complex systems. We will simply mention here, without going into detail, that in electromagnetism, besides the current and charge densities, four fields are usually defined as state variables:

E electric field;
D displacement field;
B magnetic induction field;
H magnetic field.

Maxwell's equations are then rewritten as

$$\begin{cases} \nabla \cdot \mathbf{D} = \rho \\ \nabla \cdot \mathbf{B} = 0 \\ \nabla \times \mathbf{H} = \mathbf{j} + \varepsilon_0 \dfrac{\partial \mathbf{D}}{\partial t} \\ \nabla \times \mathbf{E} = -\dfrac{\partial \mathbf{B}}{\partial t}, \end{cases} \tag{3.5.10}$$

where now to close the system one has to give the constitutive relations linking the four fields. As for continuum mechanics models, this is natural since, say, the magnetization properties will differ from material to material.

The simplest examples of constitutive laws are those dealing with ***linear materials***, e.g., vacuum, for which

$$\mathbf{D} = \varepsilon \mathbf{E} \qquad \text{and} \qquad \mathbf{B} = \mu \mathbf{H} .$$

Another example is given by the ***dispersive (isotropic) materials***, which present memory and correlation effects. These effects can be described through an integration over past time and space, respectively. The following model can then be proposed

$$\begin{cases} \mathbf{D}(t, \mathbf{x}) = \displaystyle\int_{\mathbb{R}^3} \int_{-\infty}^{t} G_E(t - s, \mathbf{x} - \mathbf{y}) \mathbf{E}(s, \mathbf{y}) \, dV \, ds \\ \mathbf{B}(t, \mathbf{x}) = \displaystyle\int_{\mathbb{R}^3} \int_{-\infty}^{t} G_B(t - s, \mathbf{x} - \mathbf{y}) \mathbf{H}(s, \mathbf{y}) \, dV \, ds . \end{cases}$$

Of course, there are materials, like the ferromagnetic materials, which require more complex constitutive laws.

3.6 Direct Simulation Models in Biology

As we have seen, the relationship between mathematics and physics was and still is characterized by success and reciprocal support and motivations. Physics has permanently stimulated mathematics, which has continuously supported physics with direct contributions to research discoveries. This remark includes, in particular, the science of mathematical modelling.

The same comment can be almost straighforwardly applied to the relationships between mathematics and technology. However this is not the case of the interplay between mathematics and the natural sciences, where several difficulties can be remarked.

As a matter of fact, the famous mathematician Eugene Wigner posed in a paper [WIG] that appeared in 1960 the problem of the unreasonable effectiveness of mathematics in natural sciences. However, despite several difficulties, some successful results can be still registered. We can mention, among others, the Nobel prize gained by Hodgkin and Huxley due to their research and, in particular, to their mathematical model on the conduction of nerve impulses, already dealt with in Section 2.2.5.

Probably, the biological sciences need a new mathematics and new computer simulation methods in this direction. Some efforts have been made, for instance, using cellular automata and complex computer architectures. However, such a pioneer activity has not yet given general results, but simply some interesting pioneer studies. Traditional modelling methods generally lead to models that can be regarded as technical generalizations of the classical ones of physical sciences. This is the case, among several others, of the celebrated Hodgkin and Huxley model (2.2.40) [HOD], which, however, still represents a pioneer step, since it deals with a space-independent situation and, therefore, there is actually no propagation of impulses along the nerve fiber. In order to model this situation one can introduce spatial derivatives in the first equation in (2.2.40) substituting the ordinary differential operator d/dt with one of the following partial differential operators

$$\begin{aligned} \mathcal{L}_1 &= \frac{\partial}{\partial t} + c\frac{\partial}{\partial x} \\ \mathcal{L}_2 &= \frac{\partial}{\partial t} - \nu\frac{\partial^2}{\partial x^2} \\ \mathcal{L}_3 &= \tau\frac{\partial^2}{\partial t^2} + \frac{\partial}{\partial t} - \nu\frac{\partial^2}{\partial x^2}\,. \end{aligned} \tag{3.6.1}$$

These three operators can be suggested on the basis of well-known properties which characterize them and make them landmarks of the literature on partial differential equations. In fact, as we shall see in the next section, they are, respectively, linked to the undamped one-way propagation, to the diffusion of a signal, and to the damped propagation of waves. The remaining three equations remain unchanged so that the model has the structure of a continuous dynamic model coupled with a discrete one.

The example we have just seen shows that too rigid structures cannot be used in mathematical modelling and that some flexibility is needed to adapt the model to the analysis of real systems. More aspects of physics and physiology of neural propagation were studied by Hille [HIL].

The model described above can be regarded as a succesful example of interaction between mathematics and biological sciences. In fact, the model was used for several generalizations and applications. An example is the simulation of inner parts of complex systems with several subsystems interacting with each other, e.g., the cardiac activity seen as an electrical network [DaN], where the propagation along each line is simulated by the model we have seen above, or the study of excitable tissues [NOB]. Mathematical aspects were studied by Mascagni [MAS].

We conclude this section by inviting the reader to be patient and to look at this model as an example offered in order to open a window to such a fascinating field of application of mathematical modelling. A deeper insight needs a specialized training and more than one section of a chapter.

3.7 Classification

Once the mathematical model has been formulated, its analysis can be developed by solving the related mathematical problems.

As we have seen in the examples given in Sections 3.3–3.6, the governing sets of equations constitute a system that is much more complex than those encountered for discrete systems. Above all, a huge variety of cases is possible. Therefore, before turning to any computation, it is better to give a careful look at the type of equation one is going to handle. This is an essential step for at least two reasons:

- To identify the proper initial and boundary conditions to be joined to the equations, in order to have a well formulated and, possibly, a well-posed problem.
- To choose a proper numerical method to deal with specific simulation problems.

3.7.1 Hyperbolic, parabolic, and elliptic equations

Because of the wide variety of partial differential equations, a complete classification will not be presented here. However, to give an idea of where to look when examining a continuous model, some examples will be given.

To start, consider the second-order partial differential equation in two dependent variables

$$A(t,x)\frac{\partial^2 u}{\partial t^2} + 2B(t,x)\frac{\partial^2 u}{\partial t \partial x} + C(t,x)\frac{\partial^2 u}{\partial x^2} = F\left(t,x;u,\frac{\partial u}{\partial t},\frac{\partial u}{\partial x}\right) . \quad (3.7.1)$$

Some important mathematical properties of the solution are essentially determined by the left side term, which contains the highest-order derivatives and, in particular, by the sign of the discriminant $B^2 - AC$. We have then the following classification:

$$\begin{aligned} B^2 - AC > 0 \quad &\Longrightarrow \quad \text{hyperbolic} \\ B^2 - AC = 0 \quad &\Longrightarrow \quad \text{parabolic} \\ B^2 - AC < 0 \quad &\Longrightarrow \quad \text{elliptic} \, . \end{aligned} \quad (3.7.2)$$

Example 3.3 Hyperbolic, parabolic, and elliptic
Referring to the models presented in Sections 3.3–3.4, one has that

- The string model (3.3.7) is hyperbolic. In fact

$$A = 1\,, \quad B = 0\,, \quad C = -c^2 \quad \Longrightarrow \quad B^2 - AC = c^2 > 0\,.$$

Actually Eq.(3.3.6) with $h = 0$

$$\frac{\partial^2 u}{\partial t^2} = c^2 \frac{\partial^2 u}{\partial x^2} \tag{3.7.3}$$

is the prototype of hyperbolic equations and is often called the ***wave equation***. If h is a non-null constant, then one has the so-called ***telegrapher's equation***. Similarly, one can see that the linear elastic model (3.4.33) is also hyperbolic and that the operator $\mathcal{L}_3$ in (3.6.1) is nothing other that the telegrapher's operator.

- The viscous incompressible model (3.4.25) and the Fourier model for the heat conduction (3.4.38) are parabolic, since only C does not vanish. Actually, Eq.(3.4.38) is the prototype of parabolic equations and is often called the ***diffusion equation*** or ***heat equation***. The operator $\mathcal{L}_2$ in (3.6.1) is then called the ***diffusion operator***.
- The steady distribution of the temperature in a rigid conductor is described by Eq.(3.4.38) after the transient has died out, that is, by

$$\nabla^2 \theta = 0 \tag{3.7.4}$$

or in two dimensions

$$\frac{\partial^2 \theta}{\partial x^2} + \frac{\partial^2 \theta}{\partial y^2} = 0\,.$$

This equation is elliptic, since

$$A = C = 1\,, \quad B = 0 \quad \Longrightarrow \quad B^2 - AC = -1 < 0\,.$$

Actually, Eq.(3.7.4) is the prototype of elliptic equations and is often called the ***Laplace's equation***. □

In general, if A, B, and C are functions of t and x, then the type of equation depends upon the local values of the coefficients.

It is important to be able to analyze the change of type of the equations related to the model. In fact, as we have already mentioned, this corresponds to different algorithms in different domains of the space variables. This may involve, as we shall see in Section 3.11, decomposition of domains techniques.

Example 3.4 Equations that change type
Take, for instance, the abstract model

$$\frac{\partial^2 u}{\partial t^2} + C(t,x)\frac{\partial^2 u}{\partial x^2} = F\left(t,x;u,\frac{\partial u}{\partial t},\frac{\partial u}{\partial x}\right), \tag{3.7.5}$$

which has $A = 1$ and $B = 0$ and therefore $B^2 - AC = -C(t,x)$.

Equation (3.7.5) is then elliptic, parabolic, or hyperbolic according to the sign of C. In particular, it is elliptic, hyperbolic, or parabolic when, respectively, C is positive, negative, or equal to zero. An example of equation of this type is the model proposed by Ferrari and Tricomi [FaT]

$$\frac{\partial^2 u}{\partial x^2} + y\frac{\partial^2 u}{\partial y^2} = 0, \tag{3.7.6}$$

which describes the space evolution of the velocity potential, in the velocity space, for inviscid flow around the sonic line for transonic flow around thin airfoils.

In fact, Eq.(3.7.6) is elliptic for $y > 0$, which corresponds to subsonic flow, parabolic for $y = 0$, which corresponds to the sonic line, and hyperbolic for $y < 0$, which corresponds to supersonic flow. □

The classification of Eq.(3.7.1) can be generalized to the scalar equation

$$\sum_{i,j=1}^{n} A_{ij}(\mathbf{x})\frac{\partial^2 u}{\partial x_i \partial x_j} = F\left(\mathbf{x};u,\frac{\partial u}{\partial x_1},\ldots,\frac{\partial u}{\partial x_n}\right) \tag{3.7.7}$$

by considering the eigenvalues of the matrix $\mathbf{A}$ in front of the second-order derivatives ($\mathbf{A}$ is symmetric and therefore its eigenvalues are real). Then, one has

$$\left.\begin{array}{l} \forall i \quad \lambda_i \neq 0 \quad \exists! \lambda_i > 0 \\ \forall i \quad \lambda_i \neq 0 \quad \exists! \lambda_i < 0 \end{array}\right\} \implies \text{hyperbolic},$$

$$\left.\begin{array}{c} \det \mathbf{A} = 0 \\ (\text{or equivalently } \exists \lambda_i = 0) \end{array}\right\} \implies \text{parabolic},$$

$$\left.\begin{array}{l} \forall i \quad \lambda_i > 0 \\ \forall i \quad \lambda_i < 0 \end{array}\right\} \implies \text{elliptic}.$$

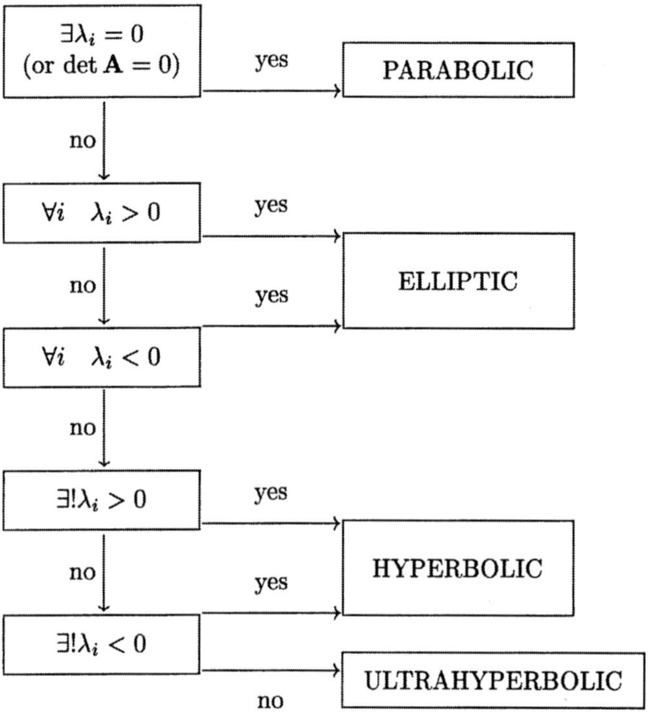

Table 3.7.1 — Classification scheme for Eq.(3.7.7).

The classification procedure is also summarized in Table 3.7.1.

We conclude by giving a partial classification of quasilinear systems of n first-order partial differential equations in two independent variables.

DEFINITION 3.7.1 Quasilinear system
The system of partial differential equations

$$\sum_j A_{ij} \frac{\partial u_j}{\partial t} + \sum_j B_{ij} \frac{\partial u_j}{\partial x} = f_i\,, \quad i = 1, \ldots, n\,, \tag{3.7.8}$$

or in vector form

$$\mathbf{A}\frac{\partial \mathbf{u}}{\partial t} + \mathbf{B}\frac{\partial \mathbf{u}}{\partial x} = \mathbf{f}\,, \tag{3.7.9}$$

where A_{ij}, B_{ij}, and f_i are functions of t, x, and $\mathbf{u}$ is said to be ***quasi-linear****. If the dependence on $\mathbf{u}$ drops in A_{ij} and in B_{ij}, then the system is said to be* ***almost linear****. If it drops also in f_i, then the system of partial differential equations is* ***linear****.*

Under the assumption that $\mathbf{A}$ is nonsingular, the classification is based on the roots of the eigenvalue problem

$$P_n(\lambda) = \det(\mathbf{B} - \lambda\mathbf{A}) = 0 \tag{3.7.10}$$

and on the number of independent eigenvectors satisfying

$$(\mathbf{B}^T - \lambda\mathbf{A}^T)\mathbf{v} = \mathbf{0}\,. \tag{3.7.11}$$

It reads as follows:

- If $P_n(\lambda)$ has n real distinct zeros $\Longrightarrow$ hyperbolic;
- If $P_n(\lambda)$ has n real zeros at least one of which is repeated and n independent eigenvectors $\Longrightarrow$ hyperbolic;
- If $P_n(\lambda)$ has n real zeros at least one of which is repeated and fewer than n independent eigenvectors $\Longrightarrow$ parabolic;
- If $P_n(\lambda)$ has no real zeros $\Longrightarrow$ elliptic.

This classification, which does not cover all possibilities, is also summarized in Table 3.7.2.

Example 3.5 Classification of first-order systems
Trivially a first-order partial differential equation

$$\frac{\partial u}{\partial t} + c(t,x;u)\frac{\partial u}{\partial x} = f(t,x;u) \tag{3.7.12}$$

is hyperbolic. Equation (3.7.12) is called, because of the properties we will see later on, a ***one-way wave equation*** or ***transport equation***.

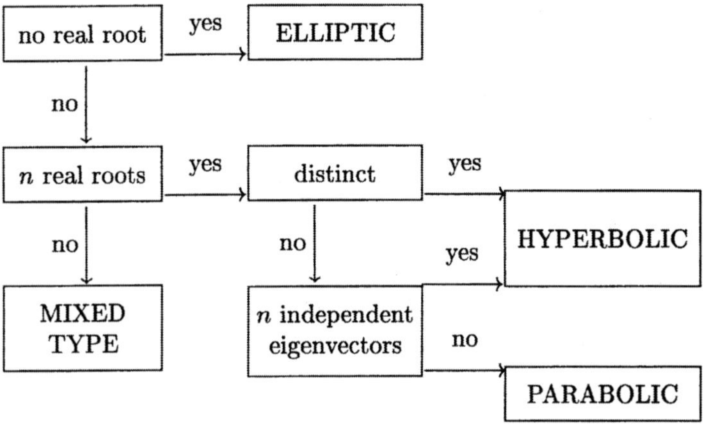

Table 3.7.2 — Classification scheme for Eq.(3.7.8).

The compressible Euler equation (3.4.22) is hyperbolic. In fact, in one dimension and for $\mathbf{b} = \mathbf{0}$, it can be rewritten as

$$\begin{cases} \dfrac{\partial \rho}{\partial t} + v\dfrac{\partial \rho}{\partial x} + \rho\dfrac{\partial v}{\partial x} = 0 \\ \dfrac{\partial v}{\partial t} + \dfrac{p'(\rho)}{\rho}\dfrac{\partial \rho}{\partial x} + v\dfrac{\partial v}{\partial x} = 0\,. \end{cases} \tag{3.7.13}$$

Hence $\mathbf{u} = \{\rho, v\}$,

$$\mathbf{A} = \begin{pmatrix} 1 & 0 \\ 0 & 1 \end{pmatrix} \qquad \text{and} \qquad \mathbf{B} = \begin{pmatrix} v & \rho \\ \dfrac{p'(\rho)}{\rho} & v \end{pmatrix}\,. \tag{3.7.14}$$

Therefore,

$$\det(\mathbf{B} - \lambda\mathbf{A}) = \begin{vmatrix} v - \lambda & \rho \\ \dfrac{p'(\rho)}{\rho} & v - \lambda \end{vmatrix} = 0$$

gives

$$\lambda = v \pm \sqrt{p'(\rho)}\,,$$

which are real since pressure is an increasing function of density. □

3.7.2 On the qualitative behavior of the solutions

What is the qualitative difference among elliptic, hyperbolic, and parabolic systems? Some preliminary indications will be given in this section. These indications only cover some introductory aspects of the general problem, which certainly deserves a deeper insight. Moreover, a relatively wider space will be given to hyperbolic and parabolic equations (with respect to the elliptic ones) as these equations generally refer to dynamic models.

Elliptic equations describe sytems in the equilibrium or steady state. They can be seen as equations describing, for instance, the final state reached by a physical system described by a parabolic equation after the transient term has died out (see Example 3.3).

Hyperbolic equations are evolution equations that describe wave-like phenomena.

The solution of a hyperbolic initial-value problem cannot be smoother than the initial data. On the other hand, it can actually develop, as time goes by, singularities even from smooth initial data, which then characterize the whole evolution and are propagated along special curves called ***characteristics***. In particular, if the solution has initially a compact support, for instance,

$$u(t=0,x)=0\,, \qquad \text{for} \qquad |x|>M_0\,,$$

then

$$\forall t\,, \quad \exists M(t) \quad \text{such that} \quad u(t,x)=0 \quad \text{for} \quad |x|>M(t)\,.$$

The rate at which the support expands can be interpreted as the ***speed of propagation*** of the effect modelled by the hyperbolic equation.

Parabolic equations are evolution equations that describe diffusion-like phenomena. In general, the solution of a parabolic problem is infinitely differentiable with respect to both space and time for any $\mathbf{x}$ and $t>0$, even if the initial data are not continuous. In other words, parabolic systems have a ***smoothing action*** and singularities can neither develop nor be maintained.

Moreover, even if the initial data have compact support, at any $t>0$ the initial condition is felt everywhere, that is

$$\forall t \quad \text{and} \quad \forall M\,, \quad \exists \widehat{x} \quad \text{with} \quad |\widehat{x}|>M \quad \text{such that} \quad u(t,\widehat{x})\neq 0\,.$$

This is usually indicated by saying that parabolic systems have an ***infinite speed of propagation.***

To better understand the differences among the types of partial differential equations, the following Example and Fig. 3.7.1 can be of help.

Example 3.6 Differences among solutions
Consider the abstract evolution scalar partial differential equation

$$\mathcal{L}u = 0 \qquad x \in \mathbb{R}$$

acting on the initial condition

$$u(t=0,x) = e^{-x^2} .$$

If

$$\mathcal{L} = \frac{\partial}{\partial t} + c\frac{\partial}{\partial x}, \qquad c > 0 \tag{3.7.15}$$

(one-way propagation or transport operator), then the solution is

$$u(t,x) = e^{-(x-ct)^2} .$$

The solution travels unchanged to the right with speed equal to c.

If

$$\mathcal{L} = \frac{\partial}{\partial t} - \nu\frac{\partial^2}{\partial x^2}, \qquad \nu > 0 \tag{3.7.16}$$

(diffusion operator), then the solution is

$$u(t,x) = \frac{1}{\sqrt{1+4\nu t}} e^{-x^2/(1+4\nu t)} .$$

The bell shape of the initial condition is preserved, but it broadens out; the maximum is always at $x = 0$, but it decreases with time.

As we shall see in Section 3.8, second-order operators (in time) need one more initial condition, say

$$\frac{\partial u}{\partial t}(t=0,x) = 0 .$$

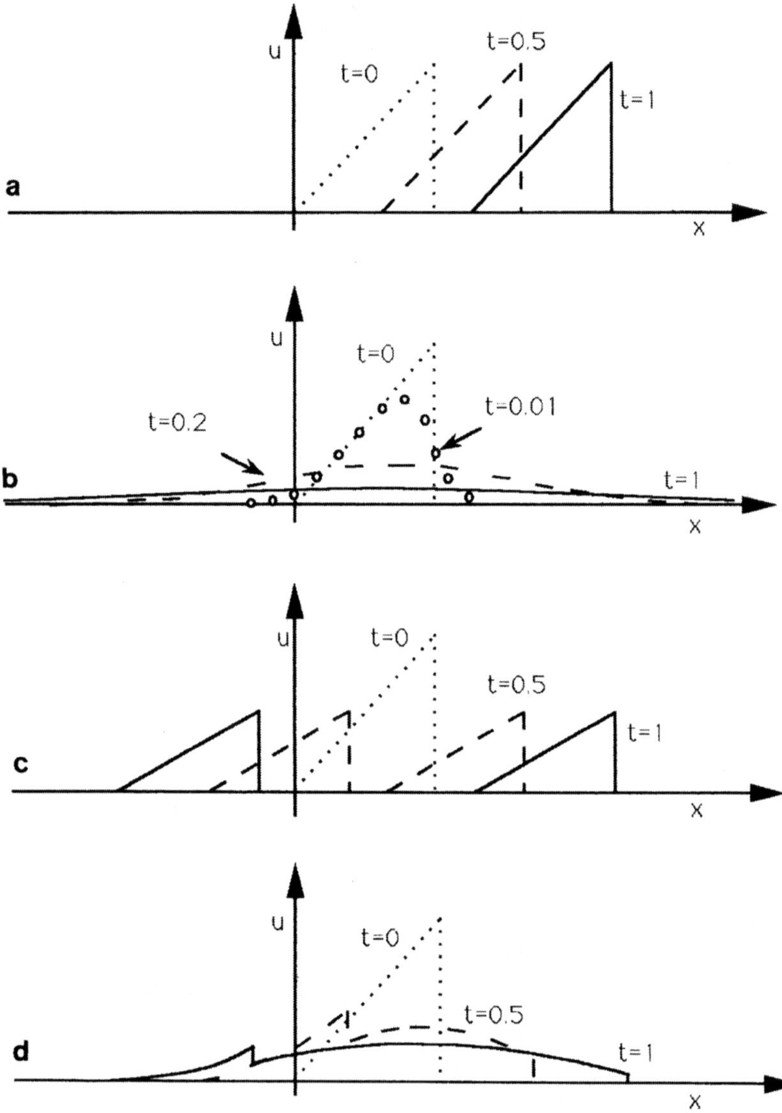

Figure 3.7.1 — Evolution of a saw-tooth initial condition under a system described by (a) Transport equation ($c = 1$); (b) Diffusion equation ($\nu = 1$); (c) Wave equation ($c = 1$); (d) Telegrapher's equation ($c = 0.2, \nu = 1$). The equations for (a), (c), and (d) are hyperbolic, while that for (b) is parabolic. As time goes by the difference between (b) and (d) becomes negligible. Observe, however, the difference between (b) at $t = 0.2$ and (d) at $t = 1$.

If now

$$\mathcal{L} = \frac{\partial^2}{\partial t^2} - c^2 \frac{\partial^2}{\partial x^2} \tag{3.7.17}$$

(wave operator), then the solution is

$$u(t,x) = \frac{1}{2}\left[e^{-(x-ct)^2} + e^{-(x+ct)^2}\right], \tag{3.7.18}$$

that is, the bell is split in two and both parts travel with speed c, one to the left and the other to the right.

Finally, if

$$\mathcal{L} = \frac{\partial^2}{\partial t^2} + \frac{1}{\tau}\frac{\partial}{\partial t} - c^2 \frac{\partial^2}{\partial x^2} \tag{3.7.19}$$

(telegrapher's operator), then the solution is

$$u(t,x) = \frac{e^{-t/2\tau}}{2}\left\{ e^{-(x-ct)^2} + e^{-(x+ct)^2} + \frac{1}{4c\tau}\int_{x-ct}^{x+ct}\left[I\left(\frac{f(z;t,x)}{2c\tau}\right) \right.\right.$$

$$\left.\left. + \frac{ct}{f(z;t,x)} I'\left(\frac{f(z;t,x)}{2c\tau}\right)\right] e^{-z^2}\, dz \right\}, \tag{3.7.20}$$

where

$$I(z) = \frac{1}{\pi}\int_{-\frac{\pi}{2}}^{\frac{\pi}{2}} e^{z\sin\varphi}\, d\varphi\,,$$

I' is its derivative and

$$f(z;t,x) = \sqrt{c^2t^2 - (z-x)^2}\,.$$

The first two terms in the curly brackets make up the solution to the wave equation; the integral represents a dispersion term. Everything is then attenuated by an exponentially decreasing factor $e^{-t/2\tau}$. □

REMARK 3.7.1 In Fig. 3.7.1 we applied the previous operators to a compact support initial condition. By looking at it, the reader can realize that if $u(t=0,x) = 0$ for $x \notin [a,b]$ and, for instance, $u(t=0,x) = u_{in}(x) > 0$ for $x \in [a,b]$, then the supports of the solutions to the different problems are

- $[a+ct, b+ct]$ for the transport problem;

- $[a - ct, b - ct] \cup [a + ct, b + ct]$ for the wave propagation problem;
- $\mathbb{R}$ for the diffusion problem;
- $[a - ct, b + ct]$ for the telegrapher's problem.

Besides the attenuation factor, an important difference between the wave propagation solution and the telegrapher's solution is in the fact that for $x \in [b - ct, a + ct]$ with $t > (b - a)/2c$, the solution to the second problem does not vanish. In fact, the first two terms in the curly brackets in Eq.(3.7.20) become $u_{in}(x - ct)$ and $u_{in}(x + ct)$, which vanish, but the integral does not. In detail, e^{-z^2} is replaced by $u_{in}(z)$ and the extrema of integration are one less than a and the other greater than b, so, because of the properies of $u_{in}(x)$, the integration interval becomes $[a, b]$.

As time goes by the difference between the solutions of the telegrapher's equation and the related diffusion equation (i.e., neglecting the term $\partial^2 u/\partial t^2$) becomes negligible. It is instructive to observe the differences between the solution in Fig. 3.7.1b at $t = 0.2$ and the solution in Fig. 3.7.1d at $t = 1$. ■

REMARK 3.7.2 The solution to (3.7.15) remains constant on the lines $x = x_0 + ct$. Similarly, the solution to (3.7.17) is made up of two terms: the first remains constant on the lines $x = x_0 + ct$, the second on the lines $x = x_0 - ct$. From this observation it should be clear that these are peculiar curves in the (t, x)-plane which deserve to be named. They are, in fact, called ***characteristics*** and, as we shall see in the next section, are very important both from the physical and the mathematical point of view. ■

3.7.3 Characteristics

To understand the concept of characteristic, consider the three-dimensional transport equation

$$\frac{\partial u}{\partial t} + \mathbf{c}(t, \mathbf{x}) \cdot \nabla u = f(t, \mathbf{x}; u)\,, \qquad \mathbf{x} \in \mathcal{D} \subseteq \mathbb{R}^3\,. \tag{3.7.21}$$

This equation is called a transport equation, since the state variable is sort of transported by the convective current $\mathbf{c}(t, \mathbf{x})$. In fact, if we consider the curves $\mathbf{x}(t)$ defined by

$$\frac{d\mathbf{x}}{dt} = \mathbf{c}(t, \mathbf{x})$$

and evaluate u on them we have that its temporal evolution is given by

$$\frac{d}{dt}u(t,\mathbf{x}(t)) = \frac{\partial u}{\partial t} + \nabla u \cdot \frac{d\mathbf{x}}{dt} \cdot$$

Recalling, then, Eq.(3.7.21) we obtain

$$\frac{d}{dt}u(t,\mathbf{x}(t)) = f\Big(t,\mathbf{x}(t); u(t,\mathbf{x}(t))\Big) .$$

The line

$$\begin{cases} \dfrac{d\mathbf{x}}{dt} = \mathbf{c}(t,\mathbf{x}) \\ \mathbf{x}(\hat{t}) = \widehat{\mathbf{x}} \end{cases} \qquad (3.7.22)$$

is the so-called ***characteristic*** through $(\hat{t},\widehat{\mathbf{x}})$.

The evolution of the solution on it is then given by

$$\begin{cases} \dfrac{d}{dt}u(t,\mathbf{x}(t)) = f\Big(t,\mathbf{x}(t); u(t,\mathbf{x}(t))\Big) \\ u(\hat{t},\widehat{\mathbf{x}}) = \widehat{u}\,, \end{cases} \qquad (3.7.23)$$

where $\widehat{u}$ is the value of u in $(\hat{t},\widehat{\mathbf{x}})$.

Thus we have reduced the solution of a first-order ***partial*** differential equation to the solution of two coupled first-order ***ordinary*** differential equations.

REMARK 3.7.3 If $f = 0$, then u remains constant along the characteristics. ■

REMARK 3.7.4 It is important to distinguish between

- Characteristics that start inside the domain $\mathcal{D}$ at time $t = 0$. In this case, one usually takes $\hat{t} = 0$, $\widehat{\mathbf{x}} \in \mathcal{D}$ and $\widehat{u} = u(t = 0,\widehat{\mathbf{x}}) = u_{in}(\widehat{\mathbf{x}})$ where u_{in} is the initial condition.
- Characteristics that start on the boundary $\partial\mathcal{D}$ of the domain. In this case, one usually takes $\widehat{\mathbf{x}} \in \partial\mathcal{D}$ and $\widehat{u} = u(\hat{t},\widehat{\mathbf{x}}) = u_{\partial\mathcal{D}}(\hat{t},\widehat{\mathbf{x}})$ where $u_{\partial\mathcal{D}}$ is the boundary condition.

This distinction is sketched in Fig. 3.7.2. ■

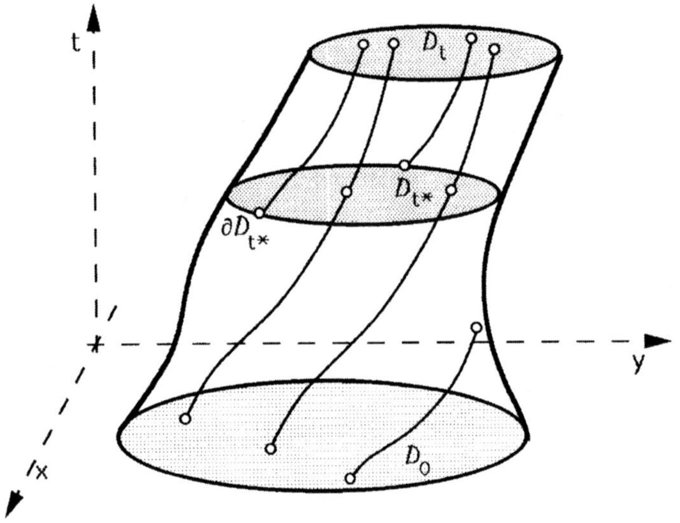

Figure 3.7.2 — The characteristic on the extreme right starts in the interior of $\mathcal{D}(t = 0)$ and meets the boundary at some $t < t^*$. The characteristic on the extreme left instead enters $\mathcal{D}$ from the boundary.

Example 3.7 Characteristics
The characteristics of the equation

$$\frac{\partial u}{\partial t} + (\alpha x + \beta t + \gamma)\frac{\partial u}{\partial x} = f(t, x; u) \tag{3.7.24}$$

can be found by solving the ordinary differential equation

$$\frac{dx}{dt} = \alpha x + \beta t + \gamma, \tag{3.7.25}$$

the solution of which is

$$x(t) = \begin{cases} Ce^{\alpha t} - \dfrac{\alpha\gamma + \beta}{\alpha^2} - \dfrac{\beta}{\alpha}t, & \text{if } \alpha \neq 0 \\ C + \gamma t + \dfrac{\beta}{2}t^2, & \text{if } \alpha = 0, \end{cases} \tag{3.7.26}$$

where C is an integration constant. Hence the characteristic through the point $(\hat{t}, \widehat{x})$ is

$$x(t) = \begin{cases} \widehat{x} e^{\alpha(t-\hat{t})} + \dfrac{\alpha\gamma + \beta}{\alpha^2}\left(e^{\alpha(t-\hat{t})} - 1\right) + \dfrac{\beta}{\alpha}\left(\hat{t} e^{\alpha(t-\hat{t})} - t\right), & \text{if } \alpha \neq 0, \\ \widehat{x} + \gamma(t - \hat{t}) + \dfrac{\beta}{2}(t^2 - \hat{t}^2), & \text{if } \alpha = 0. \end{cases} \tag{3.7.27}$$

On the characteristics the solution satisfies the initial-value problem

$$\begin{cases} \dfrac{d}{dt} u\big(t, \mathbf{x}(t)\big) = f\Big(t, x(t); u\big(t, x(t)\big)\Big) \\ u(\hat{t}, \widehat{x}) = \widehat{u}. \end{cases}$$

In particular, if $f = 0$, then the state variable remains constant along the characteristics. □

The concept of characteristics can be generalized to hyperbolic systems of first-order partial differential equations and to higher-order systems, but, as the reader can expect from the previous examples, the subject becomes too specialized to be dealt with in this volume. We just want to mention here that if Eq.(3.7.1) is hyperbolic, then there are two characteristic curves passing through a point. They are implicitly defined by $g(t, x)$ =const., where g satisfies the partial differential equation

$$A(t,x)\left(\frac{\partial g}{\partial t}\right)^2 + 2B(t,x)\frac{\partial g}{\partial t}\frac{\partial g}{\partial x} + C(t,x)\left(\frac{\partial g}{\partial x}\right)^2 = 0\,.$$

REMARK 3.7.5 As already seen, the value of the solution of a first-order partial differential equation (PDE) in a point (t, x) depends on the values on the characteristic through that point. Similarly for a second-order PDE, the two characteristics through (t, x) divide, as shown in Fig. 3.7.3, the (t, x)-plane into two regions: One upstream and one downstream. The former is called the ***domain of dependence*** since the characteristics out of it will not influence the value at (t, x). The latter is called the ***domain of influence*** since there the solution is influenced by the solution in (t, x). ■

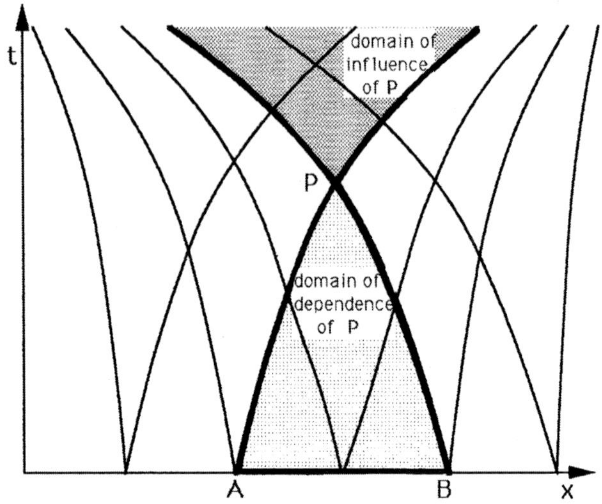

Figure 3.7.3 — Domains of dependence and of influence.

3.8 Mathematical Formulation of Problems

This section deals with the mathematical formulation of problems related to the models just presented and briefly examined in the preceding sections.

A line similar to that of the preceding chapter will be followed. In other words, the following topics will be dealt with:

- Definition of the consistency properties of a mathematical model.
- Definition of the correct formulation of a mathematical problem.
- Well-posedness of a mathematical problem.

As in the case of discrete dynamic models, it is understood that the solution of the mathematical problem can be attempted, for instance by means of the techniques to be proposed in this chapter, only if the mathematical problem is correctly formulated.

In principle, the problem should also be well posed and, as far as possible, the proof of the well-posedness of the problem should also be attempted. The reader, however, needs to be aware of the difficulties that can often arise in the proof of well-posedness of mathematical problems related to nonlinear partial differential equations.

In this section, first we will discuss the problems which have been defined above at a general level. In particular, in this section we will only introduce some examples of ill-posed problems, which are indispensable for the understanding of the numerical methods that will be dealt with later on. More on this topic can be found in Chapter 4.

As a first step, we will focus on what is meant by consistency.

DEFINITION 3.8.1 Consistency

A mathematical model defined by a system of partial differential equations is ***consistent*** *if the number of equations is equal to the dimension of the state-dependent variable.*

The definition stated above does not differ from the corresponding one, which referred to a discrete dynamic system. We have already used this concept in Section 3.4.5 to explain the need of constitutive laws in continuum mechanics.

Example 3.8 Consistency

Consider the mathematical model of the small vibrations of an elastic string in a plane in the absence of dissipative forces and with an elastic coefficient $c = c(\theta)$ modelled as depending on the temperature θ. The evolution state equation is

$$\frac{\partial^2 u}{\partial t^2} = c(\theta)\frac{\partial^2 u}{\partial x^2}, \tag{3.8.1}$$

which is not ***consistent***. In fact only one equation is given for a two-dimensional state variable: $\mathbf{u} = \{u, \theta\}$.

It is necessary, in order to define a consistent model, to link Eq.(3.8.1) to the evolution equation (model) for the variable θ. For instance, the parabolic model for the heat diffusion phenomenon can be adopted

$$\frac{\partial \theta}{\partial t} = \frac{\partial}{\partial x}\left(\kappa(\theta)\frac{\partial \theta}{\partial x}\right) + \phi(x, \theta), \tag{3.8.2}$$

where ϕ models the dispersion along the string.

The evolution equation can be written in normal vector form using the following change of variables

$$u_1 = u, \qquad u_2 = \frac{\partial u}{\partial t}, \qquad u_3 = \theta, \tag{3.8.3}$$

which yields

$$\begin{cases} \dfrac{\partial u_1}{\partial t} = u_2 \\ \dfrac{\partial u_2}{\partial t} = c^2(u_3)\dfrac{\partial^2 u_1}{\partial x^2} \\ \dfrac{\partial u_3}{\partial t} = \dfrac{\partial}{\partial x}\left(\kappa(u_3)\dfrac{\partial u_3}{\partial x}\right) + \phi(x, u_3)\,. \end{cases} \tag{3.8.4}$$

□

Once the consistency of the model has been verified, it is possible to deal with the formulation of the mathematical problem. To be more specific, consider an abstract parabolic model in one space variable

$$u = u(t,x)\ :\quad [0,T]\times[a,b] \longmapsto \mathbb{R}\,,$$

$$\frac{\partial u}{\partial t} = f\left(t, x; u, \frac{\partial u}{\partial x}, \frac{\partial^2 u}{\partial x^2}\right)\,, \tag{3.8.5}$$

which may also involve a suitable set of parameters.

It is clear that, in order to study the evolution of the system, we have to give its initial state and, sometimes, the behavior at the extrema of the space domain.

We now have to identify what information is enough to be able to solve the problem. In order to do that we introduce the following definition.

DEFINITION 3.8.2 A well-formulated problem
If the model is provided with enough initial and boundary conditions to find a solution, then the relative mathematical problem is said to be ***well-formulated.***

As we shall see in the following examples, it is difficult to give a general rule that covers all cases. What can be said is that, in most cases, the number of boundary (respectively, initial) conditions has to be equal to the order of the highest space (respectively, time) derivative.

If $\mathbf{x} \in \mathbb{R}^d$ with $d = 2, 3$, then the boundary conditions have to be assigned for all space variables. If the model is defined by a system of equations, then initial and boundary conditions have to be assigned for each equation according to the rules stated above.

REMARK 3.8.1 If both time and space derivatives appear in the mathematical model, then both initial and boundary conditions usually have to be assigned. The relative mathematical problem is then called an ***initial-boundary-value problem***. If, instead, the mathematical model is static, i.e., time independent, then, of course, no initial conditions are needed. In this case, the mathematical problem is defined as a ***boundary-value problem***. ■

The mathematical problem related to Eq.(3.8.5) is stated by linking to it one initial condition

$$u(0,x) = u_{in}(x)\,, \qquad \forall x \in [a,b] \tag{3.8.6}$$

and two boundary conditions, since in Eq.(3.8.5) a first-order time derivative appears on the left-hand side and a second-order space derivative appears as an argument of f on the right-hand side. The boundary conditions may involve the state variable, say

$$u(t,a) = u_a(t)\,, \qquad \forall t \in [0,T]\,, \tag{3.8.7}$$

or its first space derivative, say

$$\frac{\partial u}{\partial x}(t,a) = \widetilde{u}_a(t)\,, \qquad \forall t \in [0,T]\,, \tag{3.8.8}$$

or maybe a linear combination of them

$$\alpha(t)\frac{\partial u}{\partial x}(t,a) + \beta(t)u(t,a) = \widehat{u}_a(t)\,, \qquad \forall t \in [0,T]\,. \tag{3.8.9}$$

Equation (3.8.7) is called a ***Dirichlet boundary condition***, equation (3.8.8) a ***Neumann boundary condition***, and equation (3.8.9) a ***Robin boundary condition***.

Referring to the heat conduction model, giving a Dirichlet boundary condition means imposing a known temperature on the boundary, while giving a Neumann boundary condition means imposing a known heat flux through the boundary.

REMARK 3.8.2 Of course several combinations are possible. For instance, one can impose a Dirichlet boundary condition at $x = a$ and a

Neumann boundary condition at $x = b$, or impose both of them on the same boundary point. Both problems are well formulated, but it can be shown that the second is not well posed. This can be understood using the heat conduction model. In fact, imposing both conditions on the same boundary would mean prescribing on it both the temperature and the heat flux and these two physical quantities are not independent. Being more specific imposing, say, a isothermal and adiabatic boundary condition

$$u(t,a) = u_a\,, \qquad \frac{\partial u}{\partial x}(t,a) = 0 \tag{3.8.10}$$

is not physical. ■

The generalization to more space variables and to systems of partial differential equations is technical. In particular, referring to Fig. 3.8.1, the heat conduction problem

$$\frac{\partial u}{\partial t} = \kappa \nabla^2 u + f(t, \mathbf{x}; u)\,, \qquad \mathbf{x} \in \mathcal{D} \tag{3.8.11}$$

is well formulated if joined, for instance, to the initial condition

$$u(t = 0, \mathbf{x}) = u_{in}(\mathbf{x})\,, \qquad \mathbf{x} \in \mathcal{D}$$

and to the boundary conditions

$$\begin{aligned} u(t,\mathbf{x}) &= u_D(t,\mathbf{x})\,, &\quad \text{if } \mathbf{x} \in \partial\mathcal{D}_D\,,\\ \mathbf{n}(t,\mathbf{x}) \cdot \nabla u(t,\mathbf{x}) &= u_N(t,\mathbf{x})\,, &\quad \text{if } \mathbf{x} \in \partial\mathcal{D}_N\,, \end{aligned}$$

where $\partial\mathcal{D}_D$ and $\partial\mathcal{D}_N$ are a partition of the boundary $\partial\mathcal{D}$ and $\mathbf{n}$ is the normal to $\partial\mathcal{D}_N$.

The discussion of the ***well-posedness*** of mathematical problems for partial differential equations is one of the main purposes of applied mathematics and is certainly a difficult topic. An important volume in this field is the classical book by Lions and Magenes [LaM]. The reader is referred to it for a deeper insight into this topic.

The analysis needs to be organized in the framework of the classification we have just seen. Some preliminary indications will be given in the following examples.

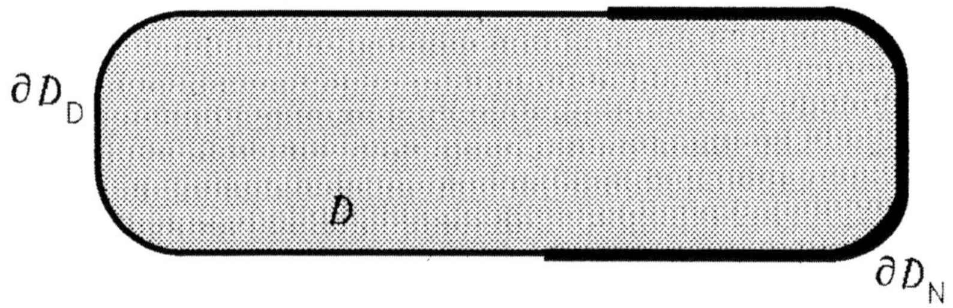

Figure 3.8.1 — Dirichlet–Neumann boundary conditions for a two-dimensional parabolic problem.

Example 3.9 Well-posedness for parabolic equations
Consider the parabolic system

$$\frac{\partial \mathbf{u}}{\partial t} = \mathbf{A}\frac{\partial^2 \mathbf{u}}{\partial x^2} + \mathbf{f}\left(t, x; \mathbf{u}, \frac{\partial \mathbf{u}}{\partial x}\right), \qquad \mathbf{u} \in \mathbb{R}^n, \tag{3.8.12}$$

which is parabolic (or Petrowski parabolic) if $\mathbf{A}$ has eigenvalues with positive real part. In order to have a well-formulated problem one needs two boundary conditions for each component. Dirichlet boundary conditions

$$\begin{aligned} \mathbf{u}(t,a) &= \mathbf{u}_a(t) \\ \mathbf{u}(t,b) &= \mathbf{u}_b(t)\,, \end{aligned} \tag{3.8.13}$$

Neumann boundary conditions

$$\begin{aligned} \frac{\partial \mathbf{u}}{\partial x}(t,a) &= \widetilde{\mathbf{u}}_a(t) \\ \frac{\partial \mathbf{u}}{\partial x}(t,b) &= \widetilde{\mathbf{u}}_b(t)\,, \end{aligned} \tag{3.8.14}$$

or Dirichlet–Neumann boundary conditions

$$\begin{aligned} \mathbf{u}(t,a) &= \mathbf{u}_a(t) \\ \frac{\partial \mathbf{u}}{\partial x}(t,b) &= \widetilde{\mathbf{u}}_b(t) \end{aligned} \tag{3.8.15}$$

all yield well-formulated and well-posed problems. On the other hand, for instance, the boundary conditions

$$\begin{aligned} &\mathbf{u}(t,a) = \mathbf{u}_a(t) \\ &\frac{\partial \mathbf{u}}{\partial x}(t,a) = \widetilde{\mathbf{u}}_a(t)\,, \end{aligned} \tag{3.8.16}$$

yield a well-formulated problem, which, as we shall see in Chapter 4, is not well posed.

In general, if $\mathbf{L}_0$ is an $n_a \times n$ matrix, $\mathbf{L}_1$ and $\mathbf{L}_2$ are $(n - n_a) \times n$ matrices, $\mathbf{R}_0$ is an $n_b \times n$ matrix, and $\mathbf{R}_1$ and $\mathbf{R}_2$ are $(n - n_b) \times n$ matrices, the boundary conditions

$$\begin{cases} \mathbf{L}_0\mathbf{u}(t,a) = \mathbf{u}_a(t) \\ \mathbf{L}_1\dfrac{\partial \mathbf{u}}{\partial x}(t,a) + \mathbf{L}_2\mathbf{u}(t,a) = \widetilde{\mathbf{u}}_a(t) \\ \mathbf{R}_0\mathbf{u}(t,b) = \mathbf{u}_b(t) \\ \mathbf{R}_1\dfrac{\partial \mathbf{u}}{\partial x}(t,b) + \mathbf{R}_2\mathbf{u}(t,b) = \widetilde{\mathbf{u}}_b(t) \end{cases} \tag{3.8.17}$$

yield a well-posed problem if the matrices

$$\begin{pmatrix} \mathbf{L}_0 \\ \mathbf{L}_1\mathbf{A}^{-1/2} \end{pmatrix} \quad \text{and} \quad \begin{pmatrix} \mathbf{R}_0 \\ \mathbf{R}_1\mathbf{A}^{-1/2} \end{pmatrix}$$

are invertible. □

As mentioned at the beginning of Section 3.7.2, elliptic problems can be seen as time independent problems that derive from parabolic problems when the transient term has died out. Then one has only to deal with boundary conditions and for the reason stated above the discussion does not differ much from that just given.

Hyperbolic problems are a lot more complex and a lot more care is needed in giving the boundary conditions. The subject is strongly related to the behavior of the characteristics near the boundaries. To understand this point consider the following one-dimensional example.

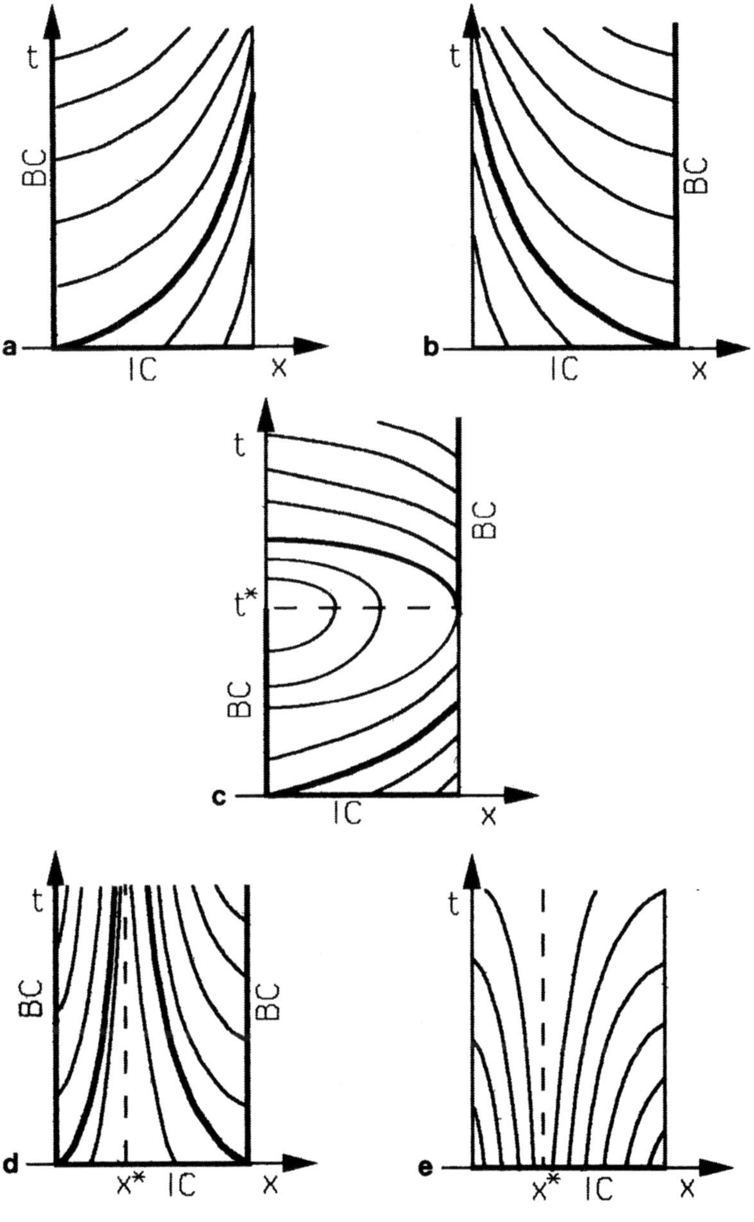

Figure 3.8.2 — Characteristics for the transport equation (3.8.18). BC indicates where to impose the natural boundary conditions. (a) $c > 0$ (b) $c < 0$ (c) c changes sign at $t = t^*$ (d) $c(a) > 0$, $c(b) < 0$ (e) $c(a) < 0$, $c(b) > 0$.

Example 3.10 Boundary conditions for hyperbolic equations
Consider the one-dimensional transport equation

$$\frac{\partial u}{\partial t} + c(t,x)\frac{\partial u}{\partial x} = 0\,, \qquad x \in [a,b]\,, \tag{3.8.18}$$

with c a smooth function of t and x.

We have already seen in Section 3.7.3 that the characteristics are given by

$$\frac{dx}{dt} = c(t,x)$$

and that the solution is constant along them. Several cases can then occur. The following is a partial scheme, but still sufficient to understand the underlying dynamics.

- $c(t,a) > 0$ and $c(t,b) > 0\,, \quad \forall t\,.$
  In this case it is natural to give (besides the initial conditions) the boundary condition on $x = a$ (Fig. 3.8.2a), so that the initial and boundary condition are propagated throughout the domain and to the other boundary. For instance, referring to Example 3.7, if $c(t,x) = c > 0$, then the solution to

$$\begin{cases} \dfrac{\partial u}{\partial t} + c\dfrac{\partial u}{\partial x} = 0 \\ u(0,x) = u_{in}(x) \\ u(t,x=a) = u_a(t) \end{cases} \tag{3.8.19}$$

  is

$$u(t,x) = \begin{cases} u_{in}(x-ct)\,, & \text{if } x > a + ct \\ u_a\left(t - \dfrac{x-a}{c}\right)\,, & \text{if } x < a + ct\,. \end{cases}$$

  Observe that if, for instance, $u_a(t=0) \neq u_{in}(x=a)$ then a discontinuity propagates along $x = a + ct$.
  If instead we try to impose $u(\hat{t}, x = b)$, say for $\hat{t} = (b-a)/2c$, and the value of u at the "other extremum" of the characteristic $u_{in}\big(x = (a+b)/2\big)$ is not equal, then we have a contradiction.
  The discussion is similar for the case

$$c(t,a) < 0 \qquad \text{and} \qquad c(t,b) < 0\,, \qquad \forall t$$

  represented in Fig. 3.8.2b.

- If, for instance,

$$c(t,a) = \begin{cases} > 0\,, & \text{if } t < \hat{t} \\ < 0\,, & \text{if } t > \hat{t} \end{cases} \qquad \text{and} \qquad c(t,b) = \begin{cases} > 0\,, & \text{if } t < \hat{t} \\ < 0\,, & \text{if } t > \hat{t} \end{cases}$$

(e.g., $f(t,x) = \hat{t} - t$), then referring to Fig. 3.8.2c, the natural boundary conditions are

$$u(t, x = a) = u_a(t)\,, \quad \text{if} \quad t < \hat{t}$$
$$u(t, x = b) = u_b(t)\,, \quad \text{if} \quad t > \hat{t}\,.$$

The discussion is similar if the signs are reversed.

- If, for instance,

$$c(t,a) > 0 \qquad \text{and} \qquad c(t,b) < 0\,, \qquad \forall t\,,$$

then referring to Fig. 3.8.2d, the natural boundary conditions are to be given, for any t, on both boundary points

$$u(t, x = a) = u_a(t)\,, \qquad u(t, x = b) = u_b(t)\,. \tag{3.8.20}$$

To better understand this case we now integrate the initial-boundary-value problem (3.8.18), (3.8.20), with $c(t,x) = x^* - x$ and $x^* \in (a,b)$. The characteristics through the points $t = 0$, $x = a$ and $t = 0$, $x = b$ are, respectively,

$$x = x^* + (a - x^*)e^{-t} \tag{3.8.21}$$

and

$$x = x^* + (b - x^*)e^{-t}\,. \tag{3.8.22}$$

The initial condition will only suffice to determine the value of u between these two characteristics. It is

$$u(t,x) = u_{in}\Big(x^* + (x - x^*)e^t\Big)$$

if

$$x \in \Big[x^* + (a - x^*)e^{-t}, x^* + (b - x^*)e^{-t}\Big]\,. \tag{3.8.23}$$

This is an example of the well-known property of first-order hyperbolic partial differential equations that the values in an interval $[\alpha, \beta]$ at time t_0 will only influence the values between the characteristics

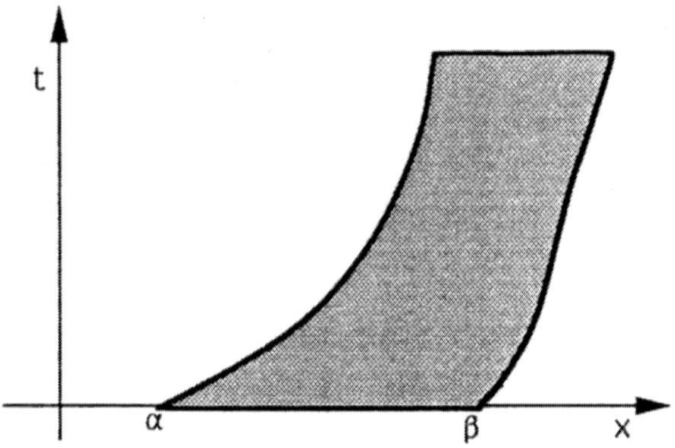

Figure 3.8.3 — Domain of influence of the interval $[\alpha, \beta]$ for a first-order hyperbolic equation.

through (t_0, α) and (t_0, β). In other words, the values of u above the characteristic through (t_0, α) cannot be determined by the values below it, as sketched in Fig. 3.8.3. Of course a similar thing occurs for the other characteristic. We will therefore say that the characteristic cannot be crossed.

The left boundary condition can only determine the values above (3.8.21)

$$u(t,x) = u_a\left(t + \log\frac{x - x^*}{a - x^*}\right), \qquad \text{if} \qquad x \in \left[a, x^* + (a - x^*)e^{-t}\right].$$

In order to compute $u(t, x)$ for $x \in [x^* + (b - x^*)e^{-t}, b]$ one needs, as already stated, the right boundary condition, too. Doing that one obtains

$$u(t,x) = u_b\left(t + \log\frac{x - x^*}{b - x^*}\right), \qquad \text{if} \qquad x \in \left[x^* + (b - x^*)e^{-t}, b\right].$$

- *Vice versa*, if

$$c(t, a) < 0 \qquad \text{and} \qquad c(t, b) > 0, \qquad \forall t,$$

then referring to Fig. 3.8.2e, one does not need any boundary condition at all. The values at the boundaries are fully determined by the initial condition.

To better understand this case we integrate the initial-boundary-value problem (3.8.18), (3.8.20), with $c(t,x) = x - x^*$ and $x^* \in (a,b)$. The characteristic through $(\hat{t}, \widehat{x})$ is

$$x(t) = x^* + (\widehat{x} - x^*)e^{t-\hat{t}}$$

and therefore for any $\hat{t}$ and $\widehat{x}$

$$\begin{aligned} u(\hat{t},\widehat{x}) &= u\Big(s, x^* + (\widehat{x} - x^*)e^{s-\hat{t}}\Big) \\ &= u\Big(0, x^* + (\widehat{x} - x^*)e^{-\hat{t}}\Big) = u_{in}\Big(x^* + (\widehat{x} - x^*)e^{-\hat{t}}\Big), \end{aligned}$$

that is,

$$u(t,x) = u_{in}\big(x^* + (x - x^*)e^{-t}\big).$$

In particular, the values at both boundaries

$$u(t,a) = u_{in}\big(x^* + (a - x^*)e^{-t}\big)$$

and

$$u(t,b) = u_{in}\big(x^* + (b - x^*)e^{-t}\big)$$

are fully determined by the initial conditions. □

Though this scheme is not exhaustive, it already gives an idea on how critical is the statement of the boundary conditions for hyperbolic partial differential equations. What can be said in general is that the number of boundary conditions at each boundary point has to be equal to the number of incoming characteristics. This can be simple to recognize for a first-order partial differential equation, but for hyperbolic systems of equations like

$$\mathbf{A}(t,x;\mathbf{u})\frac{\partial \mathbf{u}}{\partial t} + \mathbf{B}(t,x;\mathbf{u})\frac{\partial \mathbf{u}}{\partial x} = \mathbf{f}(t,x;\mathbf{u}), \tag{3.8.24}$$

it is much more complex.

Section 3.7.1 states that, in order to be hyperbolic, the eigenvalues related to (3.8.24) have to be all real and distinct. The system is then diagonalizable and can be written as

$$\frac{\partial v_i}{\partial t} + \lambda_i \frac{\partial v_i}{\partial x} = \widetilde{f}_i(t,x;\mathbf{v}), \qquad i = 1,\ldots,n, \tag{3.8.25}$$

where v_i is a linear combination of the components u_j of the state variable (i.e., the eigenvector related to the eingenvalue λ_i). If now all the eigenvalues are positive (respectively, negative), then all characteristics go to the right (respectively, left) and therefore one has to specify all v_i and therefore $\mathbf{u}$, on the left (respectively, right) boundary.

On the other hand, if some eigenvalues are positive, say $\lambda_1, \ldots, \lambda_{n'}$, and some, say $\lambda_{n'+1}, \ldots, \lambda_n$, are negative then some characteristics go to the left and some to the right. Therefore, one needs to give $v_1, \ldots, v_{n'}$ on the left boundary and $v_{n'+1}, \ldots, v_n$ on the right boundary. How to transfer this knowledges on the formulation of the proper boundary conditions for the original state variables $u_1, \ldots, u_n$ is a trivial matter.

As a final observation we note that second-order partial differential equations need two boundary conditions to be well formulated, one at each end to be well posed.

The generalization to systems or to higher-dimensional problems follows the line of thought previously outlined for parabolic equations, though in this case things are a lot more difficult.

As we have already seen for discrete systems, the proper formulation of the problem does not imply that the mathematical problem is also ***well posed***, i.e., the solution exists, is unique, and depends continuously on the initial data. The qualitative analysis of the problem can be based upon several methods and, in particular on the fixed-point theorem reported in Appendix 1. In this case the operator $\mathcal{U}$ must include both initial and boundary conditions.

Useful books that develop this topic are, among others, the classical volume by Courant and Hilbert [CaH] for the analysis of linear problems and by Lions and Magenes [LaM] for several interesting nonlinear initial-boundary-value problems. Further indications to the reader can be given with reference to the semigroup analysis of evolution equation by Belleni Morante [BEa] and [BEb] and to Martin [MAR]. In particular, [BEb] provides a concise guide which is useful and effective for several interesting applications.

It is worth mentioning that several practical situations may be such that the mathematical problem may not be well posed and may not even be well formulated. This occurs when the initial and boundary conditions cannot be properly identified. These types of problems, which are relevant in applied sciences, are dealt with in Chapter 4.

We also have to mention that the qualitative analysis of mathematical problems refers not only to the existence of solutions, but also to the related stability analysis. Several aspects of this topic were dealt with in Chapter 2 with reference to discrete dynamic systems. Some of the ideas reported in Chapter 2 can also be transferred to the analysis of continuous dynamic systems. In fact the stability analysis can be developed, similarly to the analysis of discrete systems, by a detailed insight into the spectral properties of the linearized equation or by a suitable analysis of Liapunov (energy) functionals. This topic is not, however, dealt with in this book. In fact the stability analysis of continuous models requires several technical developments that can be organized in a specialized literature.

The interested reader is referred to Galdi and Padula [GPD] for direct methods based upon detailed investigations of the structural properties of the abstract equations, to the classical book by Joseph [JOa] for the stability of fluid motion, to Dym [DYM] and Knops and Wilkes [KaW] for the stability in elasticity and to Galdi and Rionero [GaR] and Straughan [STR] for the application of energy methods in fluid dynamics and elasticity. The reader who enters into this field, must also be aware, as documented in Joseph [JOa] or in Joseph and Renardy [JaR] that very often the stability analysis must be related to a detailed investigation of the physical properties of the mathematical model. Therefore, also in this case, mathematical methods and modelling methods can find an important contact point.

3.9 Finite Difference Methods

Once the mathematical problem has been correctly formulated and studied at the qualitative level, we can turn our attention to the numerical simulation. The basic procedure to be used before massive computation is similar to that already dealt with for discrete models at the beginning of Section 2.8 and depicted in Table 2.8.1.

Also the spirit of this and of the following section is the same which motivated Section 2.8: Furnishing a *vade mecum* that serves as a starting point for the unskilled reader.

As we have mentioned several times throughout this chapter, the study and application of partial differential equations is a huge subject. As a straightforward consequence, there is a large variety of numerical methods dealing with several cases. Some of them are too specialized to be presented here. Some are improvements of the simple methods given here to particular continuous models.

This section will provide an introduction to ***finite difference methods*** for evolution equations in the same line of thought followed in the previous chapter, referring to Sewell [SEW] and Strikwerda [STI] for deeper insight.

Most of the schemes will deal with one space dimension problems. The generalization to more space dimensions will be briefly dealt with at the end of the section.

Section 3.10 will provide, in the same line of thought, a brief introduction to ***collocation methods***. Finally in Section 3.11 the spirit and need of domain decomposition methods will be dealt with.

3.9.1 Basic idea and fundamental aspects

The first step of the finite-difference methods is to divide the interval $[a, b]$ into N subintervals, usually of the same width Δx,

$$a = x_0 < x_1 < \cdots < x_{N-1} < x_N = b \, .$$

The basic idea is then to replace every partial derivative that appears in the differential equation and in the initial and boundary conditions by an appropriate approximation. Of course, this can be done in several ways. For instance, first-order space derivatives can be approximated as

$$\frac{\partial \mathbf{u}}{\partial x}(t,x) \cong \frac{\mathbf{u}(t,x+\Delta x) - \mathbf{u}(t,x)}{\Delta x} \tag{3.9.1}$$

or as

$$\frac{\partial \mathbf{u}}{\partial x}(t,x) \cong \frac{\mathbf{u}(t,x) - \mathbf{u}(t,x-\Delta x)}{\Delta x} \, . \tag{3.9.2}$$

Both of them are first-order approximations called, respectively, ***forward*** and ***backward difference***.

Better accuracy is given by the ***central difference*** approximation

$$\frac{\partial \mathbf{u}}{\partial x}(t,x) \cong \frac{\mathbf{u}(t,x+\Delta x) - \mathbf{u}(t,x-\Delta x)}{2\Delta x} \, , \tag{3.9.3}$$

which is second order.

As far as higher derivatives are concerned we recall the central differences approximations

$$\begin{aligned}
\frac{\partial^2 \mathbf{u}}{\partial x^2}(t,x) &\cong \frac{1}{\Delta x^2}\Big[\mathbf{u}(t,x+\Delta x) - 2\mathbf{u}(t,x) + \mathbf{u}(t,x-\Delta x)\Big] \\
\frac{\partial^3 \mathbf{u}}{\partial x^3}(t,x) &\cong \frac{1}{\Delta x^3}\Big[\mathbf{u}(t,x+2\Delta x) - 2\mathbf{u}(t,x+\Delta x) \\
&\quad + 2\mathbf{u}(t,x-\Delta x) - \mathbf{u}(t,x-2\Delta x)\Big] \\
\frac{\partial^4 \mathbf{u}}{\partial x^4}(t,x) &\cong \frac{1}{\Delta x^4}\Big[\mathbf{u}(t,x+2\Delta x) - 4\mathbf{u}(t,x+\Delta x) \\
&\quad + 6\mathbf{u}(t,x) - 4\mathbf{u}(t,x-\Delta x) + \mathbf{u}(t,x-2\Delta x)\Big] \, .
\end{aligned} \tag{3.9.4}$$

Once the finite difference approximations, say $\delta_x \mathbf{u}$, $\delta_{xx} \mathbf{u}$, ..., are chosen to approximate $\partial \mathbf{u}/\partial x$, $\partial^2 \mathbf{u}/\partial x^2$, ... , the system of partial differential equations, say

$$\frac{\partial \mathbf{u}}{\partial t}(t,x) = \mathbf{f}\left(t, x; \mathbf{u}, \frac{\partial \mathbf{u}}{\partial x}, \frac{\partial^2 \mathbf{u}}{\partial x^2}, \dots\right), \qquad (3.9.5)$$

is approximated by the system of ordinary differential equations

$$\frac{d\mathbf{u}}{dt}(t,x_j) = \mathbf{f}\Big(t, x_j; \mathbf{u}(t,x_j), \delta_x \mathbf{u}(t,x_j), \delta_{xx}\mathbf{u}(t,x_j), \dots\Big), \qquad (3.9.6)$$

which can now be solved using any of the methods presented in Section 2.8.

REMARK 3.9.1 The notation in (3.9.6) is slightly misleading. Despite the triviality of the following statement we feel it is essential to remark explicitly that, for instance, the expression of the fourth-order derivative of $\mathbf{u}$ in x_j in (3.9.4) is not only a function of $\mathbf{u}(t, x_j)$ but also of the neighborhood values $\mathbf{u}(t, x_{j-2})$, $\mathbf{u}(t, x_{j-1})$, $\mathbf{u}(t, x_{j+1})$, and $\mathbf{u}(t, x_{j+2})$. ∎

As in Section 2.8, the following definition can be introduced.

DEFINITION 3.9.1 Explicit, implicit, and semi-implicit
If in $\mathbf{f}$ *all quantities are evaluated at time* t_i *then one has an* ***explicit method****. If on the other hand the right-hand side of Eq.(3.9.5) is evaluated at time* t_{i+1}*, then one has an* ***implicit method****. If finally some of the terms in the right-hand side of Eq.(3.9.5) are evaluated at time* t_i *and some at time* t_{i+1} *then one has a so-called* ***semi-implicit method****.*

REMARK 3.9.2 Explicit methods are very easy to be implemented for any $\mathbf{f}$. In fact, its arguments are all known, since they are computed at time t_i. Implicit and semi-implicit methods, instead, need in general the solution of a nonlinear system of equations since the unknown values $u(t_{i+1}, x_j)$ appear also in $\mathbf{f}$. If $\mathbf{f}$ is linear in $\mathbf{u}$ and in its derivatives, then the system to be solved is linear. As mentioned in Section 2.8, implicit methods have better stability properties. ∎

Often, to advance in time, a finite difference approximation is again used and also this can be done in several ways. If, for instance, the time derivative is approximated by the forward difference approximation

$$\frac{\partial \mathbf{u}}{\partial t}(t, x_j) \cong \frac{\mathbf{u}(t + \Delta t, x_j) - \mathbf{u}(t, x_j)}{\Delta t} \tag{3.9.7}$$

and **f** is computed explicitly, we have

$$\begin{aligned} \mathbf{u}(t + \Delta t, x_j) &= \mathbf{u}(t, x_j) \\ &+ \Delta t\, \mathbf{f}\Big(t, x_j; \mathbf{u}(t, x_j), \delta_x \mathbf{u}(t, x_j), \delta_{xx} \mathbf{u}(t, x_j), \ldots\Big) . \end{aligned} \tag{3.9.8}$$

If, instead, the time derivative is approximated by the central difference approximation

$$\frac{\partial u}{\partial t}(t, x_j) \cong \frac{u(t + \Delta t, x_j) - u(t - \Delta t, x_j)}{2\Delta t}, \tag{3.9.9}$$

we have

$$\begin{aligned} \mathbf{u}(t + \Delta t, x_j) &= \mathbf{u}(t - \Delta t, x_j) \\ &+ 2\Delta t\, \mathbf{f}\Big(t, x_j; \mathbf{u}(t, x_j), \delta_x \mathbf{u}(t, x_j), \delta_{xx} \mathbf{u}(t, x_j), \ldots\Big) . \end{aligned} \tag{3.9.10}$$

DEFINITION 3.9.2 Order of the finite-difference scheme
A finite-difference approximation is said to be n-th order in space and m-th order in time if approximation of at least order n are used to approximate space derivatives and approximation of at least order m are used to approximate time derivatives.

For instance, if **f** in Eqs.(3.9.8) and (3.9.10) has derivatives up to second order and they are evaluated by central differences, then (3.9.8) will be second order in space and first order in time and (3.9.10) second order in both space and time.

The kind of approximation used strongly depends on the type of partial differential equation to be handled. The first thing to recognize is whether a system is hyperbolic, parabolic, or elliptic.

To understand the importance of this distinction we recall from Section 3.7.2 that hyperbolic problems govern propagation-like problems with possible formation of shocks and conservation of some physical quantities. Therefore the main concern is to use a method that:

- Does not smooth out discontinuities of the solutions or of its derivative;
- Is able to preserve physical quantities as required by the conservation equation;
- Is able to detect properly the propagation speed.

One may think that these requirements are trivial to be satisfied, but this is not true. In fact, as far as the first two requirements are concerned, most finite difference schemes are dissipative in the sense that they introduce (of course unwillingly) an artificial viscosity, which smooths out the solution and in some sense "dissipates". This phenomenon is called ***diffusion***. In the simulation we have to be aware of this phenomenon and control it. However, since the artificial viscosity decreases with Δx, the accuracy can be improved by decreasing Δx.

The last requirement implies that all the frequency components that make up the wave, e.g., a saw-tooth wave, travel at the same speed. Some schemes, instead, propagate the different frequencies at different speeds. This phenomenon is often referred to as ***dispersion***. One can often reduce this effect (and also diffusion) by choosing a time step near the stability limit (which, by the way, is not very restrictive for hyperbolic problems).

Parabolic problems govern diffusion-like behaviors. They smooth out the solution, therefore the introduction of an artificial viscosity or the dispersion effect is not so important. The main trouble in dealing with parabolic problems is that the stability requirements for explicit schemes are severe and therefore implicit schemes are preferred.

Elliptic problems describe systems in equilibrium or steady states. Once every derivative is replaced by its finite-difference approximation, the partial differential equation is replaced by a system of equations in the unknowns $u(x_i, y_j)$ to be solved at the same time. Therefore the problem reduces to the solution of a large system of equations. This is usually done by iterative methods, such as Gauss–Seidel, Successive Over Relaxation (SOR), Symmetric Successive Over Relaxation (SSOR), and so on. Since their presentation falls beyond the aim of this section, we refer the reader to Hageman and Young [HaY] for an analysis of these methods and, among others, to Strikwerda [STI] for their application to elliptic partial differential equations.

Another difficulty that occurs when using finite difference schemes (which is also common to collocation methods) is that in two- and three-dimensional problems the space grid will not fit, in general, the domain as shown in Fig. 3.9.1a. The same problem may even occur in one-dimensional problems with time-dependent domains as shown in

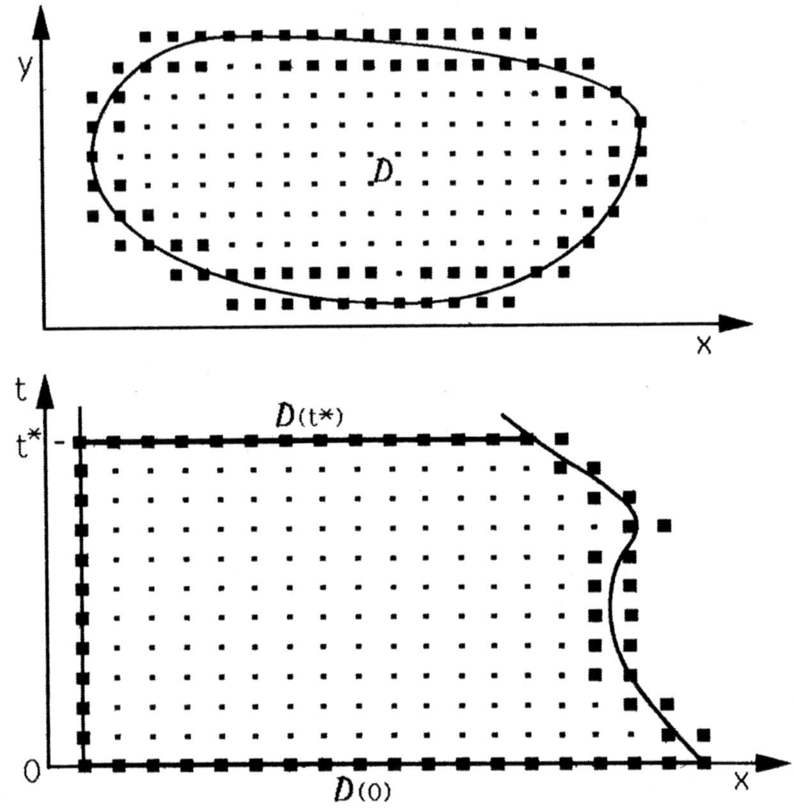

Figure 3.9.1 — Finite-difference grids.

Fig. 3.9.1b. This implies a difficulty in properly giving the boundary conditions without loosing accuracy. To overcome this difficulty one may

- Use suitable interpolation formulas to set boundary conditions;
- Map, if possible, the domain into a rectangle;
- Use, if possible, domain decomposition methods.

Anyway in this case finite element methods are much more suitable (see Akin [AKI] for an elementary introduction).

3.9.2 Finite differences for first-order hyperbolic equations

Consider the one-dimensional scalar transport equation

$$\frac{\partial u}{\partial t} + c(t,x)\frac{\partial u}{\partial x} = f(t,x;u), \qquad x \in [a,b], \tag{3.9.11}$$

joined with the initial condition

$$u(0,x) = u_{in}(x). \tag{3.9.12}$$

Consider first the case in which $c(t,x)$ is always positive. Then, as stated in Section 3.8, the natural boundary condition is to be given at the left boundary

$$u(t,a) = u_a(t). \tag{3.9.13}$$

The first step is to define a uniform grid and then to denote by

$$\begin{aligned} t_i &= i\Delta t \\ x_j &= a + j\Delta x \\ u_{i,j} &= u(t_i, x_j) \\ c_{i,j} &= c(t_i, x_j) \\ f_{i,j} &= f(t_i, x_j). \end{aligned}$$

REMARK 3.9.3 In the following the first index, usually i, will always refer to time, while the second index, usually j, will always refer to the space variable. ∎

As a first approach we may say that to compute $u_{i,j}$ we can only use the values $u_{h,k}$ with $h \le i$ and $k \le j$, otherwise we would not know how to compute the values $u_{i,N}$ at the other end of the space interval (see Fig. 3.9.2a).

The simplest conceivable scheme that satisfies this requirement is the so-called ***upwind finite-difference scheme***

$$\frac{u_{i+1,j} - u_{i,j}}{\Delta t} + c_{i,j}\frac{u_{i,j} - u_{i,j-1}}{\Delta x} = f_{i,j} \tag{3.9.14}$$

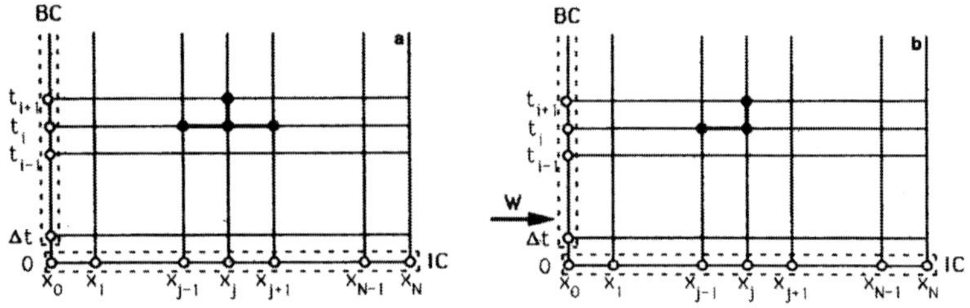

Figure 3.9.2 — (a) Central difference and (b) upwind difference stencils. BC and IC stand for the values known, respectively, by the boundary and initial conditions.

corresponding to the stencil represented in Fig. 3.9.2b. (The stencil indicates the nodes involved in the scheme.)

The initial-value problem is then rewritten as

$$\begin{cases} u_{0,j} = u_{in}(x_j) \\ u_{i+1,0} = u_a(t_{i+1}) \\ u_{i+1,j} = u_{i,j} - c_{i,j}\dfrac{\Delta t}{\Delta x}(u_{i,j} - u_{i,j-1}) + f_{i,j}\Delta t\,, \end{cases} \tag{3.9.15}$$

for $i \geq 0$ and $j = 1, \ldots, N$.

The scheme (3.9.15) is then explicit and first-order in both space and time. It can also be proved that it is stable only if

$$\Delta t \leq \frac{\Delta x}{\max|c(t,x)|}\,. \tag{3.9.16}$$

REMARK 3.9.4 As for Section 2.8 the stability conditions are usually proved for linear systems. ■

The physical meaning of the stability criterion is simple. Consider, in fact, the characteristic through $u_{i,j-1}$. As mentioned in Example 3.10 of Section 3.8, the values below that characteristic cannot properly determine values above it, that is the characteristic cannot be crossed (see Fig. 3.9.3).

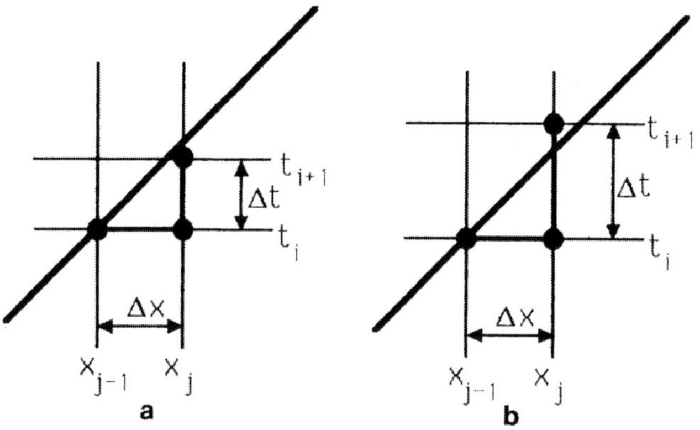

Figure 3.9.3 — Physical justification of stability limit. (a) Stable and (b) unstable time steps.

The stability condition (3.9.16) is not very restrictive, since Δt and Δx have to be of the same order anyway, because of accuracy. For this reason implicit methods are not strongly sought. In spite of this, we here recall the unconditionally stable ***implicit upwind scheme***

$$\frac{u_{i+1,j} - u_{i,j}}{\Delta t} + c_{i+1,j}\,\frac{u_{i+1,j} - u_{i+1,j-1}}{\Delta x} = f_{i+1,j}\,, \tag{3.9.17}$$

(see Fig. 3.9.4), which is often used since if f is at most linear in u, say $f(t,x;u) = F(t,x)u + g(t,x)$, it can be easily put in the explicit form

$$u_{i+1,j} = \frac{u_{i,j} + c_{i+1,j}\dfrac{\Delta t}{\Delta x}u_{i+1,j-1} + g_{i+1,j}\Delta t}{1 + c_{i+1,j}\dfrac{\Delta t}{\Delta x} - F_{i+1,j}\Delta t}\,.$$

If $c(t,x)$ is always negative then the natural boundary condition is to be given at the right boundary

$$u(t,b) = u_b(x)\,, \tag{3.9.18}$$

and the stencil to be used is that in Fig. 3.9.5, which leads to

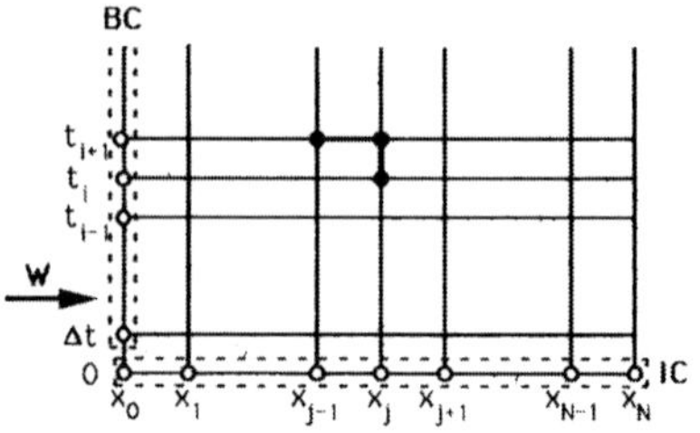

Figure 3.9.4 — Implicit upwind stencil.

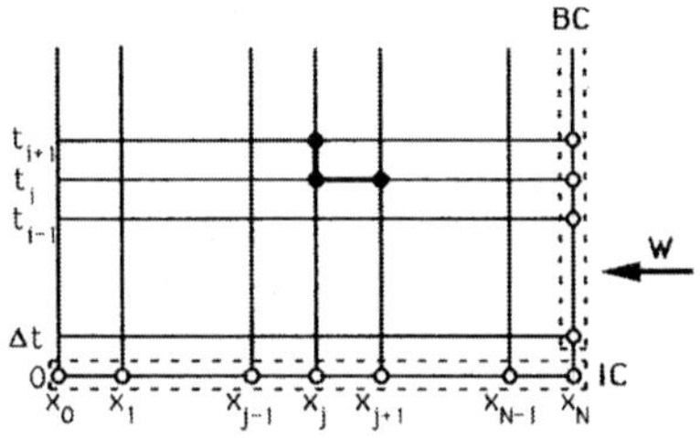

Figure 3.9.5 — Upwind stencil for $c < 0$.

$$\begin{cases} u_{0,j} = u_{in}(x_j) \\ u_{i+1,N} = u_b(t_{i+1}) \\ u_{i+1,j} = u_{i,j} - c_{i,j}\dfrac{\Delta t}{\Delta x}(u_{i,j+1} - u_{i,j}) + f_{i,j}\Delta t\,, \end{cases} \tag{3.9.19}$$

for $i \geq 0$ and $j = 0, \ldots, N-1$.

Of course, if $c(t, x)$ changes sign (as time goes by), then the method has to be properly adjusted. This requires an extra amount of care. If $c(t, x)$ changes sign with x at a fixed t, then one should change stencil according to the region.

REMARK 3.9.5 It is important to remark that using upwind differences (3.9.2) to approximate $\partial u/\partial x$ is equivalent to adding an artificial diffusion term $\frac{1}{2}|c|\Delta x \partial^2 u/\partial x^2$ to the transport equation (3.9.11) and then using the central difference scheme (3.9.4) to approximate the space derivatives of this new diffusion-convection problem. For this reason the scheme will smooth out the discontinuities or the corners of the solution. This effect is usually referred to by saying that the scheme is ***dissipative***. Of course, as $\Delta x \to 0$ the approximation will converge toward the true solution. For this reason dissipative methods work better if the solution is smooth. ■

The schemes (3.9.15) and (3.9.16) can be easily generalized to system of n first-order hyperbolic equations

$$\frac{\partial \mathbf{u}}{\partial t} + \mathbf{A}(t,x)\frac{\partial \mathbf{u}}{\partial x} = \mathbf{f}(t,x;\mathbf{u})\,, \qquad a \le x \le b\,, \tag{3.9.20}$$

in the case where the eigenvalues λ_i of $\mathbf{A}$ all have the same sign. If this is not the case, the application of the method becomes rather cumbersome, as mentioned in Section 3.8.

For instance, if all eigenvalues are positive, one can use the vector form of the upwind scheme (3.9.15)

$$\begin{cases} \mathbf{u}_{0,j} = \mathbf{u}_{in}(x_j) \\ \mathbf{u}_{i+1,0} = \mathbf{u}_a(t_{i+1}) \\ \mathbf{u}_{i+1,j} = \mathbf{u}_{i,j} - \dfrac{\Delta t}{\Delta x}\mathbf{A}_{i,j}(\mathbf{u}_{i,j} - \mathbf{u}_{i,j-1}) + \Delta t \mathbf{f}_{i,j}\,, \end{cases} \tag{3.9.21}$$

for $i \ge 0$ and $j = 1, \ldots, N$.

Equation (3.9.21) is stable if

$$\Delta t \le \frac{\Delta x}{\max_i \lambda_i}\,. \tag{3.9.22}$$

3.9.3 Higher-order schemes for first-order hyperbolic systems

The upwind scheme is only first-order accurate in space and time. A popular scheme which is second order in space and time is the ***leap-frog method*** (see Section 2.8.5), which corresponds to the central difference stencil shown in Fig. 3.9.6,

$$\frac{\mathbf{u}_{i+1,j} - \mathbf{u}_{i-1,j}}{2\Delta t} + \mathbf{A}_{i,j}\frac{\mathbf{u}_{i,j+1} - \mathbf{u}_{i,j-1}}{2\Delta x} = \mathbf{f}_{i,j} \tag{3.9.23}$$

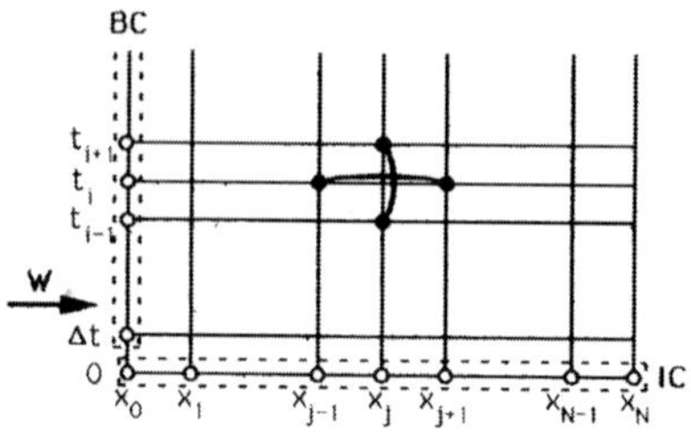

Figure 3.9.6 — Leap-frog stencil.

or

$$\mathbf{u}_{i+1,j} = \mathbf{u}_{i-1,j} - \mathbf{A}_{i,j}\frac{\Delta t}{\Delta x}(\mathbf{u}_{i,j+1} - \mathbf{u}_{i,j-1}) + 2\Delta t\mathbf{f}_{i,j}\,, \tag{3.9.24}$$

for $i > 0$ and $j = 1, \ldots, N-1$.

REMARK 3.9.6 Note that if $\mathbf{f}$ does not depend on $\mathbf{u}$, then $\mathbf{u}_{i,j}$ is not used at all in the scheme. ∎

This method has the same stability condition as the explicit upwind scheme dealt with in the previous section but it is not dissipative.

There are three main difficulties concerning its application, namely,

- A starting procedure is needed since the method involves more time levels (see Section 2.8.5 for a discussion on multistep methods).
- If $\mathbf{u}_{i,j}$ is known for $j = 0, \ldots, N$, then the scheme gives only the values $\mathbf{u}_{i+1,j}$ in the internal nodes $j = 1, \ldots, N-1$. One of the boundary values is given by the boundary condition, but the other one remains as yet unknown. A method to overcome this difficulty is to use upwind differences to compute the value of the remaining node, but this decreases the order of the method. Alternatively one can use the so-called ***quasicharacteristic extrapolations***, which, in the scalar case with $c > 0$, are written

$$u_{i+1,N} = u_{i,N-1} \tag{3.9.25a}$$

or

$$u_{i+1,N} = 2u_{i,N-1} - u_{i-1,N-2}\,. \tag{3.9.25b}$$

Anyway, special care is needed in this case, since conditions like (3.9.25) are kinds of numerical boundary conditions that may render the scheme unstable (which somewhat corresponds to rendering the problem ill posed). The easiest solution is just to try to keep in mind what was just said.

- Though the leap-frog method gives a better resolution near nonregular points, it gives a less smooth solution that presents high frequency oscillations of computational nature which are propagated without damping. We recall that, as mentioned in Section 2.8.5, the leap-frog scheme presents an extraneous solution of period $2\Delta t$ of the difference equation. A method to smooth out these oscillations is to average the new solution with that obtained at the previous step or to add a small dissipative term. Of course, neither remedy has a rigorous justification, but both essentially work.

Another popular method often used, above all for conservation equations

$$\frac{\partial \mathbf{u}}{\partial t} + \frac{\partial \mathbf{F}(\mathbf{u})}{\partial x} = \mathbf{0}, \qquad x \in [a, b], \tag{3.9.26}$$

is the Lax–Wendroff method

$$\mathbf{u}_{i+1,j} = \mathbf{u}_{i,j} - \frac{\Delta t}{2\Delta x}(\mathbf{F}_{i,j+1} - \mathbf{F}_{i,j-1}) + \left(\frac{\Delta t}{2\Delta x}\right)^2$$
$$\times \Big[(\mathbf{J}_{i,j+1} + \mathbf{J}_{i,j})(\mathbf{F}_{i,j+1} - \mathbf{F}_{i,j}) - (\mathbf{J}_{i,j} + \mathbf{J}_{i,j-1})(\mathbf{F}_{i,j} - \mathbf{F}_{i,j-1})\Big], \tag{3.9.27}$$

where $\mathbf{J}$ is the Jacobian of $\mathbf{F}$. This method is second order both in space and time.

If $\mathbf{F}(\mathbf{u}) = \mathbf{A}\mathbf{u}$, then it is possible to prove that the method is stable if

$$\Delta t \le \frac{\Delta x}{\max_i |\lambda_i|}, \tag{3.9.28}$$

where λ_i are the eigenvalues of $\mathbf{A}$. In this case the scheme simplifies to

$$\begin{aligned}\mathbf{u}_{i+1,j} = \mathbf{u}_{i,j} &- \frac{\Delta t}{2\Delta x}\mathbf{A}(\mathbf{u}_{i,j+1} - \mathbf{u}_{i,j-1}) \\ &+ \frac{\Delta t^2}{2\Delta x^2}\mathbf{A}^2(\mathbf{u}_{i,j+1} - 2\mathbf{u}_{i,j} + \mathbf{u}_{i,j-1}).\end{aligned} \tag{3.9.29}$$

If $\mathbf{F}$ is nonlinear in $\mathbf{u}$, then it is sometimes desirable to avoid the computation of the Jacobian in Eq.(3.9.27). In order to do that, several two step modifications have been devised, such as the following

$$\begin{aligned} \mathbf{u}^*_{i,j} &= \frac{1}{2}(\mathbf{u}_{i,j+1} + \mathbf{u}_{i,j}) - \frac{\Delta t}{2\Delta x}(\mathbf{F}_{i,j+1} - \mathbf{F}_{i,j}) \\ \mathbf{u}_{i+1,j} &= \mathbf{u}_{i,j} - \frac{\Delta t}{\Delta x}(\mathbf{F}^*_{i,j} - \mathbf{F}^*_{i,j-1}), \end{aligned} \tag{3.9.30}$$

where $\mathbf{F}^*_{i,j} = \mathbf{F}(\mathbf{u}^*_{i,j})$.

Both Eqs.(3.9.27) and (3.9.30) are dissipative and therefore more suitable for problems with smooth solutions and present dissipation effects. These can be reduced by choosing a time step near the stability limit. Furthermore, as for the leap-frog method, artificial boundary conditions are needed.

REMARK 3.9.7 In general, finite difference methods work efficiently in problems with smooth solutions. On the other hand, if discontinuities, or shocks, are present, greater accuracy can be accomplished by integrating along the characteristics, e.g., Eqs.(3.7.22–3.7.23). Unfortunately, this method is not always practical since in some cases it is hard to compute the characteristics. The advantage of the ***method of characteristics*** is that the solution along these curves does not change as rapidly as when it is observed for a fixed time as a function of x, or *vice versa.* In fact, since discontinuities propagate along the characteristics, if we fix t and span through x, we bump into the wave. If, instead, we follow the characteristic, we catch the wave (see Fig. 3.9.7). ■

3.9.4 Finite differences for second-order hyperbolic equations

As an example of second-order hyperbolic partial differential equation, consider the telegrapher's equation

$$\frac{\partial^2 u}{\partial t^2} + \frac{1}{\tau(t,x)}\frac{\partial u}{\partial t} = \nu(t,x)\frac{\partial^2 u}{\partial x^2} + F\left(t, x; u, \frac{\partial u}{\partial x}\right), \quad x \in [a,b], \tag{3.9.31}$$

with initial conditions

$$u(0,x) = u_{in}(x), \qquad \frac{\partial u}{\partial t}(0,x) = \tilde{u}_{in}(x) \tag{3.9.32}$$

and with mixed type boundary conditions

$$\alpha\frac{\partial u}{\partial x}(t,a) + \beta u(t,a) = u_a(t), \qquad u(t,b) = u_b(t). \tag{3.9.33}$$

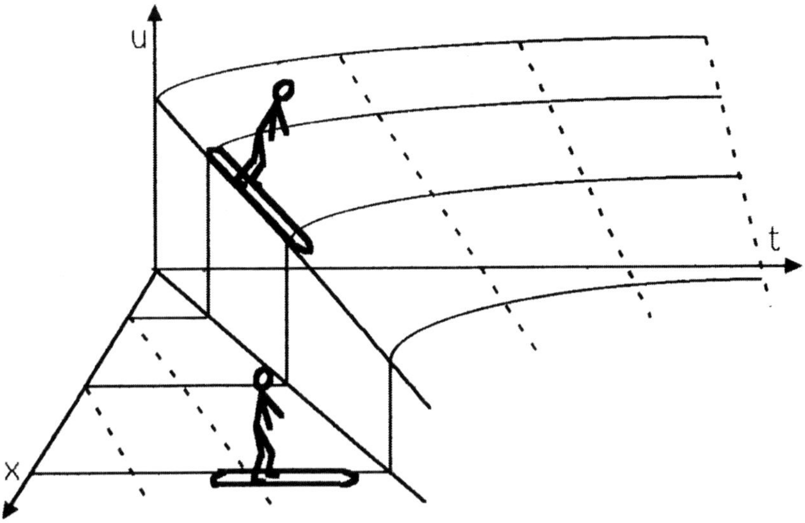

Figure 3.9.7 — Bumping into and catching a shock wave.

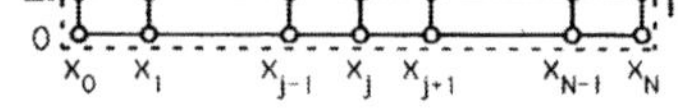

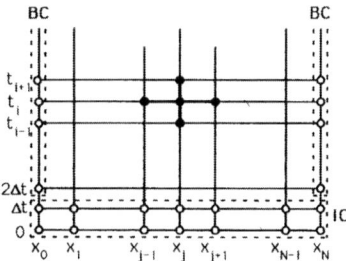

Figure 3.9.8 — Stencil for second-order hyperbolic equations.

The natural way to discretize Eq.(3.9.31) is to use central differences both in space and time (see Fig. 3.9.8), so that one has

$$\frac{u_{i+1,j} - 2u_{i,j} + u_{i-1,j}}{\Delta t^2} + \frac{1}{\tau_{i,j}} \frac{u_{i+1,j} - u_{i-1,j}}{2\Delta t}$$
$$= \nu_{i,j} \frac{u_{i,j+1} - 2u_{i,j} + u_{i,j-1}}{\Delta x^2}$$
$$+ F\left(t_i, x_j; u_{i,j}, \frac{u_{i,j+1} - u_{i,j-1}}{2\Delta x}\right), \qquad (3.9.34)$$

that is

$$u_{i+1,j} = \left(1+\frac{\Delta t}{2\tau_{i,j}}\right)^{-1}\left[2u_{i,j} - \left(1-\frac{\Delta t}{2\tau_{i,j}}\right)u_{i-1,j}\right.$$
$$+\left(\frac{\Delta t}{\Delta x}\right)^2 \nu_{i,j}(u_{i,j+1} - 2u_{i,j} + u_{i,j-1})$$
$$\left.+\,\Delta t^2 F\left(t_i, x_j; u_{i,j}, \frac{u_{i,j+1} - u_{i,j-1}}{2\Delta x}\right)\right]. \qquad (3.9.35)$$

Equation (3.9.35) gives explicitly the value of u in the internal nodes $j = 1, \dots, N-1$ and for $i \geq 1$.

In order to preserve the accuracy of the method, which is second order in both space and time, one cannot use the (first-order) forward differences to evaluate the derivative in the initial and, possibly, boundary conditions (3.9.32–3.9.33).

Using the initial condition and Taylor expansion to second order and the partial differential equation to be integrated, one can approximate

$$u(\Delta t, x) \cong u(t=0,x) + \Delta t\frac{\partial u}{\partial t}(t=0,x) + \frac{\Delta t^2}{2!}\frac{\partial^2 u}{\partial t^2}(t=0,x)$$
$$= u_{in}(x) + \Delta t\widetilde{u}_{in}(x) + \frac{\Delta t^2}{2}\times\left[-\frac{1}{\tau(0,x)}\widetilde{u}_{in}(x)\right.$$
$$\left.+\,\nu(0,x)\frac{d^2u_{in}}{dx^2}(x) + F\left(0,x;u_{in}(x),\frac{du_{in}}{dx}(x)\right)\right],$$

where the derivatives of the initial condition u_{in} can be replaced by a second-order approximation if they are difficult to compute analytically. The initializing conditions for the scheme (3.9.35) are then

$$u_{0,j} = u_{in}(x_j)$$
$$u_{1,j} = u_{in}(x_j) + \Delta t\widetilde{u}_{in}(x_j)$$
$$+\frac{\Delta t^2}{2}\left[-\frac{1}{\tau_{0,j}}\widetilde{u}_{in}(x_j) + \nu_{0,j}\frac{d^2u_{in}}{dx^2}(x_j) + F_{0,j}\right] \qquad (3.9.36)$$

with the obvious symbolism.

If Dirichlet boundary conditions are given, i.e., $\alpha = 0$ in (3.9.33), then the boundary values are given directly by the boundary conditions

$$u_{i+1,0} = \frac{u_a(t_{i+1})}{\beta}$$

$$u_{i+1,N} = u_b(t_{i+1}) .$$

If, instead, one has a Neumann ($\beta = 0$) or Robin ($\alpha, \beta \neq 0$) boundary condition, the same care is needed to keep the accuracy. In this case we can

- Introduce a fictitious node x_{-1};
- Approximate the Robin boundary condition with central differences to find

$$u_{i,-1} = u_{i,1} - \frac{2\Delta x}{\alpha}\left(u_a(t_i) - \beta u_{i,0}\right) ; \tag{3.9.37}$$

- Impose the central difference scheme (3.9.35) of the partial differential equation on the right boundary

$$\begin{aligned} u_{i+1,0} = \left(1 + \frac{\Delta t}{2\tau_{i,0}}\right)^{-1} \Bigg[& 2u_{i,0} - \left(1 - \frac{\Delta t}{2\tau_{i,0}}\right) u_{i-1,0} \\ & + \left(\frac{\Delta t}{\Delta x}\right)^2 \nu_{i,0}(u_{i,1} - 2u_{i,0} + u_{i,-1}) \\ & + \Delta t^2 F\left(t_i, a; u_{i,0}, \frac{u_{i,1} - u_{i,-1}}{2\Delta x}\right)\Bigg] ; \end{aligned} \tag{3.9.38}$$

- Use Eq.(3.9.37) to eliminate u_{-1} in (3.9.38).

REMARK 3.9.8 A similar procedure can be used for any second-order partial differential equations with Neumann or Robin boundary conditions, in particular for parabolic equations. ∎

The method just presented is stable if

$$\Delta t \leq \frac{\Delta x}{\max \nu(t, x)} ,$$

which is not a very restrictive condition, since Δt and Δx have to be of the same order anyhow because of accuracy considerations. Therefore, looking for implicit methods is not as essential as it is for parabolic

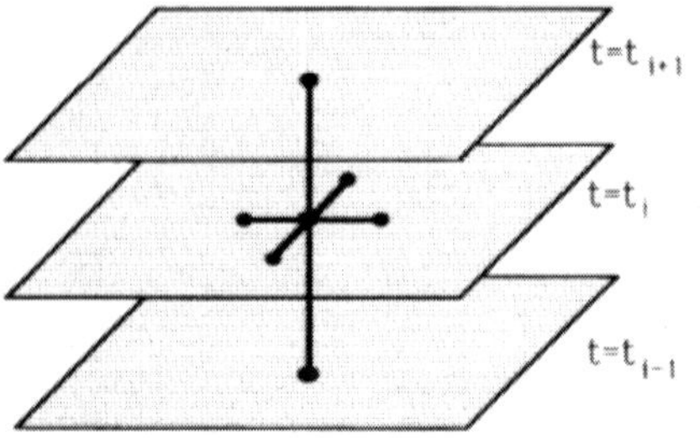

Figure 3.9.9 — **Two-dimensional stencil.**

equations. There are, however, implicit schemes that are unconditionally stable, but for the reason stated above, they are beyond the aim of this book.

REMARK 3.9.9 The finite difference scheme (3.9.35) can be technically generalized to more space dimensions

$$\frac{\partial^2 u}{\partial t^2} + \frac{1}{\tau(t,\mathbf{x})}\frac{\partial u}{\partial t} = \nu(t,\mathbf{x})\nabla^2 u + F\left(t,\mathbf{x};u,\frac{\partial u}{\partial x_1},\dots,\frac{\partial u}{\partial x_d}\right),$$

using stencils like that presented in Fig. 3.9.9 for the two-dimensional case. The stability condition in this case becomes more severe. For instance, in two-dimensional problems

$$\Delta t \le \frac{\Delta x\,\Delta y}{\sqrt{\Delta x^2 + \Delta y^2}}\,\frac{1}{\max \nu(t,x)}\,.$$

■

3.9.5 Finite differences for parabolic equations

As an example of a parabolic partial differential equation, consider the advection-diffusion equation

$$\frac{\partial u}{\partial t} = \nu(t,x)\frac{\partial^2 u}{\partial x^2} \quad c(t,x)\frac{\partial u}{\partial x} + f(t,x)u + g(t,x)\,, \quad x \in [a,b]\,, \qquad (3.9.39)$$

with initial condition

$$u(0,x) = u_{in}(x)$$

and Dirichlet boundary conditions

$$\begin{cases} u(t,a) = u_a(t) \\ u(t,b) = u_b(t) . \end{cases} \tag{3.9.40}$$

REMARK 3.9.10 The generalization to Neumann and Robin boundary conditions follows the lines given in Section 3.9.4. ■

To understand the main difference between the application of finite differences to hyperbolic and to parabolic problems, we had better follow the procedure explained in Section 3.9.1 step by step.

Using central differences for the space derivatives, Eq.(3.9.39) can be approximated by

$$\begin{aligned} \frac{du_j}{dt} &= \nu_j(t)\frac{u_{j+1} - 2u_j + u_{j-1}}{\Delta x^2} \\ &\quad - c_j(t)\frac{u_{j+1} - u_{j-1}}{2\Delta x} + f_j(t)u_j + g_j(t) , \end{aligned} \tag{3.9.41}$$

for $j = 1, \ldots, N-1$, where now $u_j(t) = u(t, x_j)$, and so on. Of course, u_0 and u_N are given by the boundary conditions (3.9.40).

Equation (3.9.41) can be written as

$$\frac{du_j}{dt} = \left(\frac{c_j}{2\Delta x} + \frac{\nu_j}{\Delta x^2}\right) u_{j-1} + \left(f_j - \frac{2\nu_j}{\Delta x^2}\right) u_j + \left(-\frac{c_j}{2\Delta x} + \frac{\nu_j}{\Delta x^2}\right) u_{j+1}$$

or in vector form as

$$\frac{d\mathbf{u}}{dt} = \mathbf{A}(t)\mathbf{u} + \mathbf{b}(t) , \tag{3.9.42}$$

where

$$\mathbf{A} = \begin{pmatrix} \beta_1 & \gamma_1 & 0 & 0 & \cdots & 0 \\ \alpha_2 & \beta_2 & \gamma_2 & 0 & \cdots & 0 \\ 0 & \alpha_3 & \beta_3 & \gamma_3 & \cdots & 0 \\ \vdots & \vdots & \ddots & \ddots & \ddots & \vdots \\ 0 & 0 & \cdots & \alpha_{N-2} & \beta_{N-2} & \gamma_{N-2} \\ 0 & 0 & \cdots & 0 & \alpha_{N-1} & \beta_{N-1} \end{pmatrix} , \tag{3.9.43}$$

$$\mathbf{u} = \begin{pmatrix} u_1 \\ u_2 \\ u_3 \\ \vdots \\ u_{N-2} \\ u_{N-1} \end{pmatrix}, \qquad \mathbf{b} = \begin{pmatrix} g_1 + \alpha_1 u_a \\ g_2 \\ g_3 \\ \vdots \\ g_{N-2} \\ g_{N-1} + \gamma_{N-1} u_b \end{pmatrix}, \tag{3.9.44}$$

and

$$\alpha_j(t) = \frac{c_j(t)}{2\Delta x} + \frac{\nu_j(t)}{\Delta x^2}$$

$$\beta_j(t) = f_j(t) - \frac{2\nu_j(t)}{\Delta x^2}$$

$$\gamma_j(t) = -\frac{c_j(t)}{2\Delta x} + \frac{\nu_j(t)}{\Delta x^2}.$$

By means of this procedure, we have approximated a partial differential equation as a system of ordinary differential equations. At this point, to integrate (3.9.42), we simply have to choose a stable and sufficiently accurate ODE solver. We recall that to check for the stability of the method one has to compute the spectrum of the linear operator related to the ordinary differential equations, in this case, the eigenvalues of $\mathbf{A}$ in (3.9.43). It is possible to prove that if ν is constant the most dangerous eigenvalue of $\mathbf{A}$ is real and about $-4\nu/\Delta x^2$.

Therefore if the forward Euler method (i.e., forward difference in time) is used to integrate (3.9.42), we have an explicit scheme that is stable only if $-4\nu\Delta t/\Delta x^2$ belongs to the stability region of the ODE solver (see Fig. 2.8.1c), that is,

$$\Delta t \le \frac{\Delta x^2}{2\nu}.$$

More precisely, one can actually prove that the explicit method is stable only if

$$\Delta x \le 2 \min \frac{\nu(t,x)}{|c(t,x)|} \tag{3.9.45}$$

and

$$\Delta t \le \frac{\Delta x^2}{2 \max \nu(t,x)}. \tag{3.9.46}$$

REMARK 3.9.11 Condition (3.9.45) gives a constraint on the space discretization of the form $\Delta x \le$ const., which becomes valuable only if the viscosity ν is very small with respect to the convection velocity c (roughly speaking we have an "almost hyperbolic system"). It must be satisfied in order for the numerical solution to behave qualitatively like that of the partial differential equation. The quantity on the left-hand side of (3.9.45) corresponds to the inverse of the Reynolds number in fluid-dynamic problems or to the Peclet number in heat transfer problems.

The oscillations that come out when this condition is violated do not grow excessively and are the result of an inadequate resolution.

One way of avoiding the restriction (3.9.45) for ν/c very small (and therefore this kind of oscillation) is to use the upwind differencing of the convective term $c(t,x_j)(u_{i,j}-u_{i,j-1})/\Delta x$, if $c>0$, which as a counterpart reduces the order of the method. In this case the stability condition for the explicit scheme becomes

$$\Delta t \leq \frac{\Delta x^2}{2\max \nu(t,x) + \Delta x \max |c(t,x)|}, \tag{3.9.47}$$

which is less restrictive than (3.9.46), especially when ν/c is very small. This method, however, introduces an artificial diffusion term similar to that encountered in Remark 3.9.5. Actually, it is this term that is responsible for the smoothing action.

On the other hand, condition (3.9.46) is of the form $\Delta t \leq \varphi(\Delta x)$ and forces one to use very small time steps to integrate the partial differential equation, since $\varphi(\Delta x)$ is quadratic in Δx. It is also important to point out that the stability condition (3.9.46) involves only the highest-order term. In fact, it is generally true that the most restrictive criterion involves only these terms. ∎

If, on the other hand, the backward Euler method (corresponding to backward differences in time) or Crank–Nicolson method is used to integrate (3.9.42), then one has an implicit scheme that is always stable.

Actually, restrictions like (3.9.45) still hold and therefore observations like those in Remark 3.9.11 can be repeated here.

For reader convenience, we explicitly write down the above mentioned schemes and represent in Fig. 3.9.10 the relative stencils.

Forward Euler

$$\begin{aligned} u_{i+1,j} &= u_{i,j} + \frac{\Delta t}{\Delta x^2}\nu_{i,j}(u_{i,j+1} - 2u_{i,j} + u_{i,j-1}) \\ &\quad - \frac{\Delta t}{2\Delta x}c_{i,j}(u_{i,j+1} - u_{i,j-1}) + \Delta t f_{i,j}u_{i,j} + \Delta t g_{i,j} \end{aligned} \tag{3.9.48}$$

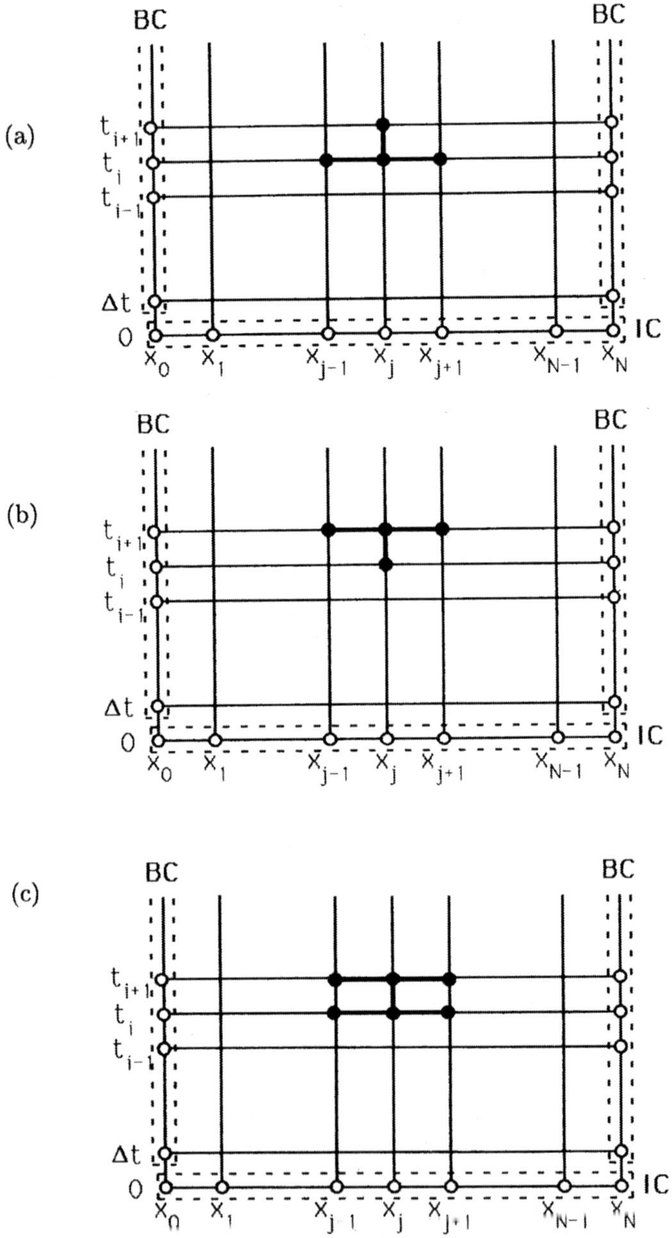

Figure 3.9.10 — Stencil for parabolic partial differential equations. Central difference in space and (a) forward Euler (b) backward Euler (c) Crank–Nicolson in time.

or, referring to Eq.(3.9.42–3.9.44)

$$\mathbf{u}_{i+1} = \big[\mathbf{I} + \Delta t \mathbf{A}(t_i)\big]\mathbf{u}_i + \Delta t \mathbf{b}(t_i)\,, \tag{3.9.49}$$

where $\mathbf{u}_i$ now stands for the vector $\mathbf{u}(t_i)$ in Eq.(3.9.44).

Backward Euler

$$\begin{aligned}
&\left(-\frac{\Delta t}{2\Delta x}c_{i+1,j} - \frac{\Delta t}{\Delta x^2}\nu_{i+1,j}\right)u_{i+1,j-1} \\
&\quad + \left(1 + \frac{2\Delta t}{\Delta x^2}\nu_{i+1,j} - \Delta t f_{i+1,j}\right)u_{i+1,j} \\
&\quad + \left(\frac{\Delta t}{2\Delta x}c_{i+1,j} - \frac{\Delta t}{\Delta x^2}\nu_{i+1,j}\right)u_{i+1,j+1} = u_{i,j} + \Delta t g_{i+1,j}
\end{aligned} \tag{3.9.50}$$

or

$$\big[\mathbf{I} - \Delta t \mathbf{A}(t_{i+1})\big]\mathbf{u}_{i+1} = \mathbf{u}_i + \Delta t \mathbf{b}(t_{i+1})\,. \tag{3.9.51}$$

Crank–Nicolson

$$\begin{aligned}
&\frac{1}{2}\left(-\frac{\Delta t}{2\Delta x}c_{i+1,j} - \frac{\Delta t}{\Delta x^2}\nu_{i+1,j}\right)u_{i+1,j-1} \\
&\quad + \left(1 + \frac{\Delta t}{\Delta x^2}\nu_{i+1,j} - \frac{\Delta t}{2}f_{i+1,j}\right)u_{i+1,j} \\
&\quad + \frac{1}{2}\left(\frac{\Delta t}{2\Delta x}c_{i+1,j} - \frac{\Delta t}{\Delta x^2}\nu_{i+1,j}\right)u_{i+1,j+1} \\
&= \frac{1}{2}\left(\frac{\Delta t}{2\Delta x}c_{i,j} + \frac{\Delta t}{\Delta x^2}\nu_{i,j}\right)u_{i,j-1} + \left(1 - \frac{\Delta t}{\Delta x^2}\nu_{i,j} + \frac{\Delta t}{2}f_{i,j}\right)u_{i,j} \\
&\quad + \frac{1}{2}\left(-\frac{\Delta t}{2\Delta x}c_{i,j} + \frac{\Delta t}{\Delta x^2}\nu_{i,j}\right)u_{i,j+1} + \frac{\Delta t}{2}(g_{i+1,j} + g_{i,j})\,,
\end{aligned} \tag{3.9.52}$$

or

$$\begin{aligned}
&\left[\mathbf{I} - \frac{\Delta t}{2}\mathbf{A}(t_{i+1})\right]\mathbf{u}_{i+1} \\
&\qquad = \left[\mathbf{I} + \frac{\Delta t}{2}\mathbf{A}(t_i)\right]\mathbf{u}_i + \frac{\Delta t}{2}[\mathbf{b}(t_i) + \mathbf{b}(t_{i+1})]\,.
\end{aligned} \tag{3.9.53}$$

REMARK 3.9.12 In both Euler methods the time derivative has been evaluated with first-order approximations, therefore both the schemes (3.9.49) and (3.9.51) are first order in time and second order in space. On the other hand, the Crank–Nicolson method (3.9.52) is second order in both space and time. ■

REMARK 3.9.13 Implicit schemes require the solution of a linear system that is tridiagonal (see the structure of **A** in Eq.(3.9.43)). This can be easily done by the algorithm presented in Appendix 2. In the case of higher-order linear partial differential equations, one still has the same structure, but with more nonvanishing diagonals, for instance a fourth-order partial differential equation yields to the solution of a pentadiagonal linear system. ■

REMARK 3.9.14 As mentioned in Section 3.9.1, the explicit scheme can be straightforwardly extended to systems of nonlinear partial differential equations, like (3.8.12). On the other hand, the simplest way to use implicit methods on nonlinear partial differential equations is to operate a sort of linearization. For instance, the model

$$\frac{\partial u}{\partial t} + u\frac{\partial u}{\partial x} = \nu(t,x;u)\frac{\partial^2 u}{\partial x^2} + k(t,x;u)\left(\frac{\partial u}{\partial x}\right)^2$$

can be approximated as

$$\begin{aligned}\frac{u_{i+1,j} - u_{i,j}}{\Delta t} &+ u_{i,j}\frac{u_{i+1,j+1} - u_{i+1,j-1}}{2\Delta x} \\ &= \nu_{i,j}\frac{u_{i+1,j+1} - 2u_{i+1,j} + u_{i+1,j-1}}{\Delta x^2} + \\ &\quad + k_{i,j}\frac{u_{i,j+1} - u_{i,j-1}}{2\Delta x}\,\frac{u_{i+1,j+1} - u_{i+1,j-1}}{2\Delta x}\,. \end{aligned} \tag{3.9.54}$$

Alternatively one can use Newton's iterative method for the solution of nonlinear systems. ■

3.9.6 Finite differences in two dimensions

Formally, there is no difficulty in generalizing the schemes (3.9.48–3.9.53) to two space dimensions. Of course, now one has to evaluate the state variable on a planar space grid, say a rectangle with $(N+1)(M+1)$ nodes, as time goes by. This might be a lengthy computation. For this reason it would be desirable to use a method with not too strong restrictions on the time step. The explicit scheme (3.9.48) is stable only if

$$\Delta t \leq \frac{\Delta x^2 \, \Delta y^2}{2\nu(\Delta x^2 + \Delta y^2)}$$

and therefore, in spite of its simplicity, it does not satisfy the just mentioned requirement.

On the other hand, implicit methods do not have significant restrictions on the time step, but imply the solution of a pentadiagonal system.

An useful method to overcome both drawbacks is the so-called ***alternating direction implicit method***, or simply ***ADI method***, which preserves the tridiagonal form of the system by first solving a sequence of one-dimensional difference equations in x and then a sequence of one-dimensional difference equations in y, and so on.

Thus, for instance, one can apply (3.9.52) in every node y_k to update the value $u^*_{i+1,j,k}$ and then apply again (3.9.52) with the index j replaced by k in every node x_j, to compute $u_{i+1,j,k}$. Referring to the two-dimensional diffusion equation

$$\frac{\partial u}{\partial t} = \nu \left(\frac{\partial^2 u}{\partial x^2} + \frac{\partial^2 u}{\partial y^2} \right),$$

one can first approximate $\partial^2 u/\partial x^2$ implicitly and $\partial^2 u/\partial y^2$ explicitly and then *vice versa*, that is, one has first to solve

$$u^*_{i+1,j,k} - \nu \frac{\Delta t}{\Delta x^2}(u^*_{i+1,j+1,k} - 2u^*_{i+1,j,k} + u^*_{i+1,j-1,k})$$
$$= u_{i,j,k} + \nu \frac{\Delta t}{\Delta x^2}(u_{i,j,k+1} - 2u_{i,j,k} + u_{i,j,k-1})$$

and then

$$u_{i+1,j,k} - \nu \frac{\Delta t}{\Delta x^2}(u_{i+1,j,k+1} - 2u_{i+1,j,k} + u_{i+1,j,k-1})$$
$$= u^*_{i,j,k} + \nu \frac{\Delta t}{\Delta x^2}(u^*_{i,j+1,k} - 2u^*_{i,j,k} + u^*_{i,j-1,k})$$

for $i \geq 0$, $j = 1, \ldots, N-1$ and $k = 1, \ldots, M-1$. The method is second order in time and space and always stable.

Before implementing the code, however, one needs to pay a particular attention to giving the boundary conditions, as is well explained in Section 7.3 of Strikwerda [STI].

3.10 The Collocation Method

As for the finite difference methods, the first step of a collocation method is to select $N+1$ nodes in the space interval $[a, b]$

$$a = x_0 < x_1 < \cdots < x_{N-1} < x_N = b\,. \tag{3.10.1}$$

The second step is then to select a set of cardinal basis functions $\Phi_j(x)$, say the Lagrangian polynomials (A.2.6), the trigonometric polynomials (A.2.25), or the fundamental splines, so that

$$\Phi_j(x_i) = \delta_{ij}\,, \qquad i, j = 0, \ldots, N\,, \tag{3.10.2}$$

where δ_{ij} is the Kronecker delta.

The solution $u(t, x)$ to the initial-boundary-value problem can then be approximated as

$$u(t, x) \cong u^N(t, x) \stackrel{\text{def}}{=} \sum_{j=0}^{N} u_j(t)\Phi_j(x)\,, \tag{3.10.3}$$

where $u_j(t) = u(t, x_j)$.

The partial derivatives of u can be naturally approximated by the partial derivatives of u^N

$$\begin{aligned}
\frac{\partial u}{\partial t}(t, x) &\cong \frac{\partial u^N}{\partial t}(t, x) = \sum_{j=0}^{N} \frac{du_j}{dt}(t)\Phi_j(x) \\
\frac{\partial u}{\partial x}(t, x) &\cong \frac{\partial u^N}{\partial x}(t, x) = \sum_{j=0}^{N} u_j(t)\frac{d\Phi_j}{dx}(x) \\
\frac{\partial^2 u}{\partial x^2}(t, x) &\cong \frac{\partial^2 u^N}{\partial x^2}(t, x) = \sum_{j=0}^{N} u_j(t)\frac{d^2\Phi_j}{dx^2}(x)\,,
\end{aligned} \tag{3.10.4}$$

and so on.

Substituting (3.10.3) and (3.10.4) into the sample advection-diffusion model

$$\frac{\partial u}{\partial t} + c(t,x;u)\frac{\partial u}{\partial x} = \nu(t,x;u)\frac{\partial^2 u}{\partial x^2} + F(t,x;u) \ , \tag{3.10.5}$$

yields

$$\begin{aligned}
&\sum_{j=0}^{N} \frac{du_j}{dt}(t)\Phi_j(x) + c\left(t,x;\sum_{j=0}^{N} u_j(t)\Phi_j(x)\right)\sum_{j=0}^{N} u_j(t)\frac{d\Phi_j}{dx}(x) \\
&= \nu\left(t,x;\sum_{j=0}^{N} u_j(t)\Phi_j(x)\right)\sum_{j=0}^{N} u_j(t)\frac{d^2\Phi_j}{dx^2}(x) \\
&+F\left(t,x;\sum_{j=0}^{N} u_j(t)\Phi_j(x)\right) \ .
\end{aligned} \tag{3.10.6}$$

The second conceptual step is to consider Eq.(3.10.6) in the collocation points x_i. Recalling Eq.(3.10.2), Eq.(3.10.6) can be rewritten as

$$\begin{aligned}
&\frac{du_i}{dt}(t) + c\big(t,x_i;u_i(t)\big)\sum_{j=0}^{N} D_{ij}u_j(t) \\
&=\nu\big(t,x;u_i(t)\big)\sum_{j=0}^{N} B_{ij}u_j(t) + F\big(t,x_i;u_i(t)\big) \ ,
\end{aligned} \tag{3.10.7}$$

where

$$\begin{aligned}
D_{ij} &= \frac{d\Phi_j}{dx}(x_i) \\
B_{ij} &= \frac{d^2\Phi_j}{dx^2}(x_i)
\end{aligned} \tag{3.10.8}$$

are the components of two matrices $\mathbf{D}$ and $\mathbf{B}$ that can be computed once and for all as the collocation points x_i and the cardinal basis functions $\Phi_j(x)$ are chosen.

There are now two choices left arbitrary

- The collocation points;
- The cardinal basis functions.

Of course, the choice depends on the properties of the solution and on the accuracy with which the derivatives are computed. Noting what is presented in Appendix 2, the reader can understand why a Chebyshev distribution of collocation points is the best choice to couple with Lagrange polynomials (in the sense that it provides a good approximation of the derivatives). If, however, we know that the solution to our problem is periodic in x, then the natural choice is to use trigonometric polynomials with equispaced nodes. The former case is referred to as the ***Chebyshev collocation method***, the latter as the ***Fourier collocation method.*** These choices are also usually preferred because the related formulas can be computationally shortened using Fast Fourier Transform algorithms (FFTs) [BRI] and because they lead to spectral accuracy for infinitely differentiable solutions.

DEFINITION 3.10.1 Spectral accuracy
The adjective ***spectral*** *is used to indicate an approximation with an error that decreases faster than any power of the number of collocation points for infinitely differentiable functions.*

3.10.1 Fourier collocation method

In Fourier collocation methods, the reference interval is, of course, $[0, 2\pi]$. The first step is then to map the periodicity interval $[a, b]$ into $[0, 2\pi]$ with a smooth transformation $\xi = f(x)$, say the linear one

$$\xi = 2\pi \frac{x-a}{b-a} \qquad \text{or} \qquad x = a + \frac{(b-a)}{2\pi}\xi .$$

It is possible to prove that the choice of equispaced collocation points

$$\xi_h = \frac{h\pi}{N}, \qquad h = 0, \ldots, 2N-1 \tag{3.10.9}$$

is equivalent to approximating u as

$$u(t,\xi) \cong u^N(t,\xi) = \sum_{k=-N}^{N-1} \widetilde{u}_k(t) e^{ik\xi}, \tag{3.10.10}$$

where, throughout this subsection, i is the ***imaginary unit*** and

$$\widetilde{u}_k(t) = \frac{1}{2N} \sum_{h=0}^{2N-1} u(t,\xi_h) e^{-ihk\pi/N}, \quad k = -N, \ldots, N-1. \tag{3.10.11}$$

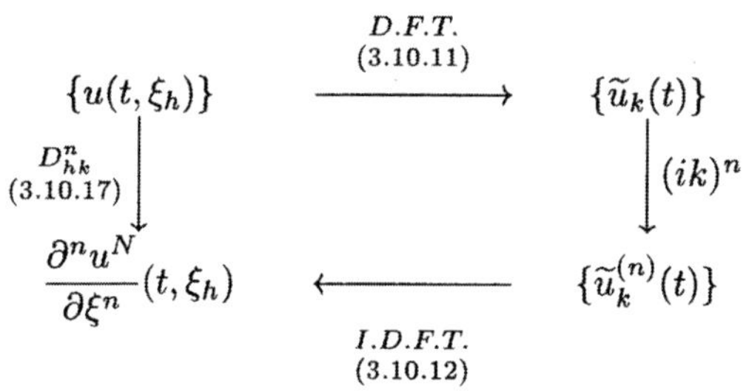

D.F.T.
(3.10.11)
$\{u(t,\xi_h)\}$
$\{\widetilde{u}_k(t)\}$
D^n_{hk}
(3.10.17)
$(ik)^n$
$\frac{\partial^n u^N}{\partial \xi^n}(t,\xi_h)$
$\{\widetilde{u}_k^{(n)}(t)\}$
I.D.F.T.
(3.10.12)

Table 3.10.1 — Fourier collocation derivatives.

From (3.10.10) it readily follows that

$$u(t,\xi_h) = u^N(t,\xi_h) = \sum_{k=-N}^{N-1} \widetilde{u}_k(t) e^{ihk\pi/N}, \tag{3.10.12}$$

for $h = 0, \ldots, 2N-1$.

The two transformations (3.10.11) and (3.10.12) are usually called ***Discrete Fourier Transform*** (DFT) and ***Inverse Discrete Fourier Transform*** (IDFT). From the computational point of view one can profit from the FFT routine to compute them and this is one of the reasons why these transformations are so popular in numerical analysis.

The values of the derivatives in the collocation points can be computed from the values of u in the collocation points by the following steps (see Table 3.10.1):

Step 1 Knowing $u(t,\xi_h)$ and using (3.10.11), compute $\widetilde{u}_k$.

Step 2 Deriving (3.10.10), obtain

$$\frac{\partial u^N}{\partial \xi}(t,\xi) = \sum_{k=-N}^{N-1} ik\widetilde{u}_k(t) e^{ik\xi}, \tag{3.10.13}$$

which says that the coefficients $\widetilde{u}_k^{(1)}$ of the derivative are

$$\widetilde{u}_k^{(1)} = ik\widetilde{u}_k = \frac{ik}{2N} \sum_{h=0}^{2N-1} u(t,\xi_h) e^{-ihk\pi/N}, \tag{3.10.14}$$

for $k = -N, \ldots, N-1$.

Step 3 Transform back with Eq.(3.10.12) to finally obtain

$$\frac{\partial u^N}{\partial \xi}(t, \xi_h) = \sum_{k=-N}^{N-1} \widetilde{u}_k^{(1)} e^{ihk\pi/N}, \qquad h = 0, \dots, 2N-1. \tag{3.10.15}$$

This approximation of the derivative is called ***Fourier collocation derivative*** and is spectrally accurate.

The formula can be easily generalized to higher-order derivatives by replacing (3.10.14) of the second step with

$$\widetilde{u}_k^{(n)} = (ik)^n \widetilde{u}_k = \frac{(ik)^n}{2N} \sum_{h=0}^{2N-1} u(t, \xi_h) e^{-ihk\pi/N} .$$

REMARK 3.10.1 Using the chain of transformations (3.10.14–3.10.15) is recomended for large N, say $N > 5$, if the Discrete Fourier Transforms are accomplished via the special form of the FFT routine, which takes advantage of the reality of u (see Canuto, et al. [CaA] for an example). If one does not exploit the reality of u in using the FFT, then round-off errors in the procedure may lead to complex value derivatives. ■

REMARK 3.10.2 The same result can be achieved by matrix multiplication. In fact, the three transformations (3.10.14–3.10.15) of Table 3.10.1 can be summarized in

$$\frac{\partial u^N}{\partial \xi}(t, \xi_h) = \sum_{k=0}^{2N-1} D_{hk} u_k(t) , \tag{3.10.16}$$

where

$$D_{hk} = \begin{cases} \dfrac{(-1)^{h+k}}{2 \sin \dfrac{(h-k)\pi}{2N}}, & \text{if } h \neq k \\ 0, & \text{if } h = k \end{cases} \tag{3.10.17}$$

or in vector form

$$\frac{\partial \mathbf{u}}{\partial \xi}(t) \cong \mathbf{D}\mathbf{u}(t) , \tag{3.10.18}$$

where

$$\mathbf{u}(t) = \{u_0(t), \ldots, u_N(t)\} \tag{3.10.19}$$

is the unknown vector of the state variable in the collocation points and

$$\frac{\partial \mathbf{u}}{\partial \xi}(t) \cong \left\{ \frac{\partial u}{\partial \xi}(t, \xi_0), \ldots, \frac{\partial u}{\partial \xi}(t, \xi_N) \right\} \tag{3.10.20}$$

is the value of the derivative of u in the collocation points. However, despite its simplicity, this procedure is computationally longer.

For higher-order derivatives

$$\frac{\partial^n \mathbf{u}}{\partial \xi^n}(t) \cong \frac{\partial^n \mathbf{u}^N}{\partial \xi^n}(t) = \mathbf{D}^n \mathbf{u} \,,$$

where $\mathbf{D}^n$ is the n-th power of the matrix $\mathbf{D}$. ∎

REMARK 3.10.3 The values of the derivatives with respect to the original independent variable x are obtained using the chain rule

$$\frac{\partial u^N}{\partial x} = \frac{\partial u^N}{\partial \xi}\frac{d\xi}{dx} = \frac{df}{dx}(x)\frac{\partial u^N}{\partial \xi} \,,$$

so that

$$\frac{\partial u^N}{\partial x}(t, x_h) = \frac{df}{dx}(x_h)\frac{\partial u^N}{\partial \xi}(t, \xi_h) \,. \tag{3.10.21}$$

For instance, if the transformation is linear, then

$$\frac{\partial u^N}{\partial x}(t, x_h) = \frac{2\pi}{b-a}\frac{\partial u^N}{\partial \xi}(t, \xi_h) \,. \tag{3.10.22}$$

Hence, recalling (3.10.16–3.10.20), it is convenient to define directly

$$\frac{\partial \mathbf{u}}{\partial x} = \widetilde{\mathbf{D}}\mathbf{u} \,, \tag{3.10.23}$$

where

$$\widetilde{\mathbf{D}} = \frac{2\pi}{b-a}\mathbf{D} \,.$$

∎

3.10.2 Chebyshev collocation method

In Chebyshev collocation methods, the reference interval is $[-1, 1]$. As before, the first step is to map the interval $[a, b]$ into $[-1, 1]$ with a smooth transformation, say the linear one

$$\xi = 1 - 2\frac{x-a}{b-a} \qquad \text{or} \qquad x = a - \frac{(b-a)}{2}(\xi - 1)\,.$$

It is possible to prove that the choice of Chebyshev collocation points

$$\xi_h = \cos\frac{h\pi}{N}\,, \qquad h = 0, \ldots, N \tag{3.10.24}$$

joined with Lagrange polynomials is equivalent to approximating u as

$$u(t, \xi) \cong u^N(t, \xi) \overset{\text{def}}{=} \sum_{k=0}^{N} \widetilde{u}_k(t) T_k(x)\,, \tag{3.10.25}$$

where $T_k(x)$ are the Chebyshev polynomials (of first kind) and the coefficients $\widetilde{u}_k$ are given by

$$\widetilde{u}_k(t) = \frac{2}{N\bar{c}_k} \sum_{h=0}^{N} \frac{1}{\bar{c}_h} u(t, \xi_h) \cos\frac{hk\pi}{N}\,, \qquad k = 0, \ldots, N\,, \tag{3.10.26}$$

where

$$\bar{c}_h = \begin{cases} 2\,, & \text{if } h = 0, N \\ 1\,, & \text{if } 1 \le h \le N-1\,. \end{cases}$$

REMARK 3.10.4 Observe that as h increases, the nodes ξ_h go from 1 to -1, for instance $\xi_0 = 1$ and $\xi_N = -1$. ■

Not only is Eq.(3.10.25) spectrally accurate, but it is also computationally convenient since

$$u(t, \xi_h) = u^N(t, \xi_h) = \sum_{k=0}^{N} \widetilde{u}_k(t) \cos\frac{hk\pi}{N}\,, \qquad h = 0, \ldots, N\,, \tag{3.10.27}$$

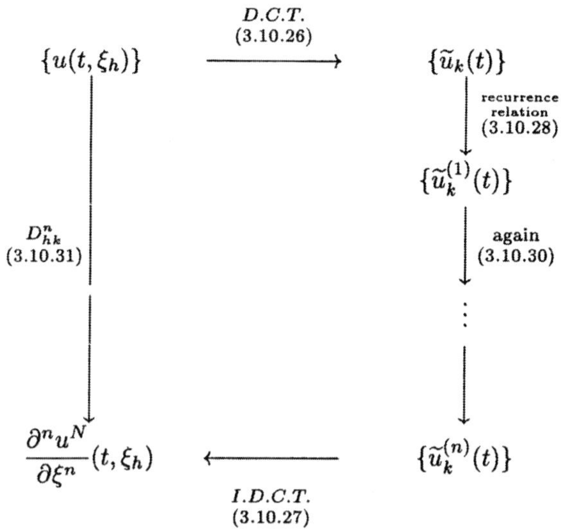

Table 3.10.2 — Chebyshev collocation derivatives.

which enables the use of FFT algorithms in computing both the ***Discrete Chebyshev Transform*** (DCT) (3.10.26) and the ***Inverse Discrete Chebyshev Transform*** (IDCT) (3.10.27) and therefore in the evaluation of the derivatives.

As for the Fourier collocation derivatives, the Chebyshev collocation derivatives can be evaluated from the values of u in the collocation points using a scheme similar to Table 3.10.1 (see Table 3.10.2).

Step 1 Knowing $u(t,\xi_h)$ and using (3.10.26), compute $\widetilde{u}_k$.

Step 2 Derive (3.10.25) to obtain the recurrence formula

$$\begin{aligned} \widetilde{u}^{(1)}_{N+1} &= 0\,, \\ \widetilde{u}^{(1)}_{N} &= 0\,, \\ \widetilde{u}^{(1)}_{k} &= \widetilde{u}^{(1)}_{k+2} + 2(k+1)\widetilde{u}_{k+1}\,, \qquad k = N-1,\ldots,1\,, \\ \widetilde{u}^{(1)}_{0} &= \frac{1}{2}\widetilde{u}^{(1)}_{2} + \widetilde{u}_1\,. \end{aligned} \tag{3.10.28}$$

(Note that the recurrence formula advances for decreasing values of k.)

Step 3 Finally, transform back with the inverse Chebyshev transform

(3.10.27), to obtain

$$\frac{\partial u^N}{\partial \xi}(t, \xi_h) = \sum_{k=0}^{N} \widetilde{u}_k^{(1)} \cos \frac{hk\pi}{N}, \qquad h = 0, \ldots, N. \tag{3.10.29}$$

This approximation of the derivative is called ***Chebyshev collocation derivative*** and is spectrally accurate.

In general, for the n-th derivative one has to repeat (3.10.28) n times, that is, the recurrence relation

$$\begin{aligned}
\widetilde{u}_{N+1}^{(j)} &= 0, \\
\widetilde{u}_{N}^{(j)} &= 0, \\
\widetilde{u}_{k}^{(j)} &= \widetilde{u}_{k+2}^{(j)} + 2(k+1)\widetilde{u}_{k+1}^{(j-1)}, \qquad k = N-1, \ldots, 1, \\
\widetilde{u}_{0}^{(j)} &= \frac{1}{2}\widetilde{u}_{2}^{(j)} + \widetilde{u}_{1}^{(j-1)},
\end{aligned} \tag{3.10.30}$$

for $j = 1, \ldots, n$. (Of course, $\widetilde{u}_k^{(0)} = \widetilde{u}_k$.)

The Chebyshev collocation derivative can also be computed using (3.10.18), where now the derivative matrix is

$$D_{hk} = \begin{cases} \dfrac{\overline{c}_h}{\overline{c}_k} \dfrac{(-1)^{h+k}}{\xi_h - \xi_k}, & \text{if } h \neq k, \\[2ex] -\dfrac{\xi_h}{2(1-\xi_h^2)}, & \text{if } 1 \leq h = k \leq N-1, \\[2ex] \dfrac{2N^2+1}{6}, & \text{if } h = k = 0, \\[2ex] -\dfrac{2N^2+1}{6}, & \text{if } h = k = N. \end{cases} \tag{3.10.31}$$

REMARK 3.10.5 It is useful to observe that computing complex derivatives such as

$$\mathcal{F}u(\xi, u) = \frac{\partial}{\partial \xi}\left[\nu(\xi, u)\frac{\partial}{\partial \xi} f(\xi, u)\right]$$

is not much harder than computing $\partial^2 u/\partial\xi^2$. In fact, in the latter case one has to operate as follows

$$u_h \xrightarrow{\mathbf{D}} \frac{\partial u}{\partial \xi}(\xi_h) \xrightarrow{\mathbf{D}} \frac{\partial^2 u}{\partial \xi^2}(\xi)\,,$$

where **D** is the operation relative to the computation of the derivative, namely, the transformation (3.10.26, 3.10.28, 3.10.29), or the matrix multiplication by (3.10.31). In the former case, one has to operate as follows

$$f(\xi_h, u_h) \xrightarrow{\mathbf{D}} \frac{\partial f}{\partial \xi}(\xi_h, u_h) \longrightarrow \nu(\xi_h, u_h)\frac{\partial f}{\partial \xi}(\xi_h, u_h) \xrightarrow{\mathbf{D}} \mathcal{F}u(\xi_h, u_h)\,.$$

■

REMARK 3.10.6 We remark again that the values of the derivatives with respect to the original independent variable x are obtained using the chain rule (3.10.2), which in the linear case is written

$$\frac{\partial u^N}{\partial x} = -\frac{2}{b-a}\,\frac{\partial u^N}{\partial \xi}\,,$$

and therefore (3.10.23) still holds with

$$\widetilde{\mathbf{D}} = -\frac{2}{b-a}\mathbf{D}\,.$$

■

3.10.3 Application of Chebyshev collocation method to the advection-diffusion model

In order to understand the main features of the collocation methods, including its advantages and disadvantages, consider as a testing ground the advection-diffusion problem in $[a, b]$

$$\begin{cases} \dfrac{\partial u}{\partial t} + c(t,x)\dfrac{\partial u}{\partial x} - \nu\dfrac{\partial^2 u}{\partial x^2} = F(t,x) \\ u(0,x) = u_{in}(x) \\ \alpha\dfrac{\partial u}{\partial x}(t,a) + \beta u(t,a) = u_a(t) \\ \gamma\dfrac{\partial u}{\partial x}(t,b) + \delta u(t,b) = u_b(t)\,. \end{cases} \tag{3.10.32}$$

In this section we will restrict our discussion to the linear case with ν constant. The collocation method leads to the semidiscrete equation

$$\begin{cases} \dfrac{du_h}{dt} + c_h \displaystyle\sum_{k=0}^{N} \widetilde{D}_{hk}u_k - \nu \sum_{k=0}^{N} \widetilde{D}^2_{hk}u_k = F_h \quad \text{for } h = 1, \ldots, N-1 \\ \alpha \displaystyle\sum_{k=0}^{N} \widetilde{D}_{0k}u_k + \beta u_0 = u_a \\ \gamma \displaystyle\sum_{k=0}^{N} \widetilde{D}_{Nk}u_k + \delta u_N = u_b\,, \end{cases} \tag{3.10.33}$$

where u_k are the values of u in the collocation points

$$x_k = a - \frac{b-a}{2}\left(\cos\frac{k\pi}{N} - 1\right),$$

and $\widetilde{D}^2_{hk}$ is the second derivative matrix, which simply is the square of the first derivative matrix $\widetilde{D}_{hk}$

$$\widetilde{D}^2_{hk} = \sum_{j=0}^{N} \widetilde{D}_{hj}\widetilde{D}_{jk}\,.$$

Of course, the initial-value problem has to be solved with the initial condition

$$u_h(0) - u_{in}(x_h)\,.$$

Before choosing the ODE solver, one should examine the eigenvalues related to the differential operators in (3.10.32) or to their discretized versions (3.10.33). Without entering into detail, we need to say that

- The eigenvalues related to the second-order derivatives with Robin boundary conditions are real and

$$-C_1 N^4 < \lambda \leq 0 \,. \tag{3.10.34}$$

Actually it can be proved that for the Dirichlet boundary conditions the eigenvalues are strictly negative.

- The eigenvalues related to the first-order derivative are complex with negative real part and

$$|\lambda| \leq C_3 N^2 \,, \qquad \text{as } N \longmapsto \infty \,. \tag{3.10.35}$$

- The advection-diffusion operator

$$-c(t,x)\frac{\partial}{\partial x} + \nu \frac{\partial^2}{\partial x^2}$$

joined with the Dirichlet boundary conditions has eigenvalues with

$$|\lambda| \leq \nu C_1 N^4 + \max|c(t,x)|\, C_2 N^2 \tag{3.10.36}$$

and

$$\Re e \lambda \leq \frac{\max|c(t,x)|}{\nu} \,.$$

In fact, for small ν and N there might be eigenvalues with positive real parts.

Once the ODE solver has been chosen, we have to set the time step Δt so that all eigenvalues are squeezed into the stability region. For instance, if the forward Euler method is used

$$u_{i+1,h} = u_{i,h} + \Delta t \left(-c_h \sum_{k=0}^{N} \widetilde{D}_{hk} u_{i,k} + \nu \sum_{k=0}^{N} \widetilde{D}^2_{hk} u_{i,k} + F_{i,h} \right) \,, \tag{3.10.37}$$

then Δt has to satisfy

$$\Delta t \leq \frac{C}{\nu N^4 + C_2 \max|c(t,x)|\, N^2} \,. \tag{3.10.38}$$

REMARK 3.10.7 If c is constant, the discretization (3.10.37) can be written in vector form as

$$\mathbf{u}_{i+1} = (\mathbf{I} - c\Delta t \widetilde{\mathbf{D}} + \nu \Delta t \widetilde{\mathbf{D}}^2)\mathbf{u}_i + \Delta t \mathbf{F}_i \,, \tag{3.10.39}$$

where $\mathbf{I}$ is the identity matrix and $\mathbf{u}_i = \{u_{i,0}, \ldots, u_{i,N}\}$. ■

REMARK 3.10.8 The stability result (3.10.38) could have been foreseen in some sense by considering that the minimum distance between Chebyshev collocation points occurs near the boundary with

$$\Delta x = 1 - \cos\frac{\pi}{N} \cong \frac{\pi^2}{2N^2}.$$

Thus the constraints (3.9.16) and (3.9.46) obtained in the finite difference case would give

$$\Delta t \leq \frac{\Delta x}{c} = \frac{C}{cN^2}$$

and

$$\Delta t \leq \frac{\Delta x^2}{2\nu} = \frac{C'}{\nu N^4}.$$

At this point one might be tempted to consider equispaced nodes, in order to soften the stability constraint to a more practical one

$$\Delta t \leq \frac{C}{\nu\, O(N^2) + c\, O(N)}, \tag{3.10.40}$$

but in this case spectral accuracy is lost (unless we are using Fourier collocation method for a periodic problem). ■

REMARK 3.10.9 The constraint (3.10.38) is prohibitive if ν/c is large. If, however, $\nu \cong c\,O(N^{-2})$ with N high enough to reach accuracy, say $N = 10$, then the stability constraint becomes $\Delta t \cong O(N^{-2})$, which is more reasonable. If, however, ν/c is very small, then the solution may present rapid changes that are not very well resolved by the method leading to Gibbs phenomenon, i.e., strong oscillations near large gradients or discontinuities. To avoid this effect the highest-order modes have to be cut out or smoothed out. ■

In order to overcome the difficulty coming from the strong stability constraint, one may consider implicit methods, say the backward Euler method

$$(\mathbf{I} + c\Delta t\mathbf{D} - \nu\Delta t\mathbf{D}^2)\mathbf{u}_{i+1} = \mathbf{u}_i + \Delta t\mathbf{F}_{i+1} \tag{3.10.41}$$

or the Crank–Nicolson method

$$\left(\mathbf{I} + \frac{c}{2}\Delta t\mathbf{D} - \frac{\nu}{2}\Delta t\mathbf{D}^2\right)\mathbf{u}_{i+1}$$
$$= \left(\mathbf{I} - \frac{c}{2}\Delta t\mathbf{D} + \frac{\nu}{2}\Delta t\mathbf{D}^2\right)\mathbf{u}_i + \frac{\mathbf{F}_i + \mathbf{F}_{i+1}}{2}, \qquad (3.10.42)$$

which are always stable.

There are essentially two problems in using these methods:

1. The inversion of the (generally full) matrix on the left-hand side, which is costly and is often done by iterative methods [HaY].
2. The treatment of nonlinearities. It is, in fact, easy to understand how to generalize explicit methods to nonlinear problems, while the treatment of nonlinearities in implicit methods is more complex. The easiest method is to freeze the coefficients at the previous time, as explained in Remark 3.9.14.

REMARK 3.10.10 Since in many models it is the convective term that is nonlinear, while the diffusive term depends only very weekly on the state variable (e.g., the Navier–Stokes equation or Fourier's heat conduction model), then one is lead to consider ***semi-implicit methods***, i.e., methods that treat implicitly the diffusive term and explicitly the convective term. One may expect, and in fact it can be proved, that only the term treated explicitly will influence the stability condition leading to estimates of the form $\Delta t = CN^{-2}$. The simplest of these methods is the ***backward-forward Euler method***

$$\left(\mathbf{I} - \Delta t\nu\mathbf{D}^2\right)\mathbf{u}_{i+1} = \left(\mathbf{I} - c\Delta t\mathbf{D}\right)\mathbf{u}_i + \Delta t\mathbf{F}_{i+1}, \qquad (3.10.43)$$

which is first order in time.

Another very popular method is the ***Adams–Bashforth Crank–Nicolson method***, which approximates with Adams–Bashforth the convective term and with Crank–Nicolson the diffusive term

$$\left(\mathbf{I} - \frac{\Delta t}{2}\nu\mathbf{D}^2\right)\mathbf{u}_{i+1} = \left(\mathbf{I} + \frac{\Delta t}{2}\nu\mathbf{D}^2 - \frac{3\Delta t}{2}c\mathbf{D}\right)\mathbf{u}_i$$
$$+ \frac{\Delta t}{2}c\mathbf{D}\mathbf{u}_{i-1} + \Delta t\frac{\mathbf{F}_{i+1} + \mathbf{F}_i}{2}. \qquad (3.10.44)$$

■

3.11 Domain Decomposition Methods

The modelling and mathematical methods presented in the preceding sections were developed by adopting the same model and the same solution techniques over the whole domain of the dependent variable.

However, this situation may be unrealistic in several physical situations. In fact, we have already seen that the partial differential equations may change type and consequently one has to adapt the solution algorithms.

In some cases, even the mathematical model may change in different domains of the dependent variables. This, in fact, occurs when different media are in contact, for instance:

- Gas-solid interactions;
- Conductors with different conductivities;
- Flow of immiscible liquids (e.g., oil and water or air and water);
- Phase changes.

In the first two cases the interface dividing the two regions is fully determined. In the last two cases it is an unknown of the problem (in most cases the most relevant one).

There are also cases in which it is simply convenient to introduce in a medium a fictitious interface that divides it into a region where a more complex model (or solution technique) has to be used and a region where an easier model can still be a good approximation. This may, in fact, be convenient to do in the following cases:

- In fluid dynamics, where one needs a viscous model, e.g., the Navier–Stokes model, near the boundary (the so-called boundary layer), while an inviscid model, e.g., the Euler model, is sufficient to provide a good description of the flow elsewhere;
- In rarefied gas dynamics, where the Boltzmann model is used only near the space ship, while the Euler or Navier–Stokes model is used in the outer environment;
- In solid mechanics, where it may be useful to distinguish between regions with large and small deformations;
- In heat transfer, where thermal diffusivity is much more important in the boundary layer than elsewhere;
- In laser technology, where to study the reaction of a body to a picosecond laser pulse, one may need a hyperbolic model (e.g., that proposed by Cattaneo) near the laser and the Fourier parabolic model away from it;
- In problems with strong inhomogeneous terms.

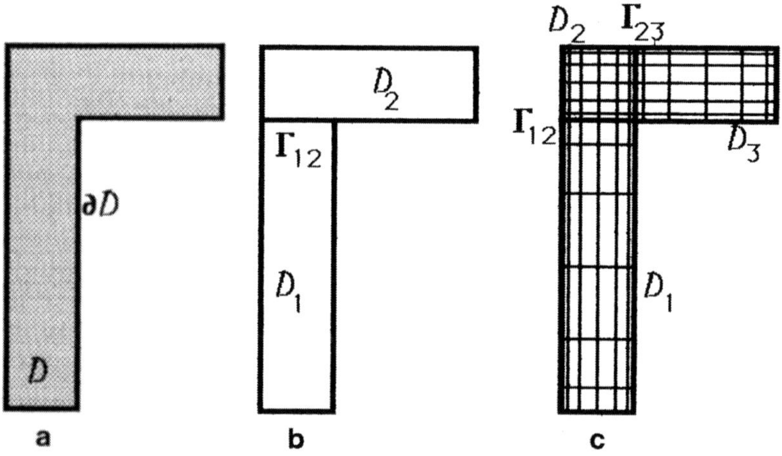

Figure 3.11.1 — Examples of decomposition of (a) a nonrectangular domain into (b) two rectangular parts and (c) three rectangular parts. In (c) a coarse Chebyshev collocation net is also indicated.

In these cases the location of the fictitious interface has to be adapted according to the solution on the basis of some error estimates.

Finally, the domain $\mathcal{D}$ might have a shape that does not possess the sufficiently regularity required for the application of the numerical scheme. Then a suitable subdivision of $\mathcal{D}$ may be organized in order to obtain a convenient applicability of the mathematical method that is adopted. For instance, as we have already mentioned, collocation methods work well in rectangular domains or in domains that can be easily mapped (i.e., with a smooth transformation) into rectangles. Then the domain in Fig. 3.11.1a can be partitioned into at least two rectangular subdomains as in Fig. 3.11.1b. Actually, because of the corner, dividing $\mathcal{D}$ into three parts, as shown in Fig. 3.11.1c, leads to better accuracy than dividing $\mathcal{D}$ in two and using more collocation points.

In summary, what it is actually done in domain decomposition methods is to cut the domain $\mathcal{D}$ into m smaller pieces $\mathcal{D}_i$ with

$$\mathcal{D} = \bigcup_{i=1}^{m} \mathcal{D}_i$$

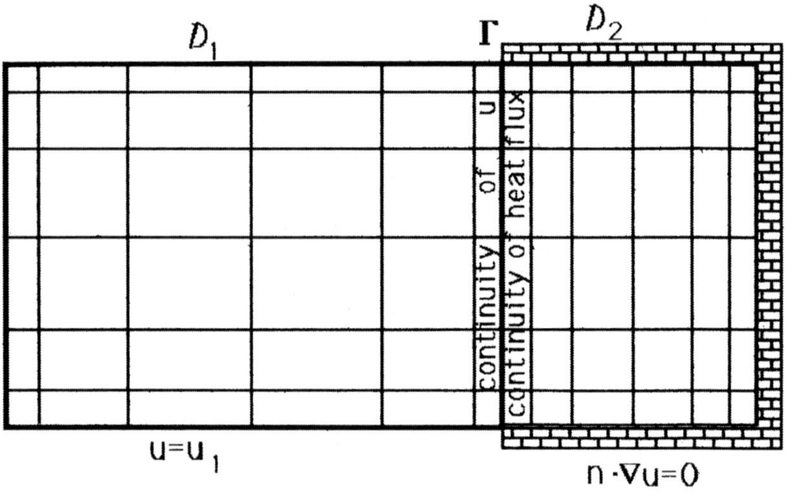

Figure 3.11.2 — Parabolic-parabolic coupling. The decomposition, the resulting collocation net, and the interface conditions are also indicated.

and to solve in each part a subproblem. At this point the reader can easily understand why with the advent of parallel computers this method is gaining more and more attention.

The crucial point is in the transmission of information from one subdomain to the adjacent ones through the real or fictitious interface. In fact, as will be clearer in the following sections, to solve these subproblems we need additional conditions on the interfaces between the subdomains. These conditions are usually imposed keeping in mind both the mathematical structure and the physical origin of the continuous model. It is in fact required that the problem is well formulated and well posed and that the scheme is stable.

In the following sections we will present several examples with the intention of giving an idea of the procedure and with no attempt to be exhaustive at all.

3.11.1 Parabolic-parabolic coupling

As a first example, consider a homogeneous solid with diffusivity κ_1 held at room temperature u_1. A second homogeneous solid with diffusivity κ_2 with all walls but one insulated is heated at temperature u_2 and put in contact with the first solid as shown in Fig. 3.11.2.

The problem can then be written as

$$\frac{\partial u}{\partial t} = \nabla \cdot \left[\kappa(x,u)\nabla u\right],$$

$$u(0,\mathbf{x}) = \begin{cases} u_1 & \text{if } \mathbf{x} \in \mathcal{D}_1\,, \\ u_2 & \text{if } \mathbf{x} \in \mathcal{D}_2\,, \end{cases} \tag{3.11.1}$$

$$u(t,\mathbf{x}) = u_1 \qquad \text{on } \partial\mathcal{D}_1 - \Gamma\,,$$

$$\mathbf{n}\cdot\nabla u(t,\mathbf{x}) = 0 \qquad \text{on } \partial\mathcal{D}_2 - \Gamma\,,$$

where

$$\kappa(x,u) = \begin{cases} \kappa_1(u) & \text{in } \mathcal{D}_1 \\ \kappa_2(u) & \text{in } \mathcal{D}_2 \end{cases}$$

or, splitting (3.11.1) in two, as

$$\begin{aligned} &\frac{\partial u^{(1)}}{\partial t} = \nabla \cdot \left[\kappa_1(u^{(1)})\nabla u^{(1)}\right] \\ &u^{(1)}(0,\mathbf{x}) = u_1 \\ &u^{(1)}(t,\mathbf{x}) = u_1 \qquad \text{on } \partial\mathcal{D}_1 - \Gamma\,, \end{aligned} \tag{3.11.2}$$

and

$$\begin{aligned} &\frac{\partial u^{(2)}}{\partial t} = \nabla \cdot \left[\kappa_2(u^{(2)})\nabla u^{(2)}\right] \\ &u^{(2)}(0,\mathbf{x}) = u_2 \\ &\mathbf{n}\cdot\nabla u^{(2)}(t,\mathbf{x}) = 0 \qquad \text{on } \partial\mathcal{D}_2 - \Gamma\,, \end{aligned} \tag{3.11.3}$$

where, of course, $u^{(i)}$ is the restriction of u on $\mathcal{D}_i$.

From what was mentioned in Section 3.7, it is clear that both (3.11.2) and (3.11.3) need one more condition on Γ to be well formulated. A condition like

$$u^{(1)}(t,\mathbf{x}) = u^{(2)}(t,\mathbf{x}) = u_\Gamma(t,\mathbf{x}) \quad \text{on} \quad \Gamma$$

would do, but we do not know u_Γ, which is actually a very important unknown of the problem. We are then left with a continuity condition, which makes sense both from the mathematical and the physical point of view.

As a second condition we might ask that also the derivative of u be continuous across Γ, but this has no physical justification. In fact, taking into account the physical background of the model, it can be easily understood that, besides the temperature, it is the heat flux that needs be continuous across Γ. This produces the interface conditions

$$\begin{cases} u^{(1)}(t,\mathbf{x}) = u^{(2)}(t,\mathbf{x})\,, \\ \kappa_1 \mathbf{n}\cdot\nabla u^{(1)}(t,\mathbf{x}) = \kappa_2 \mathbf{n}\cdot\nabla u^{(2)}(t,\mathbf{x})\,, \end{cases} \qquad \text{on } \Gamma\,, \tag{3.11.4}$$

which couples the system (3.11.2) and (3.11.3).

Since the common value of the temperature and of the heat flux is, of course, unknown, the solution has to be computed via an iterative procedure, that is, at every time step one has to

Step 1 Set a guessed interface temperature u_Γ;

Step 2 Solve (3.11.2) with $u^{(1)} = u_\Gamma$ on Γ;

Step 3 Compute the heat flux through the interface

$$\mathbf{q}_\Gamma = \kappa_1 \nabla u^{(1)}(t,\mathbf{x})\,; \tag{3.11.5}$$

Step 4 Solve (3.11.3) with

$$\kappa_2 \mathbf{n}\cdot\nabla u^{(2)} = \mathbf{n}\cdot\mathbf{q}_\Gamma \qquad \text{on } \Gamma\,; \tag{3.11.6}$$

Step 5 Set u_Γ equal to the value of $u^{(2)}$ on Γ;

Step 6 Go back to Step 2 and iterate to convergence.

Usually in the iterative procedures it is better to introduce in Steps 4 and 5 a relaxation parameter that weights the new interface condition with the previous one. The following section will furnish an explicit example of this type.

3.11.2 Elliptic-elliptic coupling

Without taking into account the physical background leading to the model, consider the abstract two-dimensional boundary-value problem

$$\begin{cases} \nabla^2 u = f(\mathbf{x};u) & \text{in } \mathcal{D} \\ u(\mathbf{x}) = u_D(\mathbf{x}) & \text{on } \partial\mathcal{D}\,, \end{cases} \tag{3.11.7}$$

where $\mathcal{D}$ is the domain in Fig. 3.11.1.

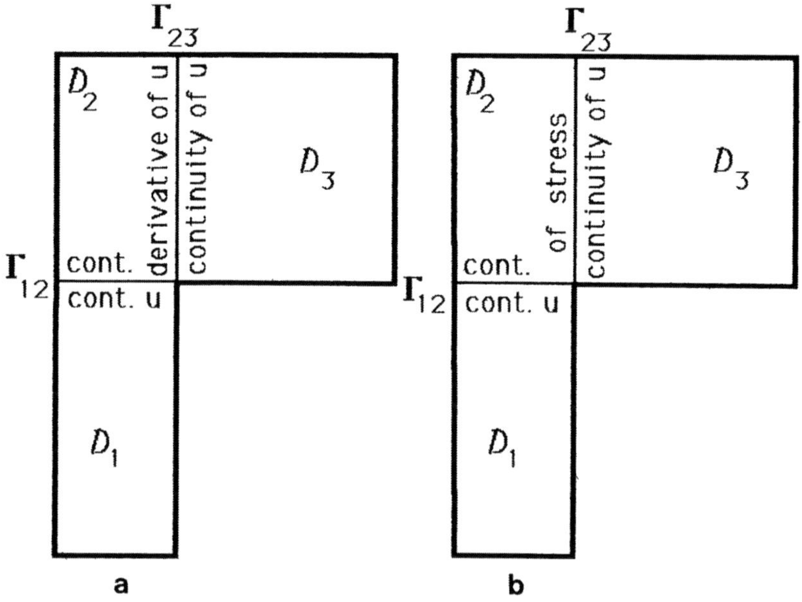

Figure 3.11.3 — Decomposition of a nonrectangular domain and related collocation net. Interface conditions for (a) Laplace's equation and for (b) Stokes' problem are also indicated.

As already stated, because of the corner, it is convenient to partition $\mathcal{D}$ into three subdomains $\mathcal{D}_1$, $\mathcal{D}_2$, and $\mathcal{D}_3$ divided, respectively, by the two interfaces Γ_{12} and Γ_{23} and to introduce in each subdomain a collocation net, as shown in Fig. 3.11.3.

In order to solve (3.11.7) in $\mathcal{D}_1$ we need to impose some conditions on Γ_{12}. As before, conditions like

$$u^{(1)}(\mathbf{x}) = a(\mathbf{x}) \qquad \text{on } \Gamma_{12}$$

would do, but we do not know $a(\mathbf{x})$.

The idea is then to impose some continuity conditions that will couple, as it is natural, the solution in the different domains. Continuity of u across the interfaces

$$\begin{aligned} u^{(1)}(\mathbf{x}) &= u^{(2)}(\mathbf{x}) \qquad \text{on } \Gamma_{12}\,, \\ u^{(3)}(\mathbf{x}) &= u^{(2)}(\mathbf{x}) \qquad \text{on } \Gamma_{23} \end{aligned} \tag{3.11.8}$$

would be sufficient if we knew the solution in $\mathcal{D}_2$. It should be clear by this argument that we need two more continuity conditions, say

$$\frac{\partial u^{(1)}}{\partial y}(\mathbf{x}) = \frac{\partial u^{(2)}}{\partial y}(\mathbf{x}) \qquad \text{on } \Gamma_{12}\,, \tag{3.11.9a}$$

and

$$\frac{\partial u^{(2)}}{\partial x}(\mathbf{x}) = \frac{\partial u^{(3)}}{\partial x}(\mathbf{x}) \qquad \text{on } \Gamma_{23}\,. \tag{3.11.9b}$$

REMARK 3.11.1 These conditions make the problem well formulated and well posed, but in principle we cannot say if the ansatz is correct till the physical origin of the model is considered. This point will be clarified in Remark 3.11.2. ■

The solution of the problem is now obtained by the following iterative procedure.

Let $u_\Gamma(\mathbf{x})$ be an initial guess on the value of u at the interfaces and solve for $n \geq 1$ (the index n refer to the n-th iteration) the boundary-value problems

$$\begin{cases} \nabla^2 u_n^{(1)} = f(\mathbf{x}; u_n^{(1)}) & \text{in } \mathcal{D}_1 \\ u_n^{(1)}(\mathbf{x}) = u_D(\mathbf{x}) & \text{on } \partial\mathcal{D}_1 - \Gamma_{12} \\ u_n^{(1)}(\mathbf{x}) = \theta_n u_{n-1}^{(2)}(\mathbf{x}) + (1-\theta_n) u_{n-1}^{(1)}(\mathbf{x}) & \text{on } \Gamma_{12} \end{cases} \tag{3.11.10}$$

and

$$\begin{cases} \nabla^2 u_n^{(3)} = f(\mathbf{x}; u_n^{(3)}) & \text{in } \mathcal{D}_3 \\ u_n^{(3)}(\mathbf{x}) = u_D(\mathbf{x}) & \text{on } \partial\mathcal{D}_3 - \Gamma_{23} \\ u_n^{(3)}(\mathbf{x}) = \Theta_n u_{n-1}^{(2)}(\mathbf{x}) + (1-\Theta_n) u_{n-1}^{(3)}(\mathbf{x}) & \text{on } \Gamma_{23}\,, \end{cases} \tag{3.11.11}$$

where θ_n and Θ_n are suitable relaxation parameters.

Then the solution in $\mathcal{D}_2$ is obtained by integrating

$$\begin{cases} \nabla^2 u_n^{(2)} = f(\mathbf{x}; u_n^{(2)}) & \text{in } \mathcal{D}_2 \\ u_n^{(2)}(\mathbf{x}) = u_D(\mathbf{x}) & \text{on } \partial\mathcal{D}_2 - (\Gamma_{23} \cup \Gamma_{12}) \\ \dfrac{\partial u_n^{(2)}}{\partial y}(\mathbf{x}) = \dfrac{\partial u_n^{(1)}}{\partial y}(\mathbf{x}) & \text{on } \Gamma_{12} \\ \dfrac{\partial u_n^{(2)}}{\partial x}(\mathbf{x}) = \dfrac{\partial u_n^{(3)}}{\partial x}(\mathbf{x}) & \text{on } \Gamma_{23}\,. \end{cases} \tag{3.11.12}$$

The solution is then iterated to convergence.

REMARK 3.11.2 Knowing the physical background of a model is, however, essential to giving physically consistent interface conditions. Every problem has its own natural interface condition. Consider, for instance, Stokes' problem in the same domain $\mathcal{D}$ shown in Fig. 3.11.3

$$\begin{cases} \nu\nabla^2\mathbf{u} = \nabla p + f(\mathbf{x}) & \text{in } \mathcal{D} \\ \nabla \cdot \mathbf{u} = 0 & \\ u(\mathbf{x}) = 0 & \text{on } \partial\mathcal{D}\,, \end{cases} \tag{3.11.13}$$

which is an elliptic problem describing the flow of a very viscous fluid (inertial terms are neglected) around the corner shown in Fig. 3.11.3. Though this system might be erroneously considered similar to (3.11.7), the method just given cannot be generalized to this case without taking into account that we are dealing with a fluid. In fact, for a continuum the relevant physical quantities one has to look at are the velocity and the stress.

In this case the natural boundary conditions needed to couple the subproblems are the continuity of velocity (as before)

$$\begin{cases} u_x^{(1)} = u_x^{(2)} \\ u_y^{(1)} = u_y^{(2)} \end{cases} \quad \text{on } \Gamma_{12}$$

$$\begin{cases} u_x^{(3)} = u_x^{(2)} \\ u_y^{(3)} = u_y^{(2)} \end{cases} \quad \text{on } \Gamma_{23}\,,$$

where $u_x^{(i)}$ and $u_y^{(i)}$ are the x and y components of the velocity $\mathbf{u}$ in $\mathcal{D}_i$ and the continuity of shear stress $\mathbf{t} \cdot \mathbf{Tn}$ and normal stresses $\mathbf{n} \cdot \mathbf{Tn}$ (instead of the continuity of the derivative) is given by

$$\begin{cases} \nu \dfrac{\partial u_y^{(2)}}{\partial y} - p^{(2)} = \nu \dfrac{\partial u_y^{(1)}}{\partial y} - p^{(1)} \\ \dfrac{\partial u_y^{(2)}}{\partial x} + \dfrac{\partial u_x^{(2)}}{\partial y} = \dfrac{\partial u_y^{(1)}}{\partial x} + \dfrac{\partial u_x^{(1)}}{\partial y} \end{cases} \quad \text{on } \Gamma_{12}$$

and

$$\begin{cases} \nu \dfrac{\partial u_x^{(2)}}{\partial x} - p^{(2)} = \nu \dfrac{\partial u_x^{(3)}}{\partial x} - p^{(3)} \\ \dfrac{\partial u_y^{(2)}}{\partial x} + \dfrac{\partial u_x^{(2)}}{\partial y} = \dfrac{\partial u_y^{(3)}}{\partial x} + \dfrac{\partial u_x^{(3)}}{\partial y} \end{cases} \quad \text{on } \Gamma_{12}\,.$$

■

3.11.3 Parabolic-hyperbolic coupling

Consider now the problem of coupling the Navier–Stokes model for viscous flow with the Euler equations in those regions where viscous effects can be neglected. As a simplified version of this problem, consider the following one-dimensional parabolic system in $[a, \Gamma]$

$$\begin{cases} \dfrac{\partial u^{\mathrm{par}}}{\partial t} + b_u \dfrac{\partial u^{\mathrm{par}}}{\partial x} = \nu \dfrac{\partial^2 u^{\mathrm{par}}}{\partial x^2} + f(t, x; u^{\mathrm{par}}, v^{\mathrm{par}}) \\ \\ \dfrac{\partial v^{\mathrm{par}}}{\partial t} + b_v \dfrac{\partial v^{\mathrm{par}}}{\partial x} = \nu \dfrac{\partial^2 v^{\mathrm{par}}}{\partial x^2} + f(t, x; u^{\mathrm{par}}, v^{\mathrm{par}}) \end{cases} \tag{3.11.14}$$

and the relative hyperbolic system obtained by neglecting the viscous terms, i.e., $\nu = 0$, in $[\Gamma, b]$

$$\begin{cases} \dfrac{\partial u^{\mathrm{hyp}}}{\partial t} + b_u \dfrac{\partial u^{\mathrm{hyp}}}{\partial x} = f(t, x; u^{\mathrm{hyp}}, v^{\mathrm{hyp}}) \\ \\ \dfrac{\partial v^{\mathrm{hyp}}}{\partial t} + b_v \dfrac{\partial v^{\mathrm{hyp}}}{\partial x} = f(t, x; u^{\mathrm{hyp}}, v^{\mathrm{hyp}})\,. \end{cases} \tag{3.11.15}$$

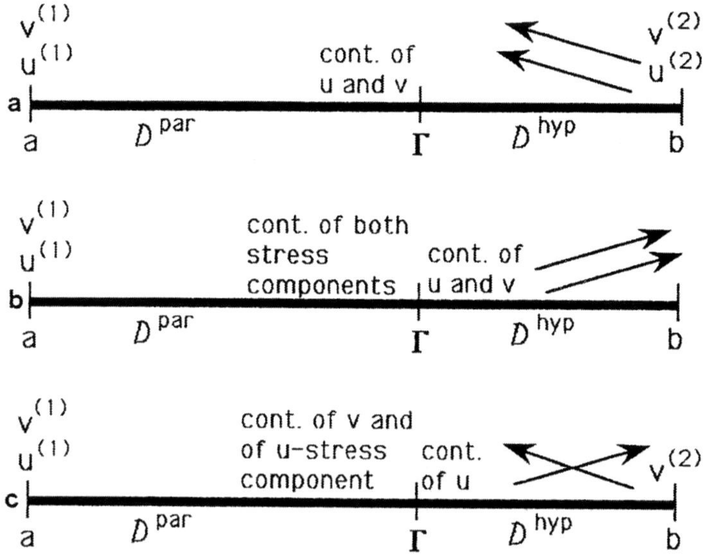

Figure 3.11.4 — Parabolic-hyperbolic coupling. Interface conditions are indicated when (a) $b_u, b_v < 0$, (b) $b_u, b_v > 0$, and (c) $b_u > 0$, $b_v < 0$. The arrows indicate the direction of the characteristics.

In order to solve (3.11.14) in $[a, \Gamma]$ we have to give, besides the initial conditions, two conditions in $x = a$, say

$$\begin{cases} u^{\text{par}}(t,a) = u_a(t) \\ v^{\text{par}}(t,a) = v_a(t) \end{cases} \tag{3.11.16}$$

and two conditions in $x = \Gamma$, one on u and one on v.

As mentioned in Section 3.7, in order to give the proper boundary conditions to (3.11.15), we have to give a look at the characteristics entering the domain. In summary, if b_u and b_v are constant, we have to give

- Two conditions in $x = b$

$$\begin{aligned} u^{\text{hyp}}(t,b) &= u_b(t) \\ v^{\text{hyp}}(t,b) &= v_b(t) \end{aligned} \tag{3.11.17}$$

if b_u and b_v are both negative (see Fig. 3.11.4a);

- Two conditions in $x = \Gamma$ if b_u and b_v are both positive (see Fig. 3.11.4b);
- u^{hyp} in $x = \Gamma$ and v^{hyp} in $x = b$, if $b_u > 0$ and $b_v < 0$ and *vice versa* (see Fig. 3.11.4c).

The first two cases are simple (and proposed as exercises to the reader). In the first case one has to

Step 1 Solve (3.11.15) with (3.11.17).

Step 2 Deduce $u^{\text{hyp}}(t,\Gamma)$ and $v^{\text{hyp}}(t,\Gamma)$.

Step 3 By the continuity of u and v, set

$$u^{\text{par}}(t,\Gamma) = u^{\text{hyp}}(t,\Gamma) \quad \text{and} \quad v^{\text{par}}(t,\Gamma) = v^{\text{hyp}}(t,\Gamma)\,. \tag{3.11.18}$$

Step 4 Solve (3.11.14) with (3.11.16) and (3.11.18).

In the second case one has to

Step 1 Solve (3.11.14) with (3.11.16) with the "continuity of stress"

$$\nu\frac{\partial u^{\text{par}}}{\partial x}(t,\Gamma) = 0 \quad \text{and} \quad \nu\frac{\partial v^{\text{par}}}{\partial x}(t,\Gamma) = 0\,.$$

Step 2 Deduce u^{par} and v^{par}.

Step 3 Solve in $\mathcal{D}_2$ as before.

The interesting case is the last one. In this case, Gastaldi and Quarteroni [GaQ] suggest to couple (3.11.14–3.11.16) through the continuity of velocity and of the "component of the stress"

$$\begin{cases} u^{\text{par}}(t,\Gamma) = u^{\text{hyp}}(t,\Gamma) \\ v^{\text{par}}(t,\Gamma) = v^{\text{hyp}}(t,\Gamma) \\ \nu\dfrac{\partial u^{\text{par}}}{\partial x}(t,\Gamma) = 0 \\ v^{\text{par}}(t,b) = v_b(t)\,, \end{cases} \tag{3.11.19}$$

if $b_u > 0$ and $b_v < 0$. If, on the contrary, $b_u < 0$ and $b_v > 0$, then the last two equations are substituted by

$$\begin{cases} \nu\dfrac{\partial v^{\text{par}}}{\partial x}(t,\Gamma) = 0 \\ u^{\text{par}}(t,b) = u_b(t)\,. \end{cases}$$

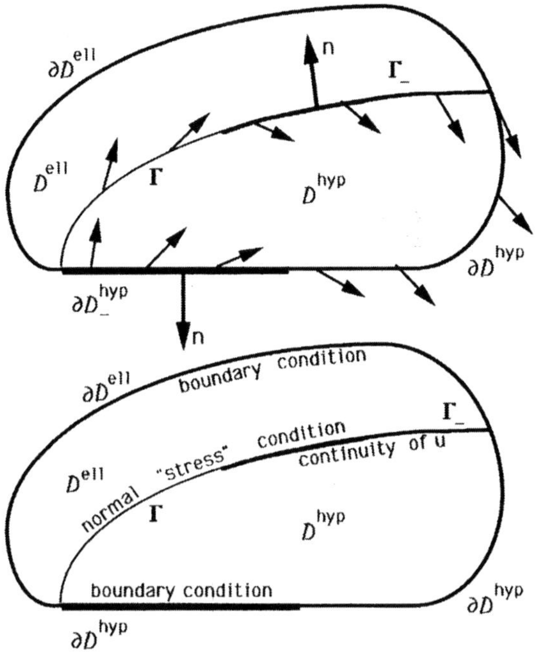

Figure 3.11.5 — Hyperbolic-elliptic coupling. The arrows indicate the direction of **c**.

Also in this case an iterative procedure is then applied.

3.11.4 Hyperbolic-elliptic coupling

As a final example and with the same aim as the previous examples, consider with reference to Fig. 3.11.5 the coupling of

$$\mathbf{c} \cdot \nabla u^{\text{ell}} = \nu \nabla^2 u^{\text{ell}} + f(t, x; u^{\text{ell}}) \quad \text{in} \quad \mathcal{D}^{\text{ell}} \subset \mathbb{R}^2 \tag{3.11.20}$$

$$\mathbf{c} \cdot \nabla u^{\text{hyp}} = f(t, x; u^{\text{hyp}}) \quad \text{in} \quad \mathcal{D}^{\text{hyp}} \subset \mathbb{R}^2 \tag{3.11.21}$$

with Dirichlet boundary conditions on $\partial\mathcal{D}$. The former equation is an elliptic model resembling the steady linearized Navier–Stokes equation. The latter equation is the hyperbolic transport model obtained again neglecting the viscous term.

Equation (3.11.20) yields a well-formulated problem when boundary conditions are given on the outer boundary $\partial\mathcal{D}^{\text{ell}} - \Gamma$ and on Γ. Equation (3.11.21), instead, needs a boundary condition only on those parts of

$\partial\mathcal{D}^{\text{hyp}} - \Gamma$ and of Γ (which we will denote, respectively, by $\partial\mathcal{D}^{\text{hyp}}_{-} - \Gamma$ and Γ_{-}) where $\mathbf{c}\cdot\mathbf{n} \leq 0$.

In summary, as far as the interface conditions are concerned, one condition is needed along the whole Γ in order to solve the elliptic problem in $\mathcal{D}^{\text{ell}}$ and a further condition only on that part of Γ where $\mathbf{c}$ is directed inside $\mathcal{D}^{\text{hyp}}$ to solve the hyperbolic problem in $\mathcal{D}^{\text{hyp}}$.

As now usual, these interface conditions are based on some continuity assumptions. In this case Gastaldi, Quarteroni, and Sacchi [GQS] suggest joining to (3.11.20) and (3.11.21) and to the relative boundary conditions, say for instance

$$\begin{aligned} u^{\text{ell}}(t,\mathbf{x}) &= u_D(t,\mathbf{x}) \qquad \text{on } \partial\mathcal{D}^{\text{ell}} - \Gamma \\ u^{\text{hyp}}(t,\mathbf{x}) &= u_D(t,\mathbf{x}) \qquad \text{on } \partial\mathcal{D}^{\text{hyp}}_{-} - \Gamma\,, \end{aligned} \tag{3.11.22}$$

the following set of interface conditions

$$\begin{aligned} &\nu\mathbf{n}\cdot\nabla u^{\text{ell}}(t,\mathbf{x}) = 0 \qquad \text{on } \Gamma_{-} \\ &\nu\mathbf{n}\cdot\nabla u^{\text{ell}}(t,\mathbf{x}) + \mathbf{c}\cdot\mathbf{n}u^{\text{ell}} = \mathbf{c}\cdot\mathbf{n}u^{\text{hyp}} \qquad \text{on } \Gamma - \Gamma_{-} \\ &u^{\text{hyp}}(t,\mathbf{x}) = u^{\text{ell}}(t,\mathbf{x}) \qquad \text{on } \Gamma_{-}\,, \end{aligned} \tag{3.11.23}$$

where $\mathbf{n}$ on Γ is directed toward $\mathcal{D}^{\text{ell}}$.

In the iterative procedure one solves first the hyperbolic problem with the continuity of velocity condition (last condition in (3.11.23)) replaced by

$$u^{\text{hyp}}_{n}(t,\mathbf{x}) = \theta_n u^{\text{ell}}_{n-1}(t,\mathbf{x}) + (1-\theta_n)u^{\text{hyp}}_{n-1}(t,\mathbf{x})\,,$$

where θ_n is a relaxation parameter and then solves the elliptic problem with the first condition in (3.11.22) and the first two in (3.11.23).

REMARK 3.11.3 We point out that the two parts of the interface, Γ_{-} and $\Gamma - \Gamma_{-}$, play separate roles in the interaction mechanism between $\mathcal{D}^{\text{ell}}$ and $\mathcal{D}^{\text{hyp}}$. In fact, the hyperbolic solution u^{hyp} is influenced by the elliptic one u^{ell} only through Γ_{-}, that is, through that part of Γ where the transport vector is directed inside $\mathcal{D}^{\text{hyp}}$, as is physically reasonable. *Vice versa*, the elliptic solution is influenced by the hyperbolic one only through $\Gamma - \Gamma_{-}$, where $\mathbf{c}$ is directed outside $\mathcal{D}^{\text{hyp}}$. ■

REMARK 3.11.4 If $\Gamma_{-} = \Gamma$ or $\Gamma_{-} = \emptyset$, then the two problems decouple. In fact, if $\Gamma_{-} = \Gamma$, one has enough conditions to solve the elliptic problem and then pass to the solution of the hyperbolic one. *Vice versa*, if $\Gamma_- = \emptyset$, one can solve the hyperbolic problem and then deduce the interface conditions to solve the elliptic one. ■

3.12 Scientific Programs

We have pointed out in Section 2.8 that a good numerical code should be efficient, economically convenient, versatile, and have reasonable storage requirements. Furthermore, we have recomended performing several tests before starting the model simulation. Applying this process to discrete models may take some time, but, in most cases, it is straightforward.

The same line of thought should also guide the choice of the numerical method and the testing procedure for continuous models. In this case, however, things become much more complex. In particular, some tests suggested in Section 2.8 might be big pieces of work by themselves, e.g., finding the spectrum of the operator, studying the stability of the system, obtaining asymptotic results, and so on. For this reason the reader has to pay particular attention to the existing literature on his model and on models similar to the one he is dealing with.

This section will present some sample applications of finite difference and collocation methods to parabolic and hyperbolic models. The same line used in Section 2.9 will be followed here. We will also refer to Section 2.9 for a full explanation of the input/output parts, since they are transferred to the following programs with only minor adjustments.

3.12.1 FDFEPAR: An explicit finite difference routine for parabolic equations

This section will describe a simple explicit routine, which integrates the convection-diffusion equation

$$\frac{\partial u}{\partial t} + c\frac{\partial u}{\partial x} = \frac{\kappa}{1+\varepsilon u}\frac{\partial^2 u}{\partial x^2} - \frac{\varepsilon\kappa}{(1+\varepsilon u)^2}\left(\frac{\partial u}{\partial x}\right)^2 \tag{3.12.1}$$

in $x \in [a, b]$ using the central finite difference method with the following initial and boundary conditions

$$\begin{aligned} u_{in}(x) &= \sin x \\ u_a(t) &= u_b(t) = 0\ . \end{aligned} \tag{3.12.2}$$

The new and most important part of the program is the finite difference solver. We will focus on it referring to Table 3.12.1a for the related flow chart.

After setting the initial conditions, the boundary conditions and the right-hand side of the equation (in the internal nodes) are computed by the central finite difference scheme. Then the forward Euler method (forward difference) is used to update the solution.

In more detail, the block 60–230 is a series of input/output commands, which, among other things, asks for the values of c, κ, and ε in Eq.(3.12.1). Then in lines 270–340, the space discretization is specified.

In lines 350–390 time is initialized and the initial condition is set. The current value of the solution is also stored in `uold` (lines 400–424), before computing the updated value. The program returns here when the updated solution is computed.

The boundary conditions are set in lines 430 and 440 and the right-hand side of the equation is computed in the internal nodes in lines 460–550. In particular, lines 470 and 480 give the central difference approximations for the spatial derivatives; lines 500–520 the terms in front of $\partial^2 u/\partial x^2$, $\partial u/\partial x$, and u, respectively; line 530 the forcing term; and finally line 540 gathers all these quantities to compute the right-hand side of Eq.(3.9.41).

Besides lines 380, 430, and 440, lines 490–540 also have to be properly adjusted if different initial conditions, boundary conditions, and/or equations are to be integrated.

The forward Euler method is then used in lines 560–580 to update the solution.

Then lines 790–810 provide the analytic solution, which might be known for certain values of the parameters (here $a = 0$, $b = 4\pi$, $c = 0$). Of course, these lines also have to be changed if different initial-boundary-value problems are being integrated (and if the analytic solution is available, of course).

After that, line 820 establishes whether the solution has to be plotted (lines 830–1020) or printed (lines 1200–1380). As already stated, these blocks are almost identical to those described in Section 2.9. The only differences are the following:

- u is plotted versus x and not t with the obvious changes.
- In line 920, the present time is printed.
- The state variable u is a scalar. The vector `u(ix)` refers to the value of u in the location `x(ix)`. Therefore, the plots are obtained by piecewise linear interpolation at each time step (lines 930–960). The same holds for the analytic solution (lines 980–1010).

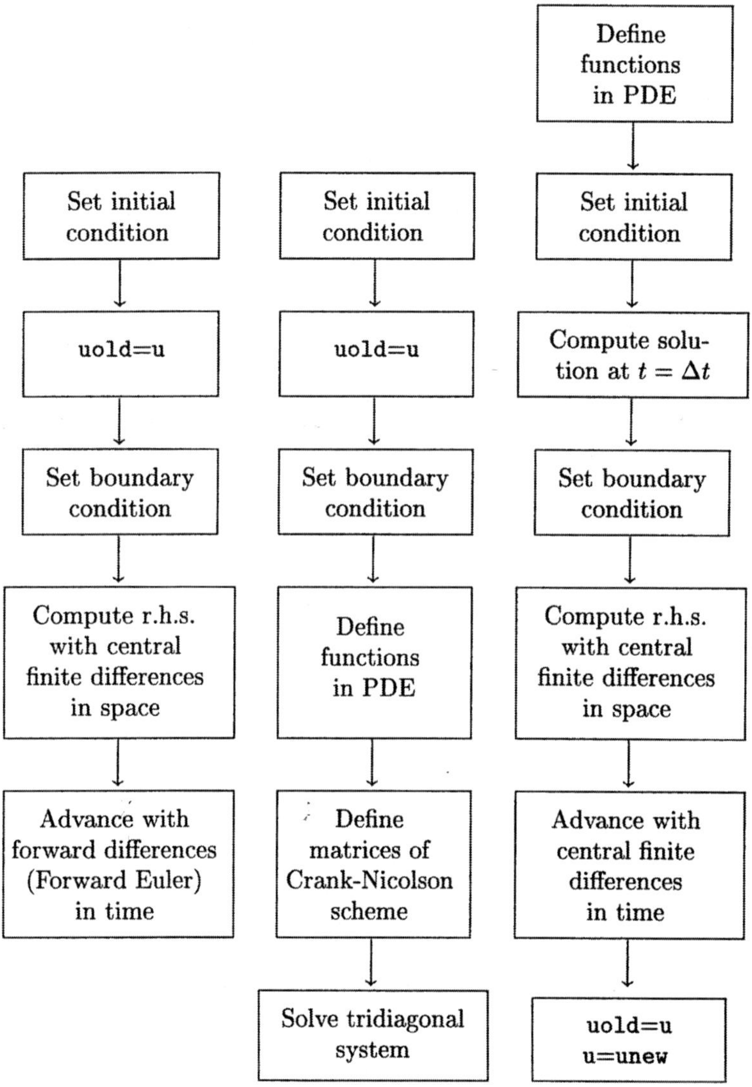

(a) (b) (c)

Table 3.12.1 — Flow charts for finite difference solvers (a) FDFEPAR (b) FDCNPAR (c) FD2HYP.

- If the analytic solution is known, the error

$$\frac{\max|\mathtt{uan(ix)} - \mathtt{u(ix)}|}{\max|\mathtt{uan(ix)}|}$$

is also computed and printed (lines 1270-1320).

After representing the solution, the program goes to line 1380 where an optional pause is inserted (this will be especially useful for taking a look at the data and examining the development of instabilities or similar phenomena). The integration is continued till the final time by going back to line 400.

FDFEPAR.BAS

```
1 REM ************************************************************
2 REM *          Solution of convection-diffusion equation       *
3 REM *          with Dirichlet boundary conditions using        *
4 REM *          explicit (central) finite differences in        *
5 REM *           space and forward Euler method in time         *
6 REM ************************************************************
7 REM ************************************************************
8 REM *                        Variables used                    *
9 REM ************************************************************
10 REM nx = number of grid points in space interval (nx1=nx-1)
11 REM x(nx) = grid points in space interval
12 REM ax, bx = extrema of space interval
13 REM dx = distance between grid points
14 REM u(nx) = state variable
15 REM uold(nx) = state variable at previous time step
16 REM uan(nx) = analytic solution
18 REM rhs(nx) = right-hand side of differential equations
19 REM der = finite difference approximation of derivative of u
20 REM der2 = finite difference approximation of second
   derivative of u
22 REM t = time
23 REM dt = time step
24 REM tfin = final time of integration
25 REM prout = print-out interval
26 REM nflag = flag indicating if the solution has to be
   printed or plotted
27 REM out$ = flag for print-out option
28 REM prstop$ = flag for inserting a pause after print-outs
29 REM iprint = flag used for switching the printing procedure
30 REM an$ = flag indicating if the analytic solution is available
```

```
31 REM pumin,pumax = minimum and maximum ordinate of graph
32 REM cc = convective coefficient
33 REM kk, eps = heat conductivity coefficients
34 REM den = used in r.h.s. of PDE
35 REM d = coefficient in front of the second derivative of u
36 REM v = coefficient in front of the first derivative of u
37 REM g = coefficient in front of u
38 REM f = forcing term
39 REM stablim = stability condition
40 DIM x(1001), u(1001), uan(1001), uold(1001), rhs(1001)
60 REM ***********************************************************
61 REM *                         Input lines                       *
62 REM ***********************************************************
70 CLS 0
80 PRINT "******************************************************"
81 PRINT "*     Solution of a convection-diffusion equation     *"
82 PRINT "*      with Dirichlet boundary conditions using       *"
84 PRINT "*          FORWARD EULER method in time and           *"
85 PRINT "*        CENTRAL FINITE DIFFERENCES in space          *"
86 PRINT "******************************************************"
90 PRINT ""
100 PRINT "When more than one quantity is requested, separate
   with a comma"
110 PRINT "Values in square brackets are the suggested values
   (not default)"
120 INPUT "Space interval [0,4*pi=12.5664]="; ax, bx
130 INPUT "Number of grid points [31]"; nx:  nx1 = nx - 1
140 INPUT "Convective coefficient c [0]"; cc
145 INPUT "Heat conductivity coefficients k/(1+eu) (e<1) [1,0]";
   kk, eps
150 stablim = ((bx - ax) / nx1) ^ 2 / (2 * kk)
155 PRINT "The stability limit is dt<"; stablim
160 INPUT "Time step and final time [.08,4]"; dt, tfin
165 INPUT "Plot [1] or print-out [0] of the solution"; ngraf
170 IF ngraf = 0 THEN INPUT "Print-out interval [.5]"; prout
180 IF ngraf = 1 THEN INPUT "Minimum and maximum ordinate of
   the graph [-1,1]"; pumin, pumax
190 IF ngraf <> 0 AND ngraf <> 1 THEN GOTO 160
200 INPUT "Do you want to stop at every time step [y/n]"; prstop$
210 IF prstop$ <> "y" AND prstop$ <> "n" THEN GOTO 200
220 INPUT "Is the analytical solution known [y]"; an$
230 IF an$ <> "y" AND an$ <> "n" THEN GOTO 220
240 REM ***********************************************************
241 REM *                      End of input lines                   *
242 REM ***********************************************************
250 CLS 0
270 dx = (bx - ax) / nx1
310 REM ***********************************************************
```

```
311 REM *                  Set space discretization                    *
312 REM **********************************************************
320 FOR ix = 1 TO nx
330 x(ix) = ax + (ix - 1) * dx
340 NEXT ix
345 REM **********************************************************
346 REM *                      Set initial data                        *
347 REM **********************************************************
350 t = 0
360 iprint = 1
370 FOR ix = 1 TO nx
380 u(ix) = SIN(x(ix))
390 NEXT ix
400 FOR ix = 1 TO nx
420 uold(ix) = u(ix)
424 NEXT ix
425 REM **********************************************************
426 REM *                  Set boundary conditions                     *
427 REM **********************************************************
430 u(1) = 0
440 u(nx) = 0
445 REM **********************************************************
446 REM *                Set coefficients of the PDE                   *
447 REM *                                                              *
448 REM * du            du            d^2u                             *
449 REM * -- + v(t,x;u) -- = d(t,x;u) --- + f(t,x;u) u + g(t,x)        *
450 REM * dt            dx            dx^2                             *
452 REM *                                                              *
453 REM * d(t,x;u) = coefficient in front of the second
    derivative of u                                                    *
454 REM * v(t,x;u) = coefficient in front of the first
    derivative of u                                                    *
455 REM * f(t,x;u) = coefficient in front of u                         *
456 REM * g(t,x) = forcing term                                        *
457 REM **********************************************************
460 FOR ix = 2 TO nx1
470 der = (uold(ix + 1) - uold(ix - 1)) / (2 * dx)
480 der2 = (uold(ix + 1) - 2 * uold(ix) + uold(ix - 1)) / dx^2
490 den = (1 + eps * uold(ix))
500 d = kk / den
510 v = cc + eps * kk / den ^ 2 * der
520 f = 0
530 g = 0
540 rhs(ix) = d * der2 - v * der + f * uold(ix) + g
550 NEXT ix
555 REM **********************************************************
556 REM *       Update solution using forward Euler method             *
557 REM **********************************************************
```

```
560 t = t + dt
565 FOR ix = 2 TO nx1
570 u(ix) = uold(ix) + rhs(ix) * dt
580 NEXT ix
780 IF an$ = "n" THEN GOTO 820
785 REM ***********************************************************
786 REM *         If known, insert analytic solution here         *
787 REM ***********************************************************
790 FOR ix = 1 TO nx
800 uan(ix) = EXP(-kk * t) * SIN(x(ix))
810 NEXT ix
820 IF ngraf = 0 THEN GOTO 1200
830 REM ***********************************************************
831 REM *                   Plotting commands                     *
832 REM ***********************************************************
840 CLS 0:  KEY OFF: SCREEN 2
850 LOCATE 14, 10:  PRINT ax
860 LOCATE 14, 70:  PRINT bx
870 LOCATE 13, 7:  PRINT pumin
880 LOCATE 2, 5:  PRINT pumax
900 VIEW (80, 10)-(560, 100), , 1
910 WINDOW (ax, pumin)-(bx, pumax)
920 LOCATE 15, 40:  PRINT "t="; t
930 PSET (ax, u(1))
940 FOR ix = 2 TO nx
950 LINE -(x(ix), u(ix))
960 NEXT ix
970 IF an$ = "n" THEN GOTO 1380
980 PSET (ax, uan(1))
990 FOR ix = 2 TO nx
1000 LINE -(x(ix), uan(ix)), , , &H7070
1010 NEXT ix
1020 GOTO 1380
1200 REM **********************************************************
1201 REM *                   Print-out commands                   *
1202 REM **********************************************************
1210 IF t < prout * iprint - dt/10 THEN GOTO 1390
1220 PRINT TAB(15); "t="; t
1230 difmax = 0:  uanmax = 0
1240 FOR i = 1 TO nx
1250 PRINT "u("; i; ")="; u(i),
1260 IF an$ = "n" THEN GOTO 1310
1270 dif = ABS(u(i) - uan(i))
1280 IF dif > difmax THEN difmax = dif
1290 IF ABS(uan(i)) > uanmax THEN uanmax = ABS(uan(i))
1300 PRINT "uan="; uan(i), "|uan-u|="; dif
1310 NEXT i
1320 IF an$ = "y" THEN PRINT "dt="; dt, "dx="; dx,
```

```
    "maximum error="; difmax / uanmax
1330 iprint = iprint + 1
1380 IF prstop$ = "y" THEN INPUT "Press RETURN to continue"; scr
1390 REM ************************************************************
1391 REM *                   Is final time passed by?                *
1392 REM ************************************************************
1410 IF t <= tfin - dt / 10 THEN GOTO 400
1420 END
```

3.12.2 FDCNPAR: An implicit finite difference routine for parabolic equations

As already remarked in Section 3.9.5, explicit schemes suffer a strong stability requirement. This difficulty can be overcome by using implicit methods. The routine presented here uses central differences in space and the Crank–Nicolson method to advance in time and applies it to the initial-boundary-value problem (3.12.1–3.12.2). In this way one has a method that is always stable and also more accurate than that used in FDFEPAR.

In this case a new difficulty arises in the evaluation of nonlinear terms. Keeping in mind Remark 3.9.14, we used the approximation

$$\kappa'(u)\left(\frac{\partial u}{\partial x}\right)^2 \cong \kappa'(\mathtt{uold(ix)})\frac{\mathtt{uold(ix+1)}-\mathtt{uold(ix-1)}}{2\Delta x} \times \frac{\mathtt{u(ix+1)}-\mathtt{u(ix-1)}}{2\Delta x}\,.$$

The structure of the program is essentially that used in FDFEPAR, but, of course, for the finite difference solver, shown in Table 3.11.1b, which is more complex and requires the solution of a tridiagonal system.

In more detail, referring to the right-hand side of Eq.(3.9.52), the coefficient in front of $u_{i,j-1}$ is computed as `aex` in line `530`, part of the second parenthesis as `bex` in line `540`, and the coefficients in front of $u_{i,j+1}$ as `cex` in line `550`.

Since the functions in the partial differential equation are independent of time, these quantities are also used in lines `560-580` to set the values of the matrix on the left-hand side of (3.9.52) (otherwise `aex`, `bex`, and `cex` should be computed again at time $t + \Delta t$).

The boundary conditions are also taken into account in lines `610` and `620`.

The tridiagonal system is finally solved by the routine provided in Eqs.(A.2.20–A.2.21), which consists of a forward substitution part (lines `640-720`) and a backward substitution part (lines `730-770`).

FDCNPAR.BAS

```
1 REM ************************************************************
2 REM *       Solution of convection-diffusion equation with      *
3 REM *     Dirichlet boundary conditions and linearization of    *
4 REM *   the nonlinear terms using central finite differences    *
5 REM *          in space and Crank-Nicolson method in time       *
6 REM ************************************************************
7 REM ************************************************************
8 REM *                      Variables used                       *
9 REM ************************************************************
10 REM nx = number of grid points in space interval (nx1=nx-1)
11 REM x(nx) = grid points in space interval
12 REM ax, bx = extrema of space interval
13 REM dx = distance between grid points
14 REM u(nx) = state variable
15 REM uold(nx) = state variable at previous time step
16 REM uan(nx) = analytic solution
18 REM d(nx) = diffusion coefficient
19 REM v(nx) = convection coefficient
20 REM f(nx) = decay coefficient
21 REM g(nx) = forcing term
22 REM t = time
23 REM dt = time step (kh2=dt/dx^2, kh=dt/dx)
24 REM tfin = final time of integration
25 REM prout = print-out interval
26 REM nflag = flag indicating whether to print or plot the
   solution
27 REM out$ = flag for print-out option
28 REM prstop$ = flag for inserting a pause after print-outs
29 REM iprint = flag used for switching the printing procedure
30 REM an$ = flag indicating if the analytic solution is available
31 REM pumin, pumax = minimum and maximum ordinate of graph
32 REM a(nx) = subdiagonal of finite difference matrix
33 REM b(nx) = diagonal of finite difference matrix
34 REM c(nx) = superdiagonal of finite difference matrix
35 REM ff(nx) = r.h.s. of linear system
36 REM cc = convective coefficient
37 REM kk, eps = heat conductivity coefficients
38 REM ratio = used in tridiagonal system solver
39 REM aex, bex, cex, sum = used in definition of coefficient
   matrix
50 DIM x(1001), u(1001), uan(1001), uold(1001)
55 DIM a(1001), b(1001), c(1001), d(1001), f(1001), g(1001),
   v(1001), ff(1001)
60 REM ************************************************************
61 REM *                       Input lines                         *
62 REM ************************************************************
```

```
70 CLS 0
80 PRINT "********************************************************"
81 PRINT "*     Solution of a convection-diffusion equation     *"
82 PRINT "*      with Dirichlet boundary conditions using       *"
84 PRINT "*          CRANK-NICOLSON method in time and          *"
85 PRINT "*         CENTRAL FINITE DIFFERENCES in space         *"
86 PRINT "********************************************************"
90 PRINT ""
100 PRINT "When more than one quantity is requested, separate
   with a comma"
110 PRINT "Values in square brackets are the suggested values
   (not default)"
120 INPUT "Space interval [0,4*pi=12.5664]="; ax, bx
130 INPUT "Number of grid points [31]"; nx
135 INPUT "Convective coefficient c [0]"; cc
140 INPUT "Heat conductivity coefficients k/(1+eu) (e<1) [1,0]";
   kk, eps
150 INPUT "Time step and final time [.5,4]"; dt, tfin
160 INPUT "Plot [1] or print-out [0] of the solution"; ngraf
170 IF ngraf = 0 THEN INPUT "Print-out interval [.5]"; prout
180 IF ngraf = 1 THEN INPUT "Minimum and maximum ordinate of
   the graph [-1,1]"; pumin, pumax
190 IF ngraf <> 0 AND ngraf <> 1 THEN GOTO 160
200 INPUT "Do you want to stop at every time step [y/n]"; prstop$
210 IF prstop$ <> "y" AND prstop$ <> "n" THEN GOTO 200
220 INPUT "Is the analytical solution known [y]"; an$
230 IF an$ <> "y" AND an$ <> "n" THEN GOTO 220
240 REM **********************************************************
241 REM *                   End of input lines                   *
242 REM **********************************************************
250 CLS 0
260 nx1 = nx - 1
270 dx = (bx - ax) / nx1
280 kh2 = dt / dx ^ 2
290 kh = dt / dx
310 REM **********************************************************
311 REM *               Set space discretization                 *
312 REM **********************************************************
320 FOR ix = 1 TO nx
330 x(ix) = ax + (ix - 1) * dx
340 NEXT ix
345 REM **********************************************************
346 REM *                   Set initial data                     *
347 REM **********************************************************
350 t = 0
360 iprint = 1
370 FOR ix = 1 TO nx
380 u(ix) = SIN(x(ix))
```

```
390 NEXT ix
400 FOR ix = 1 TO nx
420 uold(ix) = u(ix)
424 NEXT ix
425 REM ***********************************************************
426 REM *                  Set boundary conditions                  *
427 REM ***********************************************************
430 u(1) = 0
440 u(nx) = 0
445 REM ***********************************************************
446 REM *               Set coefficients of the PDE                  *
447 REM *                                                           *
448 REM * du              du              d^2u                      *
449 REM * -- + v(t,x;u) -- = d(t,x;u) --- + f(t,x;u) u + g(t,x)   *
450 REM * dt              dx              dx^2                      *
452 REM *                                                           *
453 REM * d(t,x;u) = coefficient in front of the second
   derivative of u                                                  *
454 REM * v(t,x;u) = coefficient in front of the first
   derivative of u                                                  *
455 REM * f(t,x;u) = coefficient in front of u                      *
456 REM * g(t,x) = forcing term                                     *
457 REM * ATTENTION: The u in d, v, and f is computed at step i,
   which                                                            *
458 REM * corresponds to a sort of linearization of the nonlinear
   terms                                                            *
459 REM ***********************************************************
460 FOR ix = 2 TO nx1
465 den = (1 + eps * uold(ix))
470 d(ix) = kk / den
480 v(ix) = cc + eps * kk / den ^ 2 * (uold(ix + 1) -
   uold(ix - 1)) / (2 * dx)
490 f(ix) = 0
495 REM ***********************************************************
496 REM *            In g(ix) put (g(t) + g(t+dt)) / 2              *
497 REM ***********************************************************
500 g(ix) = 0
510 NEXT ix
511 REM ***********************************************************
512 REM *               Define the matrix elements                  *
513 REM ***********************************************************
520 FOR ix = 2 TO nx1
530 aex = (d(ix) * kh2 + v(ix) * kh / 2) / 2
540 bex = -d(ix) * kh2 + f(ix) * dt / 2
550 cex = (d(ix) * kh2 - v(ix) * kh / 2) / 2
554 REM ***********************************************************
555 REM * ATTENTION: If d, v, and f explicitly depend               *
556 REM * on time, then the related terms aex, bex, and             *
```

```
557 REM * cex in 560-580 have to be computed at time          *
558 REM * t+dt (i.e., repeat 465-490 with t=t+dt)              *
559 REM ********************************************************
560 IF ix <> 2 THEN a(ix) = -aex
570 b(ix) = 1 - bex
580 IF ix <> nx1 THEN c(ix) = -cex
585 REM ********************************************************
586 REM *               Define the known vector                *
587 REM ********************************************************
590 sum = aex * uold(ix - 1) + (1 + bex) * uold(ix) + cex *
   uold(ix + 1)
600 ff(ix) = sum + g(ix) * dt
610 IF ix = 2 THEN ff(ix) = ff(ix) + aex * u(1)
620 IF ix = nx1 THEN ff(ix) = ff(ix) + cex * u(nx)
630 NEXT ix
635 REM ********************************************************
636 REM *              Tridiagonal system solver               *
637 REM ********************************************************
640 FOR i = 2 TO nx1
650 u(i) = ff(i)
660 NEXT i
670 FOR i = 3 TO nx1
680 i1 = i - 1
690 ratio = -a(i) / b(i1)
700 b(i) = b(i) + ratio * c(i1)
710 u(i) = u(i) + ratio * u(i1)
720 NEXT i
730 IF b(nx1) = 0 THEN PRINT "error in 730":  END
740 u(nx1) = u(nx1) / b(nx1)
750 FOR i = nx1 - 1 TO 2 STEP -1
760 u(i) = (u(i) - c(i) * u(i + 1)) / b(i)
770 NEXT i
775 REM ********************************************************
776 REM *            End tridiagonal system solver             *
777 REM ********************************************************
779 t = t + dt
780 IF an$ = "n" THEN GOTO 820
785 REM ********************************************************
786 REM *      If known, insert here the analytic solution     *
787 REM ********************************************************
790 FOR ix = 1 TO nx
800 uan(ix) = EXP(-kk * t) * SIN(x(ix))
810 NEXT ix
820 IF ngraf = 0 THEN GOTO 1200
830 REM ********************************************************
831 REM *                 Plotting commands                    *
832 REM ********************************************************
840 CLS 0:  KEY OFF: SCREEN 2
```

```
850 LOCATE 14, 10:  PRINT ax
860 LOCATE 14, 70:  PRINT bx
870 LOCATE 13, 7:  PRINT pumin
880 LOCATE 2, 5:  PRINT pumax
900 VIEW (80, 10)-(560, 100), , 1
910 WINDOW (ax, pumin)-(bx, pumax)
920 LOCATE 15, 40:  PRINT "t="; t
930 PSET (ax, u(1))
940 FOR ix = 2 TO nx
950 LINE -(x(ix), u(ix))
960 NEXT ix
970 IF an$ = "n" THEN GOTO 1380
980 PSET (ax, uan(1))
990 FOR ix = 2 TO nx
1000 LINE -(x(ix), uan(ix)), , , &H7070
1010 NEXT ix
1020 GOTO 1380
1200 REM ***********************************************************
1201 REM *                    Print-out commands                   *
1202 REM ***********************************************************
1210 IF t < prout * iprint - dt/10 THEN GOTO 1390
1220 PRINT TAB(15); "t="; t
1230 difmax = 0:  uanmax = 0
1240 FOR i = 1 TO nx
1250 PRINT "u("; i; ")="; u(i),
1260 IF an$ = "n" THEN GOTO 1310
1270 dif = ABS(u(i) - uan(i))
1280 IF dif > difmax THEN difmax = dif
1290 IF ABS(uan(i)) > uanmax THEN uanmax = ABS(uan(i))
1300 PRINT "uan="; uan(i), "|uan-u|="; dif
1310 NEXT i
1320 IF an$ = "y" THEN PRINT "dt="; dt, "dx="; dx,
   "maximum error="; difmax / uanmax
1330 iprint = iprint + 1
1380 IF prstop$ = "y" THEN INPUT "Press RETURN to continue"; scr
1390 REM ***********************************************************
1391 REM *      Advance time and continue till final time          *
1392 REM ***********************************************************
1410 IF t <= tfin - dt / 10 THEN GOTO 400
1420 END
```

3.12.3 FD2HYP: An explicit finite difference routine for hyperbolic equations

This section will describe a simple explicit routine, which integrates the nonlinear telegrapher's equation

$$\begin{aligned}\frac{\partial^2 u}{\partial t^2} &+ a\left(t, x; u, \frac{\partial u}{\partial x}\right)\frac{\partial u}{\partial t} \\ &= \nu\left(t, x; u, \frac{\partial u}{\partial x}\right)\frac{\partial^2 u}{\partial x^2} + f\left(t, x; u, \frac{\partial u}{\partial x}\right) \cdot \end{aligned} \tag{3.12.3}$$

In particular, in the program we chose

$$\begin{aligned} a\left(t, x; u, \frac{\partial u}{\partial x}\right) &= \varepsilon \\ \nu\left(t, x; u, \frac{\partial u}{\partial x}\right) &= \kappa \\ f\left(t, x; u, \frac{\partial u}{\partial x}\right) &= \Big\{c_1(1+\kappa-\varepsilon)e^{-t} \\ &\quad + (c_1+c_2)\big[\varepsilon\cos t - (1-\kappa)\sin t\big]\Big\}\sin x \end{aligned} \tag{3.12.4}$$

and the following initial and boundary conditions

$$\begin{aligned} u_{in}(x) &= c_1 \sin x \\ \widetilde{u}_{in}(x) &= c_2 \sin x \\ u_a(t) &= u_b(t) = 0\ . \end{aligned} \tag{3.12.5}$$

If $x \in [0, 4\pi]$, then the initial-boundary-value problem is solved by

$$u(t, x) = [c_1 e^{-t} + (c_1 + c_2)\sin t]\sin x\ .$$

If not the solution will present a propagating discontinuity.

The flow chart related to the finite difference solver is given in Table 3.12.1c. The main difference consists in the fact that special care is needed, according to Eq.(3.9.36), in giving the solution at $t = \Delta t$. For this reason the functions a, ν, and f need to be defined in advance (lines 360–420).

Line 440 defines the initial condition u_{in} and line 450 $\widetilde{u}_{in}$, while lines 460 and 470 give, respectively, the first and second derivative of

u_{in}, which are here computed analytically and are used in Eq.(3.9.36). These lines may be replaced by finite difference approximations if the derivatives are hard to compute or the initial condition is not provided as a function but in a given set of points.

The initial condition is set at lines 480-510, while the solution at $t = \Delta t$ is computed using Eq.(3.9.36) at lines 520-590.

The solution for $t > \Delta t$ is then computed imposing the boundary conditions (lines 610 and 620) and computing (3.9.35) in the internal nodes (lines 630-740). In particular, lines 660 and 670 give the central difference approximations for the spatial derivatives, lines 680-700, the present values of a, ν, and f, respectively, and finally lines 710-730 gather all these results to find the updated values of u.

We explicitly remark that unew, u, and uold are, respectively, the values of u at $t = t_{i+1}, t_i, t_{i-1}$.

At last these values are updated in lines 750-780 and the program continues as in FDFEPAR.

FD2HYP.BAS

```
1 REM ***************************************************************
2 REM *           Solution of telegrapher's equation with          *
3 REM *             Dirichlet boundary conditions using            *
5 REM *             explicit central finite differences            *
6 REM ***************************************************************
7 REM ***************************************************************
8 REM *                        Variables used                      *
9 REM ***************************************************************
10 REM nx = number of grid points in space interval (nx1=nx-1)
11 REM x(nx) = grid points in space interval
12 REM ax, bx = extrema of space interval
13 REM dx = distance between grid points (dx2=dx^2)
14 REM unew(nx) = updated value of state variable
15 REM uan(nx) = analytic solution
17 REM u(nx) = state variable at current time
18 REM uold(nx) = state variable at previous time step
19 REM uin(x) = initial condition on u
20 REM uinp(x) = initial condition on the derivative of u
21 REM duin(x) = derivative of initial condition on u
22 REM dduin(x) = second derivative of initial condition on u
23 REM nu(t,x,u,du)= function in front of second-order space
    derivative
24 REM f(t,x,u,du) = function on r.h.s. of PDE
25 REM tauinv(t,x,u,du)= function in front of time derivative
    (tauinv=1/tau)
```

```
26 REM tau0,nu0,f0 = initial value of tauinv, nu, and f
27 REM taut,nut,ft = value of tauinv, nu, and f at time t
28 REM der(nx) = f.d. approximation of spatial derivative of u
29 REM der2(nx) = f.d. approximation of second spatial
   derivative of u
30 REM t = time
31 REM dt = time step (dt2=dt^2)
32 REM tfin = final time of integration
33 REM prout = print-out interval
34 REM out$ = flag for print-out option
35 REM prstop$ = flag for inserting a pause after print-outs
36 REM iprint = flag used for switching the printing procedure
37 REM an$ = flag indicating if the analytic solution is available
38 REM pumin,pumax = minimum and maximum ordinate of graph
39 REM stablim = stability condition
40 REM p1, p2 = parentheses in Eq.(3.9.35)
41 REM c1, c2 = initial condition coefficients
42 REM cc, kk, eps = PDE constants
45 DIM x(1001), uan(1001), unew(1001), u(1001), uold(1001)
50 REM DIM der(1001), der2(1001)
60 REM ***********************************************************
61 REM *                      Input lines                          *
62 REM ***********************************************************
70 CLS 0
80 PRINT "*********************************************************"
81 PRINT "*       Solution of telegrapher's equation with        *"
82 PRINT "*         Dirichlet boundary conditions using          *"
84 PRINT "*         EXPLICIT CENTRAL FINITE DIFFERENCES          *"
86 PRINT "*********************************************************"
90 PRINT ""
100 PRINT "When more than one quantity is requested, separate
   with a comma"
110 PRINT "Values in square brackets are the suggested values
   (not default)"
120 INPUT "Space interval [0,4*pi=12.5664]="; ax, bx
130 INPUT "Number of grid points [51]"; nx:  nx1 = nx - 1
140 INPUT "PDE coefficients nu and eps [1,2 or 1,0]"; kk, eps
145 stablim = (bx - ax) / (nx1 * kk)
146 PRINT "The stability limit is dt<"; stablim
150 INPUT "Time step and final time [.25,5]"; dt, tfin
155 INPUT "Initial condition coefficients c1 and c2
   [1,0 or 0,1]"; c1, c2
160 INPUT "Plot [1] or print-out [0] of the solution"; ngraf
170 IF ngraf = 0 THEN INPUT "Print-out interval [.5]"; prout
175 cb = 2 * c1 + c2
180 IF ngraf = 1 THEN PRINT "Minimum and maximum ordinate of
   the graph [-"; cb; ","; cb; "]"; :  INPUT pumin, pumax
190 IF ngraf <> 0 AND ngraf <> 1 THEN GOTO 160
```

```
200 INPUT "Do you want to stop at every time step [y/n]";
   prstop$
210 IF prstop$ <> "y" AND prstop$ <> "n" THEN GOTO 200
220 INPUT "Is the analytical solution known [y]"; an$
230 IF an$ <> "y" AND an$ <> "n" THEN GOTO 220
240 REM ***********************************************************
241 REM *                    End of input lines                   *
242 REM ***********************************************************
250 CLS 0
280 dt2 = dt ^ 2
290 dx2 = dx ^ 2
300 dx = (bx - ax) / nx1
310 REM ***********************************************************
311 REM *                Set space discretization                 *
312 REM ***********************************************************
320 FOR ix = 1 TO nx
330 x(ix) = ax + (ix - 1) * dx
340 NEXT ix
345 REM ***********************************************************
346 REM *               Set coefficients of the PDE               *
347 REM *                                                         *
348 REM * d^2u               du  du             du d^2u         du  *
349 REM * -- + tauinv(t,x;u,--) --= nu(t,x;u,--)-- + f(t,x;u,--) *
350 REM * dt^2               dx  dt             dx dx^2         dx  *
352 REM *                                                         *
353 REM * nu = function in front of second-order space
   derivative                                                  *
354 REM * tauinv = function in front of first-order time
   derivative                                                  *
355 REM * f = function on r.h.s. of PDE                           *
357 REM ***********************************************************
360 DEF fnnu (t, xx, uu, deruu) = kk
370 DEF fntauinv (t, xx, uu, deruu) = eps
380 DEF fnf (t, xx, uu, deruu)
390 add1 = (1 - eps + kk) * c1 * EXP(-t)
400 add2 = (c1 + c2) * (eps * COS(t) - (1 - kk) * SIN(t))
410 fnf = (add1 + add2) * SIN(xx)
420 END DEF
425 REM ***********************************************************
426 REM *                   Define initial data                   *
427 REM ***********************************************************
430 iprint = 1
440 DEF fnuin (xx) = c1 * SIN(xx)
450 DEF fnuinp (xx) = c2 * SIN(xx)
454 REM ***********************************************************
455 REM *       In the next two lines the first and second        *
456 REM *        derivatives of the initial condition uin         *
457 REM *            are given.  They can be replaced by          *
```

```
458 REM *             finite difference approximations              *
459 REM ************************************************************
460 DEF fnduin (xx) = c1 * COS(xx)
470 DEF fndduin (xx) = -c1 * SIN(xx)
475 REM ************************************************************
476 REM *                     Set initial data                      *
477 REM ************************************************************
480 t = 0
490 FOR ix = 1 TO nx
500 xx = x(ix)
510 uold(ix) = fnuin(xx)
515 REM ************************************************************
516 REM *                  Compute solution at t=dt                 *
517 REM ************************************************************
519 t = dt
520 uu = fnuin(xx)
530 deruu = fnduin(xx)
540 tau0 = fntauinv(t, xx, uu, deruu)
550 nu0 = fnnu(t, xx, uu, deruu)
560 f0 = fnf(t, xx, uu, deruu)
570 par = -tau0 * fnuinp(xx) + nu0 * fndduin(xx) + f0
580 u(ix) = fnuin(xx) + dt * fnuinp(xx) + (dt2 / 2) * par
590 NEXT ix
600 GOTO 782
605 REM ************************************************************
606 REM *                  Set boundary conditions                  *
607 REM ************************************************************
610 unew(1) = 0
620 unew(nx) = 0
625 REM ************************************************************
626 REM *             Compute right-hand side of PDE by             *
627 REM *            central finite differences in space            *
628 REM ************************************************************
630 FOR ix = 2 TO nx1
640 xx = x(ix)
650 uu = u(ix)
660 der = (u(ix + 1) - u(ix - 1)) / (2 * dx)
670 der2 = (u(ix + 1) - 2 * u(ix) + u(ix - 1)) / dx ^ 2
680 taut = fntauinv(t, xx, uu, der)
690 nut = fnnu(t, xx, uu, der)
700 ft = fnf(t, xx, uu, der)
705 REM ************************************************************
706 REM *           Compute updated value of solution by            *
707 REM *             central finite differences in time            *
708 REM ************************************************************
710 p1 = 1 / (1 + taut * dt / 2)
720 p2 = 1 - taut * dt / 2
730 unew(ix) = p1 * (2 * u(ix) - p2 * uold(ix) + dt2 *
```

```
   (nut * der2 + ft))
740 NEXT ix
745 REM ***********************************************************
746 REM *                        Update values                      *
747 REM ***********************************************************
750 t = t + dt
755 FOR ix = 1 TO nx
760 uold(ix) = u(ix)
770 u(ix) = unew(ix)
780 NEXT ix
782 IF an$ = "n" THEN GOTO 820
785 REM ***********************************************************
786 REM *          If known, insert analytic solution here          *
787 REM ***********************************************************
790 FOR ix = 1 TO nx
800 uan(ix) = (c1 * EXP(-t) + (c1 + c2) * SIN(t)) * SIN(x(ix))
810 NEXT ix
820 IF ngraf = 0 THEN GOTO 1200
830 REM ***********************************************************
831 REM *                      Plotting commands                    *
832 REM ***********************************************************
840 CLS 0:  KEY OFF: SCREEN 2
850 LOCATE 14, 10:  PRINT ax
860 LOCATE 14, 70:  PRINT bx
870 LOCATE 13, 7:  PRINT pumin
880 LOCATE 2, 5:  PRINT pumax
900 VIEW (80, 10)-(560, 100), , 1
910 WINDOW (ax, pumin)-(bx, pumax)
920 LOCATE 15, 40:  PRINT "t="; t
930 PSET (ax, u(1))
940 FOR ix = 2 TO nx
950 LINE -(x(ix), u(ix))
960 NEXT ix
970 IF an$ = "n" THEN GOTO 1380
980 PSET (ax, uan(1))
990 FOR ix = 2 TO nx
1000 LINE -(x(ix), uan(ix)), , , &H7070
1010 NEXT ix
1020 GOTO 1380
1200 REM **********************************************************
1201 REM *                     Print-out commands                   *
1202 REM **********************************************************
1210 IF t < prout * iprint - dt/10 THEN GOTO 1390
1220 PRINT TAB(15); "t="; t
1230 difmax = 0:  uanmax = 0
1240 FOR i = 1 TO nx
1250 PRINT "u("; i; ")="; u(i),
1260 IF an$ = "n" THEN GOTO 1310
```

```
1270 dif = ABS(u(i) - uan(i))
1280 IF dif > difmax THEN difmax = dif
1290 IF ABS(uan(i)) > uanmax THEN uanmax = ABS(uan(i))
1300 PRINT "uan="; uan(i), "|uan-u|="; dif
1310 NEXT i
1320 IF an$ = "y" THEN PRINT "dt="; dt, "dx="; dx,
   "maximum error="; difmax / uanmax
1330 iprint = iprint + 1
1380 IF prstop$ = "y" THEN INPUT "Press RETURN to continue"; scr
1390 REM ***********************************************************
1391 REM *                Is final time passed by?                 *
1392 REM ***********************************************************
1410 IF t <= tfin + dt / 10 THEN GOTO 610
1420 END
```

3.12.4 COLLOCAT: An explicit collocation routine

The present program is a simple application of the collocation method explained in Section 3.10.

The crucial point stays in the definition of the matrices **D** and **B** defined in Eq.(3.10.8), which are used to compute the values of the derivatives in the collocation points from the values of the function itself. For pedagogic reasons this is done for a general distribution of nodes. However, for a Chebyshev distribution and, in periodic problems, for an equispaced distribution, the matrices are, respectively, given by Eq.(3.10.31) and (3.10.17) (also recalling Remarks 3.10.3 and 3.10.5). As pointed out in Section 3.10, a more efficient method would be to use FFT algorithms to compute the chain of transformations in Tables 3.10.1 and 3.10.2.

Turning now our attention to the program, the first step is to set the space discretization (lines 130–190). Uniform and Chebyshev distributions are suggested, respectively, in lines 170 and 180, but any other distribution can be inserted here.

Referring to Appendix 2, lines 210–370 compute the terms

$$\frac{d\mathcal{L}_h}{dx}(x_k) = \frac{\prod\limits_{\substack{j=1\\ j\neq h\\ j\neq k}}^{N}(x_k - x_j)}{\prod\limits_{\substack{j=1\\ j\neq h}}^{N}(x_h - x_j)},$$

while lines 380-430 compute the terms

$$\frac{d\mathcal{L}_h}{dx}(x_h) = \sum_{\substack{j=1 \\ j\neq h}}^{N} \frac{1}{(x_h - x_j)} .$$

The matrix that computes the second derivative is obtained by matrix multiplication in lines 660-730.

The remaining input lines are similar to those used in the previous programs. Then, after setting the initial data in lines 1030-1050 and defining the analytic solution in line 1110, subroutine 2000 advances in time with a Runge–Kutta solver (see Section 2.9), subroutine 1200 prints the solution, and subroutine 1700 plots the solution.

With respect to the previous programs there is a new part in the present plotting subroutine (lines 1850-1970), which optionally computes the interpolant of the solution obtained at time t (see also INTERPOL in Appendix 2).

Another important part of the program is the subroutine 3000, which is recalled by the Runge–Kutta solver to evaluate the right-hand side of the partial differential equation

$$\frac{\partial u}{\partial t} = \mathcal{F}u .$$

The boundary conditions are set here (lines 3010-3020), then the terms $\partial u/\partial x$ and $\partial^2 u/\partial x^2$ are computed in the loop 3050-3120 as du and ddu, using the matrices $\mathbf{D}$ and $\mathbf{D}^2$ previously stored.

Note that since the values of the solution at the extrema are already given by the boundary conditions, the right-hand side of the equation is computed only in the internal nodes and the Runge–Kutta solver only acts on them (see lines 2010, 2070, 2120, 2180, 2230). These lines may need additional changes if other initial-boundary-value problems are to be solved.

COLLOCAT.BAS

```
1 REM ***********************************************************
2 REM *         Solution of convection-diffusion equation       *
3 REM *          with Dirichlet boundary conditions using       *
4 REM *              a collocation method in space and          *
5 REM *           fourth-order Runge-Kutta method in time       *
6 REM ***********************************************************
```

```
7 REM ************************************************************
8 REM *                      Variables used                      *
9 REM ************************************************************
10 REM nx = number of grid points in space interval (nx1=nx-1)
11 REM x(nx) = grid points in space interval
12 REM ax, bx = extrema of space interval
13 REM distr$ = chooses the type of collocation distribution
14 REM d(nx,nx) = derivative matrix
15 REM d2(nx,nx) = second derivative matrix
16 REM pi1(nx) = PI' in (A.2.10)
17 REM u(nx) = state variable
18 REM rhs(nx) = right-hand side of differential equations
19 REM fnuan, uu = analytic solution
20 REM du = collocation derivative of u
21 REM ddu = collocation second derivative of u
22 REM t = time
23 REM dt = time step (h2=dt/2, h6=dt/6)
24 REM tfin = final time of integration
25 REM dx = distance between points of interpolant evaluation
26 REM xx = interpolation point
27 REM pL = value of interpolant in xx
28 REM prout = print-out interval
29 REM ngraf = flag indicating if the solution has to be
   printed or plotted
30 REM out$ = flag for print-out option
31 REM prstop$ = flag for inserting a pause after print-outs
32 REM iprint = flag used for switching the printing procedure
33 REM an$ = flag indicating if the analytic solution is available
34 REM pumin,pumax = minimum and maximum ordinate of graph
35 REM cc = convective coefficient
36 REM kk, eps = heat conductivity coefficients
37 REM den = used in r.h.s. of PDE
38 REM nu = term in front of the second space derivative of u
39 REM v = term in front of the first space derivative of u
40 REM g = term in front of u
41 REM f = forcing term
42 REM stablim = stability condition
43 REM pi = 3.1415926
44 REM sum = used in reinterpolation
45 REM difmax,uanmax = used in error extimate
46 REM urk(nx), k1-k4 = used in Runge-Kutta routine
50 DIM x(101), u(101), rhs(101), pi1(101)
51 DIM du(101), ddu(101), d(101, 101), d2(101, 101)
52 DIM urk(101), k1(101), k2(101), k3(101), k4(101)
60 REM ************************************************************
61 REM *                      Input lines                         *
62 REM ************************************************************
70 CLS 0
```

```
80 PRINT "*******************************************************"
81 PRINT "*     Solution of a convection-diffusion equation     *"
82 PRINT "*      with Dirichlet boundary conditions using       *"
83 PRINT "*         a COLLOCATION method in space and a         *"
84 PRINT "*       fourth-order RUNGE-KUTTA method in time       *"
85 PRINT "*******************************************************"
90 PRINT ""
100 PRINT "When more than one quantity is requested, separate
   with a comma"
110 PRINT "Values in square brackets are the suggested values
   (not default)"
120 INPUT "Space interval [0,4*pi=12.5664]="; ax, bx
125 REM ******************************************************
126 REM *                Set space discretization             *
127 REM ******************************************************
130 INPUT "Number of grid points [11]"; nx:  nx1 = nx - 1
140 INPUT "Chebyshev distribution or equispaced nodes (c/e) [c]";
   distr$
145 IF distr$ <> "c" AND distr$ <> "e" THEN GOTO 140
150 pi = 3.1415926#
160 FOR i = 1 TO nx
170 IF distr$ = "e" THEN x(i) = ax + (bx - ax) * (i - 1) / nx1
180 IF distr$ = "c" THEN x(i) = .5 * (ax + bx - (bx - ax) *
   COS((i - 1) * pi / nx1))
190 NEXT i
200 REM ******************************************************
201 REM *              Computing derivative matrix            *
202 REM ******************************************************
210 REM ******************************************************
211 REM *             Computing pi1=PI' in (A.2.10)           *
212 REM ******************************************************
220 FOR i = 1 TO nx
230 pi1(i) = 1
240 FOR k = 1 TO nx
250 IF k <> i THEN pi1(i) = pi1(i) * (x(i) - x(k))
260 NEXT k
270 NEXT i
280 FOR i = 1 TO nx
290 FOR j = 1 TO nx
300 IF i = j THEN GOTO 370
310 REM ******************************************************
311 REM *          Computing the nondiagonal elements         *
312 REM ******************************************************
320 d(j, i) = 1
330 FOR k = 1 TO nx
340 IF k <> i AND k <> j THEN d(j, i) = d(j, i) * (x(j) - x(k))
350 NEXT k
360 d(j, i) = d(j, i) / pi1(i)
```

```
370 NEXT j
380 REM ***********************************************************
381 REM *              Computing the diagonal elements            *
382 REM ***********************************************************
390 d(i, i) = 0
400 FOR k = 1 TO nx
410 IF k <> i THEN d(i, i) = d(i, i) + 1 / (x(i) - x(k))
420 NEXT k
430 NEXT i
650 REM ***********************************************************
651 REM *               Computing the matrix for the              *
652 REM *                    second derivatives                   *
653 REM ***********************************************************
660 FOR i = 1 TO nx
670 FOR j = 1 TO nx
680 d2(i, j) = 0
690 FOR k = 1 TO nx
700 d2(i, j) = d2(i, j) + d(i, k) * d(k, j)
710 NEXT k
720 NEXT j
730 NEXT i
741 REM ***********************************************************
742 REM *          End of computation derivative matrices         *
743 REM ***********************************************************
744 REM ***********************************************************
745 REM *        Inputs not involving space discretization        *
746 REM ***********************************************************
750 INPUT "Convective coefficient c [0]"; cc
760 INPUT "Heat conductivity coefficients k/(1+eu) (e<1) [1,0]";
   kk, eps
770 stablim = (x(2) - x(1)) ^ 2 / (2 * kk)
790 PRINT "The stability limit is dt<"; stablim
800 INPUT "Time step and final time [.08,4]"; dt, tfin
810 INPUT "Plot [1] or print-out [0] of the solution"; ngraf
820 IF ngraf = 0 THEN INPUT "Print-out interval [.5]";
   prout:  GOTO 870
830 IF ngraf = 1 THEN INPUT "Minimum and maximum ordinate of
   the graph [-1,1]"; pumin, pumax
840 IF ngraf <> 0 AND ngraf <> 1 THEN GOTO 810
850 PRINT "Do you want to reinterpolate the solution?"
860 INPUT "In how many points?  (0 for linear interpolation)
   [30] ", npoint
870 INPUT "Do you want to stop at every time step [y/n]"; prstop$
880 IF prstop$ <> "y" AND prstop$ <> "n" THEN GOTO 870
890 INPUT "Is the analytical solution known [y]"; an$
900 IF an$ <> "y" AND an$ <> "n" THEN GOTO 890
910 REM ***********************************************************
911 REM *                    End of input lines                   *
```

```
912 REM **********************************************************
920 CLS 0
930 h2 = dt / 2
940 h6 = dt / 6
1000 REM *********************************************************
1001 REM *                  Set initial data                     *
1002 REM *********************************************************
1010 t = 0
1020 iprint = 1
1030 FOR ix = 1 TO nx
1040 u(ix) = SIN(x(ix))
1050 NEXT ix
1100 REM *********************************************************
1101 REM *      If known, insert analytic solution here          *
1102 REM *********************************************************
1110 DEF fnuan (tt, xx) = EXP(-tt) * SIN(xx)
1120 REM *********************************************************
1121 REM *       Subroutine 2000 integrates in time with         *
1122 REM *          a fourth-order Runge-Kutta routine           *
1123 REM *********************************************************
1130 GOSUB 2000
1140 REM *********************************************************
1141 REM *          Subroutine 1200 prints the solution          *
1142 REM *           Subroutine 1700 plots the solution          *
1143 REM *********************************************************
1150 IF ngraf = 0 AND t > prout * iprint - dt/10 THEN GOSUB 1200
1160 IF ngraf = 1 THEN GOSUB 1700
1180 REM *********************************************************
1181 REM *               Is final time passed by?                *
1182 REM *********************************************************
1190 IF t <= tfin - dt / 10 THEN GOTO 1120
1195 INPUT "Do you want to solve another ibvp [y/n]"; scr$
1198 IF scr$ = "y" THEN GOTO 750
1199 END
1200 REM *********************************************************
1201 REM *                 Print-out subroutine                  *
1202 REM *********************************************************
1220 PRINT TAB(15); "t="; t
1230 difmax = 0:  uanmax = 0
1240 FOR i = 1 TO nx
1250 PRINT "u("; i; ")="; u(i),
1260 IF an$ = "n" THEN GOTO 1310
1265 uu = fnuan(t, x(i))
1270 dif = ABS(u(i) - uu)
1280 IF dif > difmax THEN difmax = dif
1290 IF ABS(uu) > uanmax THEN uanmax = ABS(uu)
1300 PRINT "uan="; uu, "|uan-u|="; dif
1310 NEXT i
```

```
1320 IF an$ = "y" THEN PRINT "dt="; dt, "maximum error=";
   difmax / uanmax
1330 iprint = iprint + 1
1380 IF prstop$ = "y" THEN INPUT "Press RETURN to continue", scr
1390 RETURN
1399 END
1700 REM ***********************************************************
1701 REM *                  Plotting subroutine                      *
1702 REM ***********************************************************
1710 CLS 0:  KEY OFF: SCREEN 2
1720 LOCATE 14, 10:  PRINT ax:  LOCATE 14, 70:  PRINT bx
1730 LOCATE 13, 7:  PRINT pumin:  LOCATE 2, 5:  PRINT pumax
1750 VIEW (80, 10)-(560, 100), , 1
1760 WINDOW (ax, pumin)-(bx, pumax)
1770 LOCATE 15, 40:  PRINT "t="; t
1780 IF an$ = "n" THEN GOTO 1830
1790 FOR ix = 1 TO 101
1800 xx = ax + (ix - 1) * (bx - ax) / 100
1810 PSET (xx, fnuan(t, xx))
1820 NEXT ix
1830 PSET (ax, u(1))
1840 IF npoint = 0 THEN FOR ix = 2 TO nx:  LINE -(x(ix), u(ix)):
   NEXT ix:  GOTO 1990
1850 REM ***********************************************************
1851 REM *          Beginning reinterpolation of solution          *
1852 REM ***********************************************************
1855 dx = (bx - ax) / npoint
1860 zz = x(1)
1870 FOR i = 1 TO nx
1880 IF xx = x(i) THEN pL = u(i):  GOTO 1950
1890 NEXT i
1900 sum = 0
1910 FOR i = 1 TO nx:  sum = sum + u(i)/(pi1(i) * (xx - x(i))):
   NEXT i
1920 pL = 1
1930 FOR i = 1 TO nx:  pL = pL * (xx - x(i)):  NEXT i
1940 pL = sum * pL
1950 LINE -(xx, pL)
1960 xx = xx + dx
1970 IF xx <= bx + dx THEN GOTO 1870
1980 REM ***********************************************************
1981 REM *            End reinterpolation of solution              *
1982 REM ***********************************************************
1990 IF prstop$ = "y" THEN INPUT "Press RETURN to continue", scr
1995 RETURN
1999 END
2000 REM ***********************************************************
2001 REM *          Fourth-order Runge-Kutta subroutine            *
```

```
2002 REM ******************************************************
2010 FOR i = 2 TO nx1
2020 urk(i) = u(i)
2040 NEXT i
2050 GOSUB 3000
2060 t = t + h2
2070 FOR i = 2 TO nx1
2080 k1(i) = rhs(i):  urk(i) = u(i) + h2 * k1(i)
2100 NEXT i
2110 GOSUB 3000
2120 FOR i = 2 TO nx1
2130 k2(i) = rhs(i):  urk(i) = u(i) + h2 * k2(i)
2150 NEXT i
2160 GOSUB 3000
2170 t = t + h2
2180 FOR i = 2 TO nx1
2190 k3(i) = rhs(i):  urk(i) = u(i) + dt * k3(i)
2210 NEXT i
2220 GOSUB 3000
2230 FOR i = 2 TO nx1
2240 k4(i) = rhs(i)
2260 u(i) = u(i) + h6 * (k1(i) + 2 * (k2(i) + k3(i)) + k4(i))
2280 NEXT i
2290 REM ******************************************************
2291 REM *             End of Runge-Kutta routine             *
2292 REM ******************************************************
2295 RETURN
2299 END
3000 REM ******************************************************
3001 REM *               Set boundary conditions              *
3002 REM ******************************************************
3010 u(1) = 0:  urk(1) = u(1)
3020 u(nx) = 0:  urk(nx) = u(nx)
3030 REM ******************************************************
3031 REM *             Set coefficients of the PDE            *
3032 REM *                                                    *
3033 REM *  du           du              d^2u                 *
3034 REM *  -- + v(t,x;u) -- = nu(t,x;u) --- + g(t,x;u) u +f(t,x) *
3035 REM *  dt           dx              dx^2                 *
3036 REM *                                                    *
3037 REM * nu(t,x;u) = function in front of the second derivative
   of u                                                     *
3038 REM * v(t,x;u) = function in front of the first derivative
   of u                                                     *
3039 REM *  g(t,x;u) = function in front of u                 *
3040 REM *  f(t,x) = forcing term                             *
3041 REM ******************************************************
3050 FOR ix = 2 TO nx1
```

```
3060 du(ix) = 0
3070 ddu(ix) = 0
3080 FOR kx = 1 TO nx
3090 du(ix) = du(ix) + d(ix, kx) * urk(kx)
3100 ddu(ix) = ddu(ix) + d2(ix, kx) * urk(kx)
3110 NEXT kx
3120 NEXT ix
3130 FOR ix = 2 TO nx1
3150 den = (1 + eps * urk(ix))
3160 nu = kk / den
3170 v = cc + eps * kk / den ^ 2 * du(ix)
3180 g = 0
3190 f = 0
3200 rhs(ix) = nu * ddu(ix) - v * du(ix) + g * urk(ix) + f
3210 NEXT ix
3290 RETURN
9299 END
```

Problems for Chapter 3

Problem 3.1

Consider the one-dimensional convection-diffusion model with a source term

$$\frac{\partial u}{\partial t} + c(u)\frac{\partial u}{\partial x} = \frac{\partial}{\partial x}\left(\kappa(u)\frac{\partial u}{\partial x}\right) + g(t,x)u\,.$$

Answer to the following questions:

1. How can this model be written in three dimensions?
2. Which are the conservation and/or equilibrium equations involved in its derivation?
3. Which are the theorems related to the derivation mentioned in question 2?
4. Derive the model in two space dimensions.

Problem 3.2

Consider the mathematical model of Problem 3.1 in two space dimensions. Answer to the following questions:

1. Provide some statements of the initial-boundary-value problem with boundary conditions on a closed boundary.
2. Which is the corresponding statement for a problem in an unbounded boundary?
3. Which is the statement of the same problem for the static model? In particular, consider the problem on the space domains $[0,1]\times[0,1]$ and $[0,1]\times[0,+\infty)$.

Problem 3.3

Compute from (3.4.13) the mechanical energy and simplify (3.4.15).

Problem 3.4

Compute $\mathbf{T}:\mathbf{D}$ for

$$\mathbf{T} = -p\mathbf{I}\,, \quad p = \rho^\gamma \quad \Longrightarrow \quad \mathbf{T}:\mathbf{D} = \frac{p'(\rho)}{\gamma} > 0$$

$$\mathbf{T} = 2\mu\mathbf{D} \qquad \text{with} \qquad \nabla\cdot\mathbf{v} = 0\,.$$

Problem 3.5
An ideal gas ($\mathbf{T} = -p\mathbf{I}$) satisfies the laws of Boyle and Gay-Lussac $p = \rho R\theta$, where R is a constant related to the universal gas constant. This equation gives a relation between the stress $\mathbf{T}$, or equivalently the pressure p, with the density and the temperature θ. If, moreover, the gas is polytropic, then the specific internal energy is proportional to the temperature

$$\epsilon = c_v\theta = \frac{R}{\gamma - 1}\theta\,,$$

where γ is a positive constant. This is a constitutive relation linking the specific internal energy ϵ with θ. Solve the following problems:

1. Deduce from (3.4.17) with $\mathbf{q} = \mathbf{0}$ the energy equation to add to (3.4.11) and (3.4.14) (use p as further state variable).
2. Prove in one dimension that the system is hyperbolic.
3. Assuming that $\mathbf{q} = -K\nabla\theta$, what can you say about the classification of the model?

Problem 3.6
Reduce Gurtin and Pipkin's model (3.4.43) to Cattaneo's model (3.4.40) when $G_\epsilon(s) = 0$ and $G_q = (K/\tau)e^{-s/\tau}$.

Note to Problem 3.6 *Introduce $s' = t - s$ and derive.*

Problem 3.7
Use the method of characteristics to solve, for $\alpha \in \mathbb{R}$, the following initial-value problem:

$$\begin{cases} \dfrac{\partial u}{\partial t} + x^\alpha \dfrac{\partial u}{\partial x} + u = 0\,, & x \in \mathbb{R}\,, \\ u(t = 0, x) = u_{in}(x)\,. \end{cases}$$

Give a well-formulated initial-boundary-value problem in $[0, 1]$.

Problem 3.8
Recalling the discussion before Remark 3.9.8, set the Robin boundary conditions for the convection-diffusion model.

Problem 3.9

Consider the program FDFEPAR and its application to the convection-diffusion model.

1. To look at the development of the numerical instability, run the program with $N = 31$ and $\varepsilon = 0$ till `tfin=33` with $\Delta t = 0.09$ and till `tfin=7` with $\Delta t = 0.1$. Observe how the instability can develop even if the solution seems to vanish for a long time.
2. To see the influence on the nonlinear term introduced in the problem, run the program with $\varepsilon = 0.5$ and $\Delta t = 0.1, 0.08, 0.07$ till `tfin=60`. You can see for $\Delta t = 0.1$ the development of an instability then smoothed out as time goes by and the formation of a standing wave at later times. These are all numerical effects that disappear for sufficiently small time steps. Try to explain the difference in the development of the temporary instability in the different cases and the reason for its smoothing out. Can you trust the result obtained after the instability has faded away?
3. Set $\kappa(u) = 1 - 0.5u$ and repeat question 2.
4. Increase c from 1 to 10 and look at the effect of the convective term. Why does the program not work for large c? (Pay attention to what happens at the right boundary!) Can you fix it?
5. Add a source term to balance the convective term so that `uan` is still the analytic solution and repeat question 4.
6. How does the system (and the numerical code for a suitable refined mesh, say $N = 301$) behave if a shock is introduced? (It is easy to do that integrating say in $[0, 10]$).

Problem 3.10

Consider the program FDCNPAR and its application to the convection-diffusion model.

1. Control the rate of convergence running the following cases

$$\Delta t = 0.25\,, \qquad N = 11, 21, 41, 81$$

$$N = 41\,, \qquad \Delta t = 1, 0.5, 0.25, 0.125, 0.0625\,,$$

   and printing the error done at $t = 1$.
2. Give a look at the difference between the analytic solution and the numerical solution stopping at every step the following runs

$$\Delta t = 0.5\,, \qquad N = 41, 36, 31, 26, 21, 16, 11, 6$$

$$N = 41\,, \qquad \Delta t = 0.25, 0.5, 1, 2, 3, 4\,.$$

3. Repeat questions 4–6 of Problem 3.9.

Problem 3.11
Consider the program FD2HYP and its application to the telegrapher's equation.

1. Observe how even lower-order terms may influence the development of the numerical instability by running the program with $\nu = 1$, $\Delta t = 0.26$, and $(\varepsilon, \texttt{tfin}) = (0, 10), (2, 20), (10, 50)$.
2. If the solution presents a discontinuity, high frequency oscillation may appear next to it. Of course, to resolve the shock, Δx has to be sufficiently small. Sometimes this oscillation can be smoothed out by choosing Δt equal to the stability limit. (Can you explain why?). To look at this effect, run the following cases in $[a, b] = [0, 10]$:

$$N = 201\ , \qquad \Delta t = 0.04, 0.05$$

$$N = 501\ , \qquad \Delta t = 0.019, 0.02, 0.021\,.$$

   It is better to zoom in on the region near the right boundary, say setting the plotting window to $x \in [9, 10]$, $u \in [-1, 1]$. Observe in the last case (in which Δt is larger than the stability limit) how the instability develops right at the shock front.
3. Integrate the wave equation with the following initial conditions

$$u_{in}(x) = \begin{cases} x & \text{if } 0 \le x \le 1 \\ 0 & \text{otherwise} \end{cases} \qquad \tilde{u}_{in}(x) = 0$$

   and

$$u_{in}(x) = \begin{cases} 1 & \text{if } 0 \le x \le 1 \\ 0 & \text{otherwise} \end{cases} \qquad \tilde{u}_{in}(x) = 0$$

   and vanishing boundary conditions. If $\nu = 1$, then for $t < 1$ you can get the analytic solution. Run then the following cases

$$N = 301\ , \qquad \Delta t = 0.01, 0.009, 0.005\ , \qquad \texttt{tfin} = 1, 3$$

   plotting the result for $u \in 0, 1.1$. Observe the smoothing of the shock front. The first initial condition is discontinuous at $x = 1$ and has a discontinuous derivative at $x = 0$. Can you explain the difference between what happens when these two discontinuities propagate?
4. Repeat the previous question for the full telegrapher's equation with $\varepsilon = 1, 3, 10$.

Problem 3.12

Consider the program COLLOCAT and its application to the convection-diffusion model.

1. Study the development of the instability for equispaced nodes.
2. Run the program with $\varepsilon = 0.5$ for both distributions. Can you explain why the equispaced distribution does not work?
3. Repeat question 4 of Problem 3.8 for both distributions. Can you explain why the equispaced distribution does not work even for moderate c?
4. Control the rate of convergence of the program for both distributions.
5. Integrate a hyperbolic system of equations.

Chapter 4

INVERSE AND STOCHASTIC PROBLEMS

4.1 Inverse Problems and Stochastic Models

Mathematical problems, in the analysis of mathematical models, are generally required to be well formulated. Moreover, the mathematical model needs to be consistent. As we have seen, a ***consistent model*** is such that

- The mathematical model is formulated by a number of linearly independent equations equal to the dimension of the state variable.
- All parameters that characterize the model are well defined.

The related mathematical problem can be defined as ***well formulated*** if

- The mathematical problem is stated with the correct number of initial and/or boundary conditions.

Specific examples were given in Chapters 2 and 3 with reference to discrete and continuous models.

Moreover, a ***well-formulated*** problem can, in some cases, be ***well posed***. This means that the solution exists, is unique, and depends continuously on the data of the problem.

Unfortunately, well posedness does not mean yet that the solution is obtained. In fact, the solution to the mathematical problems has to be sought by suitable methods and algorithms. In particular, the problem needs to be ***well conditioned*** with respect to the algorithms applied in obtaining the solution. In other words, the algorithms that operate on the approximating problem are such that convergence and stability are assured.

However, this idealistic situation does not always apply to real cases. In fact, the practical analysis of a mathematical model, or of the related mathematical problem, may be such that some information on the model and/or the problem is lacking. Conversely, suitable information on the solution of the problem may be available and may be obtained by direct measurement of the physical system. In such a case, one has an ***inverse problem***, which is a problem such that the information on the solution must be used to complete the shortage of data.

The interaction between the analysis of the inverse mathematical problem and the measurement of the real system used to solve the problem may be very useful for constructing a good model.

The following two preliminary questions arise immediately:

- Can inverse problems be well posed?
- Can the information on the solution be consistent with those needed by the mathematical problem?

In both cases, the answer is almost always negative. In fact, inverse problems are generally ill posed. In addition, the information on the real system cannot be immediately transferred into information on the model or on the problem. This is due to the fact that the ***mathematical model*** is only an approximation of the real system, which involves (among other things) a number of variables much smaller than those related to the real system.

We have already mentioned that the analysis of inverse problems is of relevant importance for mathematical modelling and, in general, for applied mathematics. With this in mind, the applied mathematician should attempt the solution of problems without artificial simplifications, which may obscure the information he hopes to obtain from the real system. Nonlinearity and stochasticity are intrinsic features in the analysis of inverse problems in mathematical modelling.

Stochasticity generally arises, as we shall see, from the fact that the information on the real system is obtained by experimental devices and, therefore, is affected by some noise. This topic, reviewed by Preziosi, Teppati, and Bellomo [PTB], will be dealt with in the second part of this chapter.

The first part of this chapter is organized in order to provide the interested reader with several tools and methods to deal with inverse ill-posed problems. The second part deals with various aspects of stochastic modelling and stochastic calculus related to the analysis of mathematical models.

In detail, the second section provides a classification of inverse-type problems. The third section deals with solution methods obtained by decomposition of domain techniques. The fourth section deals with solution methods developed via minimization techniques. The fifth section provides a general qualitative discussion on the main features of stochastic mathematical modelling. The sixth and seventh sections deal, respectively, with a classification of discrete and continuous stochastic dynamic models. The eighth section is the central one and provides some indications for the modelling and the organization of stochastic calculus for the solution of problems. The ninth section develops in a stochastic framework the content of the first part of the chapter, namely, the analysis of inverse problems is developed when some features of the mathematical model or problem are characterized by random variables. The tenth section deals with the analysis of models obtained in the framework and with the methods of statistical mechanics. A brief discussion is proposed in the eleventh section. Some conclusive discussions are proposed in Section 4.12. Finally some scientific programs are offered in the last section of this chapter.

It is important to mention that the classification of inverse problems is organized in order to direct the applied mathematician towards the solution method most useful for dealing with each class of problems. Similarly, with reference to Sections 4.6 and 4.7, the classification of stochastic models, either discrete or continuous, is useful both for the modelling in a stochastic framework and for the selection of the mathematical methods suitable to obtain quantitative results in the analysis of stochastic problems.

The analysis of this chapter is generally referred to continuous models. However, suitable references will be made in order to address the interested reader also to the solution of discrete models.

The literature on inverse problems is vast and interesting and is generally related to the analysis of problems that arise in applied mathematics and are referred to problems generated by technological applications. Without claiming to be complete and limiting our attention to books and survey papers, we cite, among the others, the book by Lavrent'ev, Reznitskaya, and Yakhno [LRY], which reports on the mathematical literature in the Soviet Union on linear inverse-type problems; the classical book by Back, Blackwell, and St. Claire [BBS] on various modelling and mathematical aspects related to the analysis of the linear heat equation; the survey paper by Payne [PAY], which provides an interesting review of the existing literature, somehow referred to a classification of inverse-type problems; and the collection of papers edited by Colton, Ewing, and Rundell [CER], which collects several important papers on the analysis of inverse problems for partial differential equations. Further information can be obtained, mainly about modelling and classification of problems, from the volumes published by Friedman [FRI] on industrial mathematics.

A complete and well-presented collection of problems and methods on inverse problems in physics is the book by Ghosh [GHO].

The bibliography on the analysis of stochastic problems is certainly vast, especially if referring to linear problems. On the other hand, the knowledge of effective methods in the analysis of nonlinear stochastic problems is quite limited. Some bibliographical references will be here reported selecting some books and papers that are, among others, close to the intellectual line reported in this chapter. In particular, the book by Nelson [NEL] provides an interesting analysis of the interactions between inner system and outer environment (the background field in Nelson's terminology) and of the related modelling aspects. The same book deals with mathematical methods referred to discrete dynamic models of Hamiltonian dynamics. The book by Sobczyk [SOa] provides a valuable presentation of mathematical methods for the analysis of discrete stochastic dynamic systems. This book has the merit of introducing the essential concepts on the identification of models by solution of inverse problems.

Various aspects of mathematical modelling and of the related mathematical methods referred to continuous dynamic models in the stochastic case are reported in the book by Bellomo, Brzezniak, and de Socio [BBD].

The analysis of inverse-type stochastic problems is reported, among others, in the papers by Preziosi and de Socio [PaD] and by Preziosi, Teppati, and Bellomo [PTB]. Some results of the two aforementioned papers will be reported and used in this chapter.

The reader is also referred to classical textbooks of probability theory, say those by Ash and Gardner [AaG] and by Papoulis [PAP]. The main concepts are reported in Appendix 3. Further references will be given throughout the chapter.

The style of this chapter is somewhat different from the preceding ones, where the contents were presented in a self-contained fashion. This chapter opens to topics that are certainly relevant to mathematical modelling, which, however, require a second additional course.

Therefore the aim of this book is to introduce the main ideas of inverse and stochastic modelling, to provide the fundamental elements for some simple quantitative analysis, and to indicate a bibliography that is useful to the reader in order to obtain a more complete knowledge of stochastic analysis related to mathematical modelling. Owing to this, some of the bibliographical indications will be organized in order to address the reader towards new frontiers of applied mathematics.

4.2 Classification of Inverse Problems

As mentioned in the Introduction to this chapter, a classification of inverse-type problems is important for several reasons. Among others we list the following:

- Organize what is dispersed in the pertinent literature;
- Refer to specific aspects of mathematical modelling;
- Organize the selection of the mathematical methods that can be applied to obtain quantitative results.

For simplicity of notation, the classification will be referred to continuous dynamic models in one space dimension such that the state variable is one dimensional, too. After providing the classification, we will present the conceivable generalizations to vector systems, or systems in several space variables. This classification can also be applied (simplified) to discrete models.

Consider then the class of models such that the ***state equation*** (the mathematical model) is of the type

$$\frac{\partial u}{\partial t} = f\left(t, x; u, \frac{\partial u}{\partial x}, \frac{\partial^2 u}{\partial x^2}; r(t,x)\right), \tag{4.2.1}$$

where

$$u = u(t,x), \quad t \in [0,T], \quad x \in [a,b] \tag{4.2.2}$$

and where

$$r = r(t,x) \tag{4.2.3}$$

is a given function of time and space.

The mathematical problem related to Eq.(4.2.1) is obtained by joining to such an equation the initial conditions

$$u(0,x) = u_{in}(x), \qquad \text{for } x \in [a,b] \tag{4.2.4}$$

and suitable boundary conditions, which in the simplest case are the Dirichlet boundary conditions

$$u(t,a) = u_a(t) \qquad \text{and} \qquad u(t,b) = u_b(t). \tag{4.2.5}$$

Of course, as we have seen in Chapter 3, more general or different boundary conditions can also be given. However this is not essential to understand either the classification or the various methods proposed in what follows.

Once more the classification will be referred to the problem described above and all conceivable generalization will be subsequently given. This will avoid tedious notations, which may obscure the substantial aspects of the matter.

In general, an inverse problem is such that some information either about the term r or about the initial and/or the boundary conditions is missing. On the other hand, some information is given about the solution to the initial-boundary-value problem.

The aim of this section is to provide a self-consistent classification of this type of problem as well as their mathematical formulation.

A first rough classification attempts simply to distinguish between well-formulated and ill-formulated problems. In addition, the first category can be subdivided into well-posed and ill-posed problems, the second one into ill-specified and inverse problems. This classification is summarized in Table 4.2.1.

Mathematical model	→	Mathematical problem	Well formulated	Well posed
				Ill posed
			Ill formulated	Ill specified
				Inverse

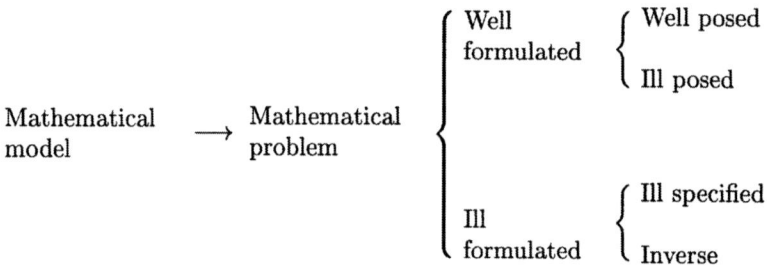

Table 4.2.1 — Classification of mathematical problems.

The scheme recalls that the mathematical model generates a mathematical problem (in our case an initial-boundary-value problem), which may be

- ***well formulated*** if the problem is stated with all initial and boundary conditions necessary for its solution;
- ***ill formulated*** when the conditions stated in the item above do not hold.

Following well-formulated problems, the scheme indicates that the problem may be well posed if the solution exists, is unique, and depends continuously upon the data of the problem, or ill posed if such a condition does not hold.

In particular, the following definitions can also be given:

DEFINITION 4.2.1 Underspecified problems
An initial-boundary-value problem is ***underspecified*** *if the information on the initial data and/or boundary conditions needed for its solution is incomplete.*

DEFINITION 4.2.2 Overspecified problems
An initial-boundary-value problem is ***overspecified*** *if further information is given in addition to the initial and/or boundary conditions needed for its solution.*

DEFINITION 4.2.3 Inverse problems
An initial-boundary-value problem is ***inverse*** *if some information on the initial and/or boundary conditions needed for its solution or/and on the parameters that characterize the model are missing and are replaced by suitable information on the solution of the mathematical problem.*

REMARK 4.2.1 The definitions above refer to continuous dynamic models. In fact, they refer to initial-boundary-value problems, which, as we have seen in Chapter 3, are generated by the aforementioned class of models. However, the same definitions include both the case of continuous static models and that of discrete dynamic models. In the first case the definitions simply involve initial and boundary conditions and parameters. In the second case, only initial conditions and parameters are involved. ■

REMARK 4.2.2 ***Underspecified*** and ***overspecified*** problems are certainly ill posed. However, one has to tackle them, whenever the mathematical formulation of the problem is imposed by practical situations. The case of ***underspecified*** problems occurs when not all initial and/or boundary conditions, needed for its solution, can be practically measured. On the other hand, the case of ***overspecified*** problems corresponds to when more measurements than those needed for their solution are given. The solution of the problem should show that the prediction of the model is close to the experimental measurements. ■

REMARK 4.2.3 The analysis, which will be developed in what follows, refers to inverse-type problems such that some information on the mathematical problem and/or the model is lacking and is substituted by suitable information on the solution to the initial-boundary-value problem. However, the reader needs to be aware that there exist ill-posed problems, which are stated with the proper initial and/or boundary conditions, but the solution is unstable and blows up. Examples of this type will be given at the end of this section. ■

Example 4.1 Ill-specified problems for the heat equation
In order to apply the definitions and remarks given above, consider Fourier's heat conduction model with variable heat conductivity

$$\frac{\partial u}{\partial t} = \frac{\partial}{\partial x}\left(\kappa(t, x, u)\frac{\partial u}{\partial x}\right), \tag{4.2.6}$$

where $u = u(t, x)$ is the temperature for $t \in [0, T]$ and $x \in [a, b]$. In addition, we refer to the well-formulated problem with initial condition

$$u(0, x) = u_{in}(x) \tag{4.2.7}$$

for $x \in [a, b]$ and boundary conditions

$$u(t, a) = u_a(t) \quad \text{and} \quad u(t, b) = u_b(t), \tag{4.2.8}$$

for $t \in [0, T]$.

Such a problem is ***underspecified*** if one of the two boundary conditions, say u_a or u_b, is not available. On the other hand, the problem becomes ***overspecified*** if in addition to conditions (4.2.7) and (4.2.8) also the heat fluxes

$$-\kappa\big(t, a, u(t, a)\big)\frac{\partial u}{\partial x}(t, a) \tag{4.2.9}$$

and/or

$$-\kappa\big(t, b, u(t, b)\big)\frac{\partial u}{\partial x}(t, b) \tag{4.2.10}$$

are prescribed.

Similar statements can be constructed for the other models introduced in Chapter 3. □

Inverse problems are generally ill posed. However, it is useful to verify if the problem is stated with a good formulation, that is, when the missing information is replaced by somewhat "equivalent" information. Say, if the unknown is a function of time and/or space, it must be replaced by additional information also a function of time or/and space.

Example 4.2 An inverse problem for the heat equation
Consider the initial-boundary-value problem (4.2.6–4.2.8). If one of the boundary conditions $u_a(t)$ or $u_b(t)$ is not known and the solution

$$u(t, x^*) = u^*(t), \quad x^* \in (a, b) \tag{4.2.11}$$

is given for $t \in [0, T]$, then one has an inverse problem. In this case we may say that the information (4.2.11) is equivalent to $u_a(t)$ or $u_b(t)$. Otherwise the problem may not be properly posed. □

We refer, for the moment, to problems such that some equivalence can be established between missing and additional information. A partial classification is proposed in Table 4.2.2.

It is, in fact, hard to provide a complete classification of mathematical problems generated by special physical situations. However, Table 4.2.2 shows a quite general framework, where a large class of problems is inserted and where the inverse formulation is referred, separately, either to the mathematical problem or to the mathematical model.

With reference to the classification reported in Table 4.2.2, the detailed statement of each problem will be given with reference to the abstract model (4.2.1). Then specific examples will be given still with reference to such a model.

Problem 4.1: Unspecified initial conditions

- The initial-boundary-value problem related to Eq.(4.2.1) is stated with the proper boundary conditions.
- The initial condition (4.2.4) is not known.
- The solution

$$u(\bar{t}, x) = \bar{u}(x) \tag{4.2.12}$$

is given for $t = \bar{t}$, where $\bar{t} \in (0, T)$, $\forall\, x \in [a, b]$.

The problem consists in finding the initial condition $u_{in}(x)$ and the solution $u(t, x)$ to the initial-boundary-value problem. □

Problem 4.2: Unspecified boundary condition

- The initial-boundary-value problem related to Eq.(4.2.1) is stated with the proper initial conditions $u_{in}(x)$ and with only one of the two boundary conditions, $u_a(t)$ or $u_b(t)$, needed for its solution.

Inverse problem	Unspecified mathematical problem	Unspecified initial conditions
		Unspecified boundary conditions
		Undefined well localized source term
		Unlocalized well defined source term
	Unspecified mathematical model	Undefined constant parameter
		Undefined time dependent term
		Undefined space dependent term
		Undefined time-space dependent term

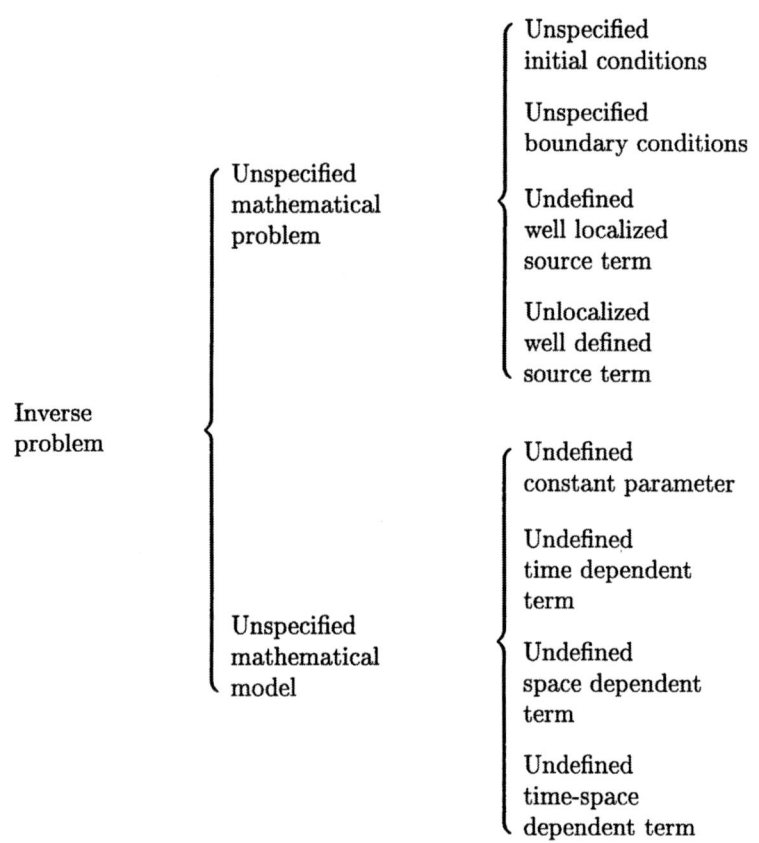

Table 4.2.2 — Classification of inverse problems.

- The solution to the initial-boundary-value problem

$$u(t, x^*) = u^*(t) \tag{4.2.13}$$

is given $\forall\, t \in [0, T]$ and $x^* \in [a, b]$.

The problem consists in determining the unknown boundary condition and the solution to the initial-boundary-value problem. □

Problem 4.3: Unspecified source term

- The mathematical model is characterized by the addition of a well-localized source/sink term, then it can be formally written as follows:

$$\frac{\partial u}{\partial t} = f\left(t, x; u, \frac{\partial u}{\partial x}, \frac{\partial^2 u}{\partial x^2}; r(t,x)\right) + s(t)\,\delta(x - x_s)\,, \qquad (4.2.14)$$

where δ is the Dirac delta function (see Appendix 3).

- The mathematical problem is stated with the proper initial and boundary conditions.
- The time-evolution of the source term $s(t)$ is not known.
- The solution

$$u(t, x^*) = u^*(t) \qquad (4.2.15)$$

to the initial-boundary-value problem is given $\forall\, t \in [0, T]$ and for some $x^* \in (a, b)$.

The problem consists in solving the initial-boundary-value problem and, in particular, in computing the source term $s(t)$. □

Problem 4.4: Unspecified parameter

- The mathematical model is characterized by a variable parameter r which may be constant, a function of time, a function of space, or a function of both time and space

$$\begin{aligned} &1.\quad r = r_0\,, \\ &2.\quad r = r(t)\,, \quad t \in [0, T]\,, \\ &3.\quad r = r(x)\,, \quad x \in [a, b]\,, \\ &4.\quad r = r(t,x)\,, \quad t, x \in [0, T] \times [a, b]\,, \end{aligned}$$

respectively.

- The mathematical problem is stated with the proper initial and boundary conditions.
- The term r is not known, however, the solution to the initial-boundary-value problem is given, with reference to the cases listed above, as follows

$$1'.\quad u = u(t^*, x^*)\,, \quad \text{for some } t^* \in (0, T] \text{ and } x^* \in (a, b)\,,$$

$2'.\quad u = u(t, x^*)\,,\quad \text{for } x^* \in (a,b) \text{ and } \forall t \in [0,T]\,,$

$3'.\quad u = u(t^*, x)\,,\quad \text{for } t^* \in (0,T] \text{ and } \forall x \in [a,b]\,,$

$4'.\quad u = u(t,x)\,,\quad \forall t \in [0,T] \text{ and } \forall x \in [a,b]\,.$

The problem consists in solving the initial-boundary-value problem and, in particular, in computing the parameter r. □

It is plain that the statements of Problems 4.1–4.4 cover only a part of all conceivable inverse-type problems. The reader must be aware that a broad range of examples can be stated with reference to the analysis of mathematical models. This concise (maybe even too concise) classification tries to indicate a logical line to follow in the statement of problems.

As stated several times throughout this book, an important property of a model is that the solution depends continuously on the initial data. Roughly speaking, this means that small errors due, for instance, to the inaccuracy of the measurements on the initial conditions or in the interpolation error committed in interpolating experimental data, should lead to small changes in the solution (of course, for the definition of small we refer to Appendix 1).

Perhaps the best way to understand the physical meaning of what was just stated is to look at the following example:

Example 4.3 Backward heat equation
Consider the following very simple initial-boundary-value problem in $(t,x) \in [0,T] \times [0,1]$

$$\begin{cases} \dfrac{\partial u}{\partial t} = \kappa \dfrac{\partial^2 u}{\partial x^2}\,, & \kappa > 0 \\ u(T,x) = u_T(x) \\ u(t,0) = u(t,1) = 0\,, \end{cases} \tag{4.2.16}$$

which is solved by

$$u(t,x) = \sum_{n=1}^{\infty} u_T^{(n)} e^{-\kappa n^2 \pi^2 (t-T)} \sin n\pi x\,,$$

where

$$u_T^{(n)} = 2\int_0^1 u_T(x)\sin n\pi x\,dx\,.$$

This implies that

$$\text{if } u_T(x) = 0 \qquad \Longrightarrow \qquad u(t,x) = 0$$

$$\text{if } u_T(x) = \sin \pi x \qquad \Longrightarrow \qquad u(t,x) = e^{\kappa\pi^2(T-t)}\sin \pi x$$

$$\text{if } u_T(x) = \frac{\sin N\pi x}{N} \qquad \Longrightarrow \qquad u(t,x) = e^{\kappa N^2\pi^2(T-t)}\frac{\sin N\pi x}{N}\,.$$

Therefore, increasing N, the initial condition becomes uniformly near $u_T = 0$, or in mathematical terms

$$\forall\varepsilon\,, \quad \exists N \quad \text{such that} \quad \left|\frac{\sin N\pi x}{N} - 0\right| < \varepsilon \quad \forall x\,.$$

In spite of this $|u(t,x)|$ grows exponentially with N, or in other words exponentially departs from the solution $u(t,x) = 0$ obtained for the initial condition $u_T(x) = 0$. Therefore, there is no continuous dependence on the initial data. □

REMARK 4.2.4 The result obtained in the previous example could have been expected. In fact, we know that parabolic operators are characterized by a smoothing action, the shorter the wavelength the stronger the action. Reversing time transforms the smoothing action into an amplifying action which is strongest for the shortest wavelengths. For this reason the previous result can be generalized to all backward parabolic systems. ■

REMARK 4.2.5 The reader has to be aware of the fact that this phenomenon is catastrophic. Refining the spatial integration mesh in the numerical code worsens the situation. In fact, if we refine the mesh, then we are able to catch finer details (i.e., higher harmonics), which blow up in a shorter time. Hence in integrating such an ill-posed system one has to use a very coarse mesh and has no hope of integrating beyond a certain interval of time. This can also explain in simple terms the difference between ill posedness and instability. In fact, the solution to the backward heat equation behaves qualitatively like the one of a linearly unstable problem, but the most dangerous perturbations have "infinitely short" wavelengths and this is physically hard to understand. In the literature this phenomenon is also termed ***Hadamard instability*** [JOb]. ■

REMARK 4.2.6 By introducing the backward time variable $\tau = T - t$, Eq.(4.2.16) can be rewritten as

$$\begin{cases} \dfrac{\partial u}{\partial \tau} = \widetilde{\kappa}\dfrac{\partial^2 u}{\partial x^2} \\ u(0,x) = u_T(x) \\ u(\tau,0) = u(\tau,1) = 0\,, \end{cases} \tag{4.2.17}$$

with $\widetilde{\kappa} = -\kappa < 0$. Hence, we can say that if the diffusivity coefficient is positive the solution $u(t,x) = 0$ is stable; if it is negative, then $u(t,x) = 0$ is Hadamard unstable. From the physical point of view, this is not a bifurcation, but a loss of well posedness. ■

We end this section by presenting two useful theorems giving some sufficient conditions for an initial-value problem to be well posed.

THEOREM 4.2.1 ***(Well Posedness of the Initial-Value Problem)***
If, applying the Fourier transform, the system of partial differential equations can be written as

$$\frac{d\widehat{\mathbf{u}}}{dt}(t,\omega) = \mathbf{A}(\omega)\widehat{\mathbf{u}}(t,\omega)$$

and all the eigenvalues $\lambda(\omega)$ of $\mathbf{A}(\omega)$ are such that

$$\Re e\,\lambda(\omega) \le \overline{\lambda}\,, \qquad \forall \omega \in \mathbb{R}\,,$$

then the initial-value problem related to the original equation is well posed. ■

THEOREM 4.2.2 ***(Well Posedness for Second-Order Equations)***
Consider the second-order partial differential equation with constant coefficients

$$\frac{\partial^2 u}{\partial t^2} = \mathcal{F}u\,, \tag{4.2.18}$$

and denote by $F(\omega)\widehat{u}$ the Fourier transform of the right-hand side. If

$$\lambda_\pm(\omega) = \pm\sqrt{F(\omega)} \le \overline{\lambda}\,, \qquad \forall \omega \in \mathbb{R}\,,$$

then the initial-value problem related to (4.2.18) is well posed. ▪

Example 4.4 Ill posedness related to Problem 4.2
Referring to the initial-boundary-value problem

$$\begin{cases} \dfrac{\partial u}{\partial t} = \dfrac{\partial^2 u}{\partial x^2} \\ u(0,x) = u_{in}(x) \\ u(t,a) = u_a(t) \\ \dfrac{\partial u}{\partial x}(t,a) = \widetilde{u}_a(t) \end{cases} \tag{4.2.19}$$

in $(t,x) \in [0,T] \times [a,b]$, consider the following initial-value problem (here t and x are switched)

$$\begin{cases} \dfrac{\partial^2 u}{\partial t^2} = \dfrac{\partial u}{\partial x} \\ u(a,x) = u_a(x) \\ \dfrac{\partial u}{\partial t}(a,x) = \widetilde{u}_a(x)\,. \end{cases} \tag{4.2.20}$$

Since $F(\omega) = i\omega$, then

$$\lambda_{\pm}(\omega) = \pm \frac{1+i}{\sqrt{2}} \sqrt{\omega}$$

(here i is the imaginary unit), which is not bounded as ω goes to infinity. Problem (4.2.20) and therefore (4.2.19) can then be (and in fact are) ill posed. □

REMARK 4.2.7 One way to render (4.2.20) well posed is to introduce a fictitious regularizing term. For instance, one can replace the partial differential equation in (4.2.20) with

$$\frac{\partial^2 u}{\partial t^2} = \frac{\partial u}{\partial x} + \varepsilon \frac{\partial^2 u}{\partial x^2}$$

and let ε go to zero. ▪

The following sections will indicate some detailed calculation methods which will be referred to the heat equation.

4.3 Solution by Decomposition of Domains

Some of the inverse-type problems stated in the preceding section can be solved by suitable mathematical methods which decompose the original problem into several properly linked, direct problems, which are well formulated.

These techniques can be considered similar to the decomposition of domain methods described briefly in Chapter 3.

In order to avoid being too general, the content of this section refers to the analysis of specific problems, in particular of Problems 4.1–4.3, and applies them to the mathematical model of the nonlinear heat equation

$$\frac{\partial u}{\partial t} = \frac{\partial}{\partial x}\left[\kappa(u)\frac{\partial u}{\partial x}\right], \qquad x \in [a, b]. \tag{4.3.1}$$

Some conceivable generalizations will be indicated in what follows.

Before describing the solution techniques to the inverse problems we report, for sake of completness, a brief summary of the solution schemes (presented in Section 3.10), referred to direct problems. These schemes are reported only for practical purposes having in mind the application proposed at the end of this chapter and the problems suggested in the last part of this book. As a matter of fact, they can be substituted by suitable alternative methods.

Scheme 1: Consider the initial-boundary-value problem related to Eq.(4.3.1) with two-point boundary conditions

$$u(t, a) = u_a(t), \qquad u(t, b) = u_b(t) \tag{4.3.2}$$

for $t \in [0, T]$ and initial conditions

$$u(0, x) = u_{in}(x), \qquad \forall x \in [a, b]. \tag{4.3.3}$$

Recalling Section 3.10, we can solve such a problem by collocating in space, i.e., by introducing the collocation points

$$a = x_0 < x_1 < \cdots < x_{N-1} < x_N = b \tag{4.3.4}$$

and then by approximating the solution with its interpolant

$$u(t,x) \cong u^N(t,x) \stackrel{\text{def}}{=} \sum_{j=0}^{N} u_j(t)\phi_j(x)\,, \qquad (4.3.5)$$

where $u_j = u(t,x_j)$, for $j = 0,\ldots,N$ and $\phi_j(x_i) = \delta_{ij}$ for $i,j = 0,\ldots,N$ are a set of cardinal basis functions.

The interpolation (4.3.5) allows us to approximate the space derivatives in the nodal points as follows

$$\begin{aligned} \frac{\partial u}{\partial x}(t,x_i) &\cong \sum_{j=0}^{N} D_{ij}u_j(t)\,, \\ \frac{\partial^2 u}{\partial x^2}(t,x_i) &\cong \sum_{j=0}^{N} B_{ij}u_j(t)\,, \end{aligned} \qquad (4.3.6)$$

where D_{ij} and B_{ij} denote, as seen in Section 3.10, the first and the second space derivative of the function ϕ_j in the nodal points x_i.

The time-evolution of the terms $u_i(t) = u(t,x_i)$ are then obtained by substituting the expression (4.3.6) into Eq.(4.3.1) and enforcing the boundary conditions with reference to the corresponding nodal points. This yields a system of nonlinear ordinary differential equations of the type

$$\frac{du_i}{dt} = \kappa(u_i)\sum_{j=0}^{N} B_{ij}u_j + \kappa_u(u_i)\left(\sum_{j=0}^{N} D_{ij}u_j\right)^2, \qquad (4.3.7)$$

for $i = 1,\ldots,N-1$, with $\kappa_u = \partial\kappa/\partial u$ and the first and last equation replaced by

$$u_0(t) = u_a(t)\,, \qquad u_N(t) = u_b(t)\,. \qquad (4.3.8)$$

In conclusion, this scheme collocates in space and integrates in time.

Scheme 2: Consider now the initial-boundary-value problem for Eq. (4.3.1) with initial conditions (4.3.3) and boundary conditions

$$u(t,a) = u_a(t)\,, \qquad \frac{\partial u}{\partial x}(t,a) = \tilde{u}_a(t)\,. \qquad (4.3.9)$$

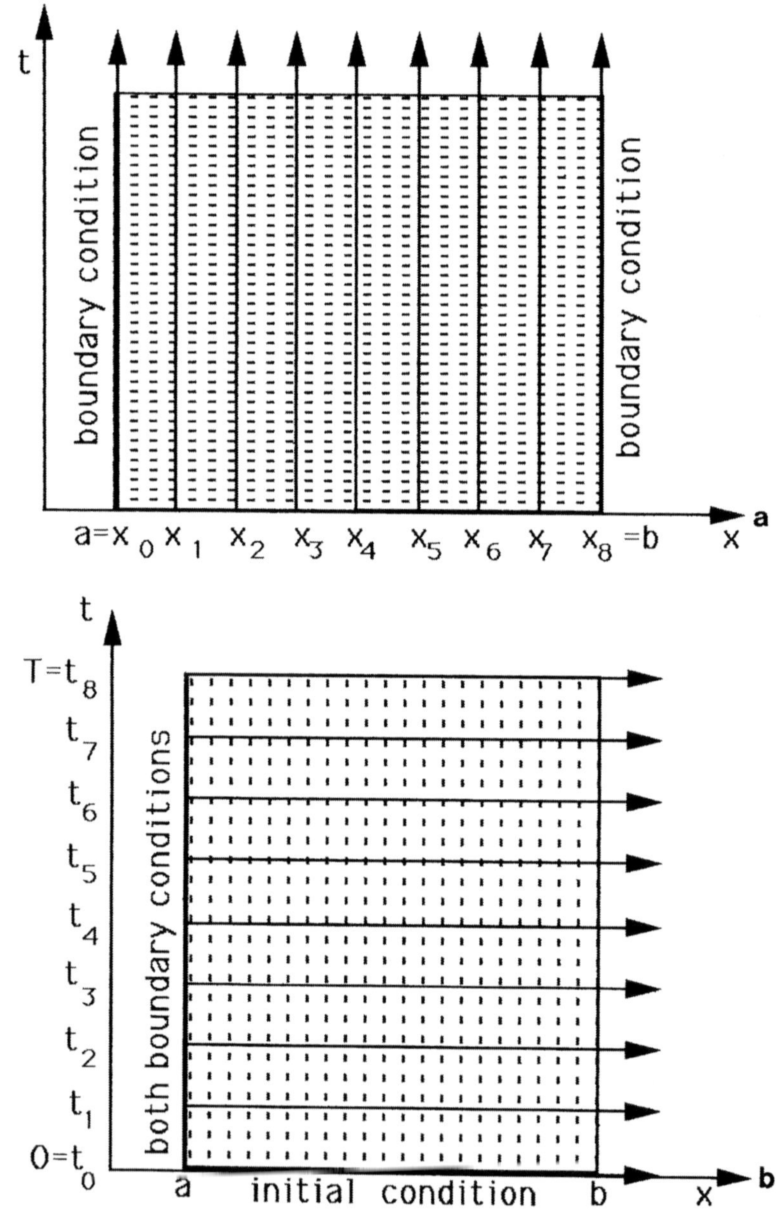

Figure 4.3.1 — Representation of (a) Scheme 1, which discretizes in space and integrates in time and (b) Scheme 2, which discretizes in time and integrates in space.

The evolution equation can be rewritten as a first-order system, with respect to the space derivatives, as follows

$$\begin{cases} \dfrac{\partial u}{\partial x} = \dfrac{1}{\kappa(u)} \dot{w} \\[2ex] \dfrac{\partial w}{\partial x} = \dfrac{\partial u}{\partial t}, \end{cases} \tag{4.3.10}$$

with the obvious meaning of symbols.

We can proceed in a fashion similar to Scheme 1, but now we collocate in time $0 = t_0 < t_1 < \cdots < t_{M-1} < t_M = T$. Hence, the state variable is approximated as follows

$$u(t,x) \cong u^M(t,x) \stackrel{\text{def}}{=} \sum_{j=0}^{M} \phi_j(t) u_j(x), \tag{4.3.11}$$

where now

$$u_j(x) = u(t_j, x), \qquad j = 0, \ldots, M. \tag{4.3.12}$$

Similar to Scheme 1, the time derivative in the nodal points can then be approximated, with the obvious meaning of symbols, as follows

$$\frac{\partial u}{\partial t}(t_i, x) \cong \sum_{j=0}^{M} D_{ij} u_j(x). \tag{4.3.13}$$

Substitution of the expressions (4.3.13) into Eq.(4.3.10) yields the following system of ordinary differential equations

$$\begin{cases} \dfrac{du_i}{dx} = \dfrac{1}{\kappa(u_i)} w_i, & \text{for } i = 1, \ldots, M \\[2ex] \dfrac{dw_i}{dx} = \displaystyle\sum_{j=0}^{M} D_{ij} u_j, & \text{for } i = 0, \ldots, M \end{cases} \tag{4.3.14}$$

which defines the evolution, in space, of u and w corresponding to the nodal points in time.

The evolution equation for u_0 in Eq.(4.3.14) is replaced by the initial condition $u(0,x) = u_{in}$, which, therefore, acts as boundary condition. The initial conditions for system (4.3.14) are the conditions at $x = a$

$$u_i(x = a) = u_a(t_i), \qquad w_i(x = a) = \kappa \widetilde{u}_a(t_i). \tag{4.3.15}$$

Namely, the boundary conditions of the original problem act as initial conditions in the new problem.

In conclusion, this scheme uses collocation in time and integration in space. The difference between the two schemes is sketched in Fig. 4.3.1.

These two schemes can be used for the solution to the inverse-type Problems 4.1–4.3 stated in Section 4.2.

REMARK 4.3.1 Several other schemes can be proposed to try to solve ill-posed problems (we recall, referring to Example 4.4, that Eq.(4.3.10) with (4.3.3), (4.3.9) acting as initial-boundary conditions, is ill posed). For instance, one may transform the problem into a well-posed one by adding a small well-behaving term. In particular, the heat diffusion model can be replaced by the following

$$\kappa \frac{\partial^2 u}{\partial x^2} = \frac{\partial u}{\partial t} + \varepsilon \frac{\partial^2 u}{\partial t^2},$$

which generates a well-posed problem, however one needs to introduce a fictitious condition on $\partial u / \partial t$. ■

The reader already possesses, after Chapter 3, sufficient knowledge to generalize the method to different types of problems or to substitute Schemes 1 and 2 with other solution techniques.

With this in mind, we can now consider the solution of the three problems mentioned above, applied to Eq.(4.3.1).

Solution of Problem 4.1: The solution of this problem is immediate and needs the application of Scheme 1.

The time interval is decomposed into two intervals

$$[0,T] = [0,\bar{t}] \cup [\bar{t},T]. \tag{4.3.16}$$

In both intervals the problem is solved by Scheme 1, but in $[0,\bar{t}]$ the time integration is reversed ($\tau = \bar{t} - t$) and $u(\bar{t},x) = \bar{u}(x)$ is used as initial condition. In particular, the ordinary differential equation to be solved is

$$\frac{du_i}{d\tau} = -\kappa(u_i) \sum_{j=0}^{N} B_{ij} u_j - \kappa_u(u_i) \left(\sum_{j=0}^{N} D_{ij} u_j \right)^2 \tag{4.3.17}$$

with the additional constraints

$$u_0(\tau) = u_a(\bar{t} - \tau), \quad \text{and} \quad u_N(\tau) = u_b(\bar{t} - \tau). \tag{4.3.18}$$

Equation (4.3.17) corresponds to the backward heat equation, which was already seen as example of an ill-posed problem. Therefore, the numerical solution of (4.3.17) needs particular attention especially when $\bar{t}$ is not small and large N are needed.

Solution of Problem 4.2: The solution of this problem needs the decomposition of the domain $\mathcal{D} = [a, b]$ of the space variable into two subdomains

$$\mathcal{D} = \mathcal{D}_1 \cup \mathcal{D}_2, \tag{4.3.19}$$

where $\mathcal{D}_1 = [a, x^*]$ and $\mathcal{D}_2 = [x^*, b]$. In order to avoid ambiguities in the application of solution algorithms, the temperature in $\mathcal{D}_1$ and $\mathcal{D}_2$ will be denoted, respectively, by $u^{(1)}$ and $u^{(2)}$

The initial-boundary-value problem in $\mathcal{D}_1$ for Eq.(4.3.1) is solved by Scheme 1. In this case $u_{in}(x)$ is the initial condition, whereas $u_a(t)$ and $u^*(t)$ act as boundary conditions.

The problem in $\mathcal{D}_2$ is solved, subsequently, by Scheme 2. In this case the initial conditions are $u^*(t)$ and the value of the space derivative of the temperature in $x = x^*$, which is obtained by the solution in $\mathcal{D}_1$ as

$$w^{(2)}(t, x^*) = \kappa\big(u^{(1)}(t, x^*)\big) \frac{\partial u}{\partial x}^{(1)}(t, x^*) = \kappa\big(u^{(1)}(t, x^*)\big) \sum_{j=0}^{N} D_{Nj} u_j^{(1)}(t). \tag{4.3.20}$$

Solution of Problem 4.3: The solution of this problem can be obtained by a suitable generalization of the solution method of the preceding problem.

In this case the domain $\mathcal{D}$ is decomposed into three subdomains

$$\mathcal{D} = \mathcal{D}_1 \cup \mathcal{D}_2 \cup \mathcal{D}_3, \tag{4.3.21}$$

where $\mathcal{D}_1 = [a, x^*]$, $\mathcal{D}_2 = [x^*, x_s]$, and $\mathcal{D}_3 = [x_s, b]$. Similar to the preceding problem, the temperature in $\mathcal{D}_3$ will be denoted by $u^{(3)}$.

The problem in $\mathcal{D}_1$ and $\mathcal{D}_2$ is solved as we have already seen for Problem 4.2 and in the visualization of Fig. 4.3.2. Among the outputs, the temperature in x_s

$$u^{(2)}(t, x_s) = u_s(t) \tag{4.3.22}$$

is known.

The problem in $\mathcal{D}_3$ is then well formulated with the proper initial and boundary conditions $u_{in}(x)$, $u_s(t)$, and $u_b(t)$. The problem in $\mathcal{D}_3$ can then be solved by Scheme 1.

The source term is then identified by the change of slope in the temperature profile

$$s(t) = -\kappa\big(u_s(t)\big)\left(\frac{\partial u}{\partial x}^{(3)}(t, x_s) - \frac{\partial u}{\partial x}^{(2)}(t, x_s)\right) \tag{4.3.23}$$

We have to point out that from the integration by Scheme 2 the space derivative of $u^{(2)}$ is known at the time collocation points t_i

$$\frac{\partial u^{(2)}}{\partial x}(t_i, x_s) = \frac{1}{\kappa\big(u_s(t_i)\big)} w^{(2)}(t_i)\,. \tag{4.3.24}$$

Therefore, to obtain its value for any t, one has to interpolate the data

$$\frac{\partial u^{(2)}}{\partial x}(t, x_s) = \sum_{i=0}^{M} \frac{1}{\kappa\big(u_s(t_i)\big)} w^{(2)}(t_i)\phi_i(t)\,. \tag{4.3.25}$$

On the other hand,

$$\frac{\partial u}{\partial x}^{(3)}(t, x_s) = \sum_{j=0}^{N} D_{0j} u_j^{(3)}(t) \tag{4.3.26}$$

is available essentially at any t, since in $\mathcal{D}_3$ we are integrating in time.

About Generalizations

As we have already mentioned, the solution schemes in this section are referred to a particular mathematical model and a limited number of mathematical problems. Several generalizations are possible. As usual, some of them are technical, some others need consistent development of mathematical methods.

Some of these generalizations are listed here for sake of completness

- Problems in several space variables;
- Problems with more general boundary conditions;

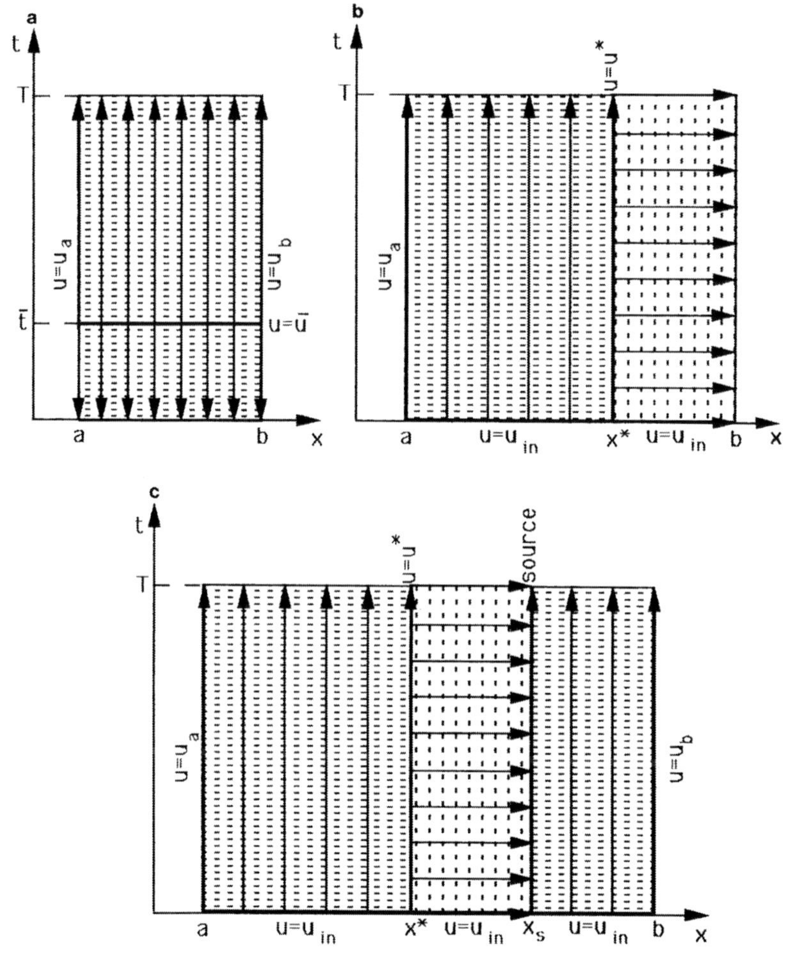

Figure 4.3.2 — Sketch of the solution procedure to solve (a) Problem 4.1, (b) Problem 4.2, and (c) Problem 4.3.

- Problems with distributed source terms.

Of course, if the problem is referred to a different mathematical model, then the generalization of the procedure may not be immediate or may even involve several additional problems.

4.4 Solution by Minimization Techniques

In order to deal with Problem 4.4, we need to introduce some concepts on minimization techniques. As known, maxima or minima of analytic functions in several variables are identified by first equating to zero the first-order partial derivatives and then studying the determinants of the second-order derivatives. In detail, let

$$g = g(x_1, \dots, x_n) \, : \, \mathcal{D} \longmapsto \mathbb{R} \tag{4.4.1}$$

be an analytic function in n dependent variables and let

$$A_m = \begin{vmatrix} \dfrac{\partial^2 g}{\partial x_1^2} & \cdots & \dfrac{\partial^2 g}{\partial x_1 \partial x_m} \\ \vdots & \ddots & \vdots \\ \dfrac{\partial^2 g}{\partial x_m \partial x_1} & \cdots & \dfrac{\partial^2 g}{\partial x_m^2} \end{vmatrix} , \tag{4.4.2}$$

with $m \leq n$, be the matrix of the first m elements of the second-order derivatives ordered with increasing j (at fixed i) along the lines and increasing i (at fixed j) along the columns.

If g is analytic, then the stationary points can be identified by solving the algebraic system

$$\frac{\partial g}{\partial x_1} = \frac{\partial g}{\partial x_2} = \cdots = \frac{\partial g}{\partial x_n} = 0 \, . \tag{4.4.3}$$

Then condition (4.4.3), which is necessary for the existence of a maximum or a minimum, is also sufficient for a minimum if

$$\det A_1 < 0 \, , \quad \det A_2 > 0 \, , \quad \det A_3 < 0 \, , \dots \, . \tag{4.4.4}$$

The application of the method just presented requires differentiability properties that may not be available in practical situations. This, for instance, is the case when g is available only by numerical calculations.

Then one needs more general methods, which may be applied to operators that map a set $\mathcal{D} \subset \mathbb{R}^n$ into $\mathbb{R}$. The simplest technique to apply, which is also easily computable, is the ***univariate search method***, see Cooper and Steinberg [CaS], Nelson [NED], and Demyanov and Rubinov [DaR].

A brief description of this method is given here. The starting point is the search of global maxima (or minima) of functions in one dependent variable, say

$$g = g(x) \, : \, [a, b] \longmapsto \mathbb{R} \, . \tag{4.4.5}$$

Considering a suitable discretization of points $\{x_q\}$, $q = 1, \ldots, n$ with $x_1 = x_a$ and $x_n = x_b$ and denoting by $g_q = g(x_q)$ the value of g corresponding to the point x_q, the presence of a maximum is then identified by the condition

$$g_i = \max_{q=1,\ldots,n} g_q \cong \max_x g \, . \tag{4.4.6}$$

In other words, one looks for the maximum of the sequence of values g_q, for $q = 1, \ldots, n$. Then one may say that this value approximates the global maximum for $g = g(x)$.

Similarly, the presence of a minimum is identified by the condition

$$g_i = \min_{q=1,\ldots,n} g_q \cong \min_x g \, . \tag{4.4.7}$$

The search provides accurate results if the discretization step is sufficiently small and if the function g has monotone behavior in each discretization interval. Otherwise, this procedure may not be reliable. In general, it is advisable to use professional scientific programs that organize the search with variable grids.

If g is a function in several variables, say

$$g = g(x_1, \ldots, x_p) \, , \tag{4.4.8}$$

then the procedure can be organized in a sequential search relaxing, sequentially, each of the p variables with fixed values of the remaining $p - 1$ variables. In detail, if x_1 is relaxed at fixed values

$$x_{20}, \ldots, x_{p0}$$

of the remaining variables, then the univariate search provides the value x_{11} corresponding to a maximum of the function

$$g = g\big(x_1; x_{20}, \ldots, x_{p0}\big) \, .$$

Then one considers the function in the independent variable x_2

$$g = g(x_2; x_{11}, x_{30}, \dots, x_{p0})$$

and looks for the point x_{21}, which identifies the presence of a maximum.

The procedure is repeated for all variables and is then iterated starting from

$$g = g(x_1; x_{21}, \dots, x_{p1})$$

until sufficient accuracy is reached.

Of course, the procedure we have just seen may not work in all cases. If, for instance, g undergoes oscillations with wavelength less than the discretization step, then the search technique does not lead to the desired result.

The method just described is the simplest to apply. However, such a method can be reasonably substituted by more efficient ones, which are reported in the pertinent literature [NEL]. We mention, among others, the ***steepest descent method***, which looks for the maxima following the steepest slopes, which are computed using a suitable set of local points to obtain the directional derivatives of the surface defined in (4.4.8).

These methods are certainly more efficient compared with the univariate method and need a shorter computational time. However, several technical aspects still need to be refined before one can technically apply the method we are talking about.

Being aware of all reliability problems, which we have mentioned above, we have sufficient information to indicate how to deal with Problem 4.4.

Consider now Problem 4.4 in the simplest case, namely when r is a scalar parameter. Then r can be identified by minimizing the distance

$$d(\mathbf{u}, \mathbf{v}; r) = \|\mathbf{u} - \mathbf{v}\|(r) \tag{4.4.9}$$

between the information $\mathbf{u}$ on the real behavior of the system and the prediction $\mathbf{v} = \mathbf{v}(r)$ of the model which will depend upon the choice of r. In this case the search of the minimum is in one variable only and the distance (4.4.9) can be the Euclidean one

$$\|\mathbf{u} - \mathbf{v}\|_2 = \left\{ \sum_{i=1}^{n} (u_i - v_i)^2 \right\}^{\frac{1}{2}} , \tag{4.4.10}$$

or

$$\|\mathbf{u} - \mathbf{v}\|_\infty = \max_{i=1,\dots,n} |u_i - v_i| , \tag{4.4.11}$$

where u_i and v_i denote the components of the vectors $\mathbf{u}$ and $\mathbf{v}$, respectively.

If one has to identify a set $\mathbf{r}$ of parameters, then the unknown is a vector and the minimization search has to be organized in several variables.

If r is a function of time, say $r = r(t)$ with $t \in [0, T]$, then a low order interpolation (say cubic splines or piecewise linear interpolant) can be used to reduce the problem to the search of the set $\mathbf{r} = \{r_j\}$ corresponding to the values of r in the collocation points.

Of course, similar techniques can be adopted if r is a function of space. In both cases one can use the minimization techniques seen above. A problem to be dealt with carefully is the control of the instabilities.

4.5 Mathematical Modelling and Stochasticity

The analysis of the preceding chapters was developed within a strictly deterministic framework. In principle, one can accept that real systems are characterized by an intrinsic deterministic behavior. On the other hand, the gap between a real system and its mathematical model may involve some stochastic behavior for various reasons which we will try to analyze in what follows.

In particular, the following sections will organize in a mathematical framework the aforementioned observations.

With this in mind, we point out, among others, the following stochasticity features in the mathematical modelling of physical systems:

- A real system, which we will call the inner system, is never fully isolated from the outer environment. If one models, in a simple way, the actions of the outer environment over the inner system, it is quite natural to do it in a stochastic manner. In other words, considering that one generally does not model both inner and outer systems, a reasonable way to model the whole system is to assume that the outer environment stochastically perturbs the behavior of the inner system.

- The parameters that characterize the physical system need to be identified and measured in order to be cast into the mathematical model. On the other hand, the evaluation of the parameters may be affected by errors, which may suggest modelling them as random variables.

  The same situation occurs when parameters and state variables are characterized at different scales. Say, the state variable is defined at a macroscopic scale: Velocity, temperature, dislocation, and so on; whereas the parameters are defined at a microscopic scale; say, the molecular scale. This situation may occur, for instance, in the modelling of materials and in the analysis of the dynamics of materials with some irregular internal behavior. Then one has to combine, in the model, the microscopic and macroscopic scales. This situation generates stochastic interactions.

- The formulation of the mathematical problems related to the analysis of models requires, as we have seen, the knowledge of suitable initial and/or boundary conditions. The physical situation, however, may be such that these quantities can be known only with uncertainties and need to be stochastically modelled. Therefore, even if the mathematical model is deterministic, the related mathematical problem may be of stochastic type.

The motivations for stochastic modelling, listed above, are not the only ones, simply the most common ones. However, sufficient indications have been given to show that dealing with stochastic calculus is of relevant importance in mathematical modelling.

The second part of this chapter is organized in order to provide the fundamental knowledge to deal with stochastic modelling: From the classification of models to the analysis of stochastic problems. The presentation of the contents is simple and concise. Certainly the whole matter needs further and deeper development. The contents will often refer to Appendix 3, where fundamental aspects of probability theory are presented in a very concise form.

4.6 Classification of Discrete Stochastic Models

Let us now deal with the class of discrete dynamic systems, namely, models such that the state variable depends upon time only

$$\mathbf{u} = \mathbf{u}(t) \; : \; [0, T] \longmapsto \mathbb{R}^n \tag{4.6.1}$$

and satisfies a system of n coupled ordinary differential equations

$$\frac{d\mathbf{u}}{dt} = \mathbf{f}(t, \mathbf{u}; \mathbf{r}) \,. \tag{4.6.2}$$

A classification of stochastic discrete dynamic models can be organized taking into account the following aspects:

- Stochasticity in the mathematical model, i.e., in the parameters;
- Stochasticity in the mathematical problem, i.e., in the initial conditions;

- Stochasticity in the external actions, i.e., in some additive noise coming from the external environment;
- Classification by the type of stochasticity: Simple random parameters or real noise.

As we shall see, these aspects are essentially the same as those to be considered in dealing with the classification of continuous dynamic systems.

If we consider the mathematical problem related to the analysis of the model as an input/output block system, then the first three types of stochasticity correspond, respectively, to the block representation given in Figs. 4.6.1–4.6.3.

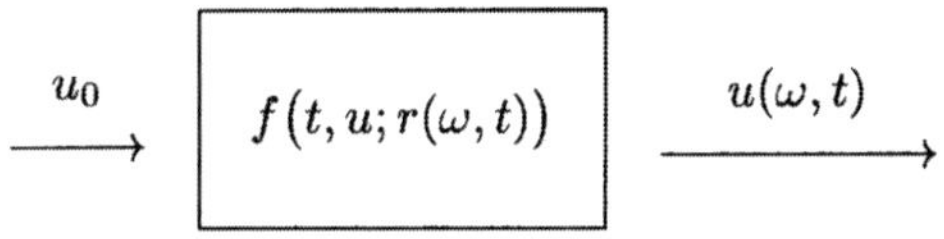

Figure 4.6.1 — Models with random parameters.

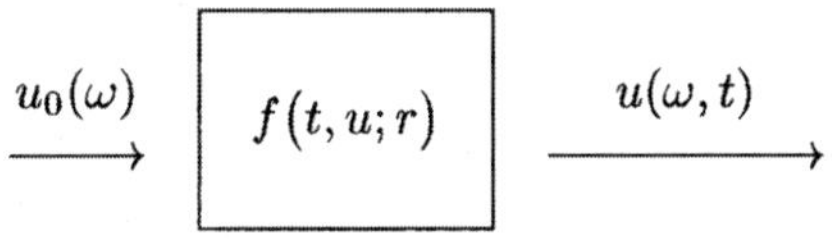

Figure 4.6.2 — Mathematical problems with random initial conditions.

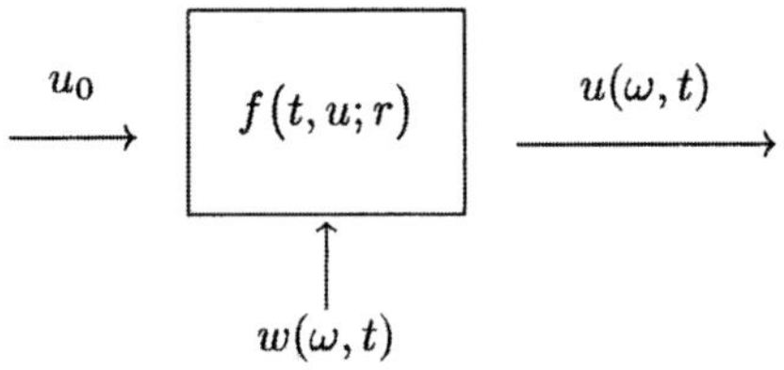

Figure 4.6.3 — Models with random additive noise.

As we can see, in the first and last representation, the mathematical model is genuinely stochastic, either in the inner system, or in the action of the outer environment. On the other hand, in the second representation the model remains strictly deterministic, but the mathematical problem is stochastic, as the input of the initial conditions is random.

In all cases, the output is a stochastic process, and, in the analysis of the model, one has to deal with the problem of the statistical analysis and representation of such a process.

An additional aspect to be taken into account is the type of stochasticity. This aspect is technical and refers to all the preceding types of stochastic modelling.

After these qualitative descriptions, some detailed definitions can be given.

DEFINITION 4.6.1 Stochastic models
A discrete dynamic model, described by a vector ordinary differential equation, is a ***stochastic discrete dynamic model*** *if some of the parameters* $\mathbf{r}$*, which characterize the model, are either random parameters* $\mathbf{r} = \mathbf{r}(\omega)$ *or stochastic processes* $\mathbf{r} = \mathbf{r}(t, \omega)$.

DEFINITION 4.6.2 Stochastically perturbed models
A discrete dynamic model, described by a vector ordinary differential equation, is a ***stochastically perturbed discrete dynamic system*** *if the evolution of the state variable is perturbed by some weighted noise.*

As we have mentioned, the output is, in any case, a stochastic process that, in general, can be regarded as a random variable with evolution in time.

It will be assumed, in order to provide a definition of ***stochastic process***, that the reader already possesses a sufficient knowledge of the foundations of probability theory, which are, for sake of completeness, summarized in Appendix 3.

We need a complete probability space (see Appendix 3) that is denoted by the triple $(\Omega, \mathcal{F}, \mathcal{P})$, where

- Ω is a nonempty set of elementary events;
- $\mathcal{F}$ is a σ-algebra of subsets of Ω;
- $\mathcal{P}$ is a probability measure on $(\Omega, \mathcal{F})$.

Then the following definitions can be proposed:

DEFINITION 4.6.3 Stochastic process
A function u defined on $[0,T] \times \Omega$, where $[0,T]$ is a time interval, is a ***stochastic process*** *if, for any $t \in [0,T]$,*

$$u(t,\cdot)\,:\quad \Omega \longmapsto \mathbb{R}$$

is a random variable.

DEFINITION 4.6.4 Realization and path
Let $t \in [0,T]$ be fixed, then the random variable $u(t;\omega)$ is called the ***realization of** u **at time** t. *For fixed $\omega \in \Omega$, the function*

$$[0,T] \longmapsto u(t;\omega)$$

is called the ***path of*** *u.*

One can denote by $F_1\big(U;t\big)$ the probability that the random variable $u(t,\omega)$ at time t, takes value not grater than U

$$F_1\big(U;t\big) = \mathcal{P}\big(\omega \in \Omega\,,\, u(t,\omega) \leq U\big)\,, \tag{4.6.3}$$

where U is a real number.

In general, if we consider a discretization of the time interval $[0,T]$

$$\big\{t_1 = 0, \ldots, t_i, \ldots, t_n = T\big\}\,, \tag{4.6.4}$$

then the set $\mathbf{u}(\omega) = \big\{u_1(\omega), \ldots, u_n(\omega)\big\}$ can be defined for each point of the discretization, where $u_i(\omega) = u(t_i,\omega)$.

In this case, the n-probability is denoted by

$$F_n\big(U_1, \ldots, U_n; t_1, \ldots, t_n\big)\,, \tag{4.6.5}$$

which defines the joint probability that $u_i \leq U_i$ for all $i = 1, \ldots, n$.

We restrict our attention to the case where F_1 or F_n has a ***probability density***, respectively, f_1 or f_n, which is defined by

$$\int_{-\infty}^{U} f_1(u;t)\,du = F_1\big(U;t\big) \tag{4.6.6}$$

and

$$\int_{-\infty}^{U_1} \cdots \int_{-\infty}^{U_n} f_n(u_1, \ldots, u_n; t_1, \ldots, t_n)\, du_1 \cdots du_n$$
$$= F_n\Big(U_1, \ldots, U_n; t_1, \ldots, t_n\Big)\,. \qquad (4.6.7)$$

If f_1 is known, then the ***first-order statistics***, i.e., all moments of u, can be computed. Similarly the ***n-order statistics***, which involves all moments and correlation of the vector $\mathbf{u} = \{u_1, \ldots, u_n\}$, can be executed if f_n is known.

In principle, a stochastic process is known if f_n is known for all types of discretizations (4.6.4).

Coming back to the case of stochastic models, U is the state variable and $u(t, \omega)$ is the related stochastic process. The mathematical method should lead to the computation of f_n. This objective is, in general, hard or impossible to obtain as documented in the classical literature [SOO], [ADO]. Often the objective is simply the first-order statistics. Some relatively simple calculations will be provided in the following sections. The reader can find in Kliemann and Namachchivaya [KaN] several useful information to further develop his knowledge.

4.7 Classification of Continuous Stochastic Models

Consider now the class of continuous dynamic models, i.e., models such that the state variable (which we will write, for sake of simplicity, in the scalar case in one space dimension) depends upon time and space

$$u = u(t, x)\ :\ [0, T] \times \mathcal{D} \longmapsto \mathbb{R} \qquad (4.7.1)$$

and satisfies a partial differential equation, for instance,

$$\frac{\partial u}{\partial t} = f\left(t, x, u, \frac{\partial u}{\partial x}, \frac{\partial^2 u}{\partial x^2}; \mathbf{r}(t, x)\right)\,. \qquad (4.7.2)$$

The types of stochasticity are essentially those listed in the preceding section, i.e.,

- Stochasticity in the mathematical model, i.e., in the parameters;
- Stochasticity in the mathematical problem, i.e., in the initial and boundary conditions;
- Stochasticity in the external inputs.

Moreover, it is necessary to distinguish (the definitions are given in Appendix 3) the type of stochasticity: Constant random parameters, stochastic processes, or random fields. In particular, the following cases can be listed:

1. Mathematical models with parameters known as random variables:

$$\mathbf{r} = \mathbf{r}(\omega) \; : \; \Omega \longmapsto \mathbb{R}\,. \tag{4.7.3}$$

2. Mathematical models with stochastic parameters known either as time stochastic processes

$$\mathbf{r} = \mathbf{r}(\omega, t) \; : \; \Omega \times [0, T] \longmapsto \mathbb{R} \tag{4.7.4}$$

or as space stochastic processes

$$\mathbf{r} = \mathbf{r}(\omega, x) \; : \; \Omega \times \mathcal{D} \longmapsto \mathbb{R}\,. \tag{4.7.5}$$

3. Mathematical models with additional weighted noise. Referring, for instance, to Eq.(4.7.2) the mathematical model is of the type

$$\frac{\partial u}{\partial t} = f\left(t, x; u, \frac{\partial u}{\partial x}, \frac{\partial^2 u}{\partial x^2}; \mathbf{r}(t, x)\right) + \varphi(t, x, u)\, \frac{dw}{dt}\,, \tag{4.7.6}$$

where w is the noise and dw/dt is its "formal derivative".

4. Mathematical problems with deterministic boundary conditions and random initial condition $u_0 = u_0(\omega, x)$.
5. Mathematical problems with deterministic initial condition and stochastic boundary conditions known as stochastic processes.

As in the case of discrete dynamic systems, the various types of stochasticities listed above can simultanously appear in a mathematical model/problem.

In all cases the output is a ***time-dependent stochastic field***

$$u = u(\omega, t, x) \; : \; \Omega \times [0, T] \times \mathcal{D} \longmapsto \mathbb{R}\,. \tag{4.7.7}$$

In detail, the following definition can be given

DEFINITION 4.7.1 Time-dependent stochastic field
*A function u defined on $[0, T] \times \mathcal{D} \times \Omega$ is a **time-dependent stochastic field** if, for any $t \in [0, T]$ and $x \in \mathcal{D}$,*

$$u(t, x, \cdot) \; : \; \Omega \longmapsto \mathbb{R}$$

is a random variable.

The concepts of realization and path are the same given in Section 4.6. The only difference is that the space variable x is an additional argument of u.

Presuming that the probability distribution has a density, one can define a first probability density distribution

$$f_1\big(u; t, x\big) \tag{4.7.8}$$

or the $n \times m$ probability density

$$f_{nm}\Big(u_{11}, \ldots, u_{n1}, \ldots, u_{1m}, \ldots, u_{nm}; t_1, \ldots, t_n, x_1, \ldots, x_m\Big), \tag{4.7.9}$$

where

$$u_{ij} = u(t_i, x_j). \tag{4.7.10}$$

The mathematical literature on methods for analyzing continuous stochastic models is limited compared with those of discrete models. Several methods for models with random parameters are reported in Bellomo, Brzezniak, and de Socio [BBD]. On the other hand, the analysis of models with real noises is limited to linear problems.

Useful references can be found when attention is focused on specific problems. For instance, in the book by Sobczyk [SOb] on the wave propagation in random media or by Pomraning [POM] on the linear transport in random media. The contents of each book is essentially restricted to the linear cases.

4.8 Modelling and Solution of Problems

The preceding two sections provided a rather simple classification of mathematical stochastic models and problems both for discrete and continuous models.

The modelling techniques do not differ, in their general lines, from those already seen in the deterministic case. The additional difficulty is that one has to deal with the identification of random variables and noises. The solution of problems requires that a statistic on the solution process is organized. This section reports about the two problems mentioned above. In detail, the first paragraph deals with the identification of random variables by means of their moments. The second paragraph describes how the statistics on the solution process can be reorganized. The content of Appendix 3 is essential to the understanding of this matter.

4.8.1 Identification of random variables

Consider the problem of identifying a random variable $r = r(\omega)$ such that its first moments

$$E\{r^p\}\,, \quad p = 1, \ldots, q\,, \tag{4.8.1}$$

are known. In addition, suppose that we also know that the variable is continuous and defined over a certain domain $\mathcal{D}_r \subseteq \mathbb{R}$, which we can properly identify.

The identification of the variable essentially consists in determining, at least in some approximation, the probability density $P = P(r)$ linked to r. We report about the method, classical in the literature [KST] and also well explained, at a tutorial level, by Adomian and Malakian [AMA]. The method essentially consists in approximating, as shown in Appendix 2, the density $P(r)$ by a sum of a finite number of orthogonal functions

$$P(r) \cong \sum_{i=0}^{N} w(r)\, C_i\, \phi_i(r)\,, \tag{4.8.2}$$

where $w(r)$ is a suitable weight function, $\phi_i(r)$ are orthonormal polynomials and C_i are suitable coefficients to be determined by

$$C_i = \langle \phi_i, P \rangle_w = \int_{D_r} \phi_i(r) P(r) w(r)\, dr\,. \tag{4.8.3}$$

It can be shown that if the polynomials ϕ_i are properly chosen, then one obtains the coefficients C_i by means of the moments of the variable r. Being practical, and referring to Appendix 2, consider the Hermite polynomials

$$\begin{aligned}
H_0(r) &= 1 \\
H_1(r) &= 2r \\
H_2(r) &= 4r^2 - 2 \\
H_3(r) &= 8r^3 - 12r \\
H_4(r) &= 16r^4 - 48r^2 + 12 \\
H_5(r) &= 32r^5 - 160r^3 + 120r \\
&\ \vdots \\
H_{i+1}(r) &= 2rH_i(r) - 2iH_{i-1}(r)\,,
\end{aligned} \tag{4.8.4}$$

which are orthogonal with respect to the weight function

$$w(r) = e^{-r^2}\,, \tag{4.8.5}$$

but not orthonormal since

$$\langle H_i, H_i \rangle_w = 2^i\, i!\, \sqrt{\pi}\,.$$

One can, of course, define the related orthonormal functions $\phi_i(r)$ in (4.8.2) as follows

$$\phi_i(r) = \frac{1}{\sqrt{2^i\, i!\, \sqrt{\pi}}} H_i(r)\,. \tag{4.8.6}$$

Performing these calculations yields

$$C_0 = 1\,,$$

$$C_1 = \sqrt{\frac{2}{\pi}}\, E\{r\}\,,$$

$$C_2 = \frac{1}{2\sqrt[4]{\pi}}\left(2E\{r^2\} - 1\right),$$

and so on.

The same calculations can be developed in more than one dependent variable as is well documented [AMA; KST]. Technically the problem is the same, however, the computational difficulty can consistently increase for multidimensional variables.

4.8.2 Statistical measures

The solution of mathematical problems related to the analysis of discrete stochastic dynamic models is, as we have mentioned, a stochastic process

$$\mathbf{u} = \mathbf{u}(\omega; t) \;:\; \Omega \times [0, T] \longmapsto \mathbb{R}^n\,. \tag{4.8.7}$$

Similarly, in the case of stochastic continuous models, $\mathbf{u}$ is a random field

$$\mathbf{u} = \mathbf{u}(\omega; t, x, y, z) \;:\; \Omega \times [0, T] \times \mathcal{D} \longmapsto \mathbb{R}^n\,, \tag{4.8.8}$$

where $\mathcal{D} \subseteq \mathbb{R}^3$

The solution of mathematical problems needs to be interpreted by use of suitable statistical measures. Therefore, it is important to report, still with reference to Appendix 3, about the technical computation of them.

Consider first the case of a discrete scalar system and assume that the first probability density

$$f_1(u;t) \,:\, \mathcal{D}_u \times [0,T] \longmapsto \mathbb{R}_+ \tag{4.8.9}$$

can be computed. Then, the ***first-order statistics*** can be organized. In fact, the knowledge of f leads to the computation of all moments of u at fixed time.

In particular, one has the p-order moments and central p-order moments

$$E\{u^p\}(t) = \int_{\mathcal{D}_u} u^p f_1(u;t)\, du \tag{4.8.10}$$

and

$$E_c\{u^p\}(t) = \int_{\mathcal{D}_u} \big|u - E(u)\big|^p f_1(u;t)\, du\,, \tag{4.8.11}$$

respectively.

Conversely, if the moments defined by (4.8.10) are known, then the probability density can be approximated by orthogonal expansions as we have seen in the preceding section.

A relatively simple case is when u depends continuously on a random variable r with known probability density $f(r)$. In this case the p-th order moments are

$$E\{u^p\}(t) = \int_{\mathcal{D}_r} u^p(t,r)\, f(r)\, dr\,, \tag{4.8.12}$$

where $\mathcal{D}_r$ is the domain of r. Similar calculations lead to central moments.

If the state variable is a vector

$$\mathbf{u}(\omega,t) \,:\, \Omega \times [0,T] \longmapsto \mathbb{R}^n\,, \tag{4.8.13}$$

then the definitions we have seen apply to each component of $\mathbf{u}$. Say, referring to the p-th order moments

$$E^p\{u_i\}(t) = \int_{\mathcal{D}_\mathbf{u}} u_i^p\, f(\mathbf{u};t)\, d\mathbf{u}\,, \tag{4.8.14}$$

and similarly in the case of random parameters.

It is useful to compute in this case the correlation matrix $[V] = [V_{ij}]$, where

$$V_{ij} = E\Big\{\big(u_i - E\{u_i\}\big)\big(u_j - E\{u_j\}\big)\Big\}\,, \tag{4.8.15}$$

which, for $i = j$, gives the variance (second-order central moment).

We have seen that, as far as the first-order statistic is concerned, the process u is regarded as a random variable at fixed time t. Therefore all computation methods of Appendix 3 can be applied.

The ***first-order statistics*** in the case of a continuous dynamic system does not differ. In fact, one deals with the density $f_1(u; t, x)$ at fixed time and space and computes the p-order moments and p-order central moments of u at fixed time and space by means of equations similar to those reported in (4.8.10–4.8.15) where the space is a further parameter.

The ***second-order statistics*** require, as we have seen, knowledge of the second-order probability density

$$f_2\big(u(t_1),\, u(t_2); t_1, t_2\big) \tag{4.8.16}$$

for a discrete model and the densities at fixed time and space, respectively, for a continuous model

$$f_2\big(u(t_1, x),\, u(t_2, x); t_1, t_2, x\big) \tag{4.8.17}$$

and

$$f_2\big(u(t, x_1),\, u(t, x_2); t, x_1, x_2\big)\,. \tag{4.8.18}$$

The knowledge of the densities defined above leads to the computation of the correlations in time and space. For instance, the autocorrelation in time

$$R(t_1, t_2; x) = E\Big\{u(t_1; x)\, u(t_2; x)\Big\} \tag{4.8.19}$$

or in space

$$R(t; x_1, x_2) = E\Big\{u(t; x_1)\, u(t; x_2)\Big\}\,, \tag{4.8.20}$$

where E is the mean value as previously defined.

In a similar way, one can compute the ***autocovariances***

$$C(t_1, t_2; x) = E\Big\{\big(u(t_1; x) - E\{u\}(t_1; x)\big)\big(u(t_2; x) - E\{u\}(t_2; x)\big)\Big\} \tag{4.8.21}$$

and

$$C(t; x_1, x_2) = E\Big\{\big(u(t; x_1) - E\{u\}(t; x_1)\big)\big(u(t; x_2) - E\{u\}(t; x_2)\big)\Big\} \tag{4.8.22}$$

Similar calculations can be developed for multidimensional state variables. In this case, the correlations and covariances have to be related also to the various components of $\mathbf{u}$.

In general, one should consider the probability density for distributions both in time and space

$$f_{22}\big(u(t_1,x_1),\, u(t_2,x_1)\,, u(t_1,x_2),\, u(t_2,x_2); t_1,t_2,x_1,x_2\big)\,, \qquad (4.8.23)$$

for t_1, $t_2 \in [0,T]$ and x_1, $x_2 \in \mathcal{D}$.

Usually, these types of calculations involve several difficulties and a detailed analysis is possible only in the case of stochasticities restricted to known random variables. The calculations are then developed as in Eq.(4.8.12).

4.9 Stochastic Aspects and Inverse Problems

The first part of this chapter brought to the reader's attention several inverse-type problems, which may arise in the analysis of mathematical models. As we have seen, these problems are such that some information needed by the statement of the problem is not given and is substituted by suitable information on the solution of the mathematical problem.

The information on the solution must be exploited to determine completely the model and/or the problem.

Several motivations suggest that a large part of inverse problems should be organized in stochastic rather than deterministic terms. In fact:

- The information on the solution to the mathematical problem is obtained by direct measurement of the real system. Consequently, stochasticity may be generated either by the inaccuracies of the measuring instruments or by the noise induced by the direct influence of the outer environment on the inner system.
- The real system involves a number of variables much higher than those taken into account in the mathematical model. Therefore the measured quantities may fluctuate because of the influence of the neglected variables.

After these preliminaries, we will now formulate some mathematical problems related to the analysis of continuous dynamic models. We refer to the contents of Sections 4.2 and 4.3 and propose a mathematical formulation such that the initial-boundary-value problem (with missing terms) is deterministic. On the other hand, the additional information obtained by measuring the real system, which refers to the solution, is of stochastic type.

The modelling of the stochasticity is achieved by the superposition of a stochastic perturbation, a noise in time or space, to the deterministic values. Some examples of stochastic inverse problems will be presented with reference to continuous dynamic models.

The following examples refer to a scalar continuous dynamic system with two-point boundary conditions. The direct formulation of the initial-boundary-value problem is one we have seen several times

$$\begin{cases} \dfrac{\partial u}{\partial t} = f\left(t, x; u, \dfrac{\partial u}{\partial x}, \dfrac{\partial^2 u}{\partial x^2}; \mathbf{r}(t,x)\right) \\ u(0,x) = u_{in}(x) \\ u(t,a) = u_a(t) \\ u(t,b) = u_b(t)\,. \end{cases} \tag{4.9.1}$$

In particular, we mention among others, the following examples.

Example 4.5 Unspecified initial condition
The mathematical model is well specified. The mathematical problem is stated with the proper boundary conditions, say

$$u(t,a) = u_a(t)\,, \qquad u(t,b) = u_b(t) \tag{4.9.2}$$

or similar ones. On the other hand, the initial conditions are not given and are replaced by the following information on the solution at some $t = \bar{t}$

$$u\big(\omega; \bar{t}, x\big) = u_d\big(\bar{t}, x\big) + \varepsilon\varphi\big(\omega; x\big)\,, \tag{4.9.3}$$

where ε is a dimensionless parameter, which can be small, and φ is a random field. □

Example 4.6 Unspecified boundary condition
The mathematical model is well specified. The mathematical problem is stated with the proper initial condition $u_{in}(x)$ and with only one boundary condition, say $u_a(t)$. On the other hand, the unspecified boundary condition is replaced by the information

$$u\big(\omega; t, x^*\big) = u_d\big(t, x^*\big) + \varepsilon\psi\big(\omega; t\big) \tag{4.9.4}$$

on the solution of the initial-boundary-value problem, where ε is a dimensionless parameter and $\psi(\omega; t)$ is a known stochastic process. □

REMARK 4.9.1 Both Examples 4.5 and 4.6 can be further particularized with reference to the heat equation, which was used in the Examples 4.1–4.3. ■

Additional problems can be stated following the same line of Section 4.3. The statement of additional problems is left to the reader as an exercise.

Referring now to the solution methods, it is important to help the reader to focus on the difficulties, which are quite hard, that these types of problems can involve. Nevertheless, if the type of stochasticity is a relatively simple one (as specified in what follows), then the problem can be dealt with by generalizing to the stochastic case the methods we have already seen in Section 4.4 with reference to deterministic problems. In particular, we note the following:

- The solution to the initial-boundary-value inverse problem, even in the case of very simple stochasticities (say simple random variables), is a time-dependent stochastic field, say $u = u(\omega; t, x)$. Therefore the interpretation-representation of u needs, as we have seen in Section 4.8, correlation both in time and space.
- The mathematical modelling of the stochasticities that characterize the problem are such that the trajectories

$$u_i = u\big(\omega_i; t, x\big)\,, \quad u_j = u\big(\omega_j; t, x\big)\,, \qquad \omega_i, \omega_j \in \Omega \tag{4.9.5}$$

  are uncorrelated (or correlated) for all $\omega_i, \omega_j \in \Omega$. In the case of uncorrelated trajectories, sample solutions are possible as we shall see in the applications. The price to be paid is a simplification in the modelling of the stochastic aspects of the model.
  If the trajectories are correlated, then solution techniques are available only in some simple and special cases. As a matter of fact, in the case of correlated trajectories the modelling of the randomness, which characterizes the mathematical model, is presumably more accurate. However the solution of the related mathematical problem may present severe difficulties.
- The solution of inverse problems by decomposition of domains can be organized with the same techniques (and difficulties) of direct problems. On the other hand, solutions by minimization techniques, as seen in Section 4.4, have to be regarded as an open problem still to be organized.

In conclusion, the message addressed to the reader is that the solution to random inverse problems is certainly relevant in mathematical modelling. Nevertheless, several technical difficulties have to be tackled and, at least in some cases, still to be solved.

4.10 Kinetic Models

The mathematical models and problems presented in the preceding sections are all characterized by stochasticities in the parameters or in the initial and/or boundary conditions related to the statement of the problem. The dependent variable is a random function of time and/or space.

It is known, in statistical mechanics, that mathematical models can be formulated by evolution equations for the probability distribution of the state variable. These types of models, sometimes called ***master equations***, can be elaborated in quite general physical situations.

These models refer to large systems of several objects that follow an evolution equation, which can be classified as a discrete dynamic system. The analysis of the whole system would need to deal with a large (or even infinite) number of differential equations. ***The statistical model*** provides, whenever it can be elaborated, the evolution of the probability distribution over the variable, which defines the state of the single objects.

This section will provide the description of some examples of models, which will be presented in order of increasing difficulty. The scientific computation proposed in the next section will show how to deal with one of them. Throughout the presentation of this example some useful bibliographical indications will be given and helpful suggestions will be provided in order to make the reader more familiar with this type of modelling.

Example 4.7 Discrete systems with random initial conditions
Consider a system made up of several elements and assume, having in mind their mathematical modelling, the following:

- The state variable for each element is a discrete-type variable $\mathbf{u}(t)$. If we refer it to each i-element, with $i = 1, \ldots, n$, then the subscript i will be applied to $\mathbf{u}$.
- The evolution of each element is described by a discrete dynamic model, which can be written, in abstract form, as

$$\frac{d\mathbf{u}}{dt} = \mathbf{g}(\mathbf{u}; \mathbf{r}), \tag{4.10.1}$$

  where $\mathbf{r}$ is a known set of parameters.
- The initial conditions $\mathbf{u}(0) = \mathbf{u}_{in}$ differ for each element. The initial state for the whole system can be simulated by a random variable $\mathbf{u}_0(\omega)$, linked to a suitable probability density $f(\mathbf{u}_0)$.
- The elements that constitute the whole system do not interact with each other.

REMARK 4.10.1 The approximation of the probability density $f(\mathbf{u}_0)$ can be organized by the moment approximation as seen in Section 4.8.

■

The analysis of the system just described can be performed by means of a mathematical method suitable to defining the time-evolution of the probability density $f(\mathbf{u};t)$. If such a density can be computed then the information on the time-evolution of the system can be obtained by computing the moments of the variable $\mathbf{u}$.

This objective can be reached by applying rather simple results of the mathematical theory of random differential equations [SOO]. The same analysis is also reported in Bellomo and Riganti [BaR] and in Sobczyk [SOa].

Consider the case of the initial-value problem relative to Eq.(4.10.1), with initial data $\mathbf{u}_0$ defined in some set $\mathcal{D}_0$, $\mathbf{u}_0 \in \mathcal{D}_0$. Moreover, assume that for every $\mathbf{u}_0 \in \mathcal{D}_0$ the solution $\mathbf{u}(t;\mathbf{u}_0)$ exists, is unique, and is defined by a one-to-one map. The evolution of the probability density is defined by the continuity of the probability measure

$$f(\mathbf{u}_0;0)\,d\mathbf{u}_0 = f(\mathbf{u};t)\,d\mathbf{u}\,, \tag{4.10.2}$$

which yields the formula

$$f(\mathbf{u};t) = f(\mathbf{u}_0)|J|(\mathbf{u} \longmapsto \mathbf{u}_0;t)\,, \tag{4.10.3}$$

where $|J|$ is the determinant of the Jacobian of the inverse mapping from the final state $\mathbf{u}$ to $\mathbf{u}_0$. The problem is then the computation of $|J|$. If the solution $\mathbf{u}(t;\mathbf{u}_0)$ is analytic, then such a problem can be easily solved. On the other hand, analytic solutions can be obtained only in simple cases. Therefore alternative methods need to be developed.

It can be shown that under suitable regularity properties of the probability measure, the density $f(\mathbf{u};t)$ satisfies the Liouville equation

$$\frac{\partial f}{\partial t} + \sum_{j=1}^{n} \frac{\partial}{\partial u_j}(g_j f) = 0\,. \tag{4.10.4}$$

This equation is a linear partial differential equation of first order, which can be solved by separation of variables

$$\frac{df}{f} = -\sum_{j=1}^{n} \frac{\partial g_j}{\partial u_j}\,dt\,. \tag{4.10.5}$$

Consequently, taking into account formula (4.10.3), one has

$$J(\mathbf{u} \longmapsto \mathbf{u}_0; t) = \exp\left\{ -\int_0^t \sum_{j=1}^n \frac{\partial g_j}{\partial u_j}(u(s))\, ds \right\}, \tag{4.10.6}$$

which also shows the positivity of J, with $J(t=0)=1$.

Derivation of Eq.(4.10.6) linked to Eq.(4.10.1) yields the system

$$\begin{cases} \dfrac{d\mathbf{u}}{dt} = \mathbf{g}(\mathbf{u}, r) \\ \dfrac{dJ}{dt} = -J \displaystyle\sum_{j=1}^n h_j(\mathbf{u}, r), \end{cases} \tag{4.10.7}$$

where

$$h_j = \frac{\partial g_j}{\partial u_j}. \tag{4.10.8}$$

The solution of Eq.(4.10.7) with the proper initial conditions yields the desired evolution model. □

Example 4.8 Population dynamics with stochastic interaction

Consider a system constituted by a large population of anonimous individuals, whose condition is defined by a factor u, called the ***state***. The evolution of the system is ruled by encounters between pairs of individuals with probabilistic type exchanges of state.

The physical meaning of ***state*** must be related to the peculiar characteristics of the population. For instance, it may refer to the social economic state or to the dominance factor of populations in competition.

A kinetic model can be constructed in order to define the time-evolution of the probability distribution over the variable u.

Let $u \in [0,1]$, which can always be obtained by suitable normalization. Then the evolution refers to the density

$$f = f(t; u), \qquad \int_0^1 f(t; u)\, du = 1, \qquad \forall t \geq 0. \tag{4.10.9}$$

We will first show a phenomenological construction of the model, then we will show how the model can refer to several interesting physical situations.

A mathematical model can be proposed on the basis of the following phenomenological assumptions:

- No individual changes its state u until an encounter with another individual with state v occurs;
- A function $\eta(u,v)$ can be properly identified (usually by phenomenological observation of the real system) in order to define the probability and rate of encounters between individuals with state u and v, respectively;
- A probability density $\psi = \psi(v,w;u)$ can be identified, always by phenomenological observations, in order to define, by

$$\psi(v,w;u)\,du \geq 0\,, \tag{4.10.10}$$

  the probability that through an encounter between a subject with state w and one with state v, the v element ends up in a state $[u, u + du]$.

Let us now forget, at least for the moment, about the actual identification of the terms η and ψ. A general criterion will be given later.

Assuming then, that η and ψ are given, the mathematical model can be derived by a conservation equation, namely, equating the derivative of f to the difference between "gain" and "loss" terms due to the interaction between individuals. With obvious meaning as symbols

$$\frac{\partial f}{\partial t} = G(f) - L(f)\,. \tag{4.10.11}$$

The definitions of η and ψ imply that

$$G(f) = \int_0^1 \int_0^1 \eta(v,w)\,\psi(v,w;u)\,f(t;v)\,f(t;w)\,dv\,dw \tag{4.10.12}$$

and

$$L(f) = f(t;u)\int_0^1 \eta(u,v)\,f(t;v)\,dv\,. \tag{4.10.13}$$

Namely, the loss term $L(f)$ refers to the individuals with state u, who loose due to encounters, ruled by $\eta(u,v)$, with the individuals with state v. The gain term $G(f)$ refers to the individuals who gain the state u, starting from v, after the encounter with the individuals with state w. The encounter is now ruled by $\eta(v,w)$ and the transition from v into u by the term $\psi(v,w;u)$.

Grouping all terms yields the model

$$\frac{\partial f}{\partial t}(t;u) = \int_0^1 \int_0^1 \eta(v,w)\, f(t;v)\, f(t;w)\psi(v,w;u)\; dv\, dw$$

$$-f(t;u) \int_0^1 \eta(u,v)\, f(t;v)\, dv\,. \qquad (4.10.14)$$

This model can be regarded as an "abstract" framework suitable to being generalized to specific modelling in physical situations, which may be apparently very far away. For instance, the model can refer to social dynamics if u takes the meaning of "social state." In this case η must model the probability and rate of interaction between different social states and ψ the possibility of changing one's state by interaction with other situations.

It is quite natural to generalize to the prediction of disease evolution due to the interaction between individuals (or cells) with pathological state classified by u. The difficulty in dealing with this type of model is certainly the identification of the terms η and ψ. On the other hand, if detailed measures can be organized, then the identification of the probability distributions can be obtained as indicated in Section 4.8.

The model (4.10.14) was originally proposed by Jager and Segel [JaS] in order to describe the evolution of a factor, called "dominance," in a population of anonimous interacting organisms. Therefore the model can be regarded as the natural evolution of the models of population dynamics presented in Chapters 1 and 2. Various studies are now developed in order to generalize the model to predict the development of pathological states due to molecular interactions.

Specific examples for the modelling of the terms η and ψ will be reported in the last section of this chapter. In general, we are interested in solving Eq.(4.10.14) with initial conditions $f(0;u) = f_0(u)$. The solution $f(t;u)$ allows us to compute the probability of finding an individual in the state $[u_1, u_2]$ by

$$\int_{u_1}^{u_2} f(t;u)\, du\,, \qquad (4.10.15)$$

where $0 \le u_1 \le u_2 \le 1$. □

The two examples we have just seen have to be regarded as simple examples among several examples that can be constructed in similar ways. We will close at this point the presentation and the sampling of this model. We simply recall that the generalization of this type of model leads to the phenomenological nonequilibrium kinetic theory and, in particular, to the celebrated Boltzmann equation. In this case the dependent variable is a distribution over time t, velocity $\mathbf{v}$, and space $\mathbf{x}$.

The interested reader is referred to Truesdell and Muncaster [TaM] for a more detailed introduction to the mathematical aspect of kinetic theory. This reference is given to stimulate the reader towards new research perspectives in this very specialized field.

4.11 Applications

This section deals with two applications: The first is related to the solution of an inverse problem for the nonlinear heat equation, the second is related to the kinetic model described in Example 4.8.

The style of this section will be more concise than that of the corresponding sections of Chapters 2 and 3. In fact the applications developed in the preceding chapters were completed with a detailed description of the scientific programs produced to obtain quantitative results. This chapter will give only suitable indications to address the reader towards the compilation of scientific programs.

In fact, the conceptual core of the scientific computation remains that proposed in Chapters 2 and 3, whereas the analysis developed in this section can be regarded as a technical development, which will be addressed to a reader, who will now be able to tackle the problem with confidence and experience.

4.11.1 An inverse problem for the heat diffusion model

Consider the nonlinear heat equation with a localized source term in one space dimension and assume that the heat diffusivity coefficient can be modelled in the following way

$$\kappa = \kappa(u) = k + \frac{\varepsilon}{u} \cdot \tag{4.11.1}$$

The evolution equation can be written as follows

$$\frac{\partial u}{\partial t} = \left(k + \frac{\varepsilon}{u}\right) \frac{\partial^2 u}{\partial x^2} - \frac{\varepsilon}{u^2} \left(\frac{\partial u}{\partial x}\right)^2 + s(t)\, \delta(x - x_s)\,, \tag{4.11.2}$$

where $t \in [0, T]$ and $x \in [0, 1]$.

In detail, we consider the mathematical problem defined by the following initial and boundary conditions

$$\begin{cases} u(0,x) = 1\,, & \forall x \in [0,1] \\ u(t,0) = 1\,, & \forall t \in [0,T] \\ u(t,1) = 1\,, & \forall t \in [0,T] \\ u^*(t) = u(t,x^*) = 1 + d\,\sin(2\pi m t)\,, & \forall t \in [0,T]\,, \end{cases} \tag{4.11.3}$$

where d and ν are positive constants.

The mathematical problem consists in computing the unspecified source term

$$s = s(t)\,, \tag{4.11.4}$$

localized, for sure, in $x = x_s$.

The solution to the problem described above can be organized by the method proposed in Section 4.3. In particular, the domain $\mathcal{D} = [0,1]$ can be decomposed into three domains

$$\mathcal{D}_1 = [0, x^*]\,, \qquad \mathcal{D}_2 = [x^*, x_s]\,, \qquad \mathcal{D}_3 = [x_s, 1]\,. \tag{4.11.5}$$

The solution of the initial-boundary-value problem in $\mathcal{D}_1$ is the classical one we saw in Chapter 3, when two-point boundary conditions are given.

Quantitative results can be obtained by several techniques, for instance, the collocation interpolation methods described in Chapter 3 and briefly summarized in Section 4.3. Finite difference schemes for parabolic equations can also be succesfully applied.

Scientific programs such as those of the last section of Chapter 3 can be applied technically to obtain quantitative results.

As we saw in Section 4.3, the solution in $\mathcal{D}_1$ provides the heat flux in x^* as a function of time

$$q^* = -\left(k + \frac{\varepsilon}{u(t,x^*)}\right)\frac{\partial u}{\partial x}(t,x^*)\,, \qquad \forall t \in [0,T]\,. \tag{4.11.6}$$

This condition, linked to the additional one $u = u(t,x^*)$, defines the initial conditions in the integration with respect to x for Eq.(4.11.2), which is written as

$$\begin{cases} \dfrac{\partial u}{\partial x} = -\dfrac{uq}{ku + \varepsilon} \\ \dfrac{\partial q}{\partial x} = -\dfrac{\partial u}{\partial t}\,. \end{cases} \tag{4.11.7}$$

For such an equation, the initial condition of the original problem acts as a boundary condition.

Once more, both collocation-interpolation methods or finite-difference schemes can be applied to obtain quantitative results. The problem in $\mathcal{D}_3$ is the same as that dealt with in $\mathcal{D}_1$. The boundary conditions are now $u(t, x_s)$ and $u(t, 1)$, where the first is obtained by the solution in $\mathcal{D}_2$ and the second is given by the statement of the original problem. The source term is then computed by Eq.(4.3.23).

A scientific program based on collocation-interpolation methods is reported in the last section of this chapter. The reader is referred to the presentation of the scientific programs of Chapter 3 for its interpretation and use. Examples of calculations of this type can be found in Preziosi and de Socio [PaD] and in Preziosi [PRE].

4.11.2 Solution of the "dominance" model

Consider now the mathematical model proposed by Jager and Segel [JaS] for the evolution of the dominance factor in a population of anonymous individuals. The structure of such a model is somewhat different from the classical models presented in Chapters 2 and 3. In fact, such a model involves an integral term related to the probabilistic structure of the model itself. It follows that quantitative results can be obtained only by means of suitable generalizations of the methods presented in Chapters 2 and 3.

This section will show the result of some calculations related to the initial-value problem. The related scientific program will also be reported and offered to the interpretation and use of the reader. Detailed calculations will be developed in the case of $\eta = 1$, namely, when all elements interact with the same rate for all values of u and v and for a specific model of the term ψ.

With this in mind, we rewrite the initial-value problem we are talking about as follows

$$\frac{\partial f}{\partial t}(t; u) + f(t; u) = \int_0^1 \int_0^1 \psi(v, w; u) f(t; v) f(t; w)\, dv\, dw\,, \qquad (4.11.8a)$$

with initial condition

$$f(0; u) = f_0(u)\,, \qquad \forall u \in [0, 1]\,. \qquad (4.11.8b)$$

The existence theory for such a problem is developed in Brzezniak and Preziosi [BaP], where it is proved that the solution exists globally in time under quite general assumptions on f_0, η, and ψ.

Calculations will be developed for the following specific model of ψ

$$\psi(v, w; u) = \frac{1}{A(v, w)} \exp\left\{ -\frac{[u - \beta(v, w)]^2}{\sigma(v, w)} \right\}, \tag{4.11.9}$$

where β defines the most probable dominance of the subject characterized by v after the encounter with the subject characterized by w. The term $A(v, w)$ is a normalizing factor related to the fact that the integral of ψ over u must be equal to one. Moreover σ determines how predictable is the outcome of the interaction. Of course, referring to the model defined in Eq.(4.10.9), several choices are possible with reference to β and σ. As far as β is concerned, calculations will be developed in the fully altruistic (maybe idealistic) behavior defined by the model

$$\beta = \frac{v + w}{2}\,. \tag{4.11.10}$$

The model is such that two individuals, after the pair interacts, equally distribute their dominance. As far as the term σ is concerned, specific calculations will be developed for $\sigma = 0.1 - 0.09|v - w|$. That is, the nearer v and w are, the more uncertain the outcome of the encounter will be.

The technical solution to the initial-value problem can be organized exploiting some ideas already used in the solution of problems referred to ordinary differential and partial differential equations. The starting point consists in the discretization of the domain of the variable u

$$u_0 = 0 < \cdots < u_i < \cdots < u_n = 1 \tag{4.11.11}$$

and interpolating $f(t; u)$ by means of Lagrange-type polynomials

$$f(t; u) \cong f^n(t; u) = \sum_{i=0}^{n} L_i(u) f_i(t)\,, \tag{4.11.12}$$

where $f_i = f(t; u_i)$ and L_i denote the Lagrange polynomials. Similar approximation can be obtained by piecewise constant functions

$$f(t; u) \cong f^n(t; u) \sum_{i=0}^{n} \chi_i(u) f_i(t)\,, \tag{4.11.13}$$

where χ_i is the characteristic function equal to one, for $u \in [u_i, u_{i+1}]$, and equal to zero otherwise.

The scientific program that can be used to obtain quantitative results is reported in Section 4.13. Some qualitative results are visualized in Fig. 4.11.1, which shows the distribution of f versus u at fixed values of t. Additional calculations can be obtained using the same program.

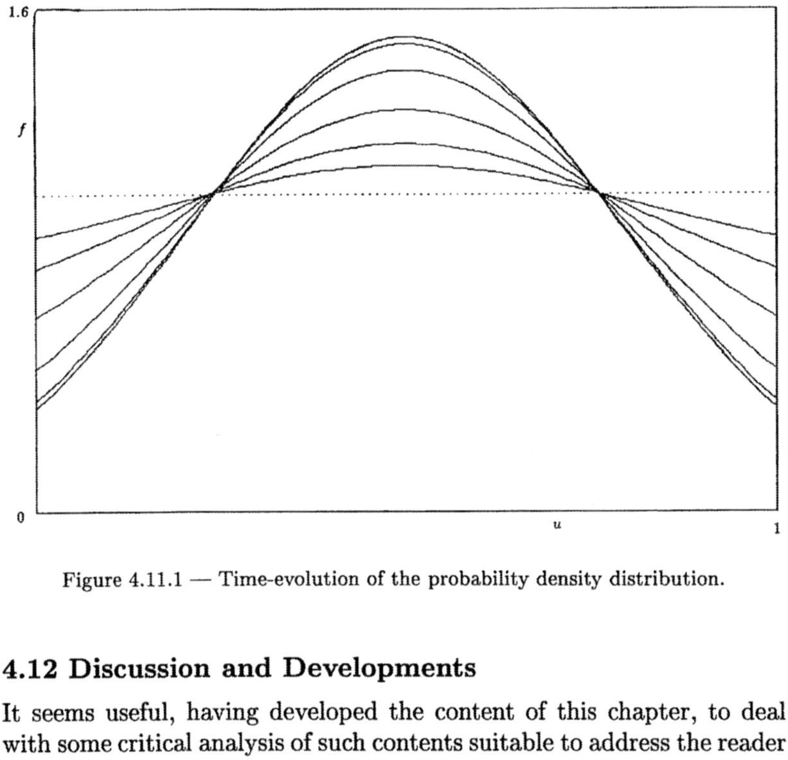

Figure 4.11.1 — Time-evolution of the probability density distribution.

4.12 Discussion and Developments

It seems useful, having developed the content of this chapter, to deal with some critical analysis of such contents suitable to address the reader towards further studies in the field of inverse and stochastic aspects in mathematical modelling.

We remind the reader that this chapter is not exhaustive. In fact, it appears to be only a bridge between the classical analysis of discrete and continuous models, presented in Chapters 2 and 3, and the inverse and stochastic type models or problems.

The following comments are offered to the attention of the reader:

- The mathematical formulation of inverse type problems for both discrete and continuous models needs to be directly related to real physical situations.
- The variety of problems that can be encountered is so vast that a complete classification cannot be exhaustively organized as we have done with reference to direct problems.

- The link between stochastic modelling and the solution of inverse problems is strong. In fact, quite often (or even almost always), the information one can obtain from the real system is affected by noise. It follows that the solution of such problems will be a stochastic process in the case of discrete systems and a time-dependent stochastic field in the case of continuous models.

- A first "rough" classification of inverse problems was proposed in this chapter: Problems that can be split, by means of suitable decomposition of domains techniques, into several direct-type problems; and genuine inverse problems that can be solved by suitable minimization techniques. This classification can be applied to a large number of problems, although some of them, in the variety of problems mentioned in the first item, may not belong to either of these two classes of problems.

- Domain decomposition methods were dealt with quite accurately in the case of one space dimension. Inverse problems in more than one space dimension cannot be regarded as a simple technical development.

These comments further indicate that this chapter has to be regarded as an introductory one. A large number of problems are left to future studies.

4.13 Scientific Programs

This section reports two scientific programs that can be used to deal with the two problems reported in Section 4.11. As we have seen, the first problem refers to the nonlinear heat equation and the second one to the kinetic dominance problem.

The first program is obtained by linking together several pieces presented in Chapter 3. The second one refers to a new problem and allows us to solve integrodifferential equations.

The reader can technically modify these programs in order to adapt them to the solution of the several problems in this chapter. In particular, he is encouraged to solve Problem 4.2 and the backward heat equation.

4.13.1 INVERSE: A program solving an inverse problem

Referring to the flow chart in Table 4.13.1, we can now describe the main steps of this program. The first one is to define the discretization of the space-like variable in the three domains. We note that, because of the switching of the role of space and time in $\mathcal{D}_2$, there the space-like variable is actually time.

In this program we have chosen the same discretization in all domains (lines 40–65). One may choose (line 45) between the Chebyshev and the equispaced distribution of nodes. The former, as explained in Appendix 2, has the advantage of being more accurate especially in computing the derivatives (which are physically related to the heat flux) at the boundary points, which is essential to transfer proper information from $\mathcal{D}_1$ to $\mathcal{D}_2$ and in computing the source term. Instead, by using the latter, a larger time step can be chosen in the integration (remember that, as explained in Chapter 3, $\Delta t <$ Const.Δx^2 and in the Chebyshev distribution Δx is very small near the boundaries).

Furthermore, the value of the temperature and of the heat flux at the source is known (after the integration in $\mathcal{D}_2$) in the nodes t_i, $i = 1, \ldots, n$, while in order to integrate the partial differential equation in $\mathcal{D}_3$, one needs $u(t, x^*)$ essentially for any t. Similarly, in order to compute the source term one needs the values of the derivative of u in x^* for any t. These values are obtained by interpolating the data obtained by integrating in $\mathcal{D}_2$. As before, we prefer the Chebyshev distribution to be more accurate in computing this interpolant, but, as already said, the need of a much smaller time step is a strong drawback, which could, however, be eliminated by using an implicit method (see Chapter 3).

After that, lines 70–510 compute the derivative matrices as explained in Chapter 3.

After some input lines (520–560), lines 570–695 set the value of some useful and recurring quantities such as the full-space discretization `zz` of the interval $[0, 1]$ (lines 660–695).

After that, the integration in $\mathcal{D}_1$ is started as in the program COLLOCAT explained in Chapter 3 (lines 700–860).

The temporal integration uses a Runge–Kutta routine as does the program RK4 in Chapter 2, but for the fact that, according to the domain, the number of equations, of internal nodes (i.e., nonboundary nodes) and the right-hand side of the equation are chosen.

When time overcomes a value of the discretization in $\mathcal{D}_2$ (line 880), the solution and the heat flux in t_i are obtained (and stored) by linear interpolation (lines 890–920) using the values `upr(j)` and `qpr` previously stored in lines 840–846. These data are then printed (lines 940–970). If the final time is passed by (line 995) the program goes to the integration in $\mathcal{D}_2$ where the role of space and time is reversed.

After setting the initial condition for problem in $\mathcal{D}_2$ (lines 1070–1110), the program goes to the integration of the system of partial differential equations, which is taken care of by the same Runge–Kutta routine of lines 2000–2299. The values of temperature and heat flux are

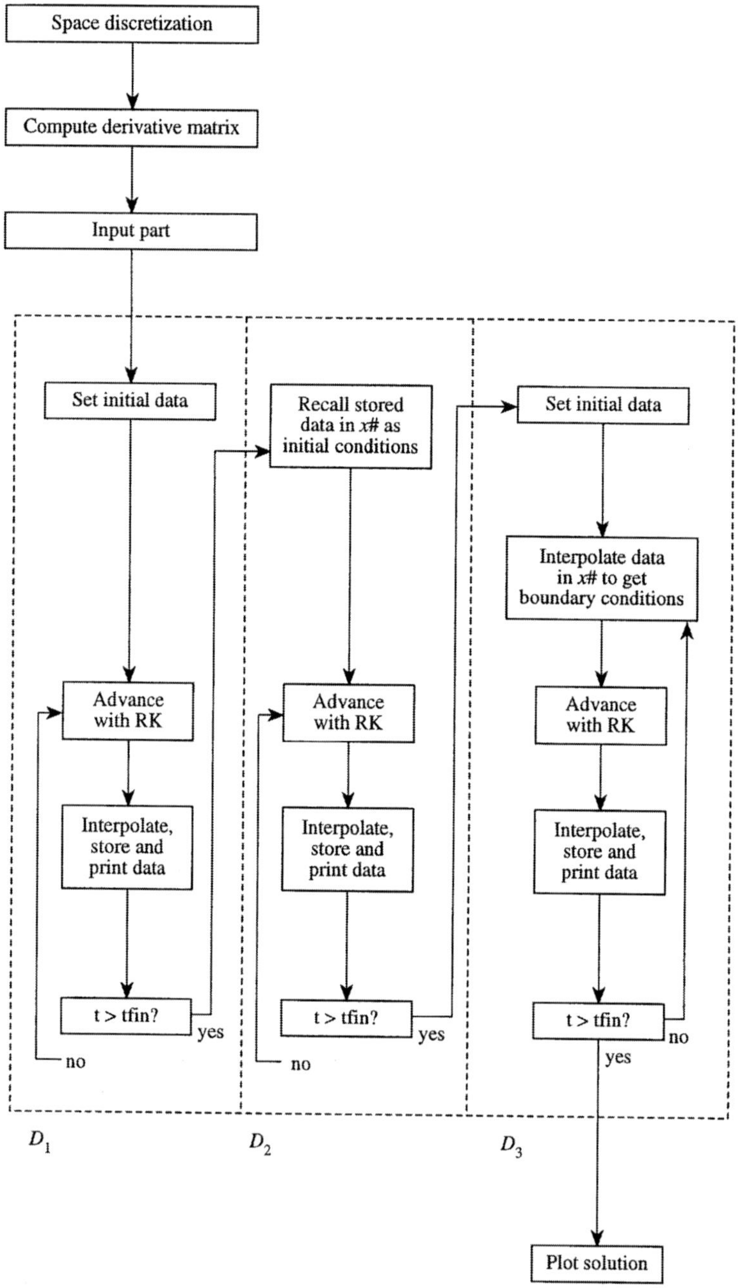

Table 4.13.1 — Flow chart of INVERSE.

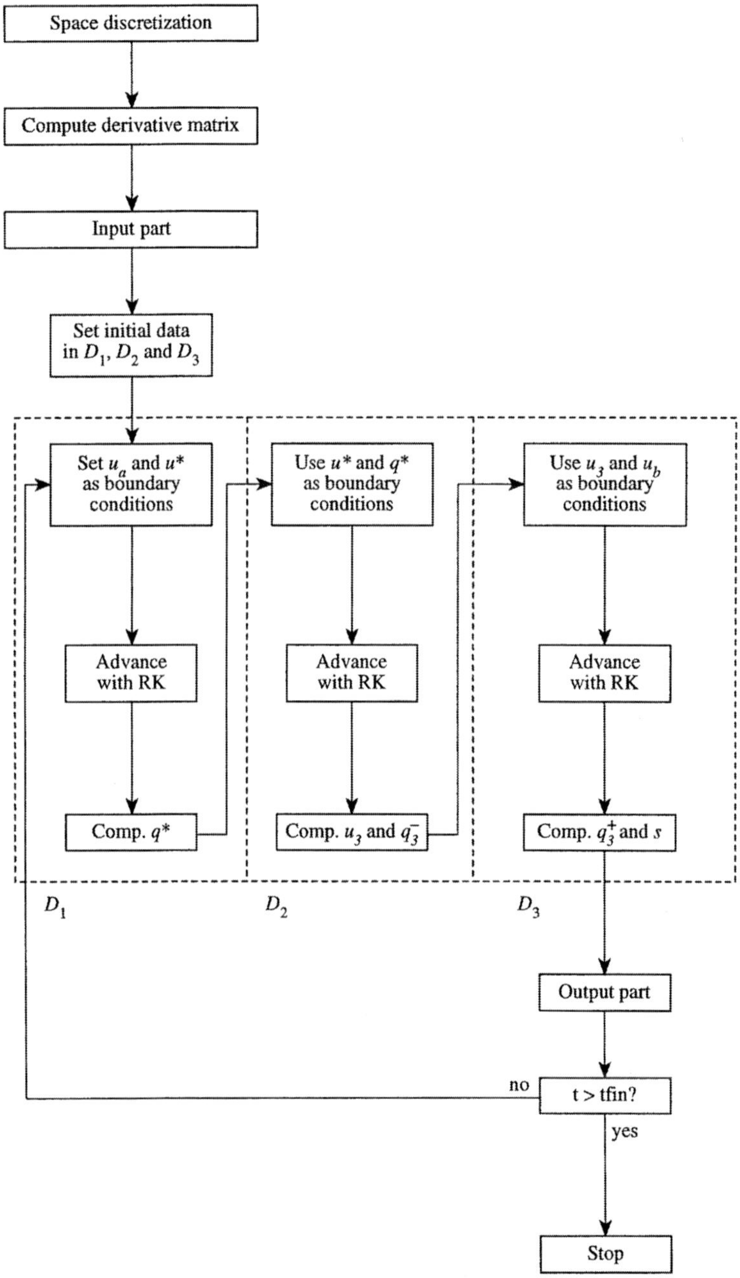

Table 4.13.2 — Alternative flow chart for an inverse problem.

stored and printed every now and then (lines 1160-1260) according to the value dx2 set in line 570. When the source location is reached (line 1260), the program passes to the integration of the initial-boundary-value problem in $\mathcal{D}_3$ (lines 1300-1440), which is done using a structure similar to that used for the integration in $\mathcal{D}_1$. There is, however, a major difference. In fact, as already stated, in order to obtain the boundary condition at the source, one has to know its temperature at any time, which is due in lines 3410-3580 by interpolating the data obtained at the end of the integration in $\mathcal{D}_2$.

When the final time is reached (line 1450), the results are plotted using the subroutine starting at line 5000. At the end of the plotting, part the program asks if more plots are desired. If not, it asks if another initial-value problem needs to be solved, otherwise it stops.

The right-hand side of the partial differential equations in $\mathcal{D}_1$, $\mathcal{D}_2$, and in $\mathcal{D}_3$ are, respectively, given by the subroutines 3000-3190, 3200-3390, and 3400-3790.

As an alternative the reader may try to modify the program avoiding the time-space switching in $\mathcal{D}_2$, as shown in Table 4.13.2. This choice allows a parallelization of the code in the sense that the program can be written so that the solution is found by advancing almost contemporarily in all domains.

INVERSE.BAS

```
1 REM ***********************************************************
2 REM *                Solution of an inverse problem           *
3 REM ***********************************************************
10 DIM u(11), urk(11), f(11), uu(11, 33), upr(11)
12 DIM q(11), qrk(11), qf(11), z(11), zz(31)
13 DIM k1(11), k2(11), k3(11), k4(11), iplot(31)
14 DIM qk1(11), qk2(11), qk3(11), qk4(11)
15 DIM d(11, 11), d2(11, 11), du(11), ddu(11), pil(11)
18 nx = 11
20 nx1 = nx - 1
25 npoint = 101
30 pi = 3.141593
35 CLS 0
36 PRINT "*******************************************************"
37 PRINT "*              Solution of an inverse problem         *"
38 PRINT "*******************************************************"
40 PRINT "When more than one quantity is requested, separate
   with a comma"
42 PRINT "Values in square brackets are the suggested values
```

```
   (not default)"
45 INPUT "Chebyshev distribution or equispaced nodes (c/e) [e]";
   distr$
50 FOR i = 1 TO nx
55 IF distr$ = "c" THEN z(i)=(1 - COS((i - 1) * pi/(nx - 1)))/2
60 IF distr$ = "e" THEN z(i) = (i - 1) / (nx - 1)
65 NEXT i
100 REM *******************************************************
101 REM *                Computing derivative matrix          *
102 REM *******************************************************
110 REM *******************************************************
111 REM *               Computing pi1=PI' in (A.2.10)         *
112 REM *******************************************************
120 FOR i = 1 TO nx
130 pi1(i) = 1
140 FOR k = 1 TO nx
150 IF k <> i THEN pi1(i) = pi1(i) * (z(i) - z(k))
160 NEXT k
170 NEXT i
180 FOR i = 1 TO nx
190 FOR j = 1 TO nx
200 IF i = j THEN GOTO 270
210 REM *******************************************************
211 REM *           Computing the nondiagonal elements        *
212 REM *******************************************************
220 d(j, i) = 1
230 FOR k = 1 TO nx
240 IF k <> i AND k <> j THEN d(j, i) = d(j, i) * (z(j) - z(k))
250 NEXT k
260 d(j, i) = d(j, i) / pi1(i)
270 NEXT j
280 REM *******************************************************
281 REM *              Computing the diagonal elements        *
282 REM *******************************************************
290 d(i, i) = 0
300 FOR k = 1 TO nx
310 IF k <> i THEN d(i, i) = d(i, i) + 1 / (z(i) - z(k))
320 NEXT k
330 NEXT i
335 REM *******************************************************
336 REM *               Computing the matrix for the          *
337 REM *                    second derivatives               *
338 REM *******************************************************
340 FOR i = 1 TO nx
350 FOR j = 1 TO nx
360 d2(i, j) = 0
370 FOR k = 1 TO nx
380 d2(i, j) = d2(i, j) + d(i, k) * d(k, j)
```

```
390 NEXT k
400 NEXT j
410 NEXT i
420 REM ***********************************************************
421 REM *          End of computation derivative matrices         *
422 REM ***********************************************************
520 REM ***********************************************************
521 REM *                        Input lines                      *
522 REM ***********************************************************
530 INPUT "Amplitude and period of measurement [.1,1]"; delu, nper
540 INPUT "Position of measurement and source/sink [.5,.6]";
   xm, xs
550 IF distr$ = "e" THEN INPUT "Final time and time steps in D1,
   D2, and D3 [2,.005,.02,.004]"; tfin, dt1, dt2, dt3
555 IF distr$ = "c" THEN INPUT "Final time and time steps in D1,
   D2, and D3 [2,.001,.005,.0005]"; tfin, dt1, dt2, dt3
560 INPUT "Heat conduction coefficients k+e/u [.2,.1]"; kk, ee
570 dx2 = .02
580 u0 = 1
590 xL = 1
610 s1 = 1 / xm
620 s2 = 1 / tfin
630 s3 = 1 / (xL - xs)
640 ns = nx + (xs - xm) / dx2
650 n1 = ns + nx1
655 nr = n1 + 1
660 FOR j = 1 TO n1
670 zz(j) = xm + (j - nx) * dx2
680 IF j <= nx THEN zz(j) = z(j) * xm
690 IF j > ns THEN zz(j) = xs + z(j - ns + 1) * (xL - xs)
695 NEXT j
700 REM ***********************************************************
701 REM *                    Integration in D1                    *
702 REM ***********************************************************
710 domain = 1
720 flag = 1
730 h = dt1
740 h2 = h / 2
750 h6 = h / 6
760 REM ***********************************************************
761 REM *                    Set initial data                     *
762 REM ***********************************************************
770 FOR j = 1 TO nx
780 u(j) = u0
785 upr(j) = u(j)
790 NEXT j
795 qq = 0
797 qpr = qq
```

```
800 t = 0
810 GOTO 880
820 REM ***********************************************************
821 REM *          The integration has to be performed on          *
822 REM *          the nodes 2,...,nx-1 since the values           *
823 REM *          u(1) and u(nx) are provided by the b.c.         *
824 REM ***********************************************************
830 i1 = 2
835 ind = nx1
840 FOR j = 1 TO nx
842 upr(j) = u(j)
844 NEXT j
846 qpr = qq
850 REM ***********************************************************
851 REM *        Subroutine 2000 is the Runge-Kutta routine        *
852 REM ***********************************************************
860 GOSUB 2000
870 REM ***********************************************************
871 REM *                       Output loop                        *
872 REM ***********************************************************
880 IF t < z(flag) * tfin THEN GOTO 820
890 FOR j = 1 TO nx
900 uu(flag, j)=u(j) - (u(j) - upr(j))*(t - z(flag) * tfin)/h
910 NEXT j
920 q(flag) = qq - (qq - qpr) * (t - z(flag) * tfin) / h
930 q(1) = 0
940 PRINT "t="; z(flag) * tfin, "q*="; q(flag)
950 FOR j = 1 TO nx
960 PRINT uu(flag, j);
970 NEXT j
980 PRINT
990 flag = flag + 1
995 IF flag <= nx THEN GOTO 820
1000 REM **********************************************************
1001 REM *                   Integration in D2                     *
1002 REM *               here t and x are switched                 *
1003 REM **********************************************************
1010 domain = 2
1020 flag = 1
1040 h = dt2
1050 h2 = h / 2
1060 h6 = h / 6
1070 REM **********************************************************
1071 REM *                   Set initial data:                     *
1072 REM *             the measured temperature and                *
1073 REM *             the computed heat flux in x*                *
1074 REM **********************************************************
1080 FOR j = 1 TO nx
```

```
1090 u(j) = uu(j, nx)
1100 REM q(j)=q(j)
1110 NEXT j
1120 t = xm
1121 REM ***********************************************************
1122 REM *        The integration has to be performed on           *
1123 REM *        the nodes 2,...,nx for u since the value          *
1124 REM *       u(1) is provided by the initial condition,         *
1125 REM *        which is now used as boundary condition,          *
1126 REM *      and on all the nodes for q.  But since q(1)         *
1127 REM *         can be found analitically from the i.c.,         *
1128 REM *       we still restrict integration to 2,...,nx          *
1129 REM ***********************************************************
1130 i1 = 2
1140 ind = nx
1144 REM ***********************************************************
1145 REM *      Subroutine 2000 is the Runge-Kutta routine          *
1146 REM ***********************************************************
1150 GOSUB 2000
1154 REM ***********************************************************
1155 REM *                      Output loop                         *
1156 REM ***********************************************************
1160 IF t < xm + dx2 * flag - h2 THEN GOTO 1150
1170 nf = nx + flag
1180 FOR j = 1 TO nx
1190 uu(j, nf) = u(j)
1200 NEXT j
1210 PRINT "x="; t
1220 FOR j = 1 TO nx
1230 PRINT "t="; z(j) * tfin, "u="; u(j), "q="; q(j)
1240 NEXT j
1250 flag = flag + 1
1260 IF t <= xs - h2 THEN GOTO 1150
1300 REM ***********************************************************
1301 REM *                   Integration in D3                      *
1302 REM ***********************************************************
1310 domain = 3
1320 flag = 1
1330 h = dt3
1340 h2 = h / 2
1350 h6 = h / 6
1360 REM ***********************************************************
1361 REM *                   Set initial data                       *
1362 REM ***********************************************************
1370 FOR j = 1 TO nx
1375 u(j) = u0
1380 upr(j) = u(j)
1385 NEXT j
```

```
1390 rs = 0
1395 rpr = rs
1400 t = 0
1410 GOTO 1450
1414 REM ***********************************************************
1415 REM *        The integration has to be performed on          *
1416 REM *        the nodes 2,...,nx-1, since the values          *
1417 REM *        u(1) and u(nx) are provided by the b.c.         *
1418 REM ***********************************************************
1420 i1 = 2
1422 ind = nx1
1424 FOR j = 1 TO nx
1426 upr(j) = u(j)
1428 NEXT j
1430 rpr = rs
1434 REM ***********************************************************
1435 REM *      Subroutine 2000 is the Runge-Kutta routine        *
1436 REM ***********************************************************
1440 GOSUB 2000
1444 REM ***********************************************************
1445 REM *                     Output loop                        *
1446 REM ***********************************************************
1450 IF t < z(flag) * tfin - h2 THEN GOTO 1420
1460 FOR j = 2 TO nx
1470 uu(flag, ns + j - 1) = u(j) - (u(j) - upr(j)) *
   (t - z(flag) * tfin) / h
1480 NEXT j
1484 REM ***********************************************************
1485 REM *           Source strength and temperature              *
1486 REM ***********************************************************
1500 uu(flag, nr) = rs - (rs - rpr) * (t - z(flag) * tfin) / h
1510 uu(1, nr) = 0
1520 PRINT "t="; z(flag) * tfin, "r="; uu(flag, nr)
1525 FOR j = 1 TO nx
1530 PRINT uu(flag, ns + j - 1);
1540 NEXT j
1550 PRINT
1560 flag = flag + 1
1570 IF flag <= nx THEN GOTO 1420
1950 GOSUB 5000
1960 INPUT "Do you want to input different data?  (y/n)"; scr$
1970 IF scr$ = "y" THEN GOTO 520
1980 IF scr$ <> "y" AND scr$ <> "n" THEN GOTO 1960
1990 END
2000 REM ***********************************************************
2001 REM *          Fourth-order Runge-Kutta subroutine           *
2002 REM ***********************************************************
2010 FOR i = i1 TO ind
```

```
2020 urk(i) = u(i)
2030 IF domain = 2 THEN qrk(i) = q(i)
2040 NEXT i
2050 ON domain GOSUB 3000, 3200, 3400
2060 t = t + h2
2070 FOR i = i1 TO ind
2080 k1(i) = f(i):  urk(i) = u(i) + h2 * k1(i)
2090 IF domain = 2 THEN qk1(i) = qf(i):  qrk(i) = q(i) + h2 *
   qk1(i)
2100 NEXT i
2110 ON domain GOSUB 3000, 3200, 3400
2120 FOR i = i1 TO ind
2130 k2(i) = f(i):  urk(i) = u(i) + h2 * k2(i)
2140 IF domain = 2 THEN qk2(i) = qf(i):  qrk(i) = q(i) + h2 *
   qk2(i)
2150 NEXT i
2160 ON domain GOSUB 3000, 3200, 3400
2170 t = t + h2
2180 FOR i = i1 TO ind
2190 k3(i) = f(i):  urk(i) = u(i) + h * k3(i)
2200 IF domain = 2 THEN qk3(i) = qf(i):  qrk(i) = q(i) + h *
   qk3(i)
2210 NEXT i
2220 ON domain GOSUB 3000, 3200, 3400
2230 FOR i = i1 TO ind
2240 k4(i) = f(i)
2250 IF domain = 2 THEN qk4(i) = qf(i)
2260 u(i) = u(i) + h6 * (k1(i) + 2 * (k2(i) + k3(i)) + k4(i))
2270 IF domain = 2 THEN q(i) = q(i) + h6 * (qk1(i) + 2 *
   (qk2(i) + qk3(i)) + qk4(i))
2280 NEXT i
2290 REM ***********************************************************
2291 REM *                 End of Runge-Kutta routine              *
2292 REM ***********************************************************
2295 RETURN
2299 END
3000 REM ***********************************************************
3001 REM *             Subroutine for integration in D1            *
3002 REM ***********************************************************
3010 REM ***********************************************************
3011 REM *                   Boundary conditions                   *
3012 REM ***********************************************************
3020 u(1) = u0
3030 u(nx) = u0 + delu * SIN(2 * pi * nper * t)
3040 urk(1) = u(1)
3050 urk(nx) = u(nx)
3060 REM ***********************************************************
3061 REM *                Derivative and r.h.s. in D1              *
```

```
3062 REM *********************************************************
3070 FOR j = i1 TO nx
3080 du(j) = 0
3090 ddu(j) = 0
3100 FOR k = 1 TO nx
3110 du(j) = du(j) + d(j, k) * urk(k)
3120 ddu(j) = ddu(j) + d2(j, k) * urk(k)
3130 NEXT k
3140 du(j) = s1 * du(j)
3150 ddu(j) = s1 ^2 * ddu(j)
3160 f(j) = ((kk + ee/urk(j))*ddu(j) - ee/urk(j) ^2*du(j) ^2)
3170 NEXT j
3180 qq = - (kk + ee / u(nx)) * du(nx)
3190 RETURN
3200 REM *********************************************************
3201 REM *          Subroutine for integration in D2            *
3210 REM *********************************************************
3220 REM *********************************************************
3230 REM *                   Boundary conditions                 *
3240 REM *********************************************************
3250 u(1) = u0
3260 urk(1) = u0
3270 q(1) = 0
3280 qrk(1) = 0
3290 REM *********************************************************
3291 REM *             Derivative and r.h.s. in D2               *
3292 REM *********************************************************
3300 FOR j = i1 TO ind
3310 du(j) = 0
3320 FOR k = 1 TO nx
3330 du(j) = du(j) + d(j, k) * urk(k)
3340 NEXT k
3350 du(j) = s2 * du(j)
3360 qf(j) = - du(j)
3370 f(j) = qrk(j) / (kk + ee / urk(j))
3380 NEXT j
3390 RETURN
3400 REM *********************************************************
3401 REM *          Subroutine for integration in D3            *
3402 REM *********************************************************
3410 REM *********************************************************
3411 REM *                   Boundary conditions                 *
3412 REM *********************************************************
3420 zz = t / tfin
3430 REM *********************************************************
3431 REM *                Reinterpolation to compute             *
3432 REM *                 left boundary conditions              *
3433 REM *********************************************************
```

```
3440 FOR i = 1 TO nx
3450 IF zz = z(i) THEN u(1) = uu(i, ns) :  phi2 = q(i) :
   GOTO 3600
3460 NEXT i
3470 sumu = 0:  sumq = 0
3480 FOR i = 1 TO nx
3490 piz = pi1(i) * (zz - z(i))
3500 sumu = sumu + uu(i, ns) / piz
3510 sumq = sumq + q(i) / piz
3520 NEXT i
3530 pL = 1
3540 FOR i = 1 TO nx:  pL = pL * (zz - z(i)):  NEXT i
3550 u(1) = sumu * pL
3560 phi2 = sumq + pL
3600 u(nx) = u0
3610 urk(1) = u(1)
3620 urk(nx) = u(nx)
3630 REM *******************************************************
3640 REM *              Derivative and r.h.s. in D3            *
3650 REM *******************************************************
3660 FOR j = 1 TO ind
3670 du(j) = 0
3680 ddu(j) = 0
3690 FOR k = 1 TO nx
3700 du(j) = du(j) + d(j, k) * urk(k)
3710 ddu(j) = ddu(j) + d2(j, k) * urk(k)
3720 NEXT k
3730 du(j) = s3 * du(j)
3740 ddu(j) = s3 ^2 * ddu(j)
3750 f(j) = ((kk + ee/urk(j))*ddu(j) - ee/urk(j) ^2*du(j) ^2)
3760 NEXT j
3764 REM *******************************************************
3755 REM *                      Source term                    *
3766 REM *******************************************************
3770 rs = - (kk + ee / u(1)) * du(1) - phi2
3780 RETURN
3790 END
5000 REM *******************************************************
5001 REM *                  Graphic subroutine                 *
5002 REM *******************************************************
5010 INPUT "Plot of source strength [0] or of temperature [1]";
   iplot
5020 IF iplot = 0 THEN plot$ = "t":  nplot = 1:  iplot(1) = nr:
   GOTO 5160
5030 INPUT "Plot versus space [x] or time [t]"; plot$
5040 IF plot$ <> "x" AND plot$ <> "t" THEN GOTO 5030
5050 PRINT "How many plots (<=";
5060 IF plot$ = "x" THEN PRINT nx; ")";
```

```
5070 IF plot$ = "t" THEN PRINT n1; ")";
5080 INPUT nplot
5090 IF plot$ = "x" THEN PRINT "At what time?  (1,...,"; nx; ")"
5100 IF plot$ = "t" THEN PRINT "Where?  (1,...,"; n1; ")", "x*=";
   nx, "xs="; ns
5110 FOR ip = 1 TO nplot
5120 INPUT iplot(ip)
5130 IF plot$ = "x" AND iplot(ip) > nx THEN PRINT
   "Impossible.  Input again":  GOTO 5120
5140 IF plot$ = "t" AND iplot(ip) > n1 THEN PRINT
   "Impossible.  Input again":  GOTO 5120
5150 NEXT ip
5160 CLS 0:  KEY OFF: SCREEN 2
5170 VIEW (30, 10)-(600, 160), , 1
5180 IF plot$ = "t" THEN GOTO 5400
5200 pmin = 0:  pmax = 1
5210 ymin = 1000:  ymax = -ymin
5220 FOR ip = 1 TO nplot
5230 FOR j = 1 TO n1
5240 IF uu(iplot(ip), j) > ymax THEN ymax = uu(iplot(ip), j)
5250 IF uu(iplot(ip), j) < ymin THEN ymin = uu(iplot(ip), j)
5260 NEXT j
5270 NEXT ip
5275 ymin = INT(ymin * 100)/100:  ymax = (INT(ymax * 100) + 1)/100
5280 WINDOW (pmin, ymin)-(pmax, ymax)
5290 LOCATE 2, 2:  PRINT ymax:  LOCATE 21, 2:  PRINT ymin
   :  LOCATE 9, 2:  PRINT "u"
5300 LOCATE 22, 4:  PRINT pmin:  LOCATE 22, 75:  PRINT pmax
5310 xxd = INT(4 + (75 - 4) * xm / pmax)
5320 xxs = INT(4 + (75 - 4) * xs / pmax)
5330 LOCATE 22, xxd:  PRINT "x*":  LOCATE 22, xxs:  PRINT "xs"
5332 FOR ii = 0 TO 20
5334 PSET (zz(nx), ymin + ii * (ymax - ymin) / 20)
5336 PSET (zz(ns), ymin + ii * (ymax - ymin) / 20)
5338 NEXT ii
5340 FOR ip = 1 TO nplot
5350 PSET (0, uu(iplot(ip), 1))
5360 FOR j = 1 TO n1
5370 LINE -(zz(j), uu(iplot(ip), j))
5380 NEXT j
5390 NEXT ip
5395 GOTO 5960
5400 pmin = 0:  pmax = tfin
5410 ymin = 1000:  ymax = -ymin
5420 FOR ip = 1 TO nplot
5430 FOR j = 1 TO nx
5440 IF uu(j, iplot(ip)) > ymax THEN ymax = uu(j, iplot(ip))
5450 IF uu(j, iplot(ip)) < ymin THEN ymin = uu(j, iplot(ip))
```

```
5460 NEXT j
5470 NEXT ip
5475 ymin = (INT(ymin * 50) - 2) / 50:  ymax =
   (INT(ymax * 50) + 2) / 50
5480 WINDOW (pmin, ymin)-(pmax, ymax)
5490 LOCATE 2, 2:  PRINT ymax:  LOCATE 21, 2:  PRINT ymin
5492 LOCATE 9, 2:  IF iplot(1) = nr THEN PRINT "r" ELSE PRINT "u"
5494 LOCATE 22, 4:  PRINT pmin:  LOCATE 22, 75:  PRINT pmax
5496 LOCATE 22, 50:  PRINT "t"
5500 FOR ip = 1 TO nplot
5510 PSET (0, uu(1, iplot(ip)))
5520 REM *********************************************************
5521 REM *          Reinterpolation of temporal evolution        *
5522 REM *            of source strength and temperature         *
5523 REM *********************************************************
5530 FOR j = 1 TO npoint
5540 t = tfin * (j - 1) / (npoint - 1)
5550 zzz = t / tfin
5560 FOR i = 1 TO nx
5580 IF zzz = z(i) THEN pL = uu(i, iplot(ip)):  GOTO 5690
5590 NEXT i
5600 sum = 0
5610 FOR i = 1 TO nx
5620 sum = sum + uu(i, iplot(ip)) / (pi1(i) * (zzz - z(i))
5630 NEXT i
5640 pL = 1
5650 FOR i = 1 TO nx:  pL = pL * (zzz - z(i)):  NEXT i
5660 pL = sum * pL
5690 LINE -(t, w)
5700 NEXT j
5720 NEXT ip
5960 INPUT "Do you want more graphs?  (y/n)"; scr$
5970 IF scr$ = "y" THEN GOTO 5000
5980 IF scr$ <> "y" AND scr$ <> "n" THEN GOTO 5960
5990 RETURN
5999 END
```

4.13.2 DEMOCRAT: A program for the population dynamics model

Consider now the initial-value problem for the model of population dynamics reported in Eq.(4.10.14). A specific example is that of Section 4.11.2, which refers to the particular case of $\eta = 1$.

We recall that the first step of the numerical simulation is the discretization of the variable u as reported in Eq.(4.11.11) so that the solution $f(t;u)$ to the initial-value problem can be approximated as shown in Eq.(4.11.12) or in Eq.(4.11.13). This interpolation can be useful for approximating the integrals in Eq.(4.10.14). In fact, substituting Eq.(4.11.12) into Eq.(4.10.14) yields

$$\int_0^1 \eta(u,v) f^n(t;v)\, dv = \sum_{j=0}^{n} I_j(u) f_j(t)\,, \tag{4.13.1}$$

where

$$I_j(u) = \int_0^1 \eta(u,v) L_j(v) dv\,. \tag{4.13.2}$$

Moreover

$$\int_0^1 \int_0^1 \eta(v,w)\psi(v,w;u) f^n(t;v) f^n(t;w)\, dv\, dw$$
$$= \sum_{j=0}^{n} \sum_{k=0}^{n} J_{jk}(u) f_j(t) f_k(t)\,, \tag{4.13.3}$$

where

$$J_{jk}(u) = \int_0^1 \int_0^1 \eta(v,w)\psi(v,w;u) L_j(v) L_k(w) dv\, dw\,. \tag{4.13.4}$$

REMARK 4.13.1 The advantage of this procedure relies on the fact that, since η and ψ are given, the terms I_j and J_{jk} can be computed, for a given discretization, once and for all. The values of I_j and J_{jk} in the collocation points will be denoted by the subscripts i

$$\begin{aligned} I_{ji} &= I_j(u_i)\,, \quad i,j = 1,\ldots,n \\ J_{jki} &= J_{jk}(u_i)\,, \quad i,j,k = 1,\ldots,n\,, \end{aligned} \tag{4.13.5}$$

so that from (4.13.2) and (4.13.4) one can explicitly write

$$I_{ji} = \int_0^1 \eta(u_i, v) L_j(v)\, dv \tag{4.13.6}$$

and

$$J_{jki} = \int_0^1 \left[\int_0^1 \eta(v,w)\, \psi(v,w;u_i)\, L_k(w)\, dw \right] L_j(v)\, dv\,. \qquad (4.13.7)$$

■

With all this in mind one obtains

$$\frac{df_i}{dt} = -f_i \sum_{j=0}^{n} I_{ji} f_j + \sum_{j=0}^{n} \sum_{k=0}^{n} J_{jki} f_j f_k\,, \quad i = 1, \ldots, n\,, \qquad (4.13.8)$$

which is a system of n ordinary differential equations to be dealt with by implementing the initial conditions

$$f_i(t=0) = f_0(u_i)\,. \qquad (4.13.9)$$

Solving (4.13.8–4.13.9) yields the time-evolution of the terms f_i, which can be used in the interpolation (4.11.12) in order to obtain $f^n(t;u)$.

In the simulation it is assumed that η is constant, which means that any individual has no prejudice in meeting individuals with different dominance. This slightly simplifies the model, since the terms I_{ji} need not be computed.

Keeping in mind the flow chart given in Table 4.13.3, we can now briefly describe the structure of the program.

The program denotes the variables u, v, and w by `x`, `y`, and `z`, respectively.

The discretization is given in lines `110-130`. For simplicity, equispaced nodes are used. In particular, different discretizations are used according to the role assumed by the parameter. In fact, `x(i)` are the values u_i in (4.11.11), while `xx`, `y`, and `z` represent finer discretizations used to compute, respectively, the normalization integral $A(v,w)$ in (4.11.9) and the v and w integrations in the double integral in Eq.(4.13.7). All these integrals are computed using Simpson's rule.

Lines `140-340` compute and store the values of the Lagrange polynomials in the integration nodes, that is

$$\texttt{pl(i,iy)} = L_i(\texttt{y(iy)})\,.$$

The computation of the coefficients J_{jki} in Eq.(4.13.7) is performed in lines `350-740` and develops as follows:

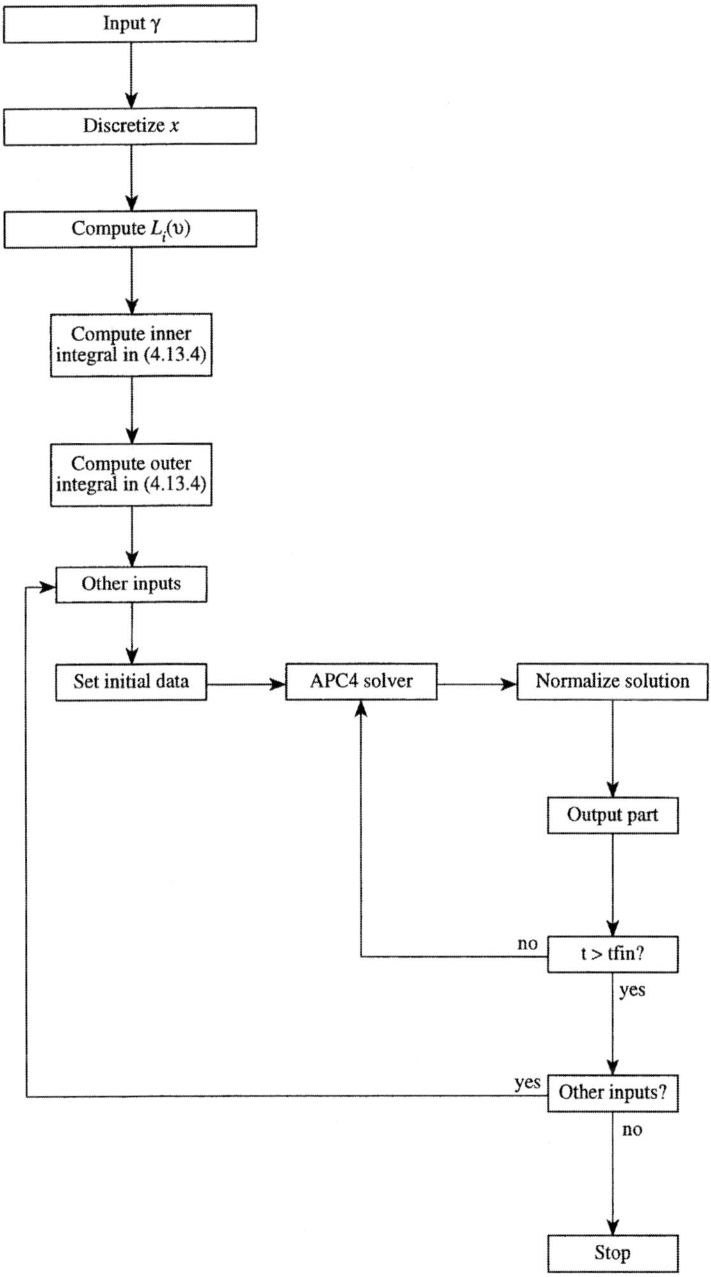

Table 4.13.3 — Flow chart of DEMOCRAT.

- The index `k` and the value `ix`, i.e., u_i are fixed (lines 360-370);
- For any `iy` (i.e., v) the inner integral is computed by evaluating the kernel ψ (lines 410-450) and the normalization function $A(v,w)$ in (4.11.9) (lines 460-540) and then using Simpson's rule (lines 550-600);
- Then for any fixed `j` (line 620), the external integral is computed (lines 630-700) and printed (line 710).

This is the part of the program that requires more time and memory, therefore it is better to store all the information to be used for different initial-value problems. In fact the program returns here (line 760) after having solved the initial-value problem. If the parameter discretization or the value of some parameters in the kernel ψ need to be changed (in this case σ), the program, of course, has to be restarted.

After some input lines (760-1050), which above all set the initial data (lines 1000-1050), the numerical integration is performed using the fourth-order predictor-corrector scheme, which was named APC4 in Chapter 2 (lines 2000-2770).

As is now usual, the right-hand side of the differential equation is given in the subroutine starting at line 3000.

The program returns to the input line 760 if one wants to repeat the integration with different data, otherwise it stops.

While integrating the differential equation, a `GOTO` statement (line 2680) sends the program to an output subroutine (lines 2800-2999), which, according to the input commands, prints the data (lines 2830-2870) or plots the solution (lines 2880-2955) with a possible interpolation (line 2935) of the data (lines 2960-2990).

Before doing that, however, on lines 2800-2825 the solution is integrated (lines 2800-2818) and normalized (lines 2820-2825), so that

$$\|f^n\|_{L_1} = \int_0^1 f^n(t;u)\,du = 1\,. \tag{4.13.10}$$

This step becomes essential (to reduce round-off errors) because of the small number of nodes used (`nx`=7, `ny`=21, `nint`=11). Increasing these values, say to `nx`=11, `ny`=101, `nint`=51, would render useless this control, but strongly increase the computation time and the memory requirements, which can be an overwhelming request for small personal computers.

The reader is encouraged to increase by a little the number of nodes, paying attention to the time required. If possible, print the value of the integral in Eq.(4.13.10) as a measure of the error and record it as the parameterization if refined (using or not the normalization of lines 2820-2825 at every time step).

Another thing to notice is how the computation is slowed down when the time step changes. This is due to the fact that in the transition the integration is performed using the Runge–Kutta part (lines 2000-2410) and the right-hand side of the ordinary differential equation is a double integral that is lengthy to compute despite the method used in the first part of the program. Therefore, the predictor-corrector method is faster that the Runge–Kutta one.

As an exercise, the reader can replace the APC4 routine with the RKF one and look at the difference in the results, in the chosen time step and in the time required to fulfill the integration.

DEMOCRAT.BAS

```
1 REM ***********************************************************
2 REM *              Program simulating the evolution of        *
3 REM *              a population of anonymous organisms        *
4 REM ***********************************************************
10 DIM u(7), f(7), f0(7), f1(7), f2(7), f3(7), ucorr(7), upred(7),
   usub(7)
20 DIM x(7), y(21), z(21), xx(11), pi0(21), pi1(7), pL(7, 21)
30 DIM inside(21), fint(7, 7, 7), rhs(7), npr(7)
   40 nx = 7:  ny = 21:  nint = 11
40 nx = 7:  ny = 21:  nint = 11
41 PRINT "*******************************************************"
42 PRINT "*            Program simulating the evolution of      *"
43 PRINT "*            a population of anonymous organisms      *"
44 PRINT "*******************************************************"
50 PRINT "When more than one quantity is requested, separate
   with a comma"
60 PRINT "Values in square brackets are the suggested values
   (not default)"
70 INPUT "gamma [.1]"; gam
80 xmin = 0:  xmax = 1
90 CLS 0
100 REM *********************************************************
101 REM *                 Parameter discretization              *
102 REM *********************************************************
110 FOR i = 1 TO nx:  x(i) = (i - 1) / (nx - 1):  NEXT i
120 FOR i = 1 TO nint:  xx(i) = (i - 1) / (nint - 1):  NEXT i
130 FOR i = 1 TO ny:  y(i) = (i - 1) / (ny - 1):  z(i) = y(i):
   NEXT i
140 REM *********************************************************
```

```
141 REM *                   Computing PI in (A.2.9)                  *
142 REM *********************************************************
150 FOR iy = 1 TO ny
160 pi0(iy) = 1
170 FOR i = 1 TO nx
180 pi0(iy) = pi0(iy) * (y(iy) - x(i))
190 NEXT i
200 NEXT iy
210 REM *********************************************************
211 REM *                  Computing PI' in (A.2.10)                 *
212 REM *********************************************************
220 FOR k = 1 TO nx
230 pi1(k) = 1
240 FOR i = 1 TO nx
250 IF k <> i THEN pi1(k) = pi1(k) * (x(k) - x(i))
260 NEXT i
270 NEXT k
280 REM *********************************************************
281 REM *                   Lagrange polynomials in                  *
282 REM *                   the integration points                   *
283 REM *********************************************************
290 FOR i = 1 TO nx
300 FOR iy = 1 TO ny
310 pL(i, iy) = 1
320 IF x(i) <> y(iy) THEN pL(i, iy) = pi0(iy) / ((y(iy) - x(i))
    * pi1(i))
330 NEXT iy
340 NEXT i
350 REM *********************************************************
351 REM *              Internal integral for any Lagrange            *
352 REM *             polynomial and integration point iy            *
353 REM *********************************************************
360 FOR ix = 1 TO nx
370 FOR k = 1 TO nx
380 FOR iy = 1 TO ny
390 inside(iy) = 0
400 FOR iz = 1 TO ny
410 REM *********************************************************
411 REM *                     Defining kernel psi                    *
412 REM *********************************************************
420 beta = (y (iy) + z(iz)) / 2
440 sigma = gam - .09 * ABS(y(iy) - z(iz))
450 psi = EXP(-(x(ix) - beta) ^2 / sigma)
460 REM *********************************************************
461 REM *              Normalization term for the kernel             *
462 REM *********************************************************
470 ayz = 0
480 FOR in = 1 TO nint
```

```
490 csimp = 2
500 IF in / 2 = INT(in / 2) THEN csimp = 4
510 IF in = 1 OR in = nint THEN csimp = 1
520 ayz = ayz + csimp * EXP(-(xx(in) - beta) ^2 / sigma)
530 NEXT in
540 ayz = ayz / (3 * (nint - 1))
550 csimp = 2
560 IF iz / 2 = INT(iz / 2) THEN csimp = 4
570 IF iz = 1 OR iz = ny THEN csimp = 1
580 inside(iy) = inside(iy) + csimp * pL(k, iz) * psi / ayz
590 NEXT iz
600 NEXT iy
610 REM ***********************************************************
611 REM *           Integral for any couple of Lagrange           *
612 REM *            polynomials j and k and point ix             *
613 REM ***********************************************************
620 FOR j = 1 TO nx
630 fint(ix, j, k) = 0
640 FOR iy = 1 TO ny
650 csimp = 2
660 IF iy / 2 = INT(iy / 2) THEN csimp = 4
670 IF iy = 1 OR iy = ny THEN csimp = 1
680 fint(ix, j, k) = fint(ix, j, k) + csimp * pL(j, iy) *
    inside(iy)
690 NEXT iy
700 fint(ix, j, k) = fint(ix, j, k) / (9 * (ny - 1) ^2)
710 PRINT "j="; j; "k="; k; "i="; ix, "integral="; fint(ix, j, k)
720 NEXT j
730 NEXT k
740 NEXT ix
750 REM ***********************************************************
751 REM *                       Input lines                       *
752 REM ***********************************************************
760 INPUT "Time step and final time [.1,8]"; h, tfin
770 INPUT "Required tolerance [.001]"; tol
780 INPUT "Plot [1] or print-out [0] of the solution "; ngraf
790 IF ngraf = 0 THEN INPUT "Print-out interval [.5]"; prout
800 IF ngraf = 1 THEN INPUT "Minimum and maximum ordinate
    of the graph [0,1.5]"; pumin, pumax
810 IF ngraf = 1 THEN INPUT "Do you want reinterpolation
    of the data [y/n]"; interp$
820 IF ngraf <> 0 AND ngraf <> 1 THEN GOTO 780
850 REM ***********************************************************
851 REM *                    End of input lines                   *
852 REM ***********************************************************
980 t = 0
990 iprint = 0
1000 REM **********************************************************
```

```
1001 REM *                       Set initial data                          *
1002 REM ************************************************************
1010 INPUT "Your own initial data [y/n]"; init$
1020 IF init$ = "n" THEN FOR ix = 1 TO nx:  u(ix) = 1:  NEXT ix:
   GOTO 2000
1030 FOR ix = 1 TO nx
1040 PRINT "u(x="; x(ix); ")="; :  INPUT u(ix)
1050 NEXT ix
1060 CLS 0
2000 REM ************************************************************
2001 REM *              Fourth-order Runge-Kutta routine                  *
2002 REM *                 to compute starting values                     *
2003 REM ************************************************************
2010 h24 = h / 24
2015 h2 = h / 2
2020 h6 = h / 6
2030 FOR i = 1 TO nx
2040 usub(i) = u(i)
2050 NEXT i
2060 GOSUB 3000
2070 FOR i = 1 TO nx
2080 f3(i) = f(i)
2090 NEXT i
2100 FOR irk = 1 TO 3
2110 GOSUB 3000
2120 t = t + h2
2130 FOR i = 1 TO nx
2140 k1(i) = f(i)
2150 usub(i) = u(i) + h2 * k1(i)
2160 NEXT i
2170 GOSUB 3000
2180 FOR i = 1 TO nx
2190 k2(i) = f(i)
2200 usub(i) = u(i) + h2 * k2(i)
2210 NEXT i
2220 GOSUB 3000
2230 t = t + h2
2240 FOR i = 1 TO nx
2250 k3(i) = f(i)
2260 usub(i) = u(i) + h * k3(i)
2270 NEXT i
2280 GOSUB 3000
2290 FOR i = 1 TO nx
2300 k4(i) = f(i)
2310 u(i) = u(i) + h6 * (k1(i) + 2 * (k2(i) + k3(i)) + k4(i))
2320 usub(i) = u(i)
2330 NEXT i
2340 GOSUB 3000
```

```
2350 FOR i = 1 TO nx
2360 IF irk = 1 THEN f2(i) = f(i)
2370 IF irk = 2 THEN f1(i) = f(i)
2380 NEXT i
2390 GOSUB 2800
2400 NEXT irk
2410 REM ***********************************************************
2411 REM *                 End Runge-Kutta routine                 *
2412 REM ***********************************************************
2420 REM ***********************************************************
2421 REM *       Fourth-order predictor-corrector routine          *
2422 REM ***********************************************************
2430 FOR i = 1 TO nx
2440 usub(i) = u(i)
2450 NEXT i
2460 GOSUB 3000
2470 REM ***********************************************************
2471 REM *                     Predictor part                      *
2472 REM ***********************************************************
2480 FOR i = 1 TO nx
2490 upred(i) = u(i) + h24 * (55 * f(i) - 59 * f1(i) + 37 * f2(i)
    - 9 * f3(i))
2500 NEXT i
2504 REM ***********************************************************
2505 REM *                     Corrector part                      *
2506 REM ***********************************************************
2510 t = t + h
2520 FOR i = 1 TO nx
2530 f0(i) = f(i)
2540 usub(i) = upred(i)
2550 NEXT i
2560 GOSUB 3000
2570 FOR i = 1 TO nx
2580 ucorr(i) = u(i) + h24 * (9 * f(i) + 19 * f0(i) - 5 * f1(i) +
   f2(i))
2582 NEXT i
2584 REM ***********************************************************
2585 REM *             End predictor-corrector routine             *
2586 REM ***********************************************************
2590 erpc = 0
2600 FOR i = 1 TO nx
2610 dif = ABS(ucorr(i) - upred(i)) * 19 / (270 * h)
2620 IF dif > erpc THEN erpc = dif
2630 NEXT i
2634 REM ***********************************************************
2635 REM *                Set new time step if er                  *
2637 REM *               is too large or too small                 *
2637 REM ***********************************************************
```

```
2640 IF erpc > tol THEN t = t - h:  h = h / 2:  GOTO 2000
2650 FOR i = 1 TO nx
2660 u(i) = ucorr(i)
2670 NEXT i
2680 GOSUB 2800
2690 IF erpc < tol / 100 THEN h = 2 * h:  GOTO 2000
2694 REM ********************************************************
2695 REM *                 Updating previous values             *
2696 REM ********************************************************
2700 FOR i = 1 TO nx
2710 f3(i) = f2(i)
2720 f2(i) = f1(i)
2730 f1(i) = f0(i)
2740 NEXT i
2764 REM ********************************************************
2765 REM *       Advance time and continue till final time      *
2766 REM ********************************************************
2770 IF t <= tfin - h / 10 THEN GOTO 2420
2780 INPUT "Different input [y/n]"; inp$
2790 IF inp$ = "y" THEN GOTO 760
2798 IF inp$ <> "y" AND inp$ <> "n" THEN GOTO 2790
2799 END
2800 REM ********************************************************
2801 REM *                    Output subroutine                 *
2802 REM ********************************************************
2804 dens = 0
2806 FOR ix = 1 TO nx
2808 csimp = 2
2810 IF ix / 2 = INT(ix / 2) THEN csimp = 4
2812 IF ix = 1 OR ix = nx THEN csimp = 1
2814 dens = dens + csimp * u(ix)
2816 NEXT ix
2818 dens = dens / (3 * (nx - 1))
2820 FOR ix = 1 TO nx
2822 u(ix) = u(ix) / dens
2825 NEXT ix
2826 REM ********************************************************
2827 REM *                    Print-out commands                *
2828 REM ********************************************************
2830 IF ngraf = 1 THEN GOTO 2880
2835 IF t < prout * iprint - .00001 THEN GOTO 2997
2840 PRINT TAB(15); "t="; t
2845 FOR i = 1 TO nx
2850 PRINT "u("; i; ")="; u(i),
2855 NEXT i
2860 iprint = iprint + 1
2870 GOTO 2997
2874 REM ********************************************************
```

```
2875 REM *                   Plotting commands                   *
2876 REM *********************************************************
2880 CLS 0:  KEY OFF: SCREEN 2
2885 ax = 0:  bx = 1
2890 LOCATE 14, 10:  PRINT ax
2895 LOCATE 14, 70:  PRINT bx
2900 LOCATE 13, 7:  PRINT pumin
2905 LOCATE 2, 5:  PRINT pumax
2910 LOCATE 15, 20
2915 VIEW (80, 10)-(560, 100), , 1
2920 WINDOW (ax, pumin)-(bx, pumax)
2925 LOCATE 15, 40:  PRINT "t="; t
2930 PSET (ax, u(1))
2935 IF interp$ = "y" THEN GOTO 2960
2940 FOR ix = 2 TO nx
2945 LINE -(x(ix), u(ix))
2950 NEXT ix
2955 GOTO 2997
2956 REM *********************************************************
2957 REM *             Beginning of reinterpolation              *
2958 REM *********************************************************
2960 FOR ix = 2 TO ny
2965 uu = 0
2970 FOR k = 1 TO nx
2975 uu = uu + pL(k, ix) * u(k)
2980 NEXT k
2985 LINE -(y(ix), uu)
2990 NEXT ix
2991 REM *********************************************************
2992 REM *                End of reinterpolation                 *
2993 REM *********************************************************
2997 RETURN
2999 END
3000 REM *********************************************************
3010 REM *          Subroutine for r.h.s. of equation            *
3060 REM *********************************************************
3070 FOR ix = 1 TO nx
3080 gain = 0
3090 FOR jx = 1 TO nx
3100 FOR kx = 1 TO nx
3110 gain = gain + fint(ix, jx, kx) * usub(jx) * usub(kx)
3120 NEXT kx
3130 NEXT jx
3140 f(ix) = gain - usub(ix)
3150 NEXT ix
3160 RETURN
3170 END
```

Problems for Chapter 4

Problem 4.1
Consider the one-dimensional model reported in Problem 1.3 and provide, for such a model, at least one example of underspecified, overspecified, and inverse problem. Provide the same indications for some mathematical models for a two population dynamics.

Note to Problem 4.1 *The reader is referred to the mathematical formulation of the initial-boundary-value problem for PDE and to Definitions 4.2.1–4.2.3.*

Problem 4.2
Consider the one-dimensional heat diffusion model with dispersion term

$$
\begin{aligned}
u = u(t,x) :& \quad [0,1] \times [0,1] \to \mathbb{R}\,,\\
\frac{\partial u}{\partial t} =& \frac{\partial}{\partial x}\left(\kappa(u)\frac{\partial u}{\partial x}\right) - x(1-x)u\,,\\
\kappa(u) =& 1 - \frac{u}{10}
\end{aligned}
$$

subject to initial condition $u_0(x) = 0$ and boundary condition at $x = 0$ of the type $u(t, x = 0) = 0$.

Let, in addition, the following information be given

$$
u^*(t) = u\left(t, x = \frac{1}{2}\right) = \frac{t^2}{10}\,.
$$

Solve the problem $\forall\, t \in [0,1]$ and $\forall\, x \in [0,1]$ and compute, in particular, the heat flow in $x = 1$.

Note to Problem 4.2 *The reader must apply the scheme "Solution to Problem 4.2" reported in Section 4.3 and use the program INVERSE.*

Problem 4.3
Consider the convection-diffusion model

$$
\frac{\partial u}{\partial t} = c\frac{\partial u}{\partial x} + \frac{\partial}{\partial x}\left(\kappa(u)\frac{\partial u}{\partial x}\right)
$$

for $t \in [0,1]$ and $x \in [0,1]$ with the following information

$$u_0(x) = u(0,x) = \frac{x}{10},$$
$$u(t) = u(t,0) = \frac{t}{5},$$
$$u^*(t) = u\left(t, \frac{1}{2}\right) = \frac{1}{20}(1-t),$$

and solve numerically the problem $\forall t \in [0,1]$ and $\forall x \in [0,1]$.

Problem 4.4
Consider Problems 4.2 and 4.3 when the heat diffusion is a random

$$\kappa(u) = 1 - \frac{r(\omega)}{10} u,$$

where $r(\omega)$ is a random variable uniformly distributed over the interval $[0,1]$. Compute the mean value and variance of u at $x = 1$.

Note to Problem 4.4 *The reader should develop a sampling of r, say $\{r_i\}_{i=1}^n$, solve the problem for each r_i, and then compute mean value and variance by suitable averaging.*

Problem 4.5
Consider the initial-value problem for the model of population dynamics with stochastic integration in the case $\eta = \psi = 1$ and with initial condition

$$f_0(u) = 1 + \varphi(u),$$

where

$$\int_0^1 \varphi(u) = 0.$$

Solve the problem by modifying the program DEMOCRAT and comparing the numerical solution with the analytic one, which can be obtained for $\eta = \psi = 1$.

Modify the program in order to obtain numerical solutions for different values of ψ and η.

Appendix 1

FUNCTION SPACES

The aim of this appendix is to provide some useful definitions of function spaces to be used in the definition of the distance between elements to be compared. The interested reader is addressed to the pertinent literature [KRA], [YOS], and [RaW] in order to obtain detailed information to complete the schematic presentation given in this appendix.

The following definitions will be given:

- Vector space;
- Normed space;
- Spaces with inner product;
- Banach space;
- Hilbert space.

The definition of ***linear vector space*** depends on the definitions of addition of elements and multiplication by a scalar element. In detail, the following definition holds:

DEFINITION A.1.1 Real linear vector space
Given a set V of elements denoted by $\mathbf{v}_1$, $\mathbf{v}_2$, $\mathbf{v}_3$, …, assume that it is possible to define in it two operations:

- *Sum, indicated by $+$, which associates to every two elements $\mathbf{v}_1, \mathbf{v}_2 \in V$ the unique element $\mathbf{v} = \mathbf{v}_1 + \mathbf{v}_2 \in V$.*
- *Product for a scalar, which associates to every $\mathbf{v} \in V$ and to any $\lambda \in \mathbb{R}$ the unique element $\mathbf{w} = \lambda\mathbf{v} \in V$.*

The set V is a vector space V if the following properties are satisfied:

$$\begin{aligned}
\mathbf{v}_1 + \mathbf{v}_2 &= \mathbf{v}_2 + \mathbf{v}_1 \,, \\
(\mathbf{v}_1 + \mathbf{v}_2) + \mathbf{v}_3 &= \mathbf{v}_1 + (\mathbf{v}_2 + \mathbf{v}_3) \,, \\
\exists \mathbf{0} \in V &: \mathbf{v} + \mathbf{0} = \mathbf{v} \,, \\
\exists \mathbf{w} \in V &: \mathbf{v} + \mathbf{w} = 0 \,, \\
\lambda_1(\lambda_2 \mathbf{v}) &= (\lambda_1 \lambda_2)\mathbf{v} \,, \\
(\lambda_1 + \lambda_2)\mathbf{v} &= \lambda_1 \mathbf{v} + \lambda_2 \mathbf{v} \,, \\
\lambda(\mathbf{v}_1 + \mathbf{v}_2) &= \lambda \mathbf{v}_1 + \lambda \mathbf{v}_2 \,,
\end{aligned}$$

for every $\mathbf{v}_1$, $\mathbf{v}_2$, $\mathbf{v}_3$, $\mathbf{v} \in V$ *and* λ_1, λ_2, $\lambda \in \mathbb{R}$.

Note that the element denoted by $\mathbf{0}$ is unique, and, for each $\mathbf{v} \in V$, the element $\mathbf{w} = -\mathbf{v}$ is unique.

Example A.1.1 Space of continuous functions
The space of the continuous functions on an interval $[a, b]$ denoted by

$$C_{[a,b]} \; : \; \{f \, : \quad [a,b] \longmapsto \mathbb{R} \quad \text{is continuous}\} \tag{A.1.1}$$

is a linear vector space if the operations of addition and multiplication for a scalar are defined as follows

$$\begin{aligned}
f, g \in C_{[a,b]} \, &: \, f + g = v \,, \quad \text{such that} \quad v(x) = f(x) + g(x) \,, \\
f \in C_{[a,b]}, \, \lambda \in \mathbb{R} \, &: \, \lambda f = v \,, \quad \text{such that} \quad v(x) = \lambda f(x)
\end{aligned}$$

for each $x \in [a, b]$. □

The generalization to functions in several dependent variables is straightforward. In fact, one simply defines a domain $\mathcal{D} \subseteq \mathbb{R}^d$ of the independent variables in a d-dimensional space and introduces the set of functions

$$f \, : \quad \mathcal{D} \longmapsto \mathbb{R} \,,$$

and the operations of addition and multiplication for a scalar in the same way.

We shall refer, from now on, to the more general case of functions defined over the domain $\mathcal{D}$.

Several spaces with the structure of a vector space can be characterized by suitable regularity and integrability properties. For instance, if it is required that

$$\int_D |f(x)|^p \, dx < \infty \,, \tag{A.1.2}$$

then one has a subspace called $L^p(\mathcal{D})$. The following spaces can be defined, among others, with sequential order of increasing smoothness:

- $\mathcal{PC}(\mathcal{D})$ is the space of the piecewise continuous functions;
- $\mathcal{C}(\mathcal{D})$ is the space of the continuous functions;
- $\mathcal{C}^n(\mathcal{D})$ is the space of the n-times continuously differentiable functions;
- $\mathcal{C}^\infty(\mathcal{D})$ is the space of the infinitely differentiable functions;
- $\mathcal{P}_N(\mathcal{D})$ is the space of the polynomials of degree N.

It is also useful introducing the following spaces, which are often used in the analysis of mathematical models:

- $\mathcal{B}^n(\mathcal{D})$ is the space of the bounded functions;
- $\tilde{\mathcal{C}}^n(\mathcal{D})$ is the space of the functions in $\mathcal{C}^n(\mathcal{D})$ that tend to zero at the boundary of $\mathcal{D}$.

All the examples reported above can be generalized to vector functions as well as to complex valued functions.

We can now define a normed space according to the following definition:

DEFINITION A.1.2 Normed space
A vector space $\mathcal{N}$ becomes a ***normed space*** *if to each element $\mathbf{v}$ of $\mathcal{N}$ a number $\|\mathbf{v}\| \in \mathbb{R}_+$ is joined such that*

$$\|\mathbf{v}\| = 0 \Longleftrightarrow \mathbf{v} = \mathbf{0} \,, \tag{A.1.3}$$

$$\|\lambda \mathbf{v}\| = |\lambda| \, \|\mathbf{v}\| \,, \tag{A.1.4}$$

$$\|\mathbf{v}_1 + \mathbf{v}_2\| \le \|\mathbf{v}_1\| + \|\mathbf{v}_2\| \,, \tag{A.1.5}$$

for every $\mathbf{v}, \mathbf{v}_1, \mathbf{v}_2 \in \mathcal{N}$ and $\lambda \in \mathbb{R}$.

As an example, if we consider the space $\mathcal{C}(\mathcal{D})$, the following norms can be defined:

DEFINITION A.1.3 ***L^∞-norm and L^p-norm***
The L^∞-norm is

$$\|v\|_\infty = \operatorname*{ess\,sup}_{\mathbf{x}\in\mathcal{D}} |v(\mathbf{x})| \,, \tag{A.1.6}$$

the L^p-norm is

$$\|v\|_p = \left(\int_{\mathcal{D}} |v(\mathbf{x})|^p \, d\mathbf{x}\right)^{\frac{1}{p}} . \tag{A.1.7}$$

In particular, if $p = 1$ one has L^1, which is often used as we have seen in Chapter 1, and for $p = 2$ one has the L^2-norm

$$\|v\|_2 = \left(\int_{\mathcal{D}} v^2(\mathbf{x}) \, d\mathbf{x}\right)^{\frac{1}{2}} . \tag{A.1.8}$$

If v is a function of time and space, then the norm can be conventionally denoted by a triple bar.

For instance

$$|||v|||_1 = \sup_{t\in[0,T]} \|v\|_1 \,, \tag{A.1.9}$$

or

$$|||v|||_2 = \sup_{t\in[0,T]} \|v\|_2 \,, \tag{A.1.10}$$

for $v = v(t, \mathbf{x})$, $t \in [0,T]$, $\mathbf{x} \in \mathcal{D}$.

If the element of the normed space is a vector, then one has to apply the norm operator to all components of the element. This means, for instance, taking either the maximum of $|v_i|$

$$|\mathbf{v}| = \max_{i=1,\dots,n} |v_i| \,, \tag{A.1.11}$$

or the square root of the sum of the square power of each component

$$|\mathbf{v}| = \left\{\sum_{i=1}^{n} v_i^2\right\}^{\frac{1}{2}} . \tag{A.1.12}$$

Then the norm is obtained by operating over time and space. Here the notation $\mathbf{v}$ is used for vector functions in $\mathbb{R}^n$.

We can now introduce the concept of distance between two elements that belong to the same function space.

DEFINITION A.1.4 Distance in norm
Let $\mathbf{u}$ *and* $\mathbf{v}$ *be two elements of the same normed space* $\mathcal{N}$, $\mathbf{u}, \mathbf{v} \in \mathcal{N}$. *Then the* ***distance*** *between* $\mathbf{u}$ *and* $\mathbf{v}$, *denoted by* $d(\mathbf{u}, \mathbf{v})$, *is the norm of the difference*

$$d(\mathbf{u}, \mathbf{v}) = \|\mathbf{u} - \mathbf{v}\| . \tag{A.1.13}$$

It is plain, by the definition given above, that we used the concept of vector space to define the difference and that of normed space to apply the norm.

To complete this appendix, also in view of the contents of the next appendix, it is useful to provide the definition of ***Banach*** and ***Hilbert*** spaces; or, in other words, of ***vector normed complete spaces*** and of ***complete spaces with inner product***, respectively.
In order to do that, we need the following definition.

DEFINITION A.1.5 Cauchy sequence
Given a normed space $\mathcal{N}$ *and a sequence*

$$\{\mathbf{v}_k\}_{k \in \mathcal{N}}$$

of elements of $\mathcal{N}$, *the sequence is a Cauchy one if for every* $\varepsilon > 0$ *there exists an integer* n_0 *such that*

$$\forall m, n > n_0 \quad \Rightarrow \quad \|\mathbf{v}_m - \mathbf{v}_n\| < \varepsilon .$$

The definition of Banach space can now be given.

DEFINITION A.1.6 Banach space
A normed space $\mathcal{B}$ *is a* ***Banach space*** *if every Cauchy sequence converges to an element of* $\mathcal{B}$.

REMARK A.1.1 The completness is an inner property of normed spaces. In particular, spaces of continuous functions endowed with the L^∞-norm (A.1.6) or the L^p-norm (A.1.7) are Banach spaces. ∎

The definition of Hilbert space requires the definition of inner product.

DEFINITION A.1.7 Inner product
Given a real linear vector space V, an ***inner product*** *is an operation that associates to any pair of elements $\mathbf{u}, \mathbf{v}$ of V a real number denoted by $\langle \mathbf{u}, \mathbf{v} \rangle$, which satisfies the following properties*

$$
\begin{aligned}
\langle \mathbf{u}, \mathbf{v} \rangle &= \langle \mathbf{v}, \mathbf{u} \rangle \,, \\
\langle \lambda_1 \mathbf{u} + \lambda_1 \mathbf{v}, \mathbf{w} \rangle &= \lambda_1 \langle \mathbf{u}, \mathbf{w} \rangle + \lambda_2 \langle \mathbf{v}, \mathbf{w} \rangle \,, \\
\langle \mathbf{u}, \mathbf{u} \rangle &\geq 0 \,, \\
\langle \mathbf{u}, \mathbf{u} \rangle &= 0 \iff \mathbf{u} = \mathbf{0} \,.
\end{aligned}
$$

for every $\mathbf{u}, \mathbf{v}, \mathbf{w} \in V$ and $\lambda_1, \lambda_2 \in \mathbb{R}$.

REMARK A.1.2 The inner product also defines a norm on V given by

$$
\|\mathbf{u}\| = \sqrt{\langle \mathbf{u}, \mathbf{u} \rangle} \,. \tag{A.1.14}
$$

■

DEFINITION A.1.8 Hilbert space
A ***Hilbert space*** *$\mathcal{H}$ is a vector space equipped with an inner product that defines a norm for which any Cauchy sequence converges to an element of $\mathcal{H}$.*

Example A.1.2 Hilbert space
Given a continuous strictly positive function $w = w(x)$ defined over $[a, b]$ (called weight function), consider the space of the functions such that

$$
\int_a^b |u(x)|^2 w(x) \, dx
$$

is finite. If the inner product

$$
\langle u, v \rangle_w = \int_a^b u(x) v(x) w(x) \, dx \,, \tag{A.1.15}
$$

is defined, then the function space, which will be denoted by $L^2_w([a,b])$, is a Hilbert space. If $w(x) = 1$, then the space is simply denoted by $L^2([a,b])$.

The subspace of L^2_w

$$H^m_w([a,b])$$
$$= \left\{ u \in L^2_w([a,b]) \quad \text{such that} \quad \frac{d^k u}{dx^k} \in L^2_w([a,b]) \quad \text{for} \quad k = 1, \dots, n \right\},$$

endowed with the inner product

$$\langle u, v \rangle_{H^m_w} = \sum_{k=0}^{m} \left\langle \frac{d^k u}{dx^k}, \frac{d^k v}{dx^k} \right\rangle_w$$

is an other example of Hilbert space called Sobolev space of order m. Here the derivatives have to be intended in distribution sense. □

Banach and Hilbert spaces can be used for existence proofs as well as for construction and/or approximation of solutions.

A few classical results are reported in this appendix in order to support the analysis developed throughout the various chapters of this book.

The first result reported here is the classical Banach fixed-point theorem, which provides existence of solution and a converging iterative scheme.

THEOREM A.1.1 ***(Banach's Fixed-Point Theorem)***
Let $\mathcal{B}$ be a Banach space and let $\mathcal{D}$ be a closed subset of $\mathcal{B}$, and consider then the equation

$$u = \mathcal{U}u\,, \tag{A.1.16}$$

where $\mathcal{U}$ is an operator that acts in $\mathcal{B}$. Then if the operator $\mathcal{U}$ is such that

$$\forall\, u \in \mathcal{D} : \mathcal{U}u \in \mathcal{D}$$
$$\forall\, u_1, u_2 \in \mathcal{D} : \|\mathcal{U}u_1 - \mathcal{U}u_2\| \le \alpha \|u_1 - u_2\|\,,$$

where $\alpha < 1$, then a unique solution exists in $\mathcal{D}$ to (A.1.16).

The solution u can be obtained by the iterative scheme

$$u_{n+1} = \mathcal{U}u_n\,, \quad n \ge 0\,, \tag{A.1.17}$$

where $u_0 \in \mathcal{D}$ is arbitrary. The rate of convergence to the solution u is given by the formula

$$\|u_n - u\| \leq \frac{\alpha^n}{1-\alpha} \|u_0 - u_1\| . \tag{A.1.18}$$

■

Theorem A.1.1 is useful in order to prove the existence of the solution, but it is not generally used to construct solutions since the rate of convergence is too slow for practical purposes. As shown in the textbook, suitable mathematical methods have to be used for such a purpose. Nevertheless, the theorem assures the construction of a solution by a sequence that converges in norm to the solution of the problem.

Consider now the problem of approximation and convergence in the mean in an L^2_w-space with the L^2_w-norm and inner product (A.1.15). The following definitions are useful.

DEFINITION A.1.9 Orthogonal functions
*Two functions u and v belonging to the L^2_w space are **orthogonals** if their inner product is equal to zero*

$$\langle u, v \rangle_w = 0 .$$

DEFINITION A.1.10 Complete system
A sequence of functions

$$u_1, u_2, \ldots \in L^2_w$$

is a system of orthogonal functions if

$$\langle u_i, u_j \rangle_w = 0 , \quad \forall\, i \neq j .$$

*An orthogonal system is **complete** if the system is infinite, all u_i are linearly independent, and $\forall\, v \in L^2_w$ there exists an expansion*

$$v = \sum_{i=1}^{\infty} c_i u_i$$

where

$$\langle u_i, u_i \rangle_w = c_i , \quad \forall\, i \geq 1 ,$$

such that

$$\sum_{i=1}^{\infty} |c_i|^2 = \|v\|^2_{L^2_w} . \tag{A.1.19}$$

In general in a Hilbert space we have

$$\lim_{n \to \infty} \|v - \sum_{i=1}^{n} c_i u_i\| = 0 . \tag{A.1.20}$$

This topic will be revisited in Appendix 2, where some examples will be given.

Classical textbooks on the topics dealt with above are those by Smart [SMA] and Alexitis [ALE].

A historical monograph is that by Banach [BAN]. Practical information can be found in the handbook compiled by Ride and Westergren [RaW].

Appendix 2

INTERPOLATION AND APPROXIMATION

Introduction

This appendix presents some techniques for the interpolation and approximation of functions. This is a fundamental topic of numerical analysis, which represents an interface between a continuous representation of physical reality and computer discrete simulation. Polynomial interpolation plays an important role, both in mathematical modelling and in model analysis. In fact, physical phenomena can often be observed only for discrete values of the independent variables (time and space). Then a continuous representation is often useful or even necessary. Moreover, as shown in Chapters 2–4, the solution of mathematical problems, related to the analysis of the mathematical models, requires the interpolation of the dependent variable over time and space.

This appendix provides a concise guide to the approximation-interpolation methods by polynomials. The reader is referred to the classical literature [DAV] for a deeper insight into the topics dealt with in this appendix. More specifically, piecewise polynomials are dealt with by Birkhoff and de Boor [BdB], splines by de Boor [deB] and Späth [SPA], Bernstein polynomials by Powell [POW], trigonometric polynomials by Zygmund [ZYG], and orthogonal polynomials by Szegö [SZE]. Finally, surface interpolation is dealt with by Lancaster and Salkauskas [LSA].

The Interpolation and the Approximation Problem

Consider a function

$$u(x) : [a, b] \longmapsto \mathbb{R} .$$

The interpolation problem consists in finding a function v in a given vector space V such that, given the points (x_h, u_h), for $h = 0, \ldots, N$, one has

$$v(x_h) = u_h . \tag{A.2.1}$$

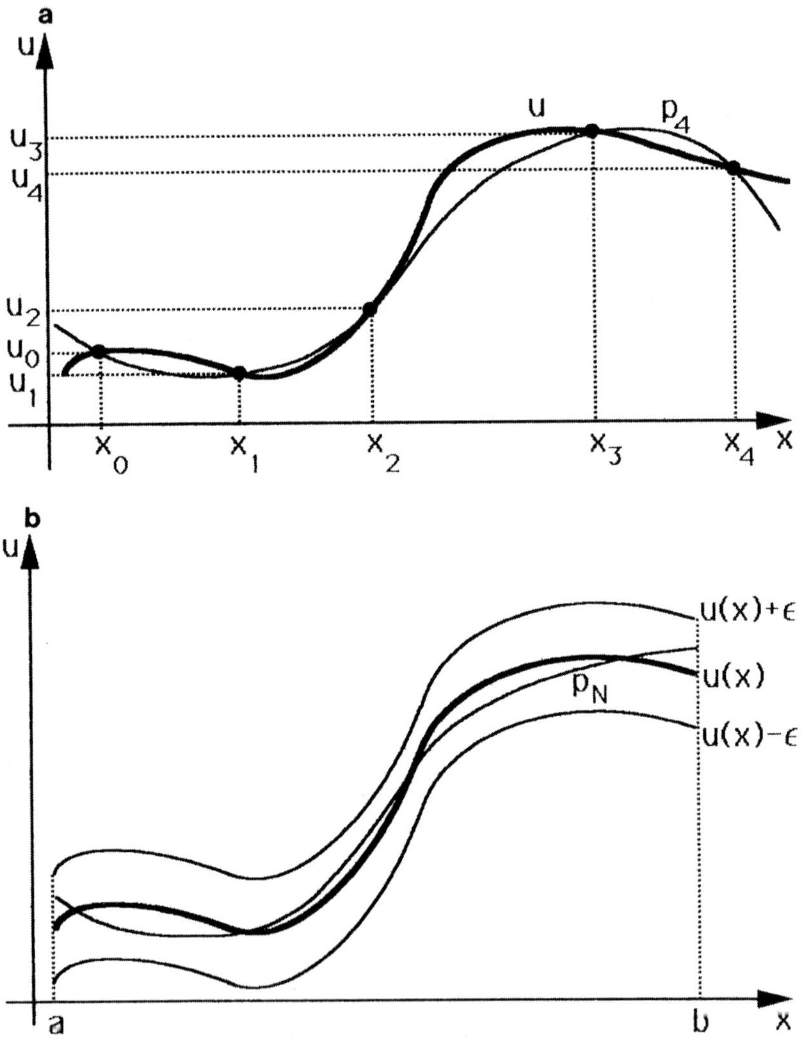

Figure A.2.1 — Depiction of (a) the interpolation problem and of (b) the approximation problem in the maximum norm. In (a) interpolation points are heavily marked.

For instance, V can be the space of polynomials, of rational functions, or of trigonometric functions. In these cases one has, respectively, the so-called polynomial, rational, and trigonometric interpolation.

The problem of approximating as closely as possible a given function by a polynomial can be approached in a slightly different way using the concept of distance between functions introduced in Appendix 1. The difference between the two problems is shown in Fig. A.2.1.

Consider, in fact, the space of functions endowed with a given norm $\|\cdot\|$ and fix an $\varepsilon > 0$, which represents the level of accuracy needed in the analysis of the model.

The problem of best approximation of a function $u(x)$ defined for $x \in [a,b]$, consists in finding a polynomial $p_N(x)$ such that

$$\big\|u(x) - p_N(x)\big\| < \varepsilon . \tag{A.2.2}$$

Namely, the distance in the chosen norm between the function and its polynomial approximation is smaller than the desired accuracy error.

The problem of interpolation and approximation can be easily generalized to continuous dynamic models. For instance, if the state variable is $u = u(x,y)$ and one knows the values u_{hk}, $h = 0,\ldots,N$, $k = 0,\ldots,M$ corresponding to $x_h \in [a,b]$ and $y_k \in [c,d]$, the interpolant is a function $v(x,y)$ such that

$$v(x_h, y_k) = u_{hk}\,, \qquad h = 0,\ldots,N\,, \quad k = 0,\ldots,M\,. \tag{A.2.3}$$

A similar generalization can be given for the problem of best approximation.

Before dealing with specific interpolation and approximation methods, it is useful to anticipate that they usually operate in a fixed domain of the independent variable, say $[-1,1]$. If the independent variable x of our model ranges between, say, x_0 and x_1, and the approximation method in $[a,b]$, it is necessary to rescale x introducing a new variable

$$\xi = a + (b-a)\frac{x - x_0}{x_1 - x_0}\,, \tag{A.2.4}$$

which now spans in the range desired by the approximation method.

In the case in which the independent variable tends to infinity, one may either choose a method working on an infinite domain or shrink the infinite interval to a finite one through a nonlinear mapping. For instance

$$\xi = \frac{b(x - x_0) + a(x_0 - \alpha)}{x - \alpha}$$

maps $[x_0, +\infty)$ into $[a, b)$ for any $\alpha < x_0$ and

$$\xi = \frac{a+b}{2} + \frac{b-a}{2} \tanh \alpha x$$

maps $(-\infty, +\infty)$ into (a, b) for any $\alpha > 0$.

The choice of the mapping is related to the properties of the function being approximated and to the approximation method used.

Lagrange Interpolation Formula

Polynomial interpolation is based on a theorem that generalizes the well-known fact that, unless for very special cases, a unique straight line passes through two points, a parabola through three, a cubic through four, and so on.

THEOREM A.2.1 *(Lagrange Interpolation Theorem)*
Given $N+1$ points $a \leq x_0 < x_1 < \cdots < x_N \leq b$ and $N+1$ values u_i, $i = 0, \ldots, N$, there exists a unique polynomial $p_N(x)$ of degree at most N such that

$$p_N(x_i) = u_i\,, \qquad i = 0, \ldots, N\,. \tag{A.2.5}$$

■

In order to write the Lagrange interpolation formula, it is useful to introduce the following Lagrange polynomials

$$L_h(x) = \frac{\prod\limits_{\substack{j=0 \\ j \neq h}}^{N} (x - x_j)}{\prod\limits_{\substack{j=0 \\ j \neq h}}^{N} (x_h - x_j)}\,, \tag{A.2.6}$$

which are such that

$$L_h(x_k) = \begin{cases} 1 & \text{if } h = k \\ 0 & \text{if } h \neq k\,. \end{cases} \tag{A.2.7}$$

Then the polynomial

$$p_N(x) = \sum_{h=0}^{N} u_h L_h(x) \tag{A.2.8}$$

is such that (A.2.5) is satisfied.

The following alternative form is also useful. It consists in introducing the ***fundamental polynomial***

$$\Pi(x) = \prod_{j=0}^{N} (x - x_j) \tag{A.2.9}$$

and the numbers

$$\Pi'_h = \frac{d\Pi}{dx}(x = x_h) = \prod_{\substack{j=0 \\ j \neq h}}^{N} (x_h - x_j) \,. \tag{A.2.10}$$

Hence from (A.2.6), for $x \neq x_h$

$$L_h(x) = \frac{\Pi(x)}{(x - x_h)\Pi'_h} \tag{A.2.11}$$

and eventually the interpolation formula (A.2.8) can be rewritten as

$$p_N(x) = \Pi(x) \sum_{h=0}^{N} \frac{u_h}{(x - x_h)\Pi'_h} \,. \tag{A.2.12}$$

The advantage of this second form is that one can compute the terms u_h/Π'_h once and for all, so that for any evaluation of the interpolating polynomial order N, operations need to be performed, while for Eq.(A.2.8) order N^2 operations are needed.

If one wants to approximate a function $u = u(x)$ with a polynomial that retains the values of the function at the nodes x_h, that is

$$p_N(x_h) = u(x_h) \qquad \text{for} \quad h = 0, \ldots, N \,,$$

the values of u_h in (A.2.8) and (A.2.12) have to be substituted with $u(x_h)$.

In the interpolation, the approximation is exact in the nodes x_h, $h = 0, \ldots, N$, however, an error is made for

$$x \in (x_h, x_{h+1}) \,, \quad \forall\, h = 0, \ldots, N-1 \,.$$

If $u(x) \in C^{N+1}_{[a,b]}$, it is possible to prove that

$$u(x) - p_N(x) = \frac{\Pi(x)}{(N+1)!} \frac{d^{N+1}u}{dx^{N+1}}(\xi), \quad \xi \in \left[\min\{x, x_0\}, \max\{x, x_N\}\right]$$

and therefore

$$\max_{x\in[a,b]} \left|u(x) - p_N(x)\right| \le \frac{1}{(N+1)!} \max_{x\in[a,b]} \left|\Pi(x)\right| \max_{x\in[a,b]} \left|\frac{d^{N+1}u}{dx^{N+1}}(x)\right| .$$

Hence, the error in the interpolation formula is, unless for a numerical constant, the product of two factors: One depending on the function $u(x)$ that is to be interpolated, the other depending on $\Pi(x)$, which, in turn, depends on the choice of the number of nodes and on their collocation in the interval $[a, b]$.

In the case of an unhappy choice of the nodes, the error can be large. For instance, if the interpolation points are crowded near one of the extrema of the interval, a large error will in general occur near the other end of the interval. For equispaced nodes, $\Pi(x)$ and the interpolation polynomial strongly oscillate near the extrema of the interpolation interval, as shown in Fig. A.2.2.

It makes sense to wonder what is the best location for the interpolation points. This question was answered by Chebyshev, who proved that if one chooses

$$x_h = \frac{b+a}{2} - \frac{b-a}{2} \cos \frac{2h+1}{2N+2} \pi , \tag{A.2.13}$$

for $h = 0, \ldots, N$, then

$$\left|\Pi(x)\right| \le \frac{(b-a)^{N+1}}{2^{2N+1}} ,$$

which might be compared with the estimate

$$\left|\Pi(x)\right| \le \frac{(b-a)^{N+1}(N-1)!}{4N^N}$$

obtained for equispaced nodes.

Note that a and b do not belong to the distribution (A.2.13). If one is compelled to include them as interpolation points, a good choice is

$$x_h = \frac{b+a}{2} - \frac{b-a}{2} \cos\left(\frac{h}{N}\pi\right) , \qquad h = 0, \ldots, N . \tag{A.2.13'}$$

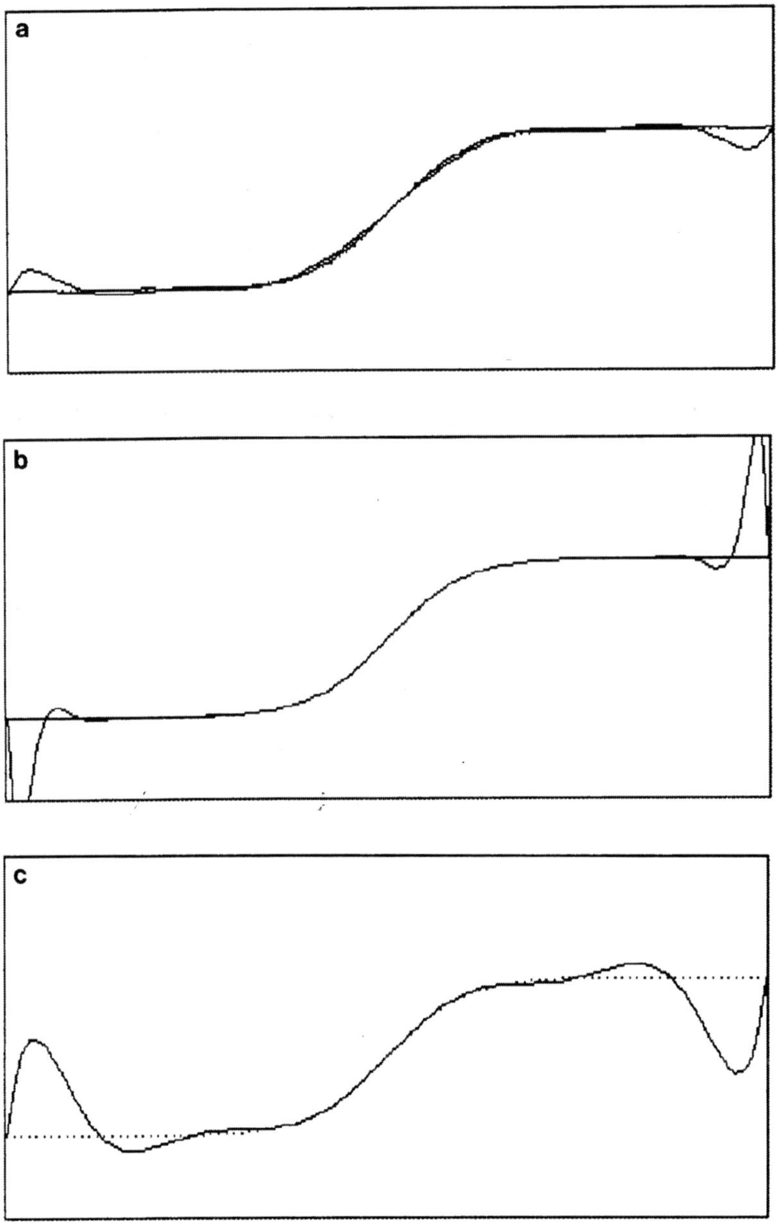

Figure A.2.2 — Lagrange interpolation of $u(x) = \tanh 5x$ in (a) 10 and (b) 20 nodes. In the figure e stands for equispaced nodes, c for Chebyshev nodes. In (c) the nodes are set randomly.

The following useful estimates hold for both the distributions (A.2.13) and (A.2.13'). If $u \in H_w^m$, for some $m \geq 1$

$$\|u(x) - p_N(x)\|_{L_w^2} \leq \frac{C(m)}{N^m}\|u\|_{H_w^m},$$

$$\|u(x) - p_N(x)\|_{L^\infty} \leq \frac{C(m)}{N^{m-\frac{1}{2}}}\|u\|_{H_w^m},$$

$$\|u(x) - p_N(x)\|_{H_w^k} \leq \frac{C(k,m)}{N^{m-2k}}\|u\|_{H_w^m}.$$

In particular

$$\left\|\frac{du}{dx}(x) - \frac{d}{dx}p_N(x)\right\|_{L_w^2} \leq \frac{C(m)}{N^{m-2}}\|u\|_{H_w^m}.$$

These estimates imply that, for these choices of interpolation points, if u is infinitely differentiable, then the interpolation error decreases faster than algebraically (i.e., spectrally).

Piecewise Interpolant

As already seen, the problem with polynomial interpolation is that if the location of the nodes is not optimal, the error might be very large. The largest error occurs near the extrema of the interval, where high degree polynomials oscillate wildly to fit all the data. In general, as shown in Fig. A.2.2, the larger N is, the higher the oscillations are.

To eliminate these oscillations one has then to keep low the degree of the interpolating polynomial. This can be done by partitioning the interval into subintervals

$$[x_0, x_N] = [x_0, x_{n_1}] \cup [x_{n_1}, x_{n_2}] \cup \cdots \cup [x_{n_m}, x_N]$$

and computing in any subinterval the relative interpolating polynomial, which will, of course, have any degree less than N (n_1, $n_2 - n_1$, ..., $N - n_m$, respectively).

The price to be paid is that the piecewise interpolant is no longer continuously differentiable in the nodes x_{n_i}, $i = 1, \ldots, m$ which are common to two intervals, as shown in Fig. A.2.3. This procedure is very useful when the function to be approximated has different character in different regions or in interpolation problems in more than one dimension.

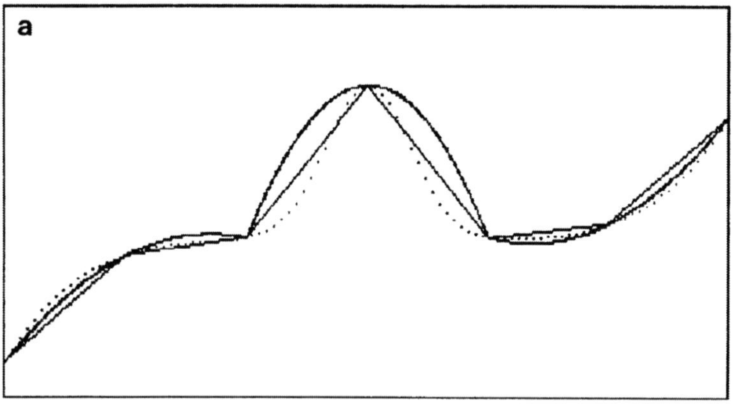

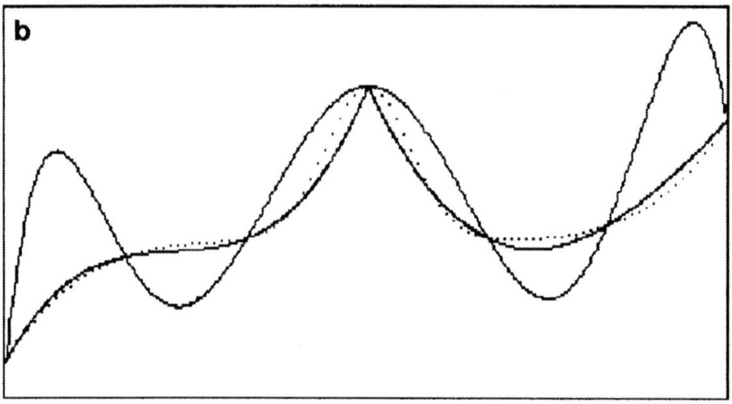

Figure A.2.3 — Piecewise interpolation of $u(x) = 10x^5 + e^{-100x^2}$ (dotted line) in six nodes. (a) Piecewise linear and quadratic interpolant (b) Piecewise cubic and full Lagrange interpolant.

Splines

A compromise between the two requirements:

- Low degree polynomials;
- Sufficient regularity

is given by the so-called splines.

DEFINITION A.2.1 Spline
*A **spline** $s(x)$ of degree m relative to the $N+1$ nodes $x_0 < \cdots < x_N$, with $a = x_0$ and $b = x_N$, is a function such that*

1. *In any interval $[x_{i-1}, x_i]$, $i = 1, \dots, N$, $s(x)$ is a polynomial of degree at most m;*
2. *$s(x) \in C^{m-1}_{[a,b]}$.*

For simplicity, we will only deal with cubic splines which, by the way, are the most commonly used because of their low degree (refer to the pertinent literature for a detailed treatment).

Item 1. of Definition A.2.1 states that in any subinterval $[x_{i-1}, x_i]$, $i = 1, \dots, N$, the cubic spline can be written as

$$s(x) = a_i + b_i x + c_i x^2 + d_i x^3\,, \qquad x \in [x_{i-1}, x_i]\,, \qquad i = 1, \dots, N\,.$$

Then one has $4N$ unknowns a_i, b_i, c_i, d_i, $i = 1, \dots, N$, to determine. The interpolating condition (A.2.5) yields the equations

$$\begin{aligned} a_i + b_i x_{i-1} + c_i x_{i-1}^2 + d_i x_{i-1}^3 &= u_{i-1}\,, \\ a_i + b_i x_i + c_i x_i^2 + d_i x_i^3 &= u_i\,, \end{aligned} \tag{A.2.14}$$

for $i = 1, \dots, N$, while the continuity of the first and second derivatives in the internal nodes gives, respectively,

$$\begin{aligned} b_i + 2c_i x_i + 3d_i x_i^2 &= b_{i+1} + 2c_{i+1} x_i + 3d_{i+1} x_i^2 \\ c_i + 3d_i x_i &= c_{i+1} + 3d_{i+1} x_i\,, \end{aligned} \tag{A.2.15}$$

for $i = 1, \dots, N-1$.

The interpolating condition (A.2.14) gives $2N$ costraints, while the continuity requirements (A.2.15) give $2(N-1)$ constraints. We have then $4N-2$ equations in $4N$ unknowns.

Hence the number of equations is not sufficient to find a unique solution. We need two more conditions, for instance on the derivatives in some interpolation point.

Usually one gives the values of either the first or of the second derivative at the extrema of the interval.

If, in particular, the second derivative of the spline at the end points is assumed to be zero, then the spline is called ***natural***.

On the other hand, if we know that the function is periodic of period $x_N - x_0$, then the values of the first and second derivatives, together with the values of the interpolant, in x_0 and x_N must be equal.

It may be proved that

THEOREM A.2.2 ***(Minimization Property of the Splines)*** *Among all the functions $f \in C^2_{[a,b]}$ that interpolate the data (A.2.5), the cubic natural spline is the one that minimizes*

$$I(f) = \int_a^b \left| \frac{d^2 f}{dx^2}(x) \right|^2 dx$$

and is the unique minimizer in $C^2_{[a,b]}$. ∎

For the actual computation of the coefficients of the spline, it is convenient to operate as follows. We know that $s(x) \in C^2_{[a,b]}$, therefore we can denote by u_i'', $i = 0, \ldots, N$, the values of the second derivative in the nodes, which are unknown except, in the case of a natural spline, for the values u_0'' and u_N''. We also know that the spline is piecewise cubic and therefore its second derivative is piecewise linear. Denoting by

$$h_i = x_i - x_{i-1}\,,$$

we can write

$$\frac{d^2 s}{dx^2}(x) = u_i'' \frac{x - x_{i-1}}{h_i} + u_{i-1}'' \frac{x_i - x}{h_i}\,, \qquad \text{for} \quad x \in [x_{i-1}, x_i]\,.$$

Integrating twice, we have

$$s(x) = u_i'' \frac{(x - x_{i-1})^3}{6h_i} + u_{i-1}'' \frac{(x_i - x)^3}{6h_i} + A_i x + B_i\,, \quad \text{for } x \in [x_{i-1}, x_i]\,,$$

where A_i and B_i are integration constants that can be evaluated by imposing the interpolation conditions at the extrema of the subinterval $[x_{i-1}, x_i]$, namely

$$s(x_{i-1}) = u_{i-1} \quad \text{and} \quad s(x_i) = u_i\,,$$

so that we obtain

$$s(x) = u_i'' \frac{(x - x_{i-1})^3}{6h_i} + u_{i-1}'' \frac{(x_i - x)^3}{6h_i}$$

$$+\left(u_{i-1}-\frac{u''_{i-1}h_i^2}{6}\right)\frac{x_i-x}{h_i}+\left(u_i-\frac{u''_i h_i^2}{6}\right)\frac{x-x_{i-1}}{h_i} \quad \text{(A.2.16)}$$

in $[x_{i-1}, x_i]$.

Imposing the continuity of the first derivative yields [LSA] the following system

$$\mu_i u''_{i-1} + 2u''_i + \lambda_i u''_{i+1} = d_i\,, \qquad i = 1, \ldots, N-1\,, \qquad \text{(A.2.17)}$$

where

$$\begin{aligned}\lambda_i &= \frac{h_{i+1}}{h_i + h_{i+1}}\,,\\ \mu_i &= \frac{h_i}{h_i + h_{i+1}}\,,\\ d_i &= \frac{6}{h_i + h_{i+1}}\left(\frac{u_{i+1}-u_i}{h_{i+1}} - \frac{u_i - u_{i-1}}{h_i}\right).\end{aligned} \qquad \text{(A.2.18)}$$

As already stated, in order to solve the linear system, we still need two conditions, which will be given setting the values of the derivatives at the extrema. For the moment, we will write these conditions in the general form

$$\begin{aligned}2u''_0 + \lambda_0 u''_1 &= d_0\,,\\ \mu_N u''_{N-1} + 2u''_N &= d_N\,,\end{aligned}$$

where the coefficients are still to be specified.

The system can then be written in matrix form as

$$\begin{pmatrix} 2 & \lambda_0 & 0 & \cdots & 0 & 0 & 0 \\ \mu_1 & 2 & \lambda_1 & \cdots & 0 & 0 & 0 \\ 0 & \mu_2 & 2 & \cdots & 0 & 0 & 0 \\ \vdots & \vdots & \vdots & \ddots & \vdots & \vdots & \vdots \\ 0 & 0 & 0 & \cdots & 2 & \lambda_{N-2} & 0 \\ 0 & 0 & 0 & \cdots & \mu_{N-1} & 2 & \lambda_{N-1} \\ 0 & 0 & 0 & \cdots & 0 & \mu_N & 2 \end{pmatrix} \begin{pmatrix} u''_0 \\ u''_1 \\ u''_2 \\ \vdots \\ u''_{N-2} \\ u''_{N-1} \\ u''_N \end{pmatrix} = \begin{pmatrix} d_0 \\ d_1 \\ d_2 \\ \vdots \\ d_{N-2} \\ d_{N-1} \\ d_N \end{pmatrix}. \qquad \text{(A.2.19)}$$

To determine λ_0, μ_N, d_0, and d_N, one can use the following rules:

1. Natural spline

$$\frac{d^2 s}{dx^2}(a) = \frac{d^2 s}{dx^2}(b) = 0 \Longleftrightarrow \lambda_0 = \mu_N = d_0 = d_N = 0\,.$$

2. Specifying the slope u'_0 and u'_N at the end points

$$\frac{ds}{dx}(a) = u'_0\,,\ \frac{ds}{dx}(b) = u'_N \Longleftrightarrow \begin{cases} \lambda_0 = \mu_N = 1 \\ d_0 = \dfrac{6}{h_1}\left(\dfrac{u_1 - u_0}{h_1} - u'_0\right) \\ d_N = \dfrac{6}{h_N}\left(u'_N - \dfrac{u_N - u_{N-1}}{h_N}\right). \end{cases}$$

If the value of the derivative at an extremum is not available, one may approximate it by interpolating the four points closest to the end of the interval with a cubic and taking its derivative at the end points.

3. Not-a-Knot spline
This choice corresponds to considering only the internal points x_i, $i = 1, \ldots, N-1$, as interpolation points and then imposing the two additional conditions of interpolation at the extrema $s(a) = u_0$ and $s(b) = u_N$. In this case, in the matrix (A.2.19) the first and last row and column are eliminated and

$$\begin{aligned} \lambda_1 &= \frac{2(h_2 - h_1)}{2h_2 + h_1}\,, \\ \mu_{N-1} &= \frac{2(h_{N-1} - h_N)}{2h_{N-1} + h_N}\,, \\ d_1 &= \frac{12}{2h_2 + h_1}\left[\frac{h_2 u_0}{h_1(h_1 + h_2)} - \frac{u_1}{h_1} + \frac{u_2}{h_1 + h_2}\right], \\ d_{N-1} &= \frac{12}{2h_{N-1} + h_N}\left[\frac{h_{N-1} u_N}{h_N(h_N + h_{N-1})} - \frac{u_{N-1}}{h_N} + \frac{u_{N-2}}{h_N + h_{N-1}}\right] \end{aligned}$$

are substituted for the values given in (A.2.18).

4. Periodic spline
Because of periodicity, in this case one has to consider Eqs.(A.2.17–A.2.18) for $i = 0, \ldots, N-1$ with

$$u''_{-1} = u''_{N-1}\,, \quad u''_N = u''_0\,, \quad u_{-1} = u_{N-1}\,, \quad u_N = u_0\,, \quad h_0 = h_N\,.$$

The linear system of N equations in the N unknowns u''_0, …, u''_{N-1} can then be written as

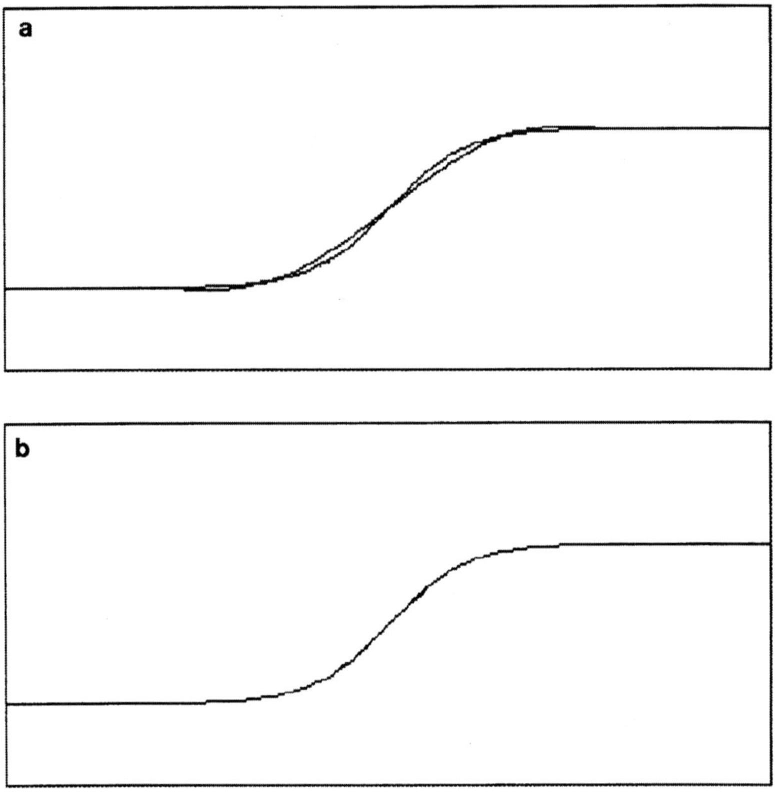

Figure A.2.4 — Spline interpolation of $f(x) = \tanh 5x$ in (a) 10 and (b) 20 nodes. In (a) e stands for equispaced nodes and c for Chebyshev nodes. In (b) the two interpolants almost coincide.

$$\begin{pmatrix} 2 & \lambda_0 & 0 & \cdots & 0 & \mu_0 \\ \mu_1 & 2 & \lambda_1 & \cdots & 0 & 0 \\ 0 & \mu_2 & 2 & \cdots & 0 & 0 \\ \vdots & \vdots & \vdots & \ddots & \vdots & \vdots \\ 0 & 0 & 0 & \cdots & 2 & \lambda_{N-2} \\ \lambda_{N-1} & 0 & 0 & \cdots & \mu_{N-1} & 2 \end{pmatrix} \begin{pmatrix} u_0'' \\ u_1'' \\ u_2'' \\ \vdots \\ u_{N-2}'' \\ u_{N-1}'' \end{pmatrix} = \begin{pmatrix} d_0 \\ d_1 \\ d_2 \\ \vdots \\ d_{N-2} \\ d_{N-1} \end{pmatrix}.$$

The success of the spline interpolation is related both to their smoothness (see Fig. A.2.4) and to the simplicity of the solution of the linear system (A.2.19). In fact, tridiagonal systems can be easily solved setting, first

$$w_i = d_i \qquad \text{for} \qquad i = 0, \dots, N$$

and then recursively computing

$$\left.\begin{aligned} p_k &= -\frac{\mu_k}{\nu_{k-1}} \\ \nu_k &= \nu_k + p_k \lambda_{k-1} \\ w_k &= w_k + p_k w_{k-1} \end{aligned}\right. \qquad k = 1, \dots, N \tag{A.2.20}$$

(in our case $\nu_k = 2$ for any k) and then proceeding backward from the value

$$u''_N = \frac{w_N}{\nu_N}$$

to obtain the others

$$u''_k = \frac{w_k - \lambda_k u''_{k+1}}{\nu_k} \qquad \text{for} \quad k = N-1, \dots, 0\,. \tag{A.2.21}$$

In conclusion, order N operations are needed to compute the coefficients u''_i of the spline and a fixed number of operations are then sufficient to evaluate the spline (A.2.16) at a point. On the other hand, Lagrange interpolation requires order N^2 operations to compute the coefficients Π'_h in (A.2.10) and order N operation to evaluate the interpolant (A.2.12) in a given point.

Hermitian Interpolation

Hermitian interpolation consists in finding a $p_N \in P_N$, such that

$$\begin{aligned} p_N(x_0) &= u_0^{(0)}\,, \quad \frac{dp_N}{dx}(x_0) = u_0^{(1)}\,, \;\dots\;, \frac{d^{\alpha_0} p_N}{dx^{\alpha_0}}(x_0) = u_0^{(\alpha_0)}\,, \\ p_N(x_1) &= u_1^{(0)}\,, \quad \frac{dp_N}{dx}(x_1) = u_1^{(1)}\,, \;\dots\;, \frac{d^{\alpha_1} p_N}{dx^{\alpha_1}}(x_1) = u_1^{(\alpha_1)}\,, \\ &\;\vdots \\ p_N(x_n) &= u_n^{(0)}\,, \quad \frac{dp_N}{dx}(x_n) = u_n^{(1)}\,, \;\dots\;, \frac{d^{\alpha_n} p_N}{dx^{\alpha_n}}(x_n) = u_n^{(\alpha_n)}\,, \end{aligned}$$

with

$$N = \left(\sum_{h=0}^{n} \alpha_h \right) + n\,.$$

To do that, set

$$\Pi_{\mathrm{Her}}(x) = \prod_{h=0}^{n} (x - x_h)^{\alpha_h + 1} \tag{A.2.22}$$

and

$$L_{hk}(x) = \frac{\Pi_{\mathrm{Her}}(x)}{k!\,(x - x_h)^{\alpha_h+1-k}} \frac{d^{\alpha_h - k}}{dx^{\alpha_h - k}} \left[\frac{(x - x_h)^{\alpha_h+1}}{\Pi_{\mathrm{Her}}(x)} \right]_{x = x_h}$$

for $h = 0, \ldots, n$ and $k = 0, \ldots, \alpha_i$. Then the desired polynomial is given by

$$p_N(x) = \sum_{h=0}^{n} \sum_{k=0}^{\alpha_h} u_h^{(k)} L_{hk}(x) \,. \tag{A.2.23}$$

When $\alpha_0 = \alpha_1 = \cdots = \alpha_n = 1$, then (A.2.23) simplifies into

$$p_N(x) = \sum_{i=0}^{n} \left[u_i^{(0)} + \left(u_i^{(1)} - \frac{\Pi_i''}{\Pi_i'} u_i^{(0)} \right) (x - x_i) \right] L_i^2(x) \,, \tag{A.2.24}$$

where Π_i' and $L_i(x)$ are, respectively, given by (A.2.10) and (A.2.11), and

$$\Pi_i'' = \frac{d^2\Pi}{dx^2}(x = x_i) = 2 \sum_{\substack{k=0 \\ k \neq i}}^{N} \prod_{\substack{j=0 \\ j \neq i \\ j \neq k}}^{n} (x_i - x_j) \,.$$

The interpolation formula (A.2.24) is called ***osculatory interpolation*** and its error can be estimated by

$$u(x) - p_N(x) = \frac{\Pi^2(x)}{(N+1)!} \frac{d^{N+1}u}{dx^{N+1}}(\xi)$$

with $\xi \in [\min\{x, x_0\}, \max\{x, x_0\}]$, and therefore

$$\max_{x \in [a,b]} \left| u(x) - p_N(x) \right| \leq \frac{1}{(N+1)!} \max_{x \in [a,b]} \left| \frac{d^{N+1}u}{dx^{N+1}}(x) \right| \max_{x \in [a,b]} \left| \Pi(x) \right|^2 .$$

Trigonometric Interpolation

If one has to interpolate a periodic function or knows that the function corresponding to the available data is periodic, then, instead of polynomials, it is convenient to use "easy" functions that take care of the periodicity. Assuming, for simplicity, that the period is 2π, which can always be done by rescaling the interpolation interval as in (A.2.4), the obvious choice is to use the trigonometric functions

$$1,\ \cos x,\ \sin x,\ \cos 2x,\ \sin 2x,\ \ldots,\ \cos Nx,\ \sin Nx\,.$$

A linear combination of these terms is known as a trigonometric polynomial of degree less than or equal to N. The corresponding linear space is usually denoted by $\mathcal{T}_N$ and, endowed with the scalar product

$$\langle u, v\rangle = \int_0^{2\pi} u(x)v(x)\,dx\,,$$

it has the structure of a Hilbert space called $L^2_{\rm per}([0,2\pi])$.

Given the $2N+1$ points

$$0 \le x_0 < x_1 < \cdots < x_{2N} < 2\pi\,,$$

one can define the ***fundamental trigonometric polynomials***

$$T_h(x) = \frac{\displaystyle\prod_{\substack{j=0\\ j\neq h}}^{2N} \sin\frac{x-x_j}{2}}{\displaystyle\prod_{\substack{j=0\\ j\neq h}}^{2N} \sin\frac{x_h-x_j}{2}}\,, \qquad h=0,\ldots,2N\,, \tag{A.2.25}$$

which, as for Lagrange fundamental polynomials, have the property that

$$T_h(x_k) = \delta_{hk}\,, \qquad \forall\, h,k = 0,\ldots,2N\,.$$

To show that the function $T_h(x)$ belongs to $\mathcal{T}_N$, it is sufficient to observe that each factor of the numerator can be written as

$$\sin\frac{x-x_j}{2} = \alpha e^{ix/2} + \beta e^{-ix/2}\,,$$

where i is the imaginary unit, for appropriate constants α and β. The product is then of the form

$$\sum_{k=-N}^{N} \alpha_k e^{ikx} ,$$

which belongs to $\mathcal{T}_N$.

The trigonometric polynomial

$$T_N(x) = \sum_{h=0}^{2N} u_h T_h(x) \tag{A.2.26}$$

interpolates the data u_k, $k = 0, \ldots, 2N$, in the given nodes.

The following useful estimates hold for equispaced distribution of nodes. If $u \in H^m$ for some $m \geq 1$, then

$$\left\| u(x) - p_N(x) \right\|_{L^2_{\text{per}}} \leq \frac{C(m)}{N^m} \left\| \frac{d^m u}{dx^m} \right\|_{L^2_{\text{per}}} ,$$

$$\left\| u(x) - p_N(x) \right\|_{L^\infty} \leq C(m) \frac{\log N}{N^m} \left\| \frac{d^m u}{dx^m} \right\|_{L^2_{\text{per}}} ,$$

$$\left\| u(x) - p_N(x) \right\|_{H^k_w} \leq \frac{C(k,m)}{N^{m-k}} \left\| \frac{d^m u}{dx^m} \right\|_{L^2_{\text{per}}} ,$$

for $k < m$. In particular

$$\left\| \frac{du}{dx} - \frac{dp_N}{dx}(x) \right\|_{L^2_{\text{per}}} \leq \frac{C(m)}{N^{m-1}} \left\| \frac{d^m u}{dx^m} \right\|_{L^2_{\text{per}}} .$$

This implies that for equispaced interpolation points, if u is infinitely differentiable, then the interpolation error decreases faster than algebraically (i.e., spectrally).

Bernstein polynomials

As far as the problem of approximation is concerned, Weierstrass proved the following theorem.

THEOREM A.2.3 ***(Best Approximation)***
Given a function $u \in C_{[a,b]}$, then for any $\varepsilon > 0$ it is possible to find a polynomial $p(x)$ of sufficiently high degree such that

$$\max_{x\in[a,b]} \left|u(x) - p(x)\right| < \varepsilon .$$

■

Bernstein proved this theorem using the polynomials

$$B_N(u;x) = \sum_{k=0}^{N} u\left(\frac{k}{N}\right)\binom{N}{k} x^k(1-x)^{N-k},$$

defined in $[0,1]$. In fact, it can be proved that if $u \in C_{[0,1]}$, then given an $\varepsilon > 0$

$$\left|u(x) - B_N(f;x)\right| \leq \varepsilon , \qquad \forall x \in [0,1],$$

for any sufficiently large N.

This is not a trivial result, since as we saw for the interpolation, increasing the degree of the Lagrange interpolant does not assure that $|u(x) - p_N(x)|$ goes to zero.

Orthogonal Polynomial Approximation

Assume that a weight function $w(x)$ has been chosen and consider the weighted space L^2_w defined in Appendix 1, endowed with the scalar product (A.1.15). Given the sequence of polynomials 1, x, $x^2, \ldots, x^N$, it is always possible to deduce from them, through a process called Gram–Schmidt orthonormalization, a sequence of polynomials

$$p_0^*(x)\,,\; p_1^*(x)\,,\; p_2^*(x)\,, \ldots \;,\; p_N^*(x)$$

orthonormal with respect to the scalar product (A.1.15) with the given weight

$$\int_a^b p_i^*(x)p_j^*(x)w(x)\,dx = \delta_{ij}\,,$$

where δ_{ij} is the Kronecker delta.

The polynomial

$$p_N(x) = \sum_{i=0}^{N} \langle f\,,\, p_i^*\rangle p_i^*$$

has the following minimum property:

THEOREM A.2.4 ***(Best Approximation in*** $\mathbf{L^2_w}$***)***
Let p_0^*, $p_1^*, \ldots, p_N^*$ *be an orthogonal system and let* u *be an arbitrary function, then*

$$\left\| u - \sum_{i=0}^{N} \langle u \,,\, p_i^* \rangle p_i^* \right\|_{L^2_w} \leq \left\| u - \sum_{i=0}^{N} C_i p_i^* \right\|_{L^2_w}$$

for any selection of the constants C_i*.* ■

The most commonly used weights and therefore orthogonal polynomials (but not orthonormal) are given in Table A.2.1.

The choice of the appropriate weight depends on what it is meant by error and on where greater accuracy is needed. That is, we might consider errors in a certain region of the interval more important than those made in other regions. For this reason we might weight more heavily errors in some regions that in others. In particular, Chebyshev weight $(1 - x^2)^{-1/2}$ is appropriate when greater accuracy is required near the ends of the interval with respect to that needed in its center. This occurs when one needs to evaluate derivatives at the extrema of the interval. The opposite occurs, for instance, for the weight $\sqrt{1 - x^2}$.

The same goal can be accomplished for periodic functions. It is, in fact, sufficient to observe that

$$\int_0^{2\pi} \sin hx \sin kx \, dx = \pi \delta_{hk} \,, \quad \forall h, k > 0 \,,$$

$$\int_0^{2\pi} \cos hx \cos kx \, dx = (1 + \delta_{h0}\delta_{k0}) \pi \delta_{hk} \,, \quad \forall h, k \geq 0 \,,$$

$$\int_0^{2\pi} \sin hx \cos kx \, dx = 0 \,, \quad \forall h, k \geq 0 \,,$$

where δ_{hk} are Kronecker deltas and therefore the set of functions

$$1, \cos x, \sin x, \cos 2x, \sin 2x, \ldots, \cos Nx, \sin Nx, \ldots$$

defines an orthogonal set in the subspace of 2π-periodic square integrable functions $L^2_{\text{per}}([0, 2\pi])$ for which the previous theorem still holds.

We conclude this topic by listing some properties of the orthogonal polynomials reported in Table A.2.1.

Name	Symbol	Interval	Weight
Legendre	$P_N(x)$	$[-1, 1]$	1
Chebyshev 1st kind	$T_N(x)$	$[-1, 1]$	$\dfrac{1}{\sqrt{1-x^2}}$
Chebyshev 2nd kind	$U_N(x)$	$[-1, 1]$	$\sqrt{1-x^2}$
Ultraspherical	$C_N^{\mu}(x)$	$[-1, 1]$	$(1-x^2)^{\mu-\frac{1}{2}}$ $\mu > -\frac{1}{2}$
Jacobi	$P_N^{(\alpha,\beta)}(x)$	$[-1, 1]$	$(1-x)^{\alpha}(1+x)^{\beta}$ $\alpha, \beta > -1$
Laguerre Standard	$L_N(x)$	$[0, +\infty)$	e^{-x}
Laguerre Generalized	$H_N^{(\alpha)}(x)$	$[0, +\infty)$	$x^{\alpha}e^{-x}$ $\alpha > -1$
Hermite	$L_N(x)$	$(-\infty, +\infty)$	e^{-x^2}

Table A.2.1 — Principal orthogonal polynomials.

Properties of Legendre Polynomials

- Explicit expression

$$P_N(x) = \frac{1}{2^N} \sum_{k=0}^{[N/2]} (-1)^k \binom{N}{k} \binom{2(N-k)}{N} x^{N-2k},$$

where $[N/2]$ is the largest integer smaller than $N/2$.

- Recurrence relation

$$P_0(x) = 1\,,$$
$$P_1(x) = x\,,$$
$$P_{N+1}(x) = \frac{2N+1}{N+1}xP_N(x) - \frac{N}{N+1}P_{N-1}(x)\,.$$

- Differential equation

$$(1-x^2)\frac{d^2u}{dx^2} - 2x\frac{du}{dx} + N(N+1)u = 0\,.$$

- Zeros

$$x_k = \left(1 - \frac{1}{8N^2} + \frac{1}{8N^3}\right)\cos\frac{4k-1}{4N+2}\pi + O\left(\frac{1}{N^4}\right)\,.$$

- Norm

$$\int_{-1}^{1} \left|P_N(x)\right|^2 dx = \frac{2}{2N+1}\,.$$

- Inequality

$$|P_N(x)| \leq 1\,.$$

Properties of Chebyshev Polynomials of the First Kind

- Explicit expression

$$T_N(x) = \frac{N}{2}\sum_{k=0}^{[N/2]}(-1)^k\frac{(N-k-1)!}{k!\,(N-2k)!}(2x)^{N-2k} = \cos\left(N\arccos x\right),$$

where $[N/2]$ is the largest integer smaller than $N/2$.

- Inverse expression

$$x^k = \begin{cases} 2^{1-k}\displaystyle\sum_{j=0}^{k/2}\binom{k}{\frac{k}{2}-j}T_{2j}(x) - \frac{1}{2^k}\binom{k}{\frac{k}{2}}, & \text{if } k \text{ is even} \\ 2^{1-k}\displaystyle\sum_{j=0}^{(k-1)/2}\binom{k}{\frac{k-1}{2}-j}T_{2j+1}(x)\,, & \text{if } k \text{ is odd.} \end{cases}$$

- Recurrence relation

$$\begin{aligned} T_0(x) &= 1\,, \\ T_1(x) &= x\,, \\ T_{N+1}(x) &= 2xT_N(x) - T_{N-1}(x)\,. \end{aligned}$$

- Differential equation

$$(1-x^2)\frac{d^2u}{dx^2} - x\frac{du}{dx} + N^2u = 0\,.$$

- Zeros

$$x_k = \cos\frac{2k-1}{2N}\,\pi\,.$$

- Norm

$$\int_{-1}^{1} \frac{|T_N(x)|^2}{\sqrt{1-x^2}}\,dx = \begin{cases} \pi\,, & \text{if } N=0\,, \\ \dfrac{\pi}{2}\,, & \text{otherwise.} \end{cases}$$

- Inequality

$$\left|T_N(x)\right| \le 1\,.$$

Properties of Chebyshev Polynomials of the Second Kind

- Explicit expression

$$\begin{aligned} U_N(x) &= \sum_{k=0}^{[N/2]} (-1)^k \frac{(k-N)!}{k!\,(N-2k)!}(2x)^{N-2k} = \frac{1}{N+1}\,\frac{dT_{N+1}}{dx}(x)\,, \\ U_N(\cos x) &= \frac{\sin(N+1)x}{\sin x}\,, \end{aligned}$$

where $[N/2]$ is the largest integer smaller than $N/2$.

- Recurrence relation

$$\begin{aligned} U_0(x) &= 1\,, \\ U_1(x) &= 2x\,, \\ U_{N+1}(x) &= 2xU_N(x) - U_{N-1}(x)\,. \end{aligned}$$

- Differential equation

$$(1-x^2)\frac{d^2u}{dx^2} - 3x\frac{du}{dx} + N(N+1)u = 0\,.$$

- Zeros

$$x_k = \cos\frac{k}{N+1}\pi\,.$$

- Norm

$$\int_{-1}^{1} \left|U_N(x)\right|^2 \sqrt{1-x^2}\,dx = \frac{\pi}{2}\,.$$

- Inequality

$$\left|U_N(x)\right| \le N+1\,.$$

Properties of Ultraspherical Polynomials

- Explicit expression

$$C_N^\mu(x) = \frac{1}{\Gamma(\mu)}\sum_{k=0}^{[N/2]}(-1)^k\frac{\Gamma(\mu+N-k)}{k!\,(N-2k)!}\,(2x)^{N-2k}\,,$$

where Γ is the gamma function and where $[N/2]$ is the largest integer smaller than $N/2$.

- Recurrence relation

$$C_0^\mu(x) = 1\,,$$

$$C_1^\mu(x) = 2\mu x\,,$$

$$C_{N+1}^\mu(x) = 2\frac{N+\mu}{N+1}xC_N^\mu(x) - \frac{N+2\mu-1}{N+1}C_{N-1}^\mu(x)\,.$$

- Differential equation

$$(1-x^2)\frac{d^2u}{dx^2} - (2\mu+1)x\frac{du}{dx} + N(N+2\mu)u = 0\,.$$

- Norm

$$\int_{-1}^{1} \left|C_N^\mu(x)\right|^2 (1-x^2)^{\mu-1/2}\,dx = \frac{\pi 2^{1-2\mu}\Gamma(N+2\mu)}{N!\,(N+\mu)[\Gamma(\mu)]^2}\,.$$

- Inequality

$$\left|C_N^\mu(x)\right| \le \begin{cases} \binom{N+2\mu-1}{N}, & \text{if } \mu > 0\,, \\ \left|C_N^\mu(\widehat{x})\right|, & \text{if } -\dfrac{1}{2} < \mu < 0\,, \end{cases}$$

where $\widehat{x} = 0$ if N is even and is equal to the maximum point nearest zero if N is odd.

Properties of Jacobi Polynomials

- Explicit expression

$$P_N^{(\alpha,\beta)}(x) = \frac{1}{2^N} \sum_{k=0}^{N} \binom{N+\alpha}{k} \binom{N+\beta}{N-k} (x-1)^{N-k}(x+1)^k\,.$$

- Recurrence relation

$$P_0^{(\alpha,\beta)}(x) = 1\,,$$

$$P_1^{(\alpha,\beta)}(x) = \frac{\alpha-\beta}{2} + \frac{2+\alpha+\beta}{2}x\,,$$

$$\begin{aligned} &2(N+1)(N+\alpha+\beta+1)(2N+\alpha+\beta)P_{N+1}^{(\alpha,\beta)}(x) \\ &= (2N+\alpha+\beta+1)\big[(\alpha^2-\beta^2) + (2N+\alpha+\beta+2)(2N+\alpha+\beta)x\big]P_N^{(\alpha,\beta)} \\ &\quad - \frac{2(N+\alpha)(N+\beta)(2N+\alpha+\beta+2)}{2(N+1)(N+\alpha+\beta+1)(2N+\alpha+\beta)}P_{N-1}^{(\alpha,\beta)}(x)\,. \end{aligned}$$

- Differential equation

$$(1-x^2)\frac{d^2u}{dx^2} + \big[\beta - \alpha - (\alpha+\beta+2)x\big]\frac{du}{dx} + N(N+\alpha+\beta+1)u = 0.$$

- Norm

$$\begin{aligned} &\int_{-1}^{1} \left|P_N^{(\alpha,\beta)}(x)\right|^2 (1-x)^\alpha (1+x)^\beta\, dx \\ &\qquad = \frac{2^{\alpha+\beta+1}\Gamma(N+\alpha+1)\Gamma(N+\beta+1)}{N!\,(2N+\alpha+\beta+1)\Gamma(N+\alpha+\beta+1)}\,, \end{aligned}$$

where Γ is the well-known gamma function.

- Inequality

$$
\left|P_N^{(\alpha,\beta)}(x)\right| \le \begin{cases} \binom{N+q}{N} \cong N^q\,, & \text{if } q \ge -\dfrac{1}{2}\,, \\ \left|P_N^{(\alpha,\beta)}(\widehat{x})\right| \cong \dfrac{1}{\sqrt{N}}\,, & \text{otherwise,} \end{cases}
$$

where $\widehat{x}$ is one of the maximum point nearest $(\beta-\alpha)/(\alpha+\beta+1)$ and $q = \max\{\alpha,\beta\}$.

Properties of Generalized Laguerre Polynomials

- Explicit expression

$$
L_N^{(\alpha)}(x) = \sum_{k=0}^{N} \frac{(-1)^k}{k!} \binom{N+\alpha}{N-k} x^k\,.
$$

- Recurrence relation

$$
\begin{aligned}
L_0^{(\alpha)}(x) &= 1\,, \\
L_1^{(\alpha)}(x) &= 1+\alpha-x\,, \\
L_{N+1}^{(\alpha)}(x) &= \frac{2N+\alpha+1-x}{N+1} L_N^{(\alpha)}(x) - \frac{N+\alpha}{N+1} L_{N-1}^{(\alpha)}(x)\,.
\end{aligned}
$$

- Differential equation

$$
x\frac{d^2u}{dx^2} + (\alpha+1-x)\frac{du}{dx} + Nu = 0\,.
$$

- Norm

$$
\int_0^{+\infty} \left|L_N^{(\alpha)}(x)\right|^2 x^\alpha e^{-x}\,dx = \frac{\Gamma(N+\alpha+1)}{N!}\,,
$$

where Γ is the well-known gamma function.

- Inequality

$$
\left|L_N^{(\alpha)}(x)\right| \le \begin{cases} \dfrac{\Gamma(N+\alpha+1)}{N!\,\Gamma(\alpha+1)} e^{x/2}\,, & \text{if } \alpha \ge 0\,, \\ \left[2 - \dfrac{\Gamma(N+\alpha+1)}{N!\,\Gamma(\alpha+1)}\right] e^{x/2}\,, & \text{if } \alpha < 0\,. \end{cases}
$$

Properties of Hermite Polynomials

- Explicit expression

$$H_N(x) = N! \sum_{k=0}^{[N/2]} \frac{(-1)^k}{k!\,(N-2k)!} (2x)^{N-2k} ,$$

where $[N/2]$ is the largest integer smaller than $N/2$.

- Recurrence relation

$$\begin{aligned} H_0(x) &= 1 , \\ H_1(x) &= 2x , \\ H_{N+1}(x) &= 2xH_N(x) - 2NH_{N-1}(x) . \end{aligned}$$

- Differential equation

$$\frac{d^2u}{dx^2} - 2x\frac{du}{dx} + 2Nu = 0 .$$

- Norm

$$\int_{-\infty}^{+\infty} \left|H_N(x)\right|^2 e^{-x^2}\, dx = \sqrt{\pi}\, 2^N\, N! .$$

- Inequality

$$\left|H_N(x)\right| \le \begin{cases} 2^N \left(\dfrac{N}{2}\right)! \left[2 - \dfrac{1}{2^N}\dbinom{N}{\frac{N}{2}}\right] e^{x^2/2} , & \text{if } N \text{ is even,} \\[2ex] \dfrac{(N+1)!}{\left(\dfrac{N+1}{2}\right)!} |x| e^{x^2/2} , & \text{if } N \text{ is odd.} \end{cases}$$

Surface Fitting

The conceptually simplest way to obtain an interpolating surface is to divide the domain in a rectangular lattice with the lines parallel to the coordinate axes and to generalize the technique previously sketched for curve fitting to two (or more) dimensions. This procedure goes under the name of ***product scheme***.

Assume that the data

$$u_{hk} , \qquad h = 0, \ldots, N , \qquad k = 0, \ldots, M ,$$

corresponding to the nodes

$$(x_h, y_k), \qquad k = 0, \ldots, N, \qquad k = 0, \ldots, M,$$

with $x_0 < x_1 < \cdots < x_N$ and $y_0 < y_1 < \cdots < y_M$, are available (for instance, $u(x_h, y_k)$).

We have already constructed some sets of so-called ***cardinal basis functions***

$$\varphi_i(x), \quad \text{such that} \quad \varphi_i(x_h) = \delta_{ih},$$
$$\psi_j(y), \quad \text{such that} \quad \psi_j(y_k) = \delta_{jk},$$

where δ_{ih} and δ_{jk} are Kronecker deltas. For instance, the Lagrange polynomials (A.2.11) or the fundamental trigonometric polynomials (A.2.25) are such functions.

One can also obtain the set of cardinal cubic splines by computing the splines $s_j(x)$ which interpolate the data $u_i = \delta_{ij}$, $i = 0, \ldots, N$, in the given nodes.

It is plain from the definition of φ_i and ψ_j that their product

$$\chi_{ij}(x, y) = \varphi_i(x)\psi_j(y)$$

is such that

$$\chi_{ij}(x_h, y_k) = \begin{cases} 1, & \text{if } h = i \text{ and } k = j, \\ 0, & \text{otherwise} \end{cases}$$

and is therefore called a ***two-dimensional cardinal basis function***.

The polynomial interpolating the data u_{ij} at (x_i, y_j) is then given by

$$p_{NM}(x, y) = \sum_{j=0}^{M} \sum_{i=0}^{N} u_{ij} \chi_{ij}(x, y); \qquad \text{(A.2.27)}$$

or

$$p_{NM}(x, y) = \sum_{j=0}^{M} \psi_j(y) \sum_{i=0}^{N} u_{ij} \varphi_i(x). \qquad \text{(A.2.28)}$$

The question now is which kind of cardinal basis function $\varphi(x)$ and $\psi(y)$ to use in this product scheme.

As we have already seen, computing the Lagrange interpolants is costly, say with respect to splines, and they might strongly oscillate near the end points. The first disadvantage might be eliminated since the two-dimensional cardinal basis functions χ_{ij} can be computed once and for all before the actual beginning of the interpolation scheme. The storage space needed can be, however, very large.

On the other hand, the strong oscillations can only be reduced by choosing adequately the location of the interpolation point or of the interpolation scheme, as it is evident in Fig. A.2.5.

In any case, all the cardinal basis functions we have encountered extend over the whole range of interpolation. It is therefore lengthy to evaluate for any (x, y) Eq.(A.2.28). To overcome this difficulty, one can use a cardinal basis function with smaller support. In this light, the use of B-spline is helpful. However, this topic is beyond the scope of this appendix and is explained very well by de Boor [deB] and by Späth [SPA].

Scientific Programs

We end this appendix with a description of two routines that perform Lagrange and spline interpolation, respectively. The former is fundamental to understanding how crucial the choice of interpolation (or collocation) points is. The latter is useful for handling the output data, though splines have also been succesfully applied in several fields, for instance to the solution of differential equations, integration, optimization, and so on.

The routines follow the same guidelines given in Chapter 2, so that anyone who has played a little with the programs given there and with the relative exercises should understand these without any trouble. Furthermore the line-by-line description is already included, in some sense, in the programs and, except for the core of the programs, does not differ much from those given in Chapter 2. Thus we will not repeat it here.

INTERPOL: A Lagrange interpolation routine

The program is made up of

- A calling program (lines `1-199`), which is essentially a long input part with several options. These are essential to guide the reader to the understanding of the importance of a correct choice of interpolation points.

- A graphic subroutine, which is essentially that used in the programs proposed in Chapter 2 (lines `800-1299`).

- The main subroutine, which performs the interpolation (lines `1300-1599`).

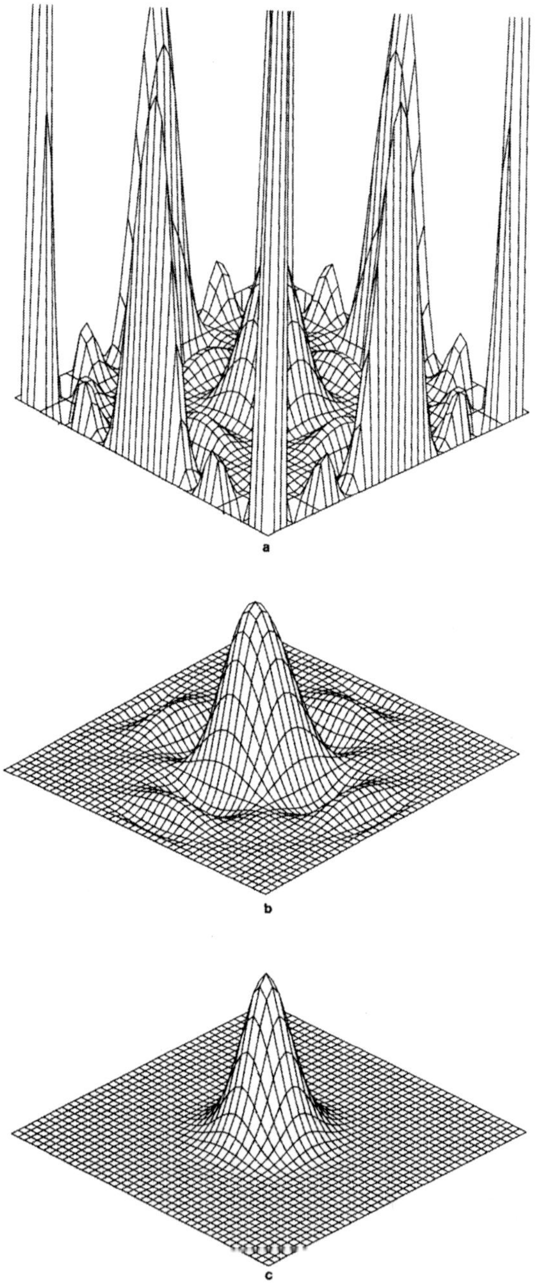

Figure A.2.5 — Surface fitting with (a) polynomial interpolation at equispaced nodes (b) polynomial interpolation at Chebyshev nodes (c) spline interpolation at equispaced nodes.

In detail, lines 150-240 are used if randomly chosen interpolation data are desired (the DO-loop 190-220 order these data) and lines 250-340 are used to input data and extrema of the plotting area.

If a function is to be interpolated, then it has to be inserted in line 240. Lines 350-480 also include input commands for the kind of interpolation points and the plotting area.

Lines 510-720 define then the interpolation points (Chebyshev, equispaced, etc.) and set the values of the functions in these points. Finally, the interpolation (and graphic) subroutine are called.

In this part, we first calculate the quantities Π'_h in (A.2.10) (lines 1310-1360) and then the constant terms appearing in the summation of Eq.(A.2.12) (line 1380). These constant terms are stored since they are independent of the point where the interpolant is evaluated. They are then used to compute the sum in Eq.(A.2.12) (lines 1470-1510). Finally, the fundamental polynomial $\Pi(x)$ in (A.2.9) is evaluated in lines 1520-1560. The multiplication of the two terms gives the value of the interpolant (line 1570).

The reader now has all the tools necessary to refer to Section 3.10 to write his own program performing the Discrete Chebyshev Transform, the Discrete Fourier Transform, and their inverses.

INTERPOL.BAS

```
1 REM ***********************************************************
2 REM *                 Lagrange interpolation                  *
3 REM ***********************************************************
4 REM
5 REM ***********************************************************
6 REM *           Main variables used in the program            *
7 REM ***********************************************************
10 REM nndd = number of interpolation points < 100 (nd=nndd-1)
11 REM x(nndd) = interpolation points
12 REM y(nndd) = interpolation data
13 REM pi1(nndd) = PI' in Eq.(A.2.10)
14 REM pL(npoint) = interpolant
15 REM sum = summation in Eq.(A.2.12)
16 REM pi0 = PI in Eq.(A.2.9)
17 REM fnf(x) = function to interpolate
18 REM c, e = parameters used in defining the function fnf(x)
19 REM a, b = extrema of interpolation
20 REM pi = 3.1415926 (used to set Chebyshev points)
21 REM npoint = number of evaluations of interpolant (<1000)
22 REM deltax,deltay = length of plotted marks
23 REM igraph = flag indicating the extrema of graph
```

```
24 REM function$ = flag indicating the type of interpolation data
25 REM ix1, ix2 = flag indicating the type of interpolation
   points (x)
26 REM kind$ = indicates the type of interpolation points
27 REM xmin, xmax, ymin, ymax = extrema of graph
28 REM ndin, ndfin, ndstep = used for automatic increase of
   number of points
30 DIM x(100), y(100), pi1(100), pL(1000)
40 CLS 0
50 REM ************************************************************
51 REM *                           Input part                             *
52 REM ************************************************************
70 PRINT "LAGRANGE INTERPOLATION"
80 PRINT "Values in parentheses are options or suggested values
   (not defaults)"
90 PRINT ""
100 INPUT "Do you want to interpolate a function (f), random data
   (r), or your own data (d)"; function$
110 IF function$ = "r" THEN GOTO 140
120 IF function$ = "d" THEN GOTO 250
130 IF function$ = "f" THEN GOTO 350
140 REM ***********************************************************
141 REM *             Setting random interpolation data           *
142 REM ***********************************************************
150 RANDOMIZE
160 nd = 10:  a = 0:  b = 1
170 PRINT " x", " y"
180 FOR i = 0 TO nd:  x(i) = RND: y(i) = RND: NEXT i
190 FOR i = 0 TO nd
200 FOR j = i + 1 TO nd
210 IF x(i) > x(j) THEN max = x(i):  x(i) = x(j):  x(j) = max
220 NEXT j:  NEXT i
230 FOR i = 0 TO nd:  PRINT x(i), y(i):  NEXT i:
   kind$ = "random nodes"
240 GOTO 310
250 REM ***********************************************************
251 REM *                  Setting your own data                  *
252 REM ***********************************************************
255 INPUT "In which interval do you want to interpolate (a,b)";
   a, b
260 INPUT "How many data points"; nndd:  nd = nndd - 1
270 FOR i = 0 TO nd
280 PRINT "datum n."; i + 1; :  INPUT "(x,y)="; x(i), y(i)
300 NEXT i
310 PRINT "Minimum and maximum value of y in plot"
320 INPUT "Default (0), Range of the interpolant (1),
   Your own (2)"; igraph
330 CLS 0
```

```
340 GOTO 730
350 REM ******************************************************
351 REM *     Interpolating the function given in line 420      *
352 REM ******************************************************
355 INPUT "How many interpolation points? (Press RETURN for
   automatic increase)"; ndd: ndin = ndd - 1
360 IF ndd <> 0 THEN ndfin = ndin: nstep = 1
370 IF ndd = 0 THEN ndin = 10: ndfin = 50: nstep = 5
380 INPUT "In which interval do you want to interpolate (-1,1)";
   a, b
390 INPUT "Coefficient of hyperbolic tangent (5)", c
410 e = 2.718281828#
420 DEF fnf (x) = (1 - e ^ (-2 * c * x)) / (1 + e ^ (-2 * c * x))
430 PRINT "Node distribution"
440 PRINT "Chebyshev (1) Chebyshev with extrema (2),"
450 INPUT "equispaced (3), denser in the middle (4),
   or your own(5)"; ix1
460 INPUT "Comparison with another distribution (no=0,
   second choice=1,2,3,4,5)"; ix2
470 PRINT "Minimum and maximum value of y in plot"
480 INPUT "Default (0), Range interpolation data (1),
   Your own (2)"; igraph
485 REM ******************************************************
486 REM *                 End of input part                    *
487 REM ******************************************************
490 CLS 0
500 FOR nd = ndin TO ndfin STEP nstep
510 REM ******************************************************
511 REM *            Setting interpolation points              *
512 REM ******************************************************
520 ix = ix1: ixx = ix2: pi = 3.1415926#
530 FOR i = 0 TO nd
540 ON ix GOTO 550, 580, 610, 640, 680
550 x(i) = a + (b - a) * (1 - COS((2 * i + 1) * pi / (2 *
   nd + 2))) / 2
560 kind$ = "Chebyshev nodes"
570 GOTO 710
580 x(i) = a + (b - a) * (1 - COS(i * pi / nd)) / 2
590 kind$ = "Chebyshev nodes with extrema"
600 GOTO 710
610 x(i) = a + (b - a) * i / nd
620 kind$ = "Equispaced nodes"
630 GOTO 710
640 xi = 2 * i / nd - 1
650 x(i) = (b + a) / 2 + (b - a) * (xi ^ 3 + xi) / 4
660 kind$ = "Nodes denser in the middle"
670 GOTO 710
680 PRINT "node n."; i + 1; : INPUT x(i)
```

```
690 kind$ = "My input"
710 y(i) = fnf(x(i))
720 NEXT i
730 REM ************************************************************
731 REM *            Calling subroutine that evaluates               *
732 REM *              the polynomial interpolant of                  *
733 REM *              (x(i),y(i)) in npoint points                   *
734 REM ************************************************************
740 npoint = 100
750 GOSUB 1300
760 REM ************************************************************
761 REM *              Calling the graphic subroutine                 *
762 REM ************************************************************
770 GOSUB 800
780 IF ixx <> 0 THEN ix = ixx:  ixx = 0:  GOTO 530
790 NEXT nd
799 END
800 REM ************************************************************
801 REM *                    Graphic subroutine                       *
802 REM ************************************************************
810 IF ix2 <> 0 AND ixx = 0 THEN GOTO 1050
820 xmin = a
830 xmax = b
840 ymin = 1000000
850 ymax = -1000000
860 REM ************************************************************
861 REM *        Choosing extrema of plotting area among              *
862 REM *         default values, minimum and maximum of              *
863 REM *         interpolation data, or your own choice              *
864 REM ************************************************************
890 ON igraph GOTO 970, 1020
900 FOR i = 0 TO nd
910 IF y(i) > ymax THEN ymax = y(i)
920 IF y(i) < ymin THEN ymin = y(i)
930 NEXT i
940 ymin = (3 * ymin - ymax) / 2
950 ymax = (3 * ymax - ymin) / 2
960 GOTO 1050
970 FOR i = 0 TO npoint
980 IF pL(i) > ymax THEN ymax = pL(i)
990 IF pL(i) < ymin THEN ymin = pL(i)
1000 NEXT i
1010 GOTO 1050
1020 PRINT "Minimum and maximum value of x and y in plot"
1030 INPUT "xmin and xmax"; xmin, xmax
1040 INPUT "ymin and ymax"; ymin, ymax
1050 IF ix2 = 0 THEN CLS 0
1060 IF ix2 <> 0 AND ixx <> 0 THEN CLS 0
```

```
1070 KEY OFF: SCREEN 2
1080 VIEW (3, 30)-(630, 180), , 1:  WINDOW (xmin, ymin)-(xmax,
   ymax)
1085 REM ********************************************************
1086 REM *                Plotting the interpolant              *
1087 REM ********************************************************
1090 PSET (a, pL(0))
1100 FOR i = 1 TO npoint
1110 xx = a + (b - a) * i / npoint
1120 LINE -(xx, pL(i))
1130 NEXT i
1140 IF function$ <> "f" THEN GOTO 1200
1150 FOR i = 0 TO 100
1160 xsub = a + (b - a) * i / 100
1170 PSET (xsub, fnf(xsub))
1180 NEXT i
1190 GOTO 1250
1195 REM ********************************************************
1196 REM *           Marking the interpolation points           *
1197 REM ********************************************************
1200 deltax = (xmax - xmin) / 40:  deltay = (ymax - ymin) / 20
1210 FOR i = 0 TO nd
1220 PSET (x(i) - deltax, y(i)):  LINE -(x(i) + deltax, y(i))
1230 PSET (x(i), y(i) - deltay):  LINE -(x(i), y(i) + deltay)
1240 NEXT i
1250 REM ********************************************************
1251 REM *                   Printing legend                    *
1252 REM ********************************************************
1255 IF ix2 = 0 THEN iloc = 1
1260 IF ix2 <> 0 AND ixx <> 0 THEN iloc = 1
1270 IF ix2 <> 0 AND ixx = 0 THEN iloc = 2
1280 LOCATE iloc, 1:  PRINT nd + 1; " "; kind$,
1285 IF iloc = 1 THEN PRINT "ymin="; ymin; " ymax="; ymax
1290 IF ixx = 0 THEN INPUT "Press RETURN to continue", scratch
1291 REM ********************************************************
1292 REM *               End of graphic subroutine              *
1293 REM ********************************************************
1299 RETURN
1300 REM ********************************************************
1301 REM *     Subroutine that evaluates in npoint points       *
1302 REM *      the polynomial interpolating (x(i),y(i))        *
1303 REM ********************************************************
1310 REM ********************************************************
1311 REM *            Computing pi1(i)=PI' in (A.2.10)          *
1312 REM ********************************************************
1320 FOR i = 0 TO nd
1330 pi1(i) = 1
1340 FOR k = 0 TO nd
```

```
1350 IF i <> k THEN pi1(i) = pi1(i) * (x(i) - x(k))
1360 NEXT k
1370 REM ***********************************************************
1371 REM *              Computing the constant terms               *
1372 REM *          appearing in the summation of (A.2.12)         *
1373 REM ***********************************************************
1380 pi1(i) = y(i) / pi1(i)
1390 NEXT i
1400 FOR ipoint = 0 TO npoint
1410 xx = a + (b - a) * ipoint / npoint
1420 REM ***********************************************************
1421 REM *       Set values of the interpolant in the nodes        *
1422 REM ***********************************************************
1430 FOR i = 0 TO nd
1440 IF xx = x(i) THEN pL(ipoint) = y(i):  GOTO 1580
1450 NEXT i
1460 REM ***********************************************************
1461 REM *      Evaluating the interpolant not in the nodes        *
1462 REM ***********************************************************
1470 REM ***********************************************************
1471 REM *            Computing the summation in (A.2.12)          *
1472 REM ***********************************************************
1480 sum = 0
1490 FOR i = 0 TO nd
1500 sum = sum + pi1(i) / (xx - x(i)):
1510 NEXT i
1520 REM ***********************************************************
1521 REM *                Computing pi0=PI in (A.2.9)              *
1522 REM ***********************************************************
1530 pi0 = 1
1540 FOR i = 0 TO nd
1550 pi0= pi0 * (xx - x(i))
1560 NEXT i
1570 pL(ipoint) = sum * pi0
1580 NEXT ipoint
1590 REM ***********************************************************
1591 REM *              End of interpolation subroutine            *
1592 REM ***********************************************************
1599 RETURN
```

SPLINTER: A spline interpolation routine

Essentially the only difference between INTERPOL and SPLINTER is, of course, in the main subroutine.

Here, first the elements of the matrix and of the right-hand side of Eq.(A.2.19) are defined in lines 1320-1410. Then the boundary elements for natural and end slope splines are, respectively, defined in lines 1430-1480 and 1490-1530. Finally the tridiagonal system (A.2.19) is solved by applying the recursive procedures (A.2.20) and (A.2.21) in lines 1680-1720. This gives the values of u_i'' in Eq.(A.2.16), which are then used in lines 1740-1870 to evaluate the spline.

The reader now has all the necessary tools and is encouraged to write his own program to compute not-a-knot and periodic splines. Finally, he can generalize all these programs to the two-dimensional case.

SPLINTER.BAS

```
1 REM ************************************************************
2 REM *                   Spline interpolation                   *
3 REM ************************************************************
4 REM
5 REM ************************************************************
6 REM *            Main variables used in the program            *
7 REM ************************************************************
8 REM ndd = number of interpolation points < 100 (nd=nndd-1)
9 REM x(ndd) = interpolation points
10 REM y(ndd) = interpolation data
11 REM h(ndd) = distance between interpolation points
12 REM a(ndd) = mu in Eq.(A.2.18)
13 REM b(ndd) = diagonal elements of the matrix (=2)
14 REM c(ndd) = lambda in Eq.(A.2.18)
15 REM d(ndd) = r.h.s. of Eq.(A.2.19), see again (A.2.18)
16 REM u(ndd) = second derivative of spline, unknown of (A.2.19)
17 REM spl(npoint) = spline interpolant
18 REM ya, yb = value of the derivative at the end points
19 REM fnf(x) = function to interpolate
20 REM c, e = parameters used in defining the function fnf(x)
21 REM a, b = extrema of interpolation
22 REM pi = 3.1415926 (used to set Chebyshev points)
23 REM npoint = number of evaluations of interpolant (<1000)
24 REM deltax,deltay = length of plotted marks
25 REM ispline = flag indicating the type of spline
26 REM igraph = flag indicating the extrema of graph
27 REM function$ = flag indicating the type of interpolation data
28 REM ix1, ix2 = flag indicating the type of interpolation
```

```
   points (x)
29 REM kind$ = indicates the type of interpolation points (x)
30 REM g1,g,add1-add3= used to evaluate splines
31 REM xmin, xmax, ymin, ymax = extrema of graph
32 REM ndin, ndfin, ndstep = used for automatic increase of
   number of points
33 REM ratio = used in tridiagonal system solver
35 DIM x(100), y(100), spl(1000)
40 DIM a(100), b(100), c(100), d(100), h(100), u(100)
45 CLS 0
50 REM ***********************************************************
51 REM *                       Input part                        *
52 REM ***********************************************************
60 PRINT " SPLINE INTERPOLATION"
70 PRINT "Values in parentheses are options or suggested values
   (not defaults)"
80 PRINT ""
90 INPUT "Do you want to interpolate a function (f), random
   data (r), or your own data (d)"; function$
95 INPUT "Which kind of spline?  Natural (0) or Fixed end
   slope (1)"; ispline
100 IF ispline = 1 THEN INPUT "Values of the slope at the
   extrema of the interval (0,0)"; ya, yb
110 IF function$ = "r" THEN GOTO 140
120 IF function$ = "d" THEN GOTO 250
130 IF function$ = "f" THEN GOTO 350
140 REM **********************************************************
141 REM *            Setting random interpolation data           *
142 REM **********************************************************
145 RANDOMIZE
150 nd = 10:  a = 0:  b = 1
160 PRINT " x", " y"
170 x(0) = a:  x(nd) = b:  y(0) = RND: y(nd) = RND
180 FOR i = 1 TO nd - 1:  x(i) = RND: y(i) = RND: NEXT i
190 FOR i = 0 TO nd
200 FOR j = i + 1 TO nd
210 IF x(i) > x(j) THEN max = x(i):  x(i) = x(j):  x(j) = max
220 NEXT j:  NEXT i
230 FOR i = 0 TO nd:  PRINT x(i), y(i):  NEXT i:
   kind$ = "random nodes"
240 GOTO 310
250 REM **********************************************************
251 REM *                 Setting your own data                  *
252 REM **********************************************************
260 INPUT "In which interval do you want to interpolate (a,b)";
   a, b
270 INPUT "How many data points"; ndd:  nd = ndd - 1
275 PRINT "ATTENTION. Extrema must be nodes"
```

```
280 FOR i = 0 TO nd
290 PRINT "datum n."; i + 1; :  INPUT "(x,y)="; x(i), y(i)
300 NEXT i
310 PRINT "Minimum and maximum value of y in plot"
320 INPUT "Default (0), Range of the interpolant (1),
   Your own (2)"; igraph
330 CLS 0
340 GOTO 730
350 REM ***********************************************************
351 REM *      Interpolating the function given in line 420       *
352 REM ***********************************************************
360 INPUT "How many interpolation points?  (Press RETURN for
   automatic increase)"; ndd:  ndin = ndd - 1
370 IF ndd <> 0 THEN ndfin = ndin:  nstep = 1
380 IF ndd = 0 THEN ndin = 10:  ndfin = 50:  nstep = 5
390 INPUT "In which interval do you want to interpolate (-1,1)";
   a, b
400 INPUT "Coefficient of hyperbolic tangent (10)", c
410 e = 2.718281828#
420 DEF fnf (x) = (1 - e ^ (-2 * c * x)) / (1 + e ^ (-2 * c * x))
430 PRINT "Node distribution"
440 PRINT "Chebyshev with extrema (1), equispaced (2), denser
   in the middle (3),"
450 INPUT "or your own(4)"; ix1
460 INPUT "Comparison with another distribution (no=0,
   second choice=1,2,3,4)"; ix2
470 PRINT "Minimum and maximum value of y in plot"
480 INPUT "Default (0), Range of the interpolant (1),
   Your own (2)"; igraph
485 REM ***********************************************************
486 REM *                  End of input part                      *
487 REM ***********************************************************
490 CLS 0
500 FOR nd = ndin TO ndfin STEP nstep
510 REM ***********************************************************
511 REM *             Setting interpolation points                *
512 REM ***********************************************************
520 ix = ix1:  ixx = ix2:  pi = 3.1415926#
530 FOR i = 0 TO nd
540 ON ix GOTO 580, 610, 640, 680
580 x(i) = a + (b - a) * (1 - COS(i * pi / nd)) / 2
590 kind$ = "Chebyshev nodes with extrema"
600 GOTO 710
610 x(i) = a + (b - a) * i / nd
620 kind$ = "Equispaced nodes"
630 GOTO 710
640 xi = 2 * i / nd - 1
650 x(i) = (b + a) / 2 + (b - a) * (xi ^ 3 + xi) / 4
```

```
660 kind$ = "Nodes denser in the middle"
670 GOTO 710
680 PRINT "node n."; i + 1; :  INPUT x(i)
690 kind$ = "My input"
710 y(i) = fnf(x(i))
720 NEXT i
730 REM ***********************************************************
731 REM *              Calling subroutine that evaluates           *
732 REM *                the polynomial interpolant of             *
733 REM *                (x(i),y(i)) in npoint points              *
734 REM ***********************************************************
740 npoint = 100
750 GOSUB 1300
760 REM ***********************************************************
761 REM *               Calling the graphic subroutine             *
762 REM ***********************************************************
770 GOSUB 800
780 IF ixx <> 0 THEN ix = ixx:  ixx = 0:  GOTO 530
790 NEXT nd
799 END
800 REM ***********************************************************
801 REM *                     Graphic subroutine                   *
802 REM ***********************************************************
810 IF ix2 <> 0 AND ixx = 0 THEN GOTO 1050
820 xmin = a
830 xmax = b
840 ymin = 1000000
850 ymax = -1000000
860 REM ***********************************************************
861 REM *        Choosing extrema of plotting area among           *
862 REM *        default values, minimum and maximum of            *
863 REM *        interpolation data, or your own choice            *
864 REM ***********************************************************
890 ON igraph GOTO 970, 1020
900 FOR i = 0 TO nd
910 IF y(i) > ymax THEN ymax = y(i)
920 IF y(i) < ymin THEN ymin = y(i)
930 NEXT i
940 ymin = (3 * ymin - ymax) / 2
950 ymax = (3 * ymax - ymin) / 2
960 GOTO 1050
970 FOR i = 0 TO npoint
980 IF spl(i) > ymax THEN ymax = spl(i)
990 IF spl(i) < ymin THEN ymin = spl(i)
1000 NEXT i
1010 GOTO 1050
1020 PRINT "Minimum and maximum value of x and y in plot"
1030 INPUT "xmin and xmax"; xmin, xmax
```

```
1040 INPUT "ymin and ymax"; ymin, ymax
1050 IF ix2 = 0 THEN CLS 0
1060 IF ix2 <> 0 AND ixx <> 0 THEN CLS 0
1070 KEY OFF: SCREEN 2
1080 VIEW (3, 30)-(630, 180), , 1:  WINDOW (xmin, ymin)-(xmax,
   ymax)
1085 REM ***********************************************************
1086 REM *                  Plotting the interpolant               *
1087 REM ***********************************************************
1090 PSET (a, spl(0))
1100 FOR i = 1 TO npoint
1110 xx = a + (b - a) * i / npoint
1120 LINE -(xx, spl(i))
1130 NEXT i
1140 IF function$ <> "f" THEN GOTO 1200
1150 FOR i = 0 TO 100
1160 xsub = a + (b - a) * i / 100
1170 PSET (xsub, fnf(xsub))
1180 NEXT i
1190 GOTO 1250
1195 REM ***********************************************************
1196 REM *              Marking the interpolation points            *
1197 REM ***********************************************************
1200 deltax = (xmax - xmin) / 40:  deltay = (ymax - ymin) / 20
1210 FOR i = 0 TO nd
1220 PSET (x(i) - deltax, y(i)):  LINE -(x(i) + deltax, y(i))
1230 PSET (x(i), y(i) - deltay):  LINE -(x(i), y(i) + deltay)
1240 NEXT i
1250 REM ***********************************************************
1251 REM *                     Printing legend                     *
1252 REM ***********************************************************
1255 IF ix2 = 0 THEN iloc = 1
1260 IF ix2 <> 0 AND ixx <> 0 THEN iloc = 1
1270 IF ix2 <> 0 AND ixx = 0 THEN iloc = 2
1280 LOCATE iloc, 1:  PRINT nd + 1; " "; kind$,
1285 IF iloc = 1 THEN PRINT "ymin="; ymin; " ymax="; ymax
1290 IF ixx = 0 THEN INPUT "Press RETURN to continue", scratch
1291 REM ***********************************************************
1292 REM *                End of graphic subroutine                *
1293 REM ***********************************************************
1299 RETURN
1300 REM ***********************************************************
1301 REM *        Subroutine that evaluates in npoint points       *
1302 REM *         the polynomial interpolating (x(i),y(i))        *
1303 REM ***********************************************************
1310 REM ***********************************************************
1311 REM *         Defining coefficients in matrix (A.2.19)        *
1312 REM *       h(i) = distance between interpolation points      *
```

```
1313 REM *              a(i) = mu(i) in Eq.(A.2.18)                *
1314 REM *      b(i) = diagonal elements of the matrix (=2)        *
1315 REM *             c(i) = lambda(i) in Eq.(A.2.18)             *
1316 REM *     d(i) = r.h.s. of (A.2.19), see again (A.2.18)       *
1317 REM *********************************************************
1320 FOR i = 1 TO nd:  h(i) = x(i) - x(i - 1):  NEXT i
1330 FOR i = 1 TO nd - 1
1340 i1 = i + 1
1345 hh = h(i) + h(i1)
1350 a(i) = h(i1) / hh
1360 b(i) = 2
1370 c(i) = h(i) / hh
1380 d(i) = ((y(i1) - y(i)) / h(i1) - (y(i) - y(i - 1)) / h(i))
   * 6 / hh
1390 NEXT i
1400 b(0) = 2
1410 b(nd) = 2
1420 IF ispline = 1 THEN GOTO 1490
1430 REM *********************************************************
1431 REM *                     Natural spline                      *
1432 REM *********************************************************
1440 c(0) = 0
1450 a(nd) = 0
1460 d(0) = 0
1470 d(nd) = 0
1480 GOTO 1600
1490 REM *********************************************************
1491 REM *                    End-slope spline                     *
1492 REM *********************************************************
1500 c(0) = 1
1510 a(nd) = 1
1520 d(0) = ((y(1) - y(0)) / h(1) - ya) * 6 / h(1)
1530 d(nd) = (yb - (y(nd) - y(nd - 1)) / h(nd)) * 6 / h(nd)
1590 REM *********************************************************
1591 REM *             Solving a tridiagonal system                *
1592 REM *********************************************************
1593 REM *********************************************************
1594 REM *             Forward substitution (A.2.20)               *
1595 REM *********************************************************
1600 FOR i = 0 TO nd
1610 u(i) = d(i)
1620 NEXT i
1630 FOR i = 1 TO nd
1635 i1 = i - 1
1640 ratio = -a(i) / b(i1)
1650 b(i) = b(i) + ratio * c(i1)
1660 u(i) = u(i) + ratio * u(i1)
1670 NEXT i
```

```
1673 REM ********************************************************
1674 REM *             Backward substitution (A.2.21)           *
1675 REM ********************************************************
1680 IF b(nd) = 0 THEN GOTO 140
1690 u(nd) = u(nd) / b(nd)
1700 FOR i = nd - 1 TO 0 STEP -1
1710 u(i) = (u(i) - c(i) * u(i + 1)) / b(i)
1720 NEXT i
1730 i = 1
1740 FOR ipoint = 0 TO npoint
1750 xx = a + (b - a) * ipoint / npoint
1760 REM
1770 IF xx > x(i) THEN i = i + 1
1780 i1 = i - 1
1790 REM ********************************************************
1791 REM *          Evaluating the spline, see (A.2.16)         *
1800 REM ********************************************************
1810 g1 = xx - x(i1)
1820 g = x(i) - xx
1830 add1 = (u(i) * g1 ^ 3 + u(i1) * g ^ 3) / (6 * h(i))
1840 add2 = (y(i1) / h(i) - u(i1) * h(i) / 6) * g
1850 add3 = (y(i) / h(i) - u(i) * h(i) / 6) * g1
1860 spl(ipoint) = add1 + add2 + add3
1870 NEXT ipoint
1890 REM ********************************************************
1891 REM *            End of interpolation subroutine           *
1892 REM ********************************************************
1899 RETURN
1900 END
```

Appendix 3

RANDOM VARIABLES

Introduction

The aim of this Appendix is to provide a brief summary of some concepts of stochastic calculus that are useful to the analysis of stochastic models dealt with in Chapter 4.

We assume that the reader already possesses the fundamental knowledge of probability theory and of random variables, in particular. Starting from this knowledge, this appendix will provide a concise survey of the following topics:

- Concept and definitions of random variables;
- Moment calculations;
- Functions of random variables;
- Examples of random variables.

The content of this appendix is limited to the analysis of random variables, and will be mainly devoted to continuous variables. Discrete variables will often be considered as special cases.

The conceptual description of stochastic processes and random fields is briefly surveyed in Chapter 4. The reader who is interested in this topic can start with the classical books by Papoulis [PAP] and by Ash and Gardner [AaG], who provide also several examples and applications, and go on to a mathematically more sophisticated book such as that by Ivanov and Leonenko [IaL]. Chapter 2 of the book by Adomian [ADO] gives several intuitive physical interpretations of stochastic processes. The book by Soize [SOI] is an interesting example of the strong link between probability theory and applied sciences. In particular, such a book deals with signal analysis.

Additional references can be found in the bibliographies included in the various books cited above.

Random Variables

In probability theory one deals with random experiments with outcomes called ***events***.

The collection of all possible events is the ***sample space*** Ω. Every subset $A \subset \Omega$ of the space is an ***observable event***, whose collection constitutes a Borel field $\mathcal{B}$. It implies that the events that come from countable unions and intersections of observable events are also observable and are contained in $\mathcal{B}$.

For a given random experiment it is possible to define a finite number $\mathcal{P}(A)$, which defines the ***probability*** that the output of the experiment falls in the subset $A \subset \Omega$. The triple $(\Omega, \mathcal{B}, \mathcal{P})$ defines the ***probability space***.

Moment Calculation

Consider a one-dimensional random variable

$$r(\omega) : \Omega \longmapsto \mathbb{R}. \tag{A.3.1}$$

The quantity

$$F(R) = \mathcal{P}(r \leq R) \tag{A.3.2}$$

defines the probability that r is less than or equal to R, i.e., $r \in (-\infty, R)$.

As known, $F(r)$ is called the ***distribution function*** of the random variable $r(\omega)$. It is non-negative and nondecreasing from 0 to 1.

If the distribution function is a continuous function of r and differentiable, then $r(\omega)$ is a ***continuous random variable***. On the other hand, if $F(r)$ has a finite or countably finite number of discontinuity poits, then $r(\omega)$ is a ***discrete random variable***.

Let r be continuous, then the derivative $P(r)$ of the distribution function is the ***probability density function***.

As known, the probability density function $P = P(r)$ is positive defined and characterized by the following properties:

$$\int_a^b P(r)\, dr = \mathcal{P}\big(r \in [a,b]\big)\,, \qquad \int_{-\infty}^{\infty} P(r)\, dr = 1\,. \tag{A.3.3}$$

The n-th order moment of a random variable $r(\omega)$ with probability density $P(r)$ is defined by

$$E\{r^n\} = \int_{-\infty}^{\infty} r^n\, P(r)\, dr\,, \tag{A.3.4}$$

so that the first-order moment

$$E\{r\} = \int_{-\infty}^{\infty} r\,P(r)\,dr \tag{A.3.5}$$

is called the ***mean value*** of $r(\omega)$ and gives the statistical average of the random variable.

The ***n-th central moment*** of $r(\omega)$ is the moment of order n, calculated with respect to its mean value

$$E\Big\{\big(r - E(r)\big)^n\Big\} = \int_{-\infty}^{\infty} \big(r - E(r)\big)^n\, P(r)\,dr\,. \tag{A.3.6}$$

The second-order central moment gives the dispersion of the probability measures about the mean value. Such a central moment is called the ***variance*** of the random variable

$$v(r) = E\Big\{\big(r - E(r)\big)^2\Big\} = \int_{-\infty}^{\infty} \big(r - E(r)\big)^2\, P(r)\,dr\,. \tag{A.3.7}$$

The positive square root of the variance is the ***standard deviation***.

The concepts we have just seen for a one-dimensional random variable can be easily generalized to multidimensional variables. Consider, for simplicity of notation, the two-dimensional random variable $\mathbf{r} = \{r_1, r_2\}$ and let

$$P(r_1, r_2) = P(\mathbf{r}) : \mathbb{R}^2 \longmapsto \mathbb{R}_+ \tag{A.3.8}$$

be the probability density function linked to $\mathbf{r}$.

The probability density is joined to the distribution function by

$$F(r_1, r_2) = \int_{-\infty}^{r_1} \int_{-\infty}^{r_2} P(\alpha, \beta)\,d\alpha\,d\beta\,. \tag{A.3.9}$$

The marginal densities are given by

$$P(r_1) = \int_{-\infty}^{\infty} P(r_1, r_2)\,dr_2\,, \qquad P(r_2) = \int_{-\infty}^{\infty} P(r_1, r_2)\,dr_2\,. \tag{A.3.10}$$

One says that the two component r_1 and r_2 of $\mathbf{r}$ are statistically independent if

$$P(\mathbf{r}) = P(r_1)\,P(r_2)\,. \tag{A.3.11}$$

Moment definitions can now be given simply by generalizing what we have already seen in the one-dimensional case.

The ***joint moment of order*** (n,m) of two random variables is defined by

$$E\{r_1^n r_2^m\} = \int_{-\infty}^{\infty}\int_{-\infty}^{\infty} r_1^n r_2^m P(r_1,r_2)\,dr_1\,dr_2\,. \tag{A.3.12}$$

If r_1 and r_2 are statistically independent, one simply has

$$E\{r_1^n\}E\{r_2^m\} = E\{r_1^n r_2^m\}\,. \tag{A.3.13}$$

Similarly, the ***joint central moment*** of order (n,m) is given by

$$E\Big\{\big(r_1 - E(r_1)\big)^n \big(r_2 - E(r_2)\big)^m\Big\}$$
$$= \int_{-\infty}^{\infty}\int_{-\infty}^{\infty} \big(r_1 - E(r_1)\big)^n \big(r_2 - E(r_2)\big)^m P(r_1,r_2)\,dr_1\,dr_2\,.$$

In particular, the second-order joint central moment is called the ***covariance*** of r_1 and r_2, and can be calculated as

$$E\Big\{\big(r_1 - E(r_1)\big)\big(r_2 - E(r_2)\big)\Big\} = E\{r_1 r_2\} - E\{r_1\}E\{r_2\}\,. \tag{A.3.14}$$

The second-order joint moment $E\{r_1 r_2\}$ is called ***correlation***.

The two random variables are said to be ***uncorrelated*** if their covariance is equal to zero.

The generalization of the definitions given above to multidimensional random variables is simply a matter of additional notation.

If the random variable is a discrete one, then r can only attain a finite number of values $r_1, \ldots, r_n$ with probability $\mathcal{P}_i = \mathcal{P}(r = r_i)$. Then the probability density function can be expressed as follows

$$P(r) = \sum_{i=1}^{n} \mathcal{P}_i \delta(r - r_i)\,, \tag{A.3.15}$$

where δ denotes the Dirac delta function.

Calculations can be performed as in the case of the continuous variable. For this purpose it is useful to recall the following properties of the delta function

$$\delta(r - r_0) = \begin{cases} 0\,, & \text{for } r \neq r_0 \\ \infty\,, & \text{for } r = r_0\,, \end{cases}$$

$$\int_{-\infty}^{\infty} \delta(r - r_0)\, dr = 1\,,$$

and

$$\int_{-\infty}^{\infty} f(r)\delta(r - r_0) = f(r_0)\,.$$

Functions of random variables

Let $\mathbf{r}_1(\omega)$ and $\mathbf{r}_2(\omega)$ be two n-dimensional random variables related by a deterministic map

$$\mathbf{r}_2 = \mathbf{g}(\mathbf{r}_1) : \mathcal{D}_1 \longmapsto \mathcal{D}_2; \quad \mathcal{D}_1, \mathcal{D}_2 \subset \mathbb{R}^n_+\,. \tag{A.3.16}$$

It is useful to obtain the statistical properties of $\mathbf{r}_2$ from those of $\mathbf{r}_1$. With this in mind, we assume the following:

- The probability density $P(\mathbf{r}_1)$ linked to the variable $\mathbf{r}_1$ is known;
- The vector function $\mathbf{g}$ defines a one-to-one map from $\mathcal{D}_1$ into $\mathcal{D}_2$;
- Each component g_i of $\mathbf{g}$ has continuous partial derivatives with respect to all components of $\mathbf{r}_1$.

Then the conservation of probability measure

$$P(\mathbf{r}_2)\, d\mathbf{r}_2 = P(\mathbf{r}_1)\, d\mathbf{r}_1 \tag{A.3.17}$$

yields the following formula

$$P(\mathbf{r}_2) = P\Big(\mathbf{r}_1 = \mathbf{g}^{-1}(\mathbf{r}_2)\Big)\Big|J(\mathbf{r}_2 \longmapsto \mathbf{r}_1)\Big|\,, \tag{A.3.18}$$

where $|J|$ is the determinant of the Jacobian of the inverse mapping

$$g^{-1}(\mathbf{r}_2) : \mathcal{D}_2 \longmapsto \mathcal{D}_1\,. \tag{A.3.19}$$

In explicit form

$$\big|J\big| = \det\big[g_{ij}^{-1}\big]\,, \tag{A.3.20}$$

where

$$g_{ij}^{-1} = \frac{\partial g_i^{-1}}{\partial r_{2j}}\,. \tag{A.3.21}$$

For one-dimensional variables, one simply has

$$P(r_2) = P\big(r_1 = g^{-1}(r_2)\big)\frac{dr_1}{dr_2}\,. \tag{A.3.22}$$

Examples of continuous variables

This appendix is completed by reporting some examples of probability densities. These examples can be useful in technical calculations and in the analysis of specific mathematical models.

The examples reported in what follows will first provide the expression of the probability density with its domain of definition and then the parameters that characterize such a distribution.

The examples will refer to continuous variables in this section and to discrete variables in the last section of this appendix.

Uniform distribution

The random variable r is defined over an interval $\mathcal{D} = [a, b]$ and the probability density is constant over such interval

$$P(r) = \frac{1}{b-a}\,, \qquad r \in [a, b]\,. \tag{A.3.23}$$

This distribution is characterized only by the domain $\mathcal{D} = [a, b]$. The mean value and variance are

$$E(r) = \frac{b+a}{2}\,, \qquad v(r) = \frac{(b-a)^2}{12}\,.$$

Beta distribution

The random variable is defined over the interval $\mathcal{D} = [0, 1]$ and the probability density is

$$P(r) = \frac{\Gamma(\alpha+\beta)}{\Gamma(\alpha)\Gamma(\beta)} r^{\alpha-1}(1-r)^{\beta-1}\,, \tag{A.3.24}$$

where α, $\beta > 1$, and Γ denotes the classical gamma function.

This distribution is characterized by two parameters α and β, which define mean value and variance of the random variable by the expression

$$\begin{aligned} E(r) &= \frac{\alpha}{\alpha+\beta}\,, \\ v(r) &= \frac{\alpha\beta}{(\alpha+\beta)^2(\alpha+\beta+1)}\,. \end{aligned} \tag{A.3.25}$$

Maxwell distribution
The random variable is defined over the semi-infinite interval $D = [0, \infty)$ and the probability density is

$$P(r) = \frac{r^2}{\alpha^3 \sqrt{\pi/2}} e^{-\frac{r^2}{2\alpha^2}}. \tag{A.3.26}$$

This distribution is characterized by one parameter α, which is related to the mean value by the relation

$$E(r) = 2\alpha\sqrt{\frac{2}{\pi}}. \tag{A.3.27}$$

The maximum of $P(r)$ is located at $r = \sqrt{2}\alpha$.

Rayleigh distribution
The random variable is defined over the semi-infinite interval $D = [0, \infty)$ and the probability density is

$$P(r) = \frac{r}{\alpha^2} e^{-\frac{r^2}{2\alpha^2}}, \tag{A.3.28}$$

where the parameter $\alpha > 0$ is related to the mean value by

$$E(r) = \alpha\sqrt{\frac{\pi}{2}}. \tag{A.3.29}$$

The maximum of $P(r)$ is located at $r = \alpha$.

Normal distribution
The random variable is defined on the whole line, $r \in \mathbb{R}$, and the probability density is

$$P(r) = \frac{1}{\sqrt{2\pi v(r)}} e^{-\frac{(r-E(r))^2}{2v(r)}}. \tag{A.3.30}$$

As it is shown by Eq.(A.5.7), this distribution is characterized by two parameters: Mean value and variance.

The maximum of the $P(r)$ is localized at $r = E(r)$.

Examples of discrete variables

Discrete random variables are characterized, as we have seen in (A.3.15), by probability density distribution of the type

$$P(r) = \sum_i \lambda_i \delta(r - r_i)\,, \tag{A.3.31}$$

where δ denotes, as we have already seen, the delta function. Two examples follow.

Poisson distribution

The probability density function is

$$P(r) = e^{-\lambda} \sum_{i=0}^{\infty} \frac{\lambda^i}{i\,!} \delta(r - i)\,. \tag{A.3.32}$$

Mean value and variance are equal to the parameter λ

$$E(r) = v(r) = \lambda\,. \tag{A.3.33}$$

Binomial distribution

The probability density function is

$$P(r) = \sum_{i=1}^{n} \binom{n}{i} p^i q^{n-i} \delta(r - i)\,, \tag{A.3.34}$$

where $p + q = 1$.

Mean value and variance of such a distribution are equal to

$$E(r) = np\,, \qquad v(r) = npq\,. \tag{A.3.35}$$

REFERENCES

[AaG] Ash R. and Gardner M., **Topics in Stochastic Processes**, Academic Press (1975).

[AaM] Astarita G. and Marrucci G., **Principles of Non-Newtonian Fluid Mechanics**, McGraw-Hill (1974).

[ADO] Adomian G., **Stochastic Systems**, Academic Press (1983).

[AIK] Aiken R.C., **Stiff Computation**, Oxford Univ. Press (1985).

[AKI] Akin J.E., **Finite Element Analysis for Undergraduates**, Academic Press (1986).

[ALE] Alexitis G., **Convergence Problems on Orthogonal Series**, Pergamon Press (1961).

[AMA] Adomian G. and Malakian M., Stochastic analysis, *Math. Model.*, **1** (1983), 211–235.

[AMR] Ascher U.M., Mattheij R.M.M., and Russell R.D., **Numerical Solution of Boundary Value Problems for Ordinary Differential Equations**, Prentice-Hall (1988).

[ARN] Arnold V.I., **Ordinary Differential Equations**, Cambridge MIT Press (1973).

[ATP] Anderson D.A., Tannehill J.C., and Pletcher R.H., **Computational Fluid Mechanics and Heat Transfer**, Hemisphere Press (1984).

[BAN] Banach S., **Théorie des Opérations Lineares**, Monografie Matematyczne (1932).

[BaP] Brzezniak Z. and Preziosi L., On the Cauchy problem for a biological model on the distribution of dominance in a population of interacting organisms, *Math. Modelling and Sci. Comp.*, **1** (1993).

[BaR] Bellomo N. and Riganti R., **Nonlinear Stochastic Problems in Physics and Mechanics**, World Scientific (1987).

[BBD] Bellomo N., Brzezniak Z., and de Socio L.M., **Nonlinear Stochastic Problems in Applied Sciences**, Kluwer (1992).

[BBS] Back I.V., Blackwell B., and St. Claire C.R., **Inverse Heat Conduction**, Wiley (1985).

[BdB] Birkhoff G. and de Boor C., Piecewise Polynomial Interpolation and Approximation, in **Approximation of Functions**, Garabedian ed., Elsevier, 164–190 (1965).

[BEa] Belleni Morante A., **Applied Semigroup and Evolution Equations**, Oxford Univ. Press (1980).

[BEb] Belleni Morante A., **A Concise Guide to Semigroups and Evolution Equations**, World Scientific (1994).

[BEL] Beltrami E., **Mathematics for Dynamic Modelling**, Academic Press (1987).

[BEN] Benton E.R., Ordinary Differential and Difference Equations, in **Handbook of Applied Mathematics: Selected Results and Methods**, Pearson C.E. ed., Van Nostrand Reinhold (1990).

[BRI] Brigham E.O., **The Fast Fourier Transform**, Prentice-Hall (1974).

[BRO] Birkhoff G. and Rota G.C., **Ordinary Differential Equations**, Wiley (1989).

[BSW] Bailey P.B., Shampine L.F., and Waltman P.E., **Nonlinear Two-Point Boundary Value Problems**, Academic Press (1968).

[BUT] Butcher J.C., **The Numerical Analysis of Ordinary Differential Equations**, Wiley (1985).

[CaA] Canuto C., Hussaini M.Y., Quarteroni A., and Zang T.A., **Spectral Methods in Fluid Dynamics**, Springer (1989).

[CaG] Cottingham W.N. and Greenwood D.A., **Electricity and Magnetism**, Cambridge Univ. Press (1991).

[CaH] Courant R. and Hilbert D., **Methods of Mathematical Physics**, Wiley (1953).

[CaM] Chorin A.J. and Marsden J.F., **Mathematical Introduction to Fluid Mechanics**, Springer (1979).

[CAR] Carrier G.F., Perturbation Methods, in **Handbook of Applied Mathematics: Selected Results and Methods**, Pearson C.E. Ed., Van Nostrand Reinhold (1990).

[CaS] Cooper L. and Steinberg D., **Introduction to Methods of Optimization**, Saunders Publ. (1970).

[CER] Colton D., Ewing R., and Rundell W., **Inverse Problems in Partial Differential Equations**, SIAM Publ. (1990).

[CHA] Chan T.F. et al. eds., **Domain Decomposition Methods**, SIAM, Vol. 1 (1989), Vol. 2 (1990).

[CRO] Cronin J., **Mathematical Aspects of Hodgkin and Huxley Neural Theory**, Cambridge Univ. Press (1987).

[DaI] Dym C.L. and Ivey E.S., **Principles of Mathematical Modeling**, Academic Press (1990).

[DaN] Di Francesco D. and Noble D., A model of cardiac activity incorporating ionic pumps and concentration changes, *Phil. Trans. R. Soc. London*, **B 307** (1985), 353–398.

[DaR] Demyanov V.F. and Rubinov A.M., **Approximate Methods in Optimization Problems**, Elsevier (1970).

[DAV] Davis P.J., **Interpolation and Approximation**, Dover Publ. (1963).

[DaZ] Duchateau P. and Zachmann D.W., **Theory and Problems of Partial Differential Equations**, McGraw-Hill (1986).

[deB] de Boor C., **A Practical Guide to Splines**, Appl. Math. Series, **27**, Springer (1978).

[DYM] Dym C.L., **Stability Theory and its Applications to Structural Mechanics**, Noordhoff Int. (1974).

[EaM] Edwards D. and Hamson M., **Guide to Mathematical Modelling**, CRC Press (1990).

[FaT] Ferrari C. and Tricomi F., **Transonic Aerodynamics**, McGraw-Hill (1962).

[FaW] Forsythe G.E. and Wasow W.R., **Finite Difference Methods for Partial Differential Equations**, Wiley (1960).

[FIT] FitzHugh R., Impulses and physiological states in theoretical models of nerve membrane, *Biophys. J.*, **1** (1961), 445–466.

[FOX] Fox L., **The Numerical Solution of Two-Point Boundary Value Problems in Ordinary Differential Equations**, Oxford Univ. Press (1957).

[FRI] Friedman A., **Mathematics in Industrial Problems**, Springer, **I** (1988); **II** (1989); **III** (1990).

[GaP] Gurtin M.E. and Pipkin A.C., A general theory of heat conduction with finite wave speeds, *Arch. Rat. Mech. Anal.*, **31** (1968), 113–126.

[GaQ] Gastaldi F. and Quarteroni A., On the coupling of hyperbolic and parabolic systems: Analytic and numerical approach, *Appl. Num. Math.*, **6** (1989), 3–31.

[GaO] Gottlieb D. and Orszag S.A., **Numerical Analysis of Spectral Methods: Theory and Applications**, SIAM-CBMS (1977).

[GaR] Galdi G.P. and Rionero S., **Weighted Energy Methods in Fluid-Dynamics and Elasticity**, Springer Lecture Notes in Mathematics, **1134** (1985).

[GEA] Gear C.W., **Numerical Initial Value Problems in Ordinary Differential Equations**, Prentice-Hall (1971).

[GHO] Ghosh R., **Methods of Inverse Problems in Physics**, CRC Press (1991).

[GPD] Galdi G.P. and Padula M., A new approach to energy theory in the stability of fluid motion, *Arch. Rat. Mech. Anal.*, **110** (1990), 187–286.

[GQS] Gastaldi F., Quarteroni A., and Sacchi Landriani G., On the coupling of two-dimensional hyperbolic and elliptic equations: Analitical and numerical approach, in Chan T.F. et al. eds., **Domain Decomposition Methods**, SIAM, Vol. 2 (1990), 22–63.

[GUR] Gurtin M.E., **Theory of Elasticity**, *Handbuck der Physics*, **VI**, Springer (1971).

[GUT] Gurtin M.E., **An Introduction to Continuum Mechanics**, Academic Press (1981).

[HAB] Haberman R., **Mathematical Models**, Prentice-Hall (1877).

[HaH] Hodgkin A.L. and Huxley A.F., A qualitative description of membrane current and its application to conduction and excitation in nerves, *J. Physiol.*, **117** (1952), 500–544.

[HaN] Hasler M. and Neirynck J., **Nonlinear Circuits**, Artech Publ. (1987).

[HaY] Hageman L.A. and Young D.M., **Applied Iterative Methods**, Academic Press (1981).

[HER] Hernandez D.B., On some guiding principles in mathematical modelling with special emphasis on determinism, *Math. Model.*, **2** (1981), 179–190.

[HIL] Hille B., **Hionic Channels of Excitable Membranes**, Sinauer (1984).

[HIN] Hindmarsh A.C., A systematic collection of ODE solvers, in **Numerical Methods for Differential Equations**, Lapidus L. and Schiener W.E. eds., Academic Press (1983).

[HIR] Hirsh C., **Numerical Computation of Internal and External Flows**, Wiley (1990).

[HKW] Hassard B.D., Kazarinoff N.D., and Wan Y.H., **Theory and Applications of Hopf Bifurcation**, Cambridge Univ. Press (1981).

[HOD] Hodgkin A.L., **The Conduction of Nerve Impulse**, The Sherrington Lectures VII, Liverpool Univ. Press (1964).

[IaJ] Iooss G. and Joseph D.D., **Elementary Stability and Bifurcation Theory**, Springer (1980).

[IaL] Ivanov A.V. and Leonenko N.N., **Statistical Analysis of Random Fields, Mathematics and Its Applications** (Soviet Series), Hazewinkel M. ed., Kluwer (1989).

[IMS] IMSL Library, Edition 8, International Statistical and Mathematical Libraries, Houston, Texas (1982).

[INC] Ince E.L., **Ordinary Differential Equations**, Dover Publ. (1926).

[JaJ] Jeffreys H. and Jeffreys B., **Methods of Mathematical Physics**, Cambridge Univ. Press (1966).

[JaP] Joseph D.D. and Preziosi L., Heat Waves, *Rev. Mod. Phys.*, **61** (1989), 41–74.

[JaR] Joseph D.D. and Renardy Y., **Fundamentals of Two-Fluid Mechanics**, Springer (1993).

[JaS] Jager E. and Segel L., On the distribution of dominance in a population of interacting anonimous organisms, *SIAM J. Appl. Math.*, **52** (1993), 1442–1468.

[JEF] Jeffrey A., **Linear Algebra and Ordinary Differential Equations**, Blackwell Scientific (1990).

[JNT] Jack J.J.B., Noble D., and Tsien R.W., **Electric Current Flow in Excitable Cells**, Oxford Univ. Press, (1975).

[JOa] Joseph D.D., **Stability of Fluid Motions**, Springer (1976).

[JOb] Joseph D.D., **Fluid Dynamics of Viscoelastic Liquids**, Springer (1990).

[KaN] Kliemann W. and Namachchivaya S., **Stochastic and Nonlinear Dynamics: Application to Mechanical Systems**, CRC Press (1993).

[KaW] Knops R.J. and Wilkes E.W., **Elastic Stability**, *Handbuck der Physics*, **VI a/3**, Springer (1973).

[KEL] Keller H.B., **Numerical Methods in Bifurcation Problems**, Tata Institute, India (1987).

[KLA] Klamkin M. S., **Mathematical Modelling: Classroom Notes in Applied Mathematics**, SIAM Publ. (1987).

[KRA] Krantz S.G., **Real Analysis and Foundations**, CRC Press (1991).

[KST] Kustnezov P., Stratonovich L. and Tichonov I., Quasi-moment functions in the theory of random processes, *Prob. Theory Appl.*, **5** (1960), 80–97.

[LaM] Lions J.L. and Magenes E., **Problémes aux Limites Nonhomogenes et Applications**, Dunod, Vol. 1 (1968), Vol. 2 (1969), Vol. 3 (1970).

[LaS] Lin C.C. and Segel L.A., **Mathematics Applied to Deterministic Problems in the Natural Sciences**, SIAM Publ. (1988).

[LRY] Lavrent'ev M.M., Reznitskaya K.G., and Yakhno V.G., **One-dimensional Inverse Problems of Mathematical Physics**, American Mathematical Society Translations, Series 2, Vol. 130, AMS Publ. (1985).

[LSA] Lancaster P. and Salkauskas K., **Curve and Surface Fitting**, Academic Press (1986).

[MaM] Marsden J.E. and Mc Cracken M., **The Hopf Bifurcation and its Applications**, Springer (1976).

[MAR] Martin C., **Nonlinear Operators and Differential Equations in Banach Spaces**, Wiley (1976).

[MAS] Mascagni M., An initial-boundary-value problem of physiological significance for equation of nerve conduction, *Comm. Pure Appl. Math.*, **42** (1989), 213–227.

[MAY] May R.M., **Stability and Complexity in Model Ecosystems**, Princeton Univ. Press (1973).

[MSM] Maynard Smith J., **Models in Ecology**, Cambridge Univ. Press (1974).

[MUL] Müller C., **Foundations of Mathematical Theory of Electromagnetic Waves**, Springer (1969).

[MUR] Murray J.D., **Mathematical Biology**, Springer (1993).

[NAG] NAG Library, Mark 10, Numerical Analysis Group, NAG Central Office, Oxford (1983).

[NED] Nelson Dorny C., **Optimization a Vector Space Approach to Models and Optimization**, Wiley (1975).

[NEL] Nelson E., **Quantum Fluctuations**, Princeton Univ. Press (1985).

[NOB] Noble D., Application of Hodgkin-Huxley equation to excitable tissue, *Physiol. Rev.*, **46** (1966), 1–50.

[PaD] Preziosi L. and de Socio L., Nonlinear inverse phase transition problems for the nonlinear heat equation, *Math. Models Meth. Appl. Sci.*, **1** (1991), 167–182.

[PAP] Papoulis A., **Probability Random Variables and Stochastic Processes**, McGraw-Hill (1965).

[PAY] Payne L.E., **Improperly Posed Problems in Partial Differential Equations**, Reg. Conf. Series in Appl. Math., SIAM Publ. (1975).

[POM] Pomraning G., **Linear Transport Theory in Stochastic Media**, World Scientific (1991).

[POW] Powell M.J.D., **Approximation Theory and Methods**, Cambridge Univ. Press (1981).

[PRE] Preziosi L., A source-sink inverse problem for the nonlinear heat equation, *Comp. Math. Model.*, **17** (1993), 3–11.

[PTB] Preziosi L., Teppati G., and Bellomo N., Modelling and solution of stochastic inverse problems in mathematical physics, *Comp. Math. Model.*, **16** (1992), 37–51.

[QUA] Quarteroni A., Domain decomposition techniques using spectral methods, *Surveys on Mathematics for Industry*, **1** (1991), 75–118.

[RaW] Rade L. and Westergren B., **BETA Mathematics Handbook**, CRC Press (1992).

[RIC] Richtmeyer R.D., **Principles of Advanced Mathematical Physics**, Springer (1978).

[ROS] Rosen R., **Foundations of Mathematical Biology**, Academic Press (1972).

[SEG] Segel L.A., **Modelling Dynamic Phenomena in Molecular and Cellular Biology**, Cambridge Univ. Press (1989).

[SEW] Sewell G., **The Numerical Solution of Ordinary and Partial Differential Equations**, Academic Press (1988).

[SEY] Seydel R., **From Equilibrium to Chaos: Practical Bifurcation and Stability Analysis**, Elsevier (1988).

[SMA] Smart D.S., **Fixed Point Theorems**, Cambridge Univ. Press (1974).

[SOa] Sobczyk K., **Random Differential Equations and Applications**, Kluwer (1991).

[SOb] Sobczyk K., **Stochastic Wave Propagation**, Elsevier (1985).

[SOI] Soize C., **Méthodes Mathématiques en Analyse du Signal**, Masson (1993).

[SOO] Soong T.T., **Random Differential Equations in Science and Engineering**, Academic Press (1973).

[SPA] Späth H., **Splines Algorithms for Curves and Surfaces**, Utilitas Mathematica (1974).

[SPW] Sparrow C., **The Lorentz Equation: Bifurcation, Chaos and Strange Attractors**, Springer (1982).

[STI] Strikwerda J.C., **Finite Difference Schemes and Partial Differential Equations**, Wandsworth Brooks (1989).

[STO] Stoker J.J., **Nonlinear Vibrations**, Interscience (1950).

[STR] Straughan B., **The Energy Method, Stability and Nonlinear Convection**, Springer (1992).

[SZE] Szegö G., **Orthogonal Polynomials**, American Mathematical Society (1959)

[TaM] Truesdell C. and Muncaster R.G., **Fundamentals of Maxwell's Kinetic Theory of a Simple Monoatomic Gas**, Academic Press (1980).

[TaS] Tichonov A.N. and Samarskii A.A., **Equations of Mathematical Physics**, In Russian, Nauka, Moscow (1977).

[TRU] Truesdell C., **A First Course in Rational Continuum Mechanics**, Academic Press (1977).

[WIG] Wigner E., The unreasonable effectiveness of mathematics in the natural sciences, *Comm. Pure Appl. Math.*, **13** (1960), 1–14.

[YAN] Yan-Qian Y., **Theory of Limit Cycles**, *Translations of Mathematical Monographs*, Vol. 66, Amer. Math. Soc. Press (1986).

[YOS] Yoshida K., **Functional Analysis**, Springer (1970).

[ZEE] Zeeman E.C., **Catastrophe Theory**, Addison-Wesley (1977).

[ZYG] Zygmund A., **Trigonometric Series**, Cambridge Univ. Press (1959).

SUBJECTS INDEX